MPH- CLASS PREP

Ann Weir 292 Giltner, M.S.U.

# Methods in Enzymology

## Volume LVIII
## CELL CULTURE

# METHODS IN ENZYMOLOGY

EDITORS-IN-CHIEF

Sidney P. Colowick  Nathan O. Kaplan

*Methods in Enzymology*

*Volume LVIII*

# *Cell Culture*

EDITED BY

*William B. Jakoby*

ROOM 9N-119, BUILDING 10
NATIONAL INSTITUTES OF HEALTH
BETHESDA, MARYLAND

*Ira H. Pastan*

ROOM B27, BUILDING 37
NATIONAL INSTITUTES OF HEALTH
BETHESDA, MARYLAND

ACADEMIC PRESS
A Subsidiary of Harcourt Brace Jovanovich, Publishers
New York London Toronto Sydney San Francisco

ACADEMIC PRESS, INC.
111 Fifth Avenue, New York, New York 10003

*United Kingdom Edition published by*
ACADEMIC PRESS, INC. (LONDON) LTD.
24/28 Oval Road, London NW1 7DX

**Library of Congress Cataloging in Publication Data**

Main entry under title:

Cell Culture.

(Methods in enzymology ; v. 58)
Bibliography.
Includes index.
1. Cell Culture. I. Jakoby, William B., Date
II. Pastan, Ira H. III. Series.
QP601.M49 vol. 58 [QH585] 574.1'925'08s [574.8'028]
ISBN 0-12-181958-2 78-24360

PRINTED IN THE UNITED STATES OF AMERICA

83 84 85 9 8 7 6 5 4 3

# Table of Contents



## Section I. Basic Methods

### A. Laboratory Requirements and Media



### B. General Cell Culture Techniques

## Section II. Specialized Techniques

### A. Metabolism

## B. Genetics, Hybridization, and Transformation



## C. Virus Preparation



## Section III. Specific Cell Lines

# Contributors to Volume LVIII

Article numbers are in parentheses following the names of contributors.
Affiliations listed are current.


RONALD T. ACTON (17, 18), *Department of Microbiology and the Diabetes Research and Training Center, University of Alabama in Birmingham, University Station, Birmingham, Alabama 35294*

DOLPH O. ADAMS (43), *Department of Pathology, Duke Medical Center, Durham, North Carolina 27710*

W. FRENCH ANDERSON (44), *Laboratory of Molecular Hematology, National Heart, Lung, and Blood Institutes, National Institutes of Health, Bethesda, Maryland 20014*

TSUKASA ASHIHARA (20), *Department of Pathology, Shiga Medical College, Moriyama-cho, Moriyama City, Shiga 524, Japan*

W. EMMETT BARKLEY (4), *Building 13, Room 2E47, National Institutes of Health, Bethesda, Maryland 20014*

PAUL A. BARSTAD (17), *Department of Microbiology and the Diabetes Research and Training Center, University of Alabama in Birmingham, University Station, Birmingham, Alabama 35294*

RENATO BASERGA (20), *Department of Pathology and Fels Research Institute, Temple University School of Medicine, Philadelphia, Pennsylvania 19140*

MARK M. BASHOR (9), *Letterman Army Institute of Research, Presidio of San Francisco, San Francisco, California 94129*

SHELBY L. BERGER (42), *Section of Cellular and Molecular Physiology, Laboratory of Pathophysiology, National Cancer Institute, National Institutes of Health, Bethesda, Maryland 20014*

JANE BOTTENSTEIN (6), *Department of Biology, University of California, San Diego, La Jolla, California 92093*

NÖEL BOUCK (24), *Department of Microbiology, University of Illinois at the Medical Center, Chicago, Illinois 60680*

ROBERT B. CAMPENOT (25), *Department of Neurobiology, Harvard Medical School, Boston, Massachusetts 02115*

WILLIAM CARLISLE (54), *The Salk Institute, P.O. Box 1809, San Diego, California 92112*

P. COFFINO (19), *Departments of Medicine and Microbiology, University of California, San Francisco, California 94143*

LEWIS L. CORIELL (3), *Institute for Medical Research, Copewood Street, Camden, New Jersey 08103*

ROBERT T. DELL'ORCO (1), *Biochemical Division, The S. R. Noble Foundation, Route 1, Ardmore, Oklahoma 73401*

GIAMPIERO DI MAYORCA (24), *Department of Microbiology, University of Illinois at the Medical Center, Chicago, Illinois 60680*

WILLIAM H. J. DOUGLAS (1,10), *Department of Anatomy, Tufts University School of Medicine, Boston, Massachusetts 02115*

CATHERINE DUFF (27), *Genetics Department and Research Institute, The Hospital for Sick Children, Toronto, Ontario, Canada*

ROBERT M. FRIEDMAN (23), *Laboratory of Experimental Pathology, National Institute of Arthritis, Metabolism, and Digestive Diseases, National Institutes of Health, Bethesda, Maryland 20014*

T. V. GOPALAKRISHNAN (44), *Laboratory of Molecular Hematology, National Heart, Lung, and Blood Institutes, National Institutes of Health, Bethesda, Maryland 20014*

J. W. GRAY (19), *Biomedical Sciences, Lawrence Livermore Laboratory, University of California, P. O. Box 808, Livermore, California 94143*

MAURICE GREEN (36), *Institute for Molecular Virology, St. Louis University School of Medicine, St. Louis, Missouri 63110*

JEFFREY GRUBB (38), *Department of Pediatrics, Division of Medical Genetics, Washington University School of Medicine, St. Louis, Missouri 63110*

P. M. GULLINO (14), *Laboratory of Pathophysiology, National Cancer Institute, National Institutes of Health, Bethesda, Maryland 20014*

RICHARD G. HAM (5), *Department of Molecular, Cellular, and Developmental Biology, University of Colorado, Boulder, Colorado 80309*

EDWARD HAWROT (53), *Department of Neurobiology, Harvard Medical School, Boston, Massachusetts 02115*

IZUMI HAYASHI (6), *Department of Biology, University of California, San Diego, La Jolla, California 92093*

W. FRED HINK (39), *Department of Entomology, Ohio State University, Columbus, Ohio 43210*

BHARATI HUKKU (13), *The Child Research Center of Michigan, Children's Hospital of Michigan, Detroit, Michigan 48201*

ERIC HUNTER (32), *Department of Microbiology, The Medical Center, University of Alabama in Birmingham, Birmingham, Alabama 35294*

SHARON HUTCHINGS (6), *Department of Biology, University of California, San Diego, La Jolla, California 92093*

ROGER H. KENNETT (28), *Department of Human Genetics, The Human Genetics Cell Center, University of Pennsylvania School of Medicine, Philadelphia, Pennsylvania 19104*

GEORGE KHOURY (34), *Laboratory of DNA Tumor Viruses, National Cancer Institute, National Institutes of Health, Bethesda, Maryland 20014*

MICHAEL KLAGSBRUN (50), *Departments of Surgical Research and Biological Chemistry, Children's Hospital Medical Center, Harvard Medical School, Boston, Massachusetts 02115*

R. A. KNAZEK (14), *Laboratory of Pathophysiology, National Cancer Institute, National Institutes of Health, Bethesda, Maryland 20014*

K. S. KOCH (47), *The Salk Institute, Post Office Box 1809, San Diego, California 92112*

IRWIN R. KONIGSBERG (45), *Department of Biology, University of Virginia, Charlottesville, Virginia 22901*

CHING-JU LAI (34), *Laboratory of DNA Tumor Viruses, National Cancer Institute, National Institutes of Health, Bethesda, Maryland 20014*

H. L. LEFFERT (47), *The Salk Institute, Post Office Box 1809, San Diego, California 92112*

DAVID W. LEVINE (15), *The Cell Culture Center, Massachusetts Institute of Technology, Cambridge, Massachusetts 02139*

JAMES A. MCATEER (10), *W. Alton Jones Cell Science Center, Old Barn Road, Lake Placid, New York 12946*

DON B. MCCLURE (6), *Department of Biology, University of California, San Diego, La Jolla, California 92093*

GERALD J. MCGARRITY (2, 37), *Institute for Medical Research, Copewood Street, Camden, New Jersey 08103*

WALLACE L. MCKEEHAN (5), *Department of Molecular, Cellular, and Developmental Biology, University of Colorado, Boulder, Colorado 80309*

WILLIAM F. MCLIMANS (16), *Roswell Park Memorial Cancer Institute, New York State Department of Health, Buffalo, New York 14263*

HIDEO MASUI (6), *Department of Biology, University of California, San Diego, La Jolla, California 92093*

JENNIE MATHER (6), *Department of Biology, University of California, San Diego, La Jolla, California 92093*

T. MORAN (47), *The Salk Institute, Post Office Box 1809, San Diego, California 92112*

TOSHIO MURASHIGE (41), *Department of Plant Science, University of California, Riverside, California 92521*

SUGAYUKI OHASA (6), *Department of Biology, University of California, San Diego, La Jolla, California 92093*

IRA PASTAN (30), *Room B27, Building 37, National Institutes of Health, Bethesda, Maryland 20014*

MANFORD K. PATTERSON, JR. (11), *The Samuel Roberts Noble Foundation, Route One, Ardmore, Oklahoma 73401*

PAUL H. PATTERSON (53), *Department of*

*Neurobiology, Harvard Medical School, Boston, Massachusetts 02115*

JOHN PAWELEK (51), *Department of Dermatology, Yale University, School of Medicine, New Haven, Connecticut 06510*

D. PERLMAN (7), *School of Pharmacy, University of Wisconsin, Madison, Wisconsin 53706*

WARD D. PETERSON, JR. (13), *The Child Research Center of Michigan, Children's Hospital of Michigan, Detroit, Michigan 48201*

LOLA C. M. REID (12, 21), *Department of Molecular Pharmacology and Liver Research Center, Departments of Medicine and Biochemistry, Albert Einstein College of Medicine, Bronx, New York 10461*

JOHN F. REYNOLDS (41), *Department of Plant Science, University of California, Riverside, Riverside, California*

ANGIE RIZZINO (6), *Department of Biology, University of California, San Diego, La Jolla, California 92093*

MARCOS ROJKIND (21), *Department of Molecular Pharmacology and Liver Research Center, Departments of Medicine and Biochemistry, Albert Einstein College of Medicine, Bronx, New York 10461*

ALBERT W. RUESINK (29), *Department of Biology, Indiana University, Jordan Hall 138, Bloomington, Indiana 47401*

MILTON H. SAIER, JR. (49), *The Department of Biology, The John Muir College, University of California, San Diego, La Jolla, California 92093*

GORDON SATO (6), *Department of Biology, University of California, San Diego, La Jolla, California 92093*

BERNARD P. SCHIMMER (52), *Banting and Best Department of Medical Research, University of Toronto, Toronto, Ontario M5G 1L6, Canada*

DAVID SCHUBERT (54), *The Salk Institute, P.O. Box 1809, San Diego, California 92112*

GINETTE SERRERO (6), *Department of Biology, University of California, San Diego, La Jolla, California 92093*

CHARLES J. SHERR (35), *Laboratory of Viral Carcinogenesis, National Cancer Institute, National Institutes of Health, Bethesda, Maryland 20014*

SEUNG-IL SHIN (31), *Department of Genetics, Albert Einstein College of Medicine, Bronx, New York 10461*

WILLIAM F. SIMPSON (13), *The Child Research Center of Michigan, Children's Hospital of Michigan, Detroit, Michigan 48201*

WILLIAM S. SLY (38), *Department of Pediatrics, Division of Medical Genetics, Washington University School of Medicine, St. Louis, Missouri 63110*

RALPH E. SMITH (33), *Departments of Microbiology and Immunology, Duke University Medical Center, Durham, North Carolina 27710*

GRETCHEN H. STEIN (22), *Department of Molecular, Cellular, and Developmental Biology, University of Colorado, Boulder, Colorado 80309*

ARMEN H. TASHJIAN, JR. (46), *Laboratory of Toxicology, Harvard School of Public Health, and Department of Pharmacology, Harvard Medical School, Boston, Massachusetts 02115*

MARY TAUB (49), *The Department of Biology, The John Muir College, University of California, San Diego, La Jolla, California 92093*

WILLIAM G. THILLY (15), *Genetic Toxicology Group, Department of Nutrition and Food Science, Massachusetts Institute of Technology, Cambridge, Massachusetts 02139*

E. BRAD THOMPSON (48), *Laboratory of Biochemistry, National Cancer Institute, National Institutes of Health, Bethesda, Maryland 20014*

LARRY H. THOMPSON (26), *Biomedical Sciences Division L-452, Lawrence Livermore Laboratory, University of California, Livermore, California 94550*

GEORGE J. TODARO (35), *Laboratory of Viral Carcinogenesis, National Cancer Institute, National Institutes of Health, Bethesda, Maryland 20014*

M. WILLIAMS (47), *The Salk Institute, Post Office Box 1809, San Diego, California 92112*

KIM S. WISE (18), *Department of Microbiology, and the Diabetes Research and*

*Training Center, University of Alabama in Birmingham, University Station, Birmingham, Alabama 35294*

WILLIAM S. M. WOLD (36), *Institute for Molecular Virology, St. Louis University School of Medicine, St. Louis, Missouri 63110*

KEN WOLF (8, 40), *National Fisheries Center-Leetown, Fish and Wildlife Service, Department of the Interior, Route 3, Box 41, Kearneyville, West Virginia 25430*

RICHARD WOLFE (6), *Department of Biology, University of California, San Diego, La Jolla, California 92093*

RONALD G. WORTON (27), *Genetics Department and Research Institute, The Hospital for Sick Children, Toronto, Ontario, Canada*

REEN WU (6), *Department of Biology, University of California, San Diego, La Jolla, California 92093*

ROSALIND YANISHEVSKY (22), *Department of Molecular, Cellular and Developmental Biology, University of Colorado, Boulder, Colorado 80309*

ROBERT K. ZWERNER (17, 18), *Department of Microbiology and the Diabetes Research and Training Center, University of Alabama in Birmingham, University Station, Birmingham, Alabama 35294*

# Preface

Many of the problems that have caught the interest and imagination of biochemists are studied best with cultured cells. We offer in this volume, in a format familiar to investigators in biochemistry, the general techniques necessary for working with cells in culture and illustrate such general methods with specific examples from the large variety of cells that have been cultivated.

The tools and methods for cell culture are presented in Part I. Part II provides a group of specialized techniques that are useful for many of the applications that biochemists and other investigators with their widely different approaches may require. Part III is concerned with specific methods for specific cell types that have been chosen to represent the wide range of cells that may now be prepared.

There is some duplication in the presentations. For example, portions of certain methods are repeated in one or another form both in Part I and Part III. We believe that this repetition is necessary to convey faithfully to the reader a complete method of proven effectiveness. Additionally, we hope that a heuristic effect will be achieved that will enable investigators unfamiliar with cell culture to assess what is available and to predict what might be most suitable for their own purposes.

William B. Jakoby
Ira H. Pastan

# METHODS IN ENZYMOLOGY

EDITED BY

Sidney P. Colowick and Nathan O. Kaplan

VANDERBILT UNIVERSITY
SCHOOL OF MEDICINE
NASHVILLE, TENNESSEE

DEPARTMENT OF CHEMISTRY
UNIVERSITY OF CALIFORNIA
AT SAN DIEGO
LA JOLLA, CALIFORNIA

I. Preparation and Assay of Enzymes
II. Preparation and Assay of Enzymes
III. Preparation and Assay of Substrates
IV. Special Techniques for the Enzymologist
V. Preparation and Assay of Enzymes
VI. Preparation and Assay of Enzymes *(Continued)*
Preparation and Assay of Substrates
Special Techniques
VII. Cumulative Subject Index

# METHODS IN ENZYMOLOGY

EDITORS-IN-CHIEF

Sidney P. Colowick     Nathan O. Kaplan

VOLUME VIII. Complex Carbohydrates
*Edited by* ELIZABETH F. NEUFELD AND VICTOR GINSBURG

VOLUME IX. Carbohydrate Metabolism
*Edited by* WILLIS A. WOOD

VOLUME X. Oxidation and Phosphorylation
*Edited by* RONALD W. ESTABROOK AND MAYNARD E. PULLMAN

VOLUME XI. Enzyme Structure
*Edited by* C. H. W. HIRS

VOLUME XII. Nucleic Acids (Parts A and B)
*Edited by* LAWRENCE GROSSMAN AND KIVIE MOLDAVE

VOLUME XIII. Citric Acid Cycle
*Edited by* J. M. LOWENSTEIN

VOLUME XIV. Lipids
*Edited by* J. M. LOWENSTEIN

VOLUME XV. Steroids and Terpenoids
*Edited by* RAYMOND B. CLAYTON

VOLUME XVI. Fast Reactions
*Edited by* KENNETH KUSTIN

VOLUME XVII. Metabolism of Amino Acids and Amines (Parts A and B)
*Edited by* HERBERT TABOR AND CELIA WHITE TABOR

VOLUME XVIII. Vitamins and Coenzymes (Parts A, B, and C)
*Edited by* DONALD B. MCCORMICK AND LEMUEL D. WRIGHT

VOLUME XIX. Proteolytic Enzymes
*Edited by* GERTRUDE E. PERLMANN AND LASZLO LORAND

VOLUME XX. Nucleic Acids and Protein Synthesis (Part C)
*Edited by* KIVIE MOLDAVE AND LAWRENCE GROSSMAN

VOLUME XXI. Nucleic Acids (Part D)
*Edited by* LAWRENCE GROSSMAN AND KIVIE MOLDAVE

VOLUME XXII. Enzyme Purification and Related Techniques
*Edited by* WILLIAM B. JAKOBY

VOLUME XXIII. Photosynthesis (Part A)
*Edited by* ANTHONY SAN PIETRO

VOLUME XXIV. Photosynthesis and Nitrogen Fixation (Part B)
*Edited by* ANTHONY SAN PIETRO

VOLUME XXV. Enzyme Structure (Part B)
*Edited by* C. H. W. HIRS AND SERGE N. TIMASHEFF

VOLUME XXVI. Enzyme Structure (Part C)
*Edited by* C. H. W. HIRS AND SERGE N. TIMASHEFF

VOLUME XXVII. Enzyme Structure (Part D)
*Edited by* C. H. W. HIRS AND SERGE N. TIMASHEFF

VOLUME XXVIII. Complex Carbohydrates (Part B)
*Edited by* VICTOR GINSBURG

VOLUME XXIX. Nucleic Acids and Protein Synthesis (Part E)
*Edited by* LAWRENCE GROSSMAN AND KIVIE MOLDAVE

VOLUME XXX. Nucleic Acids and Protein Synthesis (Part F)
*Edited by* KIVIE MOLDAVE AND LAWRENCE GROSSMAN

VOLUME XXXI. Biomembranes (Part A)
*Edited by* SIDNEY FLEISCHER AND LESTER PACKER

VOLUME XXXII. Biomembranes (Part B)
*Edited by* SIDNEY FLEISCHER AND LESTER PACKER

VOLUME XXXIII. Cumulative Subject Index Volumes I–XXX
*Edited by* MARTHA G. DENNIS AND EDWARD A. DENNIS

Section I

# Basic Methods

A. Laboratory Requirements and Media
*Articles 1 through 8*

B. General Cell Culture Techniques
*Articles 9 through 18*

# [1] Physical Aspects of a Tissue Culture Laboratory

*By* WILLIAM H. J. DOUGLAS and ROBERT T. DELL'ORCO

## I. Introduction

The material included in this section is intended to present the basic requirements necessary for the introduction of cell and tissue culture techniques into a biochemistry laboratory. The information will be presented as the space and equipment needs for performing the routine operations that are necessary for cell culture production regardless of the size of the proposed facility. These will be considered under four headings:

1. Cleaning and sterilization facilities
2. Media preparation and storage facilities
3. Work area for aseptic manipulation of cell cultures
4. Equipment for routine cell maintenance

These four topics are generally applicable to any type of proposed cell culture; however, this presentation deals exclusively with the culture of mammalian cells. While it will offer a suitable starting point, certain modifications will be necessary for the cultivation of cells from other sources, such as invertebrates and plants. More detailed information on the requirements for these systems can be obtained in recently published reviews.[1,2]

Regardless of the cell system to be employed, the scope of the laboratory facilities will depend largely upon the role planned for cell culture procedures in the individual investigative program. When only a minor role is planned, a minimum of space will be dedicated to cell production and support facilities. When a more active role is anticipated, however, space requirements will be increased and more elaborate facilities may be deemed necessary. Thus, the facilities could all be compressed into one laboratory or separated into individual laboratories each performing only one function. Whether or not a major involvement is planned, another factor to be considered in overall space and equipment requirements is the type of investigations that will be done. For example, cells grown for the harvest of a biological product, such as a hormone, or for the purification of a particular enzyme would require large quantities of cells and the necessary space and equipment requirements for mass culture

[1] O. L. Gamborg and L. R. Wetter, "Plant Tissue Culture Methods." Prairie Reg. Lab., Nat. Res. Counc. Can., Saskatoon, 1975. See also this volume [41].

[2] K. Maramorosch, ed., "Invertebrate Tissue Culture: Research Applications." Academic Press, New York, 1976. See also this volume [39].




ISBN 0-12-181958-2

capabilities. In contrast, most routine biochemical procedures, such as enzyme assays, can be performed with relatively little cellular material and proportionately less space and equipment. Therefore, the types of specialized equipment that are necessary for particular programs can be predetermined with some degree of accuracy.

Although some of the facilities needed for an adequate cell culture laboratory can be used for nothing else, much of the required facilities and equipment need not be dedicated exclusively to this purpose. By sharing equipment through collaborative efforts, the initial investment necessary to begin a tissue culture laboratory can be reduced. Central services and already available resources such as sterility testing, glassware washing, and animal handling areas should be utilized whenever possible. Also, the shared use of ancillary equipment, e.g., microscopes, pH meters, centrifuges, etc., within the same laboratory or with other laboratories should be considered. However, certain precautions, to be detailed later, must be taken when the sharing of facilities is contemplated.

It is hoped that this article will cover most of the main requirements for setting up and running a functional cell culture laboratory. Additional information concerning not only laboratory set-up but also detailed cell culture techniques can be found in several well-written books.[3-6] These texts should be referred to before introducing cell and tissue culture technology into any laboratory.

## II. Cleaning and Sterilization Facilities

Although this area of a cell culture laboratory remains critically important, some of the impact of poor laboratory practices has been lost in recent years due to the ready availability of sterile, disposable labware and commercially prepared media and reagents. With the exception of very small operations in which it is financially feasible to purchase all materials in a prepackaged, disposable form, at least some cleaning and sterilization of glassware is necessary in almost every laboratory. Because of this and because cells in culture can be nutritionally fastidious, investigators should be aware of the care that must be taken in the proper handling of glassware which not only comes into direct contact with the

[3] P. F. Kruse, Jr. and M. K. Patterson, Jr., eds., "Tissue Culture: Methods and Applications." Academic Press, New York, 1973.

[4] J. Paul, "Cell and Tissue Culture," 5th ed. Churchill-Livingstone, Edinburgh and London, 1975.

[5] R. C. Parker, "Methods of Tissue Culture," 3rd ed. Harper (Hoeber), New York, 1961.

[6] G. Penso and D. Balducci, "Tissue Cultures in Biological Research." Elsevier, Amsterdam, 1963.

cells but also with such things as pipettes which are used to transfer culture media. This is of particular importance in connection with toxic substances introduced into glassware by normal processing. While everyone is aware of the problems associated with microbial contamination, little thought may be given to toxic organic products introduced during the manufacture of certain items, inorganic residues from detergent washes, or contamination by metal ions sloughed from pipes. Therefore, proper procedures for cleaning and sterilization of glassware should be carefully followed; several such procedures are available in the literature.[3,5-7] Also it is recommended that glassware used in cell culture procedures be employed exclusively for this purpose and not mixed with glassware used for other purposes. This includes not only the culture vessels themselves but flasks, pipettes, and other miscellaneous items. This precaution insures that all material used in culturing techniques has been subjected to the same vigorous cleaning and that difficult-to-remove reagents do not contaminate the culture systems.

A complete separation of the cleaning facilities from the preparation and the aseptic areas is the ideal situation; however, because of space limitations, the preparation area can be combined with the cleaning area if the glassware is not routinely contaminated with viruses or bacteria. If at all possible, the aseptic areas should be maintained in a location isolated from the cleaning area. The general size of the cleanup area is largely dependent upon the quantity of material to be handled, but a laboratory of 100 to 150 $ft^2$ will accommodate the maximum amount of equipment that would be needed. In laboratories where acid cleaning is to be employed,[5,7] a fume hood with sufficient ventilation and safety features should be incorporated into the overall design.

In general the layout of the laboratory will be determined by the location of the sink. The sources of hot and cold tap water will dictate the placement of the washing equipment, i.e., decontamination and soaking buckets, water purification system, and pipette washer. If the volume of glassware is sufficiently large, a built-in glass washer would be advantageous. Several commercially available models are acceptable; however, an adequate supply of purified water is necessary for the final rinses. Water of suitable purity for these final rinses can be obtained by a single glass distillation, demineralization, or reverse osmosis. The choice of purification method depends on such factors as the condition of the untreated source water, the quantity of water needed, and the amount of space available for the necessary equipment. It may be practical to employ a single water purification system for all laboratory needs if the system is

[7] F. M. Price and K. K. Sanford, *Tissue Cult. Assoc. Man.* **2,** 379 (1976).

capable of producing ultrapure, reagent-grade water. Such systems are discussed in greater detail in Section III which deals with the media preparation area of the laboratory.

After final rinsing, glassware is ready for drying and prepared for sterilization. Large, bulky items may be drained and dried at room temperature on a drying rack. Most glassware is dried at elevated temperatures in a drying oven. It is also convenient to use a specially designed dryer for pipettes since large numbers of pipettes are frequently used in cell culture procedures. While the drying ovens and the sterilizing equipment may be located in a separate room or area, we have found it more convenient for the drying apparatus to be located close to the washing area. After drying, all glass vessels should be covered with paper or aluminum foil and stored in a covered area to prevent dust accumulation. Pipettes should be plugged with cotton and immediately stored in drawers. For larger operations an apparatus for automatically plugging pipettes is available. Adequate bench and storage space needs to be allotted for the handling of glassware; a 10- to 15-foot bench area with overhead and under-counter storage is sufficient for most medium-size laboratories. Since glassware is seldom being washed and prepared for sterilization at the same time, this bench area can be used for both functions.

Sterilization of glassware and other materials can be accomplished by either dry heat with a sterilizing oven or with moist heat by autoclaving. Both pieces of equipment are necessary, and their size and subsequent location in the laboratory design depends upon the projected amount of use. Because of obvious problems with heat and ventilation, it would be better to locate this equipment in a separate room which is readily accessible to the cleanup area. If smaller units are suitable for the intended traffic, they can be placed in the same room as the washing facilities. All sterilizing units should be equipped with temperature-recording charts to maintain a complete record of sterilizing time and temperature. Because of loading and/or air circulation problems within any unit, certain articles in any one load may not reach the desired temperature. It is, therefore, a good practice to label each article with a heat-sensitive indicator tape which changes when the proper temperature has been reached and maintained for the proper time.

Almost all glassware, except that containing rubber tubing connections, may be sterilized by dry heat. The method also is used for material such as silicone grease which cannot be effectively sterilized by moist heat. However, dry heat sterilization is time-consuming and more difficult to control even when a forced-air circulation system maintains uniform conditions within the oven. Because of this, most sterilization procedures are carried out with moist heat by autoclaving.

Since autoclaving is the most commonly used method, a word of caution is necessary about the quality of steam used to supply the autoclave. When the steam is heavily contaminated with impurities, they settle on the surfaces during autoclaving and the advantages of careful washing and rinsing procedures are lost. This may be a major problem where larger, shared facilities are supplied by house steam. Whatever the situation, however, it is recommended that any autoclave using house steam be equipped with a filtering device to remove contaminating material. A better solution to the problem is to use an autoclave that has provisions for its own steam generation. The water for such a unit can be obtained from a purified source thereby eliminating the contaminants at the source.

In addition to the major items of equipment mentioned in this section, several minor ones have proven useful and should be considered when outfitting the cleaning and sterilizing facilities. These include carts to facilitate the transfer of articles between the different areas of the laboratory. These are almost a necessity when the different functional units are very widely separated. Also, provisions should be made for disposal containers in the cleanup area. Ideally, these should be closed containers which would serve as receptacles for used disposable labware, wrappings from sterilized items, and the like. Other items are pipette jars for soaking pipettes before washing, a liquid detergent dispenser, and an ultrasonic cleaning bath for hard-to-clean glassware.

## III. Media Preparation and Storage Facilities

As with the other functional units described in this article, it would be ideal if a separate area were set aside exclusively for media and reagent preparation. If laboratory space is available, a 100 to 150 $ft^2$ room would be adequate to handle the equipment and to provide the bench space for the necessary operations. However, as noted in the previous section, this area can be conveniently combined with that designated for cleanup and sterilization. Although media and other reagents can be purchased as sterile, ready-to-use material, most investigators formulate at least part of what they use. The operations involved in preparing any reagent for use in cell culture are extremely critical, and several things, including a suitable water source, high-quality chemicals, good filtration equipment, and proper storage facilities, are essential for the successful maintenance of cell populations *in vitro*.

The major component of media and other reagents for the propagation of cells in culture is water. Although completely chemically defined media are not yet possible, it is necessary to know within reasonable limits what has been added to media. It is also necessary to eliminate toxic elements of both an organic and an inorganic nature. For these reasons, only water of the highest possible purity should be used to formulate cell culture reagents.[8] The selection of the proper system for generating ultrapure water is largely based on the quality of water coming into laboratory and the quantity of water needed. In many cases it is feasible to employ a relatively high output unit so that the pure product can be used for all laboratory purposes. Even when the water entering the laboratory has been treated by a central purification system, it is advisable to purify it again before use in the reagents.

Treatment of tap water with a mixed-bed ion exchanger followed by glass distillation or two to three successive glass distillations will supply enough ultrapure water for reagent purposes for an average cell culture operation. A second method utilizes water previously treated by a central facility and is capable of producing enough reagent-grade water for all laboratory purposes (final rinsing in glassware washers, autoclave uses, and media preparation). The water is first filtered to remove particulate material, and then it is passed over activated charcoal to absorb organic contaminents, subjected to a mixed-bed ion-exchange resin, and, finally, passed through a submicron filter. A third method can use tap water as its primary source and provides high-purity water for all laboratory uses. This system pre-filters the water and then, in successive steps, subjects it to reverse osmosis, mixed-bed deionization, and submicron filtration. Regardless of the method chosen to produce ultrapure water, provisions should be made for storage of limited quantities of the product water as well as for periodic testing of the water. It is not advisable to store reagent-grade water for long periods of time. However, when it is necessary to store water in a rapid turnover situation, tightly closed, borosilicate glass carboys are recommended. With respect to testing, conductivity is the easiest method for measuring product water, but it only indicates the amount of dissolved ions with no indication of particulate or organic contamination. Since it is usually impractical to test for the other contaminants, and since conductivity at least gives a guide to the efficiency of ion exchange, it should be used routinely. Usually conductivity meters are incorporated into most purification systems; if not, relatively inexpensive units can be purchased for testing purposes.

With few exceptions, essentially all liquid reagents for cell culture are sterilized by filtration. Several types of bacteriological filtering systems

[8] R. W. Pumper, *in* "Tissue Culture: Methods and Applications" (P. F. Kruse, Jr. and M. K. Patterson, Jr., eds.), p. 674. Academic Press, New York, 1973.

have been employed for cell culture purposes including porcelain, asbestos, sintered glass, and membrane filters. Because of their uniform characteristics, ready availability, and superior filtering qualities, membrane filters with a 0.22 $\mu$m pore size are most widely used for filter sterilization in cell culture laboratories. These filters can be obtained in sizes ranging from 13 to almost 300 mm in diameter with the subsequent capability of sterilizing from a few milliliters to several liters of material. The filter holders are constructed of noncorrosive material and, like the filters, come in various sizes and configurations, some of which are prepackaged in a sterile disposable form. It is necessary to have a variety of sizes on hand. The most useful are syringe adapters which take the 13 and 25 mm diameter filters for small volumes; the pressure filter holder for the 47 mm filter which is used for volumes up to 1 liter; and at least one of the larger sizes for sterilizing liter quantities of media. All membrane filtration should be accomplished under positive pressure, thereby reducing foaming, by use of either 95% air–5% $CO_2$ or nitrogen. Larger quantities of media may be placed in specially designed stainless-steel pressure tanks and passed through the filter into a receiving vessel fitted with a filling bell for dispensing into bottles for storage. The entire filtering assembly and receiving vessel can be sterilized as a unit by autoclaving.

Both before and after cell culture media and related reagents are formulated and sterilized, specific types of storage facilities are required. At room temperature, closed, dust-free space is necessary for the storage of stable chemicals, pipettes and other glassware, and filtering apparatus. Again, the size of this space depends upon the size of the operation and the amount of formulation that is anticipated. In addition, a refrigerator, $-20°$ freezer, and $-70°$ ultrafreezer space are required. The refrigerator is used to store a medium once it has been sterilized. Filtered media are usually dispensed into 100 or 500 ml bottles and are stable at refrigerator temperatures for several weeks. The $-20°$ freezer is needed for the short-term storage of labile biochemicals, stock solutions of amino acids and vitamins, and serum. For longer term storage of these materials, a $-70°$ ultrafreezer is necessary.

In addition to the equipment that has been mentioned as being required for a media preparation and storage area, several more items are recommended to complete the facility. These include an analytical balance, a pH meter, an osmometer, a hot plate, a magnetic stirrer, and a full range of glassware, including a good supply of bottles for media storage. It is possible to use much of this equipment in other laboratory areas; but certain instruments, e.g., the osmometer, should be maintained in the media preparation area.

## IV. Work Area for Aseptic Manipulation of Cell Cultures

Many cell culture manipulations may be routinely conducted on a clean bench in an area free from drafts and removed from heavy traffic areas. For more rigorous work or when hazardous agents are used, many different shields and hoods are available which provide the several degrees of protection required by the guidelines of the National Cancer Institute (see this volume [4]). These protective devices provide a sheltered working area and should be equipped with a germicidal lamp and be accessible to gas, vacuum, and electrical outlets. The work surface in a hood must be constructed of stainless steel or Formica. These surfaces will not readily retain dust, and they can be repeatedly cleaned with disinfectants. A minimum of 9 $ft^2$ of work surface should be available in the hood.

The use of filtered laminar air ventilation was introduced to help prevent airborn contaminants in work with cell cultures. Units are available in a variety of configurations which range from whole-room units to models which can be conveniently placed on laboratory bench tops. The working area within a laminar air-flow hood is vented by a continuous displacement of air that has passed through a high-efficiency particle (HEPA) filter. HEPA filters remove particulates (diameter greater than 0.3 $\mu$m) from the air and thus provide a clean environment for manipulation of cell cultures. Filtered air enters the hood from the top and exits at the bottom. Thus aerosols generated during handling of the cells are rapidly removed from the area as they ride the airstream to the exit. Horizontal flow hoods that direct the air toward the operator should not be used for cell culture work. Bunsen burners or hot plates must be avoided in laminar flow hoods since the generation of thermal and convection currents will destroy the laminar flow patterns and reduce the effectiveness of the unit. The performance of a laminar flow unit (HEPA filters, air balance, and air flow) should be checked at least three times a year; a detailed description of these procedures has recently been published.[9]

In order to ensure a clean work surface within the hood, it is important to establish and maintain daily cleaning procedures. The walls and work surface of the hood should be washed with disinfectant (Chlorofen, Zephiran chloride) at the beginning and end of each work day. Additionally, it is important to clean the work surface with disinfectant after completing work with each cell strain. If this practice is rigidly adhered to, the likelihood of cross-contamination of cell strains is greatly reduced.

Sterile materials and other items routinely used in the aseptic work areas should be stored in close proximity to the tissue culture hood in

[9] G. J. McGarrity, *Tissue Cult. Assoc.* **1,** 181 (1975).

closed, dust-free cabinets. Sterile, reusable glassware should be replenished frequently and a schedule of resterilization of unused material should be maintained. Those items that are to be used in a specific cell culture procedure should be placed in the hood prior to work. This step is particularly important when working in a laminar flow system. If the materials are placed in the unit 10 to 15 min before work commences then the laminar flow system has ample time to remove particulates from the work area.

## V. Equipment for Routine Cell Maintenance

Basic tissue culture operations require an incubator, microscope, cell repository materials, and culture vessels in which to grow the cells.

### A. Incubators

Incubators may range from temperature-regulated boxes to elaborate units which control temperature, humidity, and carbon dioxide levels in the atmosphere. The units should be maintained on emergency power circuits in case of an electrical power failure and should be placed close to the aseptic work area. If the buffering system in the tissue culture medium requires equilibration with $CO_2$, e.g., Earle's salts, one will want to use a $CO_2$ incubator. When selecting a $CO_2$ incubator for purchase, consideration should be given to the type of air flow patterns that exist in the unit. An even temperature distribution within the incubator is essential, and experience indicates that horizontal air flow units provide a more uniform distribution of both temperature and $CO_2$. Some incubators are water-jacketed but this is not a necessity. When tissue culture medium containing Hanks' salt is used or when closed culture systems are employed, a $CO_2$ incubator is not necessary and one may simply use a unit that provides a constant and even temperature distribution throughout the chamber.

### B. Microscopes

Microscopes are essential since it is important to examine cultures daily to observe their morphology and rate of growth. Many investigators grow cells, but do not take the time to become familiar with the morphological features of their cultures. In the last few years many laboratories have encountered problems of contamination of their cell cultures with other cell types (see this volume [2]). Frequently, cultures have become contaminated with HeLa cells; these cells replicate rapidly and become the dominant cell type in the culture. If an investigator is familiar with the morphological appearance of the cells in use, it is easier to detect the introduction of a second cell type into the culture system. An inverted microscope equipped with phase-contrast optics is essential for morphological analysis of monolayer cultures and is ideally placed close to the incubator facility. In addition, it is advantageous to have a dissecting microscope and a conventional, compound microscope. One may also want to consider utilization of ultraviolet optics for fluorescence.

### C. Cell Repository Materials

Techniques for long-term preservation of cultured cells by storage in liquid nitrogen are now available, and it is essential for cell culture work to have access to a cryogenic storage facility. A cell repository permits one to bank cultured cells at low population-doubling levels and to store these cells in essentially the same state for a number of years. The availability of a cell repository also eliminates the need to constantly subculture cells in order to maintain cell strains in the laboratory. (See this volume [3])

Monodisperse cell suspensions in tissue culture medium and a cryoprotective agent (glycerol or dimethylsulfoxide) may be frozen at a controlled rate.[10] For most mammalian cells a freezing rate of 1°–3°/min is used until a temperature of −70° is reached. The cells are then transferred to liquid nitrogen storage at −196°, or they may be stored in liquid nitrogen vapor phase at −156°. The choice of the type of controlled-rate freezer that would be most practical for individual needs can be made from several commercially available models. Prior to controlled-rate freezing, cells are placed in glass ampules or plastic screw-cap vials. When glass ampules are used, the neck of the container is sealed by using an ampule sealer and oxygen torch. Plastic vials are sealed by a screw-cap lid. However, this does not provide a tight seal, and the containers must not be immersed in liquid nitrogen. Plastic vials are routinely stored in liquid nitrogen vapor phase.

### D. Culture Vessels

Many types of glass and plastic cell culture vessels are commercially available for use in the cell culture laboratory. These generally are either borosilicate, soda glass, or polystyrene plastic vessels. The polystyrene plastic and soda glass containers are inexpensive and can be routinely discarded after one use. The borosilicate vessels are more expensive, but

[10] J. E. Shannon and M. L. Macy, *in* "Tissue Culture: Methods and Applications" (P. F. Kruse, Jr. and M. K. Patterson, Jr., eds.), p. 712. Academic Press, New York, 1973.

they are resistant to heat and can be washed and sterilized repeatedly without adverse effect on utility. The selection of vessel is usually based partly on personal preference and on the availability of a properly equipped and functioning glassware washing facility.

When borosilicate glassware is reused, care must be taken during the washing procedure to remove all serum and cellular residues from the glassware. In addition, washed glassware must be adequately rinsed to insure complete removal of all traces of solvents and detergents. Because of the problems encountered in washing and rinsing tissue culture glassware, many laboratories routinely use commercially available, sterile, disposable tissue culture plasticware for their monolayer culture work. Sterile, polystyrene petri dishes are available in a variety of sizes (35, 60, 100, and 150 mm diameter). It is important to order tissue-culture-grade plastic petri dishes because cultured cells generally do not attach to bacteriological-grade petri dishes. Various sizes of plastic tissue culture flasks are also available (T-25, T-75, T-150 flasks). These vessels provide 25, 75, or 150 $cm^2$ of surface area, respectively, for growing cells. In addition, plastic tubes or multiwell plates can also be used as substrates for growing monolayer cultures.

For many biochemical analyses a large number of cells are required. The method selected for growing mass cultures is clearly related to the particular cell strain or cell line to be used. If diploid, anchorage-dependent cells are employed, then monolayer cell culture techniques are required. In contrast, if heteroploid or other nonanchorage-dependent cell types are selected then suspension culture techniques most probably can be utilized. Culture systems that permit growth of large numbers of cells with either monolayer or suspension culture techniques are commercially available.

For large-scale growth of anchorage-dependent cells in monolayer culture, many laboratories use roller bottles. These culture vessels are available in a variety of sizes in both borosilicate glass or polystyrene plastic, and they permit harvests of up to $2 \times 10^7$ viable cells from one 730-$cm^2$ vessel. In 1970 Kruse[11] introduced a roller bottle perfusion apparatus that permits harvest of $1.9 \times 10^8$ normal diploid cells or $2 \times 10^9$ heteroploid cells from a single 730 $cm^2$ borosilicate glass roller bottle. This high cell density is achieved by the perfusion of fresh medium into the culture vessels at programmed intervals. The system permits control of pH and influent nutrient levels in a more precise manner than is possible in monolayer cultures fed at 2 to 3 day intervals.

Polystyrene roller bottles containing a spiral of Melinex are also commercially available. The Melinex polyester sheet provides 8000 $cm^2$ of

[11] P. F. Kruse, Jr., L. N. Keen, and W. L. Whittle, *In Vitro* **6**, 75 (1970).

culture surface, and from a single roller bottle $2 \times 10^9$ heteroploid cells can be routinely harvested.[12]

Suspension culture techniques are available for the large-scale production of nonanchorage-dependent cells. Cells can be maintained in suspension by use of a rotary shaker or by stirring. The shaker technique was introduced by Earle *et al.*[13] and is particularly suited to studies requiring replicate cultures. The cells are suspended in medium containing methylcellulose (15 cps) in Erlenmeyer or round-bottom flasks and placed in a rotary shaker. Currently the most popular method for suspension cultures uses a magnetic stirrer to keep the cells constantly agitated in tissue culture medium supplemented with methylcellulose.[14] A wide range of sizes of spinner culture vessels are available. The stirring rate varies depending on the cell type used, but it must be slow enough to avoid foaming.

A recent modification of the spinner system is the spin filter culture device.[15] This unit was developed to grow cells to high population densities and to provide for removal of drugs from the tissue culture medium at accelerated rates. Medium is removed through a rotating filter, the design of which prevents the accumulation of cells on or within the filter. The volume of medium used in this unit can vary from 200—1000 ml, and cells routinely can be grown to densities of $10^8$ cells/ml.

Two other culture systems have recently been developed for specific purposes, but they warrant mention because of their applicability to a variety of research systems. The dual-rotary circumfusion system developed by Rose permits maintenance of cells for periods of several months in a differentiated, functional state *in vitro*.[16] Such cells can be readily examined throughout their *in vitro* culture period by high-resolution microscopy, and replicate cultures are possible. The cells are grown in Rose chambers under a cellophane membrane. In many instances the monolayer cultures are initiated from explants. The cells in this culture system undergo only limited replication but remain in a viable and differentiated state for several months *in vitro*.[16]

The artificial capillary system introduced by Knazek and Gullino[17] permits cells to grow to tissue-like densities that often approximate the *in*

[12] W. House, *in* "Tissue Culture: Methods and Applications" (P. F. Kruse, Jr. and M. K. Patterson, Jr., eds.), p. 338. Academic Press, New York, 1973.

[13] W. R. Earle, E. L. Schilling, J. C. Bryant, and V. J. Evans, *J. Natl. Cancer Inst.* **14,** 1159 (1954).

[14] R. S. Kuchler and D. J. Merchant, *Proc. Soc. Exp. Biol. Med.* **92,** 803 (1956).

[15] P. Hemmelfarb, P. S. Thayer, and H. E. Martin, *Science* **164,** 555 (1969).

[16] G. G. Rose, *in* "Tissue Culture: Methods and Applications" (P. F. Kruse, Jr. and M. K. Patterson, Jr., eds.), p. 283. Academic Press, New York, 1973.

[17] R. A. Knazek and P. M. Gullino, *in* "Tissue Culture: Methods and Applications" (P. F. Kruse, Jr. and M. K. Patterson, Jr., eds.), p. 321. Academic Press, New York, 1973. See this volume [14].

*vivo* state. The artificial capillaries are semipermeable cellulose acetate or polycarbonate membranes through which tissue culture media is constantly perfused. Diffusable cell products can be retrieved from the perfusate, and in several instances hormone-producing cells synthesize and secrete at much greater rates in the artificial capillary system than comparable numbers of cells growing in monolayer culture.

## VI. Conclusion

The material in this article, including the partial list of commercial sources provided in the table, outlines in general terms the equipment and

A Partial List of Suppliers of Cell Culture Materials

*Media & Biologicals*

Associated Biomedical Systems
872 Main St.
Buffalo, NY 14203

Biofluids, Inc.
1146 Taft St.
Rockville, MD 20852

Cappel Laboratories, Inc.
Downingtown, PA 19335

Colorado Serum Company
4950 York St.
Denver, CO 80216

Connaught Laboratories Ltd.
1755 Steeles Ave. W.
Willowdale, Ontario
CANADA M2N 5T8

Difco Laboratories, Inc.
P. O. Box 1058A
Detroit, MI 48201

Flow Laboratories, Inc.
1710 Chapman Ave.
Rockville, MD 20852

GIBCO
3175 Staley Rd.
Grand Island, NY 14072

HEM Research, Inc.
5451 Randolph Rd.
Rockville, MD 20852

ICN Pharmaceuticals, Inc.
Life Science Group
26201 Miles Rd.
Cleveland, OH 44128

International Biological
481 S. Stonestreet Ave.
Rockville, MD 20852

International Scientific Industries, Inc.
P. O. Box 9
Cary, IL 60013

Irvine Scientific Sales Co., Inc.
P. O. Box 4492
Irvine, CA 92664

K. C. Biologicals, Inc.
P. O. Box 5441
Lenexá, KS 66215

Microbiological Associates, Inc.
4733 Bethesda Ave.
Bethesda, MD 20014

Miles Laboratories
Elkhart, IN 46514

(*continued*)

A PARTIAL LIST OF SUPPLIERS OF CELL CULTURE MATERIALS (*continued*)

*Media & Biologicals* (continued)

Mogul Corp.
GIBCO Diagnostics
Laboratory Park
Chagrin Park, OH 44022

North American Biologicals, Inc.
15960 NW 15th Ave.
Miami, FL 33169

Pacific Biological Co.
2400 Wright Ave.
Richmond, CA 94804

Reheis Biochemical Dept.
Armour Pharmaceutical Co.
P. O. Box 511
Kankakee, IL 60901

*Plasticware*

Cooke Laboratory Products
900 Slaters Lane
Alexandria, VA 22314

Corning Glass Works
Science Products Div.
Corning, NY 14830

Costar
Div. Data Packaging Corp.
205 Broadway
Cambridge, MA 02139

Falcon
Div. Becton-Dickinson
P. O. Box 243
Cockeysville, MD 20130

Lab-Tek Products
Div. Miles Laboratories
30 W. 475 N. Aurora Rd.
Naperville, IL 60540

Linbro Scientific
681 Dixwell Ave.
New Haven, CT. 06511

Lux Scientific Corp.
1157 Tourmaline Cr.
Newbury Park, CA 91320

Vangard International
Box 3112
Red Bank, NJ 07701

York Scientific Ltd.
Ogdensburg, NY 13669

*Incubators*

Forma Scientific
P. O. Box 649
Marietta, OH 45750

Hotpack Corp.
5086A Cottman Ave.
Philadelphia, PA 19135

Lab-Line Instruments, Inc.
Bloomingdale Ave.
Melrose Park, IL 60160

National Appliance Co.
Heinicke Instruments Co.
P. O. Box 23008
Portland, OR 97223

Percival Refrigeration & MFG. Co. Inc.
P. O. Box 249
Boone IA 50036

Wedco, Inc.
P. O. Box 223
Silver Spring, MD 20907

*Filtration Systems*

Bioquest
Div. of Becton-Dickinson & Co.
P. O. Box 243
Cockeysville, MD 21030

Gelman Instrument Co.
15 Chestnut St.
Sussex, NJ 07461

*Filtration Systems* (continued)

Millipore Corporation
Ashby Road
Bedford, MA 07130

Nalge Co., Div. Sybron Corp
Nalge Labware Div.
75 Panorama Creek Dr.
Rochester, NY 14602

Nuclepore Corp.
7035 Commerce Circle
Pleasanton, CA 94566

*Water Systems*

Barnstead Sybron Corp.
225 Rivermoor St.
Boston, MA 02132

Bellco Glass Co.
340 Edwardo Rd.
Vineland, NJ 08360

Corning Glass Works
Science Products Div.
Corning, NY 14830

Culligan USA
1 Culligan Parkway
Northbrook, IL 60062

Hydro Service & Supplies, Inc.
P. O. Box 2891
Durham, NC 27705

Ultrasciences, Inc.
2504 Gross Point Rd.
Evanston, IL 60201

*Microscopes*

American Optical Corp.
Scientific Instrument Div.
Sugar & Eggert Rds.
Buffalo, NY 14215

Bausch & Lomb, Inc.
Scientific Instrument Div.
Depot Rd., R.D. #6
Auburn, NY 13021

E. Leitz
Link Drive
Rockleigh, NJ 07647

Nikon, Inc.
Instrument Group
EPOI, 623 Stewart Ave.
Garden City, NY 11530

Olympus Corp.
Precision Instrument Div.
2 Nevada Dr.
New Hyde Park, NY 11040

Swift Instruments, Inc.
1190 N. 4th St.
San Jose, CA 95112

Wild Heerbrugg Instruments, Inc.
465 Smith St.
Farmingdale, NY 11735

Carl Zeiss, Inc.
444 Fifth Ave.
New York, NY 10018

*Cryogenic Supplies*

Forma Scientific
Box 649-T
Marietta, OH 45750

Kelvinator Commercial Products, Inc.
621 Quay St.
Manitowoc, WI 54220

Revco, Inc.
Scientific and Industrial Div.
1188 Memorial Dr.
West Columbia, SC 29169

Union Carbide
Linde Division
270 Park Avenue
New York, NY 10017

space required to initiate a cell culture program in a biochemistry laboratory. Few specific recommendations were given because the absolute requirements for any individual program will depend upon the existing facilities, the projected scope of the program, as well as the type of cells to be cultured and their intended use. Although the specifics will change according to the individual circumstances, the basic principles and problems considered here will remain unchanged regardless of the situation. Whether coverslip cultures or mass cultures are employed, or whether the entire operation is contained in one room or a suite of laboratories, cleanliness of glassware, purity of water and other reagents, and maintenance of sterility will ultimately determine the reliability of any cell culture system. The information presented here is a guide to what is necessary to achieve these basic goals.

## [2] Detection of Contamination

*By* GERARD J. MCGARRITY

The presence of adventitious agents in cell cultures is incompatible with the concept of standardized, defined systems. Presence of extraneous agents may produce either gross turbidity and rapid destruction of the culture, or no turbidity and little to moderate cytopathic effects on the culture. The latter type can remain undetected for prolonged periods, and have profound effects on the cell culture and experimental results.

Contamination may originate in the tissue specimen used to initiate the cell culture, in media, especially in bovine serum, or in the general environment. The contamination may be bacterial, mycoplasmal, fungal, viral, or cellular. Detection methods described here will effectively monitor microbiological contamination in cell cultures and media. These should be viewed as part of an overall quality-control program. Methods to prevent and control contamination are described elsewhere.[1-4]

The limitations of detection methods must be appreciated. The major limitations for detection of bacterial, mycoplasmal, and fungal organisms are sample size, level of contamination, type of growth media utilized, and presence of antibiotics in the sample. Contamination in bovine sera

[1] G. J. McGarrity and L. L. Coriell, *In Vitro* **6**, 257 (1971).

[2] G. J. McGarrity, *Tissue Cult. Assoc. Manual* **1**, 167 (1975).

[3] G. J. McGarrity, *Tissue Cult. Assoc. Manual* **1**, 181 (1975).

[4] G. J. McGarrity, V. Vanaman, and J. Sarama, *in* "Mycoplasma Infection of Cell Cultures" (G. J. McGarrity, D. G. Murphy, and W. W. Nichols, eds.), p. 213. Plenum, New York, 1978.



ISBN 0-12-181958-2

and biologicals is generally more difficult to detect because of the low frequency of contamination (1% or less of a lot) and the low concentration of organisms in contaminated units, 1–10 organisms per milliliter or less. Difficulties in detecting contaminants in cell cultures are frequently due to improper selection of microbiological media and presence of antibiotics in the sample. Antibiotics can mask contamination and yield false negatives. The United States Pharmacopeia states that there is a possibility that low levels of contaminants may remain undetected even in properly designed sterility tests.[5]

## Detection of Bacterial and Fungal Contaminants

Three factors influence the results of sterility tests. First, no single microbiological medium will detect all possible contaminants. A compromise is necessary for practical quality-control testing. Second, the sterility of the unit is based on testing a small portion of the unit. Third, presence of antibiotics can yield false negative results.

When contamination is introduced into antibiotic free cell cultures, the organisms typically will grow rapidly and produce gross turbidity in 18–24 hr. Antibiotics in the medium will kill sensitive organisms and select resistant ones. Resistant organisms may produce rapid turbidity, e.g., pseudomonads and molds, or they may be more fastidious and slower growing. In the latter case, they can remain undetected even when subjected to routine testing. In this laboratory, whenever fastidious organisms have been isolated from cell cultures, they have been in cultures containing antibiotics. L forms, mycobacteria, anaerobes, species of *Hemophilus,* fastidious diphtheroids, and yeast have been isolated from cell cultures containing antibiotics.

### A. Bovine Serum

A large number of bacterial, fungal, mycoplasmal, and viral agents have been isolated from bovine serum.[4] In this Institute, samples of each lot of bovine serum are tested before purchase. After purchase, each bottle is tested for bacteria and fungi before use. Testing before purchase includes bacterial, fungal, mycoplasmal, and growth promotion. For tests before purchase, the serum supplier should be requested to supply a sample of the lot. If an entire lot is purchased, three bottles can be requested. If less than a lot is purchased, a 100–200 ml aliquot can be

[5] "The United States Pharmacopeia," 19th ed. Mack Publ. Co., Easton, Pennsylvania, 1975.

requested. Bacterial and growth promotion tests are described below. Tests for mycoplasma and viruses are described elsewhere in this article.

### *1. Testing for Bacteria and Fungi before Purchase*

Prepare duplicate racks containing one each of the following tubes of culture media.

a. Fluid thioglycolate broth (Bioquest, Inc., Cockeysville, Maryland), 5 ml/tube
b. Trypticase soy broth (Bioquest, Inc.) 5 ml/tube
c. Sabouraud dextrose broth (Bioquest, Inc.), 5 ml/tube
d. Sterile short disposable tube
e. Two blood agar plates

Aseptically open serum bottle in a mass air flow room or laminar air flow cabinet. Carefully flame the top of the bottle. Using a sterile pipette, remove a 10-ml aliquot for testing. Add 3.0 ml of serum to each of the short disposable tubes. Add 0.5 ml to each of the culture tubes of fluid thioglycolate broth, trypticase soy broth, and Sabouraud dextrose broth. Inoculate 0.1 ml onto each blood agar plate. Incubate one rack and one blood agar plate at 37° and the other set at 30°. Check the tubes and plates daily for 2 weeks for evidence of contamination.

If present, contamination is generally evident within 3–4 days. Contamination in the tubes of broth must be distinguished from crystal formation and precipitation. This can be done by making a transfer to another tube of broth and blood agar plate and by making a gram stain on the broth or a centrifuged pellet of the broth. If possible, contaminants should be identified to aid in tracing possible sources. To prevent false positives, introduction of environmental contaminants during the testing must be minimized. One episode of bacterial contamination of cell cultures has been traced to environmental contamination that occurred during sterility tests.[4]

### *2. Testing for Growth Promotion of Cell Cultures*

Select two standard cell cultures of the type for which the serum is to be used as a growth supplement.

Prepare culture medium as it will be used for these cell cultures and add 20% of the test bovine serum. As a control, use medium with 20% bovine serum of proven growth promotion ability.

Seed a T25 plastic flask with 500,000 cells of each cell culture and add 5 ml of culture medium.

Incubate at 37° for 1 week or until culture is confluent. Harvest the cells by the usual technique and pass to three T25 flasks.

Repeat this procedure once a week for four passages and observe for growth, morphology, full sheeting of cells after 1 week incubation, and for evidence of toxicity, granularity, and vacuolization as compared to growth of the same cells in a standard nontoxic serum.

By the end of 1 month data are available on sterility tests and growth promotion and on whether the serum lot can be accepted or rejected. Note: The above time and cell concentration references may have to be modified for different cell cultures.

*3. Testing Individual Bottles of Serum*

Contaminated bottles of bovine serum have occasionally been detected in a lot which passed all sterility tests before purchase. For this reason we require that users pretest each bottle of serum before it is added to cell cultures media. This test is carried out in the same way as listed under paragraph 1 above, and observed for 2 weeks before use.

### B. Cell Culture Media

All purchased single-strength media, as well as media and solutions formulated in-house from basic ingredients or powdered base, are aseptically opened and tested as in paragraph 1 above except that the broths are incubated only at 37° and observed for 10 days.

### C. Cell Cultures

Cell cultures free of antibiotics can be tested for bacterial and fungal contamination by procedures described in paragraph 1 above except that broths are incubated only at 37° and observed for 10 days.

## Mycoplasmal Testing

Mycoplasmas are particularly well suited as infectants of cell cultures. They grow to high concentrations in cell cultures, $10^6$ to $10^8$ colony forming units (CFU) per milliliter of supernatant medium being representative; they are resistant to many antibiotics; they usually produce no gross turbidity in the cultures they infect; they are relatively difficult to detect; and they have profound effects on the cultures they infect. Reviews of the effects mycoplasmas have on cell cultures and other effects are available.[6,7]

[6] E. Stanbridge, *Bacteriol. Rev.* **35,** 206 (1971).

[7] G. J. McGarrity, D. G. Murphy, and W. W. Nichols, eds., "Mycoplasma Infection of Cell Cultures." Plenum, New York, 1978.

Many techniques are available for detection of mycoplasmas. These are based on microbiological culture,[8,9] fluorescence methods,[10,11] electron microscopy,[12] and biochemical methods.[13-17] Each approach has certain limitations; the relative sensitivity of each is unknown. Most of the data available on detection techniques have been based on fibroblast cell cultures. Some methods and criteria may have to be modified when working with epithelial cells, lymphocytes, endothelial cultures, and other cells with specialized functions. Positive and negative controls are essential for each test method. This may necessitate the use of an indicator cell culture for some detection methods.

### A. Bovine Serum

Mycoplasmal testing is performed on bovine serum before purchase. If mycoplasmas are present, they are likely to be in low concentrations. Therefore, test samples are performed with large volumes, following the procedure of Barile and Kern.[18]

#### *1. Media*

Base medium is prepared in bulk and stored in 75-ml amounts until use. Base medium consists of mycoplasma broth base (Bioquest, Inc.) supplemented with 1 g dextrose, 1 g L-arginine monohydrochloride (Eastman Kodak Co., Rochester, New York), and 1 ml of 0.2% phenol red in 1000 ml distilled water. After autoclaving and cooling, the medium is dispensed in 75-ml aliquots. Add 5 ml yeast extract (Flow Laboratories, Rockville, Maryland), 20 ml horse serum (Flow Laboratories), and 1 ml of a 0.2% DNA stock solution (Thymic DNA, Miles Labs, Kankakee, Illinois). Broth tubes are dispensed in 5-ml aliquots in screw-capped tubes and refrigerated until used. If mycoplasma testing will be done on a lim-

[8] G. J. McGarrity, *Tissue Cult. Assoc. Man.* **1**, 133 (1975).

[9] M. F. Barile, H. E. Hopps, M. W. Grabowski, D. B. Riggs, and R. A. DelGiudice, *Ann. N.Y. Acad. Sci.* **225**, 251 (1973).

[10] T. R. Chen, *Exp. Cell Res.* **104**, 255 (1977).

[11] R. A. DelGiudice and H. E. Hopps, *in* "Mycoplasmal Infection of Cell Cultures" (G. J. McGarrity, D. G. Murphy, and W. W. Nichols, eds.), p. 57. Plenum, New York, 1978.

[12] D. M. Phillips, *in* "Mycoplasmal Infection of Cell Cultures" (G. J. McGarrity, D. G. Murphy, and W. W. Nichols, eds.), p. 105. Plenum, New York, 1978.

[13] E. M. Levine, *Methods Cell Biol.* **8**, 229 (1974).

[14] E. L. Schneider, E. J. Stanbridge, and C. J. Epstein, *Exp. Cell Res.* **84**, 311 (1974).

[15] A. G. Perez, J. H. Kim, A. S. Gelbard and B. Djordjevic, *Exp. Cell Res.* **70**, 301 (1972).

[16] E. M. Levine, L. Thomas, D. McGregor, L. Hayflick, and H. Eagle, *Proc. Natl. Acad. Sci. U.S.A.* **60**, 583 (1968).

[17] G. J. Todaro, S. A. Aaronson, and E. Rands, *Exp. Cell Res.* **65**, 256 (1971).

[18] M. F. Barile and J. Kern, *Proc. Soc. Exp. Biol. Med.* **138**, 432 (1971).

ited basis, the volume of medium prepared at one time can be reduced accordingly.

For agar medium, 9 g of Noble agar (Difco Labs, Detroit, Michigan) are added per liter of base medium, autoclaved, cooled, and dispensed in 75-ml aliquots in containers that will conveniently hold 100 ml. These bottles are stored in the refrigerator until use. For preparation of agar plates, heat the base medium to 96° to dissolve the agar. Place the bottles of base medium, horse serum, yeast extract, and DNA into a water bath at 50°. Allow all components to equilibrate at 50°. Add 5 ml yeast extract, 20 ml horse serum, and 1 ml of a 0.2% DNA solution to 75-ml aliquots. Dispense into 60 × 15 mm Petri dishes (Falcon Plastics, Oxnard, California), approximately 10 ml per plate. Final mixing of media components above 50° can result in cloudy plates that are more difficult to read. Mixing and dispensing should be done quickly since agar will solidify at 45°. Agar plates are refrigerated and used within 1–2 weeks. Wrapping stacks of plates in aluminum foil or plastic will reduce dehydration. The gel strength of different lots of Noble agar may vary. The lowest concentration that yields a gel should be used. All media should be autoclaved as soon after mixing as possible. Final pH should be 7.2 ± 0.2.

### 2. *Testing Serum*

Testing serum for mycoplasma is performed by inoculation of 25 ml of the serum under test into each of two bottles containing 100 ml of mycoplasma broth and incubation at 37° under aerobic and anaerobic conditions. Anareobic atmospheres can be generated easily in a Gas Pak system (Bioquest, Inc.) Aliquots of 0.1 ml are subcultured from the broths to agar plates after 4 and 7 days incubation. Agar plates are examined microscopically (×100) for colonies for at least 3 weeks. Aerobic plates should be taped to reduce drying.

### 3. *Testing Cell Cultures*

*a. Monolayer Cultures.* Monolayer cultures should be at least 60% confluent for testing and be in antibiotic-free medium. Since proteolytic enzymes may have an adverse effect on mycoplasma, cells should be scraped from the container surface.

Remove and discard all but 3 ml of the antibiotic-free medium. Scrape some cells from the monolayer surface with a rubber policeman or glass rod. Place the suspension of cells and spent medium in a Wasserman or tube of similar size. Cap. To test the suspension cultures, gently shake flask and remove 3 ml for assay.

Inoculate a mycoplasma broth tube with 0.5 ml and mycoplasma agar plate with 0.1 ml of the test specimen. Incubate the plate and broth anaerobically in a Gas Pak or equivalent system. The cap on the broth tube should be loose and the plate unsealed. Monitor the anaerobic state with a methylene blue oxidation-reduction indicator (Bioquest, Inc.) or equivalent. After 7 days of incubation inoculate an agar plate with 0.1 ml of the inoculated broth.

Microscopically examine plates at least weekly for presence of typical mycoplasma colonies using $\times 100$ magnification. Colonial morphology may vary among different species, especially on primary isolation. However, the "fried egg" appearance will predominate.

Keep plates 3 weeks before recording negative results. Mycoplasma colonies must be distinguished from crystal formation or "pseudocolonies." Mycoplasmal isolates can be identified by the growth inhibition test[19] or by epi-fluorescence.[20]

Microbiological culturing may be difficult to set up unless a microbiology lab or microbiologist is available. Quality control of various components should be instituted. It should be recognized that microbiological culture will not detect all strains of *Mycoplasma hyorhinis*.

*b. Limitations of Microbiological Culture for Mycoplasmas.* Hopps *et al.*[21] have shown that certain strains of *M. hyorhinis,* a frequent cell culture isolate, will not propagate on cell-free media. DelGiudice and Hopps[11] have shown that 62% of *M. hyorhinis* cell culture isolates did not propagate on agar. To detect these strains and to obtain a rapid test result, we inoculate 0.1 ml of the test specimen into a plastic Leighton tube containing a plastic coverslip (Costar Industries, Cambridge, Massachusetts) that has been inoculated 1–2 days previously with 20,000 3T-6 mouse embryo fibroblast cells as an indicator culture. This system is incubated for 4 days. At the end of that time, the plastic coverslip is removed, cut with sterile scissors, and two fluorescent stains are performed as enumerated in sections c and d, below. These are an immunofluorescent sera specific for *M. hyorhinis* and Hoechst fluorochrome compound 33258. The Hoechst stain binds to DNA, staining both cell nuclei and, if present, mycoplasmal DNA.

*c. Immunofluorescent Staining for M. hyorhinis.* The procedure is essentially that reported by DelGiudice and Hopps.[11] Coverslips are rinsed twice in phosphate-buffered saline (PBS), pH 7.4, fixed in acetone for 2 min, rinsed in PBS, and stained with immuno-fluorescent antisera specific for *M. hyorhinis,* strain BTS-7. Stock antisera and antigens for preparing

[19] W. A. Clyde, Jr., *J. Immunol.* **92,** 958 (1964).

[20] R. A. DelGiudice, N. Robillard, and T. R. Carski, *J. Bacteriol.* **93,** 1205 (1966).

[21] H. E. Hopps, B. C. Meyer, M. F. Barile, and R. A. DelGiudice, *Ann. N.Y. Acad. Sci.* **225,** 265 (1973).

working solutions are available from Research Resources Branch, National Institute of Allergy and Infectious Diseases, NIH, Bethesda, Maryland 20014. The stain is allowed to incubate for 30 min at room temperature and is rinsed with PBS. Buffered glycerol (Bioquest, Inc.) is used as a mounting medium.

*d. DNA Stain.* This procedure has been reported by Chen.[10] We have incorporated the modifications of DelGiudice and Hopps into our procedures.[11]

Prepare stock solution of stain to a concentration of 50 $\mu$g/ml in distilled water. This is stored at 4° in light-protected bottles. Thimersol (1:10,000) or similar disinfectant is added to retard microbial growth. Working solutions are prepared fresh for each application by dilution in potassium phospate–citric acid, pH 5.5, to make a final stain concentration of 0.05 ug/ml.

Keep the coverslip submerged in 1 ml of medium and add 1 ml of fixative, acetic acid methanol (1:3). Aspirate after 2 min. Air dry and stain for 10 min; wash twice with distilled water, and mount in buffered glycerol (Bioquest, Inc.).

*e. Microscopy.* Preparations can be viewed with either transmitted or incident illumination. Incident illumination has been used in this laboratory (Leitz Orthoplan plus Ploem Illumination). In immunofluorescent preparations, individual *M. hyorhinis* organisms will fluoresce. This is particulary noticeable on plasma membranes of the cultured cells. With the DNA stain, appearance of mycoplasma and other prokaryotic organisms can be clearly observed as extranuclear fluorescent particles.

Controlled studies have indicated that the DNA stain is effective in detecting mycoplasmas[11] although standardization, proper controls, and confirmation testing by other techniques are necessary.

*f. Biochemical Tests.* A variety of biochemical tests are available. These are based on properties of mycoplasmas not shared by mammalian cells in culture. Depending on the capabilities and facilities, these procedures might be more suitable than the techniques described above. The limitations of each test method must be acknowledged. Appropriate controls must be included in each test.

*g. Uridine Phosphorylase Assay.* This procedure was developed by Levine and is based on the presence in all mycoplasma[13] tested of uridine phosphorylase activity.[22] Although mammalian tissues *in vivo* have this activity, cells in culture do not. Activity is detected by paper chromatography.

[22] E. M. Levine and B. G. Becker, *in* "Mycoplasma Infection of Cell Cultures" (G. J. McGarrity, D. G. Murphy, and W. W. Nichols, eds.), p. 87. Plenum, New York, 1978.

A total of $2 \times 10^7$ cultured cells are used and are lysed with 3 ml of phosphate–Triton buffer.[23] Levine states that as few as $3 \times 10^6$ cells in 0.5 ml buffer can be used. [$^{14}$C]uridine (40 $\mu$Ci/$\mu$mol) is added as the nucleoside substrate, and the mixture is incubated at 37° for 180 min. After incubation, uridine is separated from uracil at 37° by paper chromatography. The solvent system consists of 70 ml boric acid (4%, ca. 0.65 *M*), 1 ml concentrated ammonium hydroxide, and 430 ml *n*-butanol. Some phase separation occurs at room temperature but is cleared by warming to 37° and shaking.

Samples are obtained at 30 and 180 min. Uridine and uracil spots are located on the developed chromatogram with a short-wave UV light. The radioactivity in these spots is determined. The conversion rate at 180 min in uninfected cells is generally lower than 10%; mycoplasma-infected cells have much higher rates, typically 20–90%.

*h. Uridine/Uracil Uptake.* This procedure was developed by Schneider and Stanbridge,[14,24] and based on the marked alterations in the incorporations of free bases and nucleosides into the nucleic acid of mycoplasma-infected cells. Mammalian cells incorporate relatively large amounts of uridine and only negligible amounts of uracil. Mycoplasma can incorporate uracil. Mycoplasmal uridine phosphorylase can convert uridine to uracil, preventing uptake of uridine by cultured cells. The net effect is that the ratio of uptake of uridine to uptake of uracil is significantly reduced in mycoplasma-infected cells.

Cells to be tested are inoculated into each of six plastic 25 cm$^2$ flasks, $2.5 \times 10^5$ cells per flask. After 24 hr incubation at 37°, three flasks are inoculated with [$^3$H]uridine, 5 $\mu$Ci/ml; three flasks are inoculated with [$^3$H]uracil, 5 $\mu$Ci/ml. After 18–24 hr additional incubation at 37°, the labeled nucleic acids are extracted and counted. Cells are removed from the flask by trypsinization and pelleted at 200 g for 10 min in a refrigerated centrifuge. The pellet is washed twice with 1 ml of phosphate-buffered saline (PBS, pH 7.4) free of calcium and magnesium. Centrifuge at 200 g for 10 min. Decant and suspend the pellet in 1 ml 5% trichloracetic acid (TCA). Precipitate the nucleic acids by placing the tube in ice for 30 min. Centrifuge at 900 g for 15 min. Decant supernatant liquid and add 1 ml of

[23] The phosphate–Triton buffer is made from three stock solutions. A 0.5 *M* solution of $Na_2HPO_4$ is prepared and adjusted to pH 8.1 with 1 *N* HCl (approximately 2.9 ml/100 ml). Triton X-100 (Rohm and Haas, Inc., Philadelphia, Pennsylvania) is prepared as a 10% (v/v) solution in water. Nonradioactive uracil is prepared in a 0.01 *M* water solution for use as a tracer in the chromatography system. All solutions are stored at 4°; $Na_2HPO_4$ must be warmed before use. The final buffer is prepared as follows: 10 ml $Na_2NPO_4$, 10 ml uracil, 5 ml Triton X-100, 75 ml water. The final buffer is stored at 4°.

[24] E. L. Schneider and E. J. Stanbridge, *Methods Cell Biol.* **10,** 277 (1975).

perchloric acid (PCA). Incubate in 90° water bath for 15 min to hydrolyze the nucleic acids. Remove and retain the supernatant fluid. Add 0.1 ml of each of the PCA supernatants to separate vials containing Bray scintillation fluid or equivalent. Count in a liquid scintillation counter.

Ratio of uptake of uridine to uracil is determined by dividing the mean of uridine incorporation by the mean of uracil incorporation. According to Stanbridge and Schneider, ratios of 1000 for cell lines and 400 for human skin fibroblasts are considered negative; values below 100 have always been found to indicate mycoplasma.[14,24] An alternate method is available using double labeling.[25]

*i. Other Biochemical Procedures.* An adenosine phosphorylase assay was developed similar to uridine phosphorylase by Hatanaka *et al.*[26] This is similar to the uridine phosphorylase described above, although mycoplasma strains have been encountered that have lacked the enzyme.

Todaro *et al.* developed a rapid detection method using density gradient separation of tritiated nucleic acids.[17] Cell cultures are incubated for 18–20 hr with either tritiated uridine or uracil, supernatant fluids are precipitated with ammonium sulfate, layered on a 15–60% linear sucrose density gradient, and centrifuged at 40,000 rpm in an SW-41 rotor for 90 min. The gradient is collected dropwise, precipitated in trichloracetic acid, and counted in a scintillation counter. Mycoplasmal infected cultures yield a clearly defined peak at 1.22–1.24 gm/cm$^3$; clean cultures lack this peak. This technique has not been extensively used with different cell cultures and mycoplasma species.

Mycoplasma ribosomal RNA can be distinguished from mammalian ribosomal RNA by electrophoretic mobilities.[16] Polyacrylamide gel electrophoresis has been used to detect mycoplasma 16 S and 23 S RNA in infected cultures. Levine has noted that some mycoplasmal strains will not incorporate uridine tracers, leading to false negatives.

Autoradiography has been used to detect mycoplasma infection.[15] The difference between infected and noninfected cultures can be striking. Uninfected cultures will incorporate tritiated thymidine only over the nuclear areas. Infected cultures, however, will incorporate thymidine (cleaved to thymine by mycoplasma phosphorylase) around the plasma membrane of cells, giving the appearance of cytoplasmic grains. It has been reported that certain strains of mycoplasma cannot incorporate precursors and may result in false negatives.[22]

[25] E. J. Stanbridge and E. L. Schneider, *Tissue Cult. Assoc. Manual* **2,** 371 (1976).

[26] M. Hatanaka, R. A. DelGiudice, and C. Long, *Proc. Natl. Acad. Sci. U.S.A.* **72,** 1401 (1975).

### B. Electron Microscopy

Since the concentration of mycoplasmas in infected cultures is usually on the order of $10^6$ to $10^8$ per milliliter, electron microscopy can be used as a method of detection.[12] True infection must be distinguished from artifacts that can render diagnosis difficult. Use of an indicator cell culture known to be free of mycoplasma can minimize this difficulty.

Although transmission electron microscopy (TEM) is generally more available, sample preparation is more prolonged. In the examination of sections, only one plane through a cell is examined. Therefore, multiple sections are required. Because of the geometry of sectioning and the level of mycoplasma adsorbed to cells, false negatives are possible.

Scanning electron microscopy (SEM) is more suitable for routine screening. Van Diggelen *et al.*[27] and Phillips[12] have used this to detect mycoplasma infection. Cells are grown on glass coverslips for 3–4 days and fixed in 2.5% glutaraldehyde buffered at pH 7.4 with 0.2 *M* Collidine. Cells are dehydrated in ethanol series to absolute alcohol and transferred to acetone. Coverslips are critical point dried with liquid $CO_2$ in a Sorvall Critical Point Drying System, coated with gold using an Edwards 306 coater, and viewed in a scanning electron microscope.

Cultures examined by SEM are usually so heavily infected that interpretation is easy. Mycoplasma species vary in their ability to cytadsorb onto cultured cells. In cultures infected with *M. hyorhinis,* most mycoplasma organisms are cytadsorbed onto cells. Little cytadsorption occurs in A-9 mouse cells infected with *Acholeplasma laidlawii,* with most of the mycoplasmas attached to the coverslip.

SEM has worked well in examination of fibroblast cultures. Some difficulty has been observed in interpretation of lymphocyte cultures grown in suspension (D. M. Phillips and G. J. McGarrity, unpublished results). As with other test systems, use of an indicator cell culture may alleviate some problems in interpretation.

### Virus Testing

Bovine serum has been shown to contain several bovine viruses.[28,29] It is not always feasible for individual laboratories to screen for them unless specialized facilities are available. The viruses most commonly isolated from bovine sera are bovine virus, diarrhea virus, infectious bovine rhinotracheitis virus, parainfluenze virus, and bovine herpes virus.

[27] O. P. Van Diggelen, S. Shin, and D. M. Phillips, *Exp. Cell Res.* **106,** 191 (1977).

[28] C. W. Molander, A. Paley, C. W. Boone, A. J. Kniazeff, and D. T. Imagawa, *In Vitro* **4,** 148 (1968).

[29] C. W. Molander, A. J. Kniazeff, C. W. Boone, A. Paley, and D. T. Imagawa, *In Vitro* **7,** 168 (1972).

Many suppliers of serum test their product for presence of bovine viruses. Requests can be made for specific test methods and results. Procedures are available for concentration of pooled sera,[29] and these preparations can be viewed under electron microscopy and/or inoculated in embryonic bovine tracheal cells grown in virus-free serum. These cultures can be compared to uninoculated controls for cytopathic effect and can be viewed by electron microscopy. Alternately, preparations can be histologically stained and, if desired, stained with appropriate fluorescent antibodies.

### Comments

Each laboratory must decide which quality-control procedure is appropriate and feasible for its own use. However, a minimum of testing of serum for sterility and growth promotion and assay of serum and cell cultures for mycoplasma are essential. Culturing for bacteria and fungi is relatively easy. Mycoplasma assays are somewhat difficult to perform, and most laboratories use two methods. Appropriate controls are essential in each test. Testing for viruses is more complex and, to the best of our knowledge, not a routine in most laboratories.

Special note should be taken of bovine serum as an experimental variable. In addition to the organisms isolated from bovine serum, considerable variation exists in concentrations of chemicals in serum.[30] In this study, the concentration of free fatty acids was the only variable that affected growth of human fibroblasts although other factors would be expected to affect the outcome of biochemical studies.

In addition to the detection techniques described here, publications are available listing other aspects of quality control.[1–4] Techniques are available for monitoring contamination by HeLa and other cell types.[31] All of these measures are relatively easy and highly cost-effective in the establishment of axenic, standardized cell cultures and experimental and diagnostic procedures.

[30] C. W. Boone, N. Mantel, T. D. Caruso, E. Kazam, and R. E. Stevenson, *In Vitro* **7**, 174 (1972).

[31] W. A. Nelson-Rees and R. R. Flandermeyer, *Science* **195**, 1343 (1977).

# [3] Preservation, Storage, and Shipment

*By* LEWIS L. CORIELL

Mammalian cells serially cultivated *in vitro* are in an artificial environment which differs in many known and unknown respects from the normal habitat *in vivo*. No culture medium can duplicate the *in vivo* environment,



ISBN 0-12-181958-2

and we have learned by experience that cells taken from different organs require different culture media for growth and/or preservation of specialized function. Some cell types inevitably die out in any given culture medium, and the cells which persist and continue to proliferate also lose or gain characteristics which were not apparent *in vivo*. Examples of changes frequently observed include: microbial contamination, chromosome abnormalities, loss of antigens, loss of ability to proliferate, and death of the cell culture. Many schemes have been employed with varied success to avoid or delay these undesired changes. Use of primary cell cultures and the use of organ cultures are examples which function well for selected short-term studies. Incubation of cell cultures at reduced temperature slows the metabolic rate and the frequency of refeeding and subculture. Storage of cell cultures in the frozen state following the general procedures used commercially for storage of bovine semen[1] was reported by Scherer and Hoogasian in 1954[2] and by Swim *et al.* in 1958.[3] These techniques used glycerol as an adjuvant with storage in Dry Ice at −70°. Viable cell cultures could be recovered for many months, but there was gradual loss of viability when stored at −70°. Storage of frozen cell cultures in liquid nitrogen (−196°) is the accepted procedure at this time, and when properly carried out it seems to be compatible with prolonged preservation of viability and other characteristics of mammalian cell cultures.[4–6]

Without the presence of an adjuvant, freezing is lethal to most mammalian cells. Damage is caused by mechanical injury by ice crystals, concentration of electrolytes, dehydration, pH changes, denaturation of proteins, and other factors not well understood. These lethal effects are minimized by: (1) adding an adjuvant such as glycerol or dimethylsulfoxide (DMSO) which lowers the freezing point; (2) a slow cooling rate which permits water to move out of the cells before it freezes; (3) storage at a temperature below −130° which retards the growth of ice crystals; and (4) rapid warming at the time of recovery so that the frozen cell culture passes rapidly through the temperature zone between −50–0° where most cell damage is believed to occur.

Different cell strains may vary widely in their ability to withstand physical and chemical insults during the freezing and thawing process.

[1] C. Polge, A. U. Smith, and A. S. Parks, *Nature* **164,** 666 (1949).
[2] W. F. Scherer and A. Hoogasian, *Proc. Soc. Exp. Biol. Med.* **87,** 480 (1954).
[3] H. E. Swim, R. F. Hall, and R. F. Parker, *Cancer Res.* **18,** 711 (1958).
[4] H. T. Meryman, *Science* **124,** 515 (1956).
[5] H. T. Meryman, *Fed. Proc., Fed. Am. Soc. Exp. Biol.* **22,** 81 (1963).
[6] J. Nagington and R. I. Greaves, *Nature (London)* **194,** 933 (1962).

Epithelial cells of human skin survive best in 20–30% glycerol adjuvant, whereas fibroblasts in the same tissue survive best in 10% glycerol.[7] We have frozen and recovered mosquito cell cultures with high viability in 1% glycerol.[8] The optimal cooling rate for mouse embryo cells is 1°/min,[9] for marrow stem cells 2–3°/min,[10] for yeast cells 10°/min,[11] and for human red blood cells in Polyvinylpyrrolidone (PVP) the optimal freezing rate is 300°/min.[12] These extreme examples are mentioned here to caution the reader that no single suspending medium and procedure will be ideal for processing and cryogenic storage of all cell cultures. However, most investigators have had excellent success in preserving many types of mammalian cells with 5–10% glycerol or dimethylsulfoxide as the adjuvant and a cooling rate of 1–10°/min. The procedures recommended in this article are designed for preservation of human fibroblast and lymphocyte cell cultures. They are equally satisfactory for many other human and animal normal and malignant cell cultures, but if they prove less than satisfactory for a given cell culture, the reader should titrate each variable to arrive at the optimal procedure for the cell under study.

## Preservation

### *Nutrition*

Cell cultures survive cryogenic storage best if they are frozen in a good state of nutrition and maximal viability. This is achieved as follows: Select cells in the logarithmic phase of growth and remove the culture medium. Assess sterility by culture on a blood agar plate. Replace with fresh medium and harvest 24 hr later for storage in liquid nitrogen. The cell culture medium should be the one which best supports growth and preservation of desired characteristics of the cell. Cell cultures grown in synthetic medium without serum have been successfully frozen and recovered.[13,14] However, most workers employ a culture medium for preservation containing 5–25% fetal bovine serum which provides unknown

[7] B. Athreya, E. Grimes, H. Lehr, A. Greene, and L. Coriell, *Cryobiology* **5**, 262 (1969).
[8] L. Coriell, unpublished.
[9] D. G. Wittinghorn, S. P. Liebo, and P. Mazur, *Science* **178**, 411 (1972).
[10] P. Mazur and J. J. Schmidt, *Cryobiology* **5**, 1 (1966).
[11] P. Mazur and J. Schmidt, *Cryobiology* **5**, 1 (1968).
[12] A. P. Rinfret, *Fed. Proc., Fed. Am. Soc. Exp. Biol.* **22**, 94 (1963).
[13] V. Evans, H. Montes de Oca, J. C. Bryant, E. L. Schilling, and J. E. Shannon, *J. Natl. Cancer Inst.* **29**, 749 (1962).
[14] C. Waymouth and D. Varnum, *Tissue Cult. Assoc.* **2**, 311 (1976).

growth factors and acts as an adjuvant for protection against damage during cryogenic storage.

*Harvest*

Suspension cell cultures are counted to determine the number of viable cells, centrifuged at 800–1000 g for 10 min, and suspended in inactivated fetal bovine serum (56° for 15 min) containing 5% dimethylsulfoxide. Dilute to 10 million cells/ml, test for sterility, dispense into ampules, and process as described below.

For monolayer cultures each flask is inspected under the binocular microscope before harvest. Any flasks containing cells of abnormal appearance are discarded. It is desirable to release the cells from the monolayer surface with a minimum of chemical and physical trauma. Add to each T75 flask, 5 ml of 0.02% ethylenediaminetetraacetic acid (EDTA) and let stand for 8 min. Remove the liquid and discard it. Add 3 ml of EDTA in dilute trypsin[15] and observe under the binocular miscroscope until the cell sheet starts to lift from the surface around the edges. Tap the flask gently to loosen all the cells. This process usually requires 1 to 7 min. Stop the action of trypsin–EDTA by adding 3 ml of growth medium containing 20% fetal calf serum. Transfer these cells to a common pool in a centrifuge bottle that is chilled in a cracked-ice bath. Mix thoroughly, count in a haemocytometer, and dilute with growth medium to $5 \times 10^5$ viable cells per milliliter as determined by trypan blue exclusion. Culture aliquots on a blood agar plate, trypticase soy broth, Sabouraud dextrose broth, fluid thioglycolate broth, mycoplasma agar and broth, and by one of the indirect methods to detect *Mycoplasma hyorhinis* (see this volume [2]). Sediment the cells in a refrigerated centrifuge at 800–1000 g, discard the supernatant, and resuspend in complete culture medium with 20% fetal calf serum plus 10% sterile glycerol. Dispense the chilled cell suspension into 1.2 ml borosilicate glass ampules with an automatic syringe. The ampules should be previously marked with printer's ink and sterilized in the dry air oven to insure a label that will not come off when stored in liquid nitrogen. The ampules must be closed in an oxygen flame with a pulled seal. To detect leaks, ampules are placed on aluminum canes[16] and tested by immersion in a cylinder of alcohol stained with methylene blue and chilled to 4°. If a pinhole leak exists the contents of an ampule will be stained blue and should be discarded. Sealed ampules should be kept at 4° and frozen as soon as possible. Significant loss of viability of WI-38 cells was observed when stored at 4° for more than 6 hr.[17]

[15] T. Puck, *J. Exp. Med.* **104,** 615 (1956).

[16] Available from McCracken & Sons, Inc., 636 N 13 Street, Philadelphia, Pennsylvania, 19123.

[17] A. Greene, B. H. Athreya, H. B. Lehr, and L. L. Coriell, *Cryobiology* **6,** 552 (1970).

### Freezing

Place the sealed ampules on labeled canes and position in a controlled cooling rate apparatus. Set the controls to achieve a cooling rate of 1–2° per minute. When the heat of fusion is released at about −10° the cooling rate should be increased briefly to exchange the heat of fusion and to reestablish the cooling rate. When the temperature reaches −25° the cooling rate can be increased to 5–10° per minute. When the temperature of the specimen reaches −100° the ampules can be transferred quickly to a liquid nitrogen refrigerator for storage in the vapor or in the liquid phase. Success with cell cultures has been achieved by many different slow freezing techniques.[10,17,18,19] When large numbers of ampules are frozen conditions and cooling rate settings must be adjusted to provide the desired cooling rates. The optimal cooling rate and procedures should be determined for each cell type. Mouse embryos, for example, survive poorly if cooled faster than 1° per minute[9] whereas many fibroblast or epithelial cell cultures tolerate a faster cooling rate[10]. As a general guide to the adequacy of the whole process, the trypan blue viability of the cell culture after recovery from liquid nitrogen storage should not be more than 5% below the viability determined before storage. If it consistently exceeds 10% loss of viability every step in the process should be critically evaluated.

### Storage

Permanent storage should be in liquid nitrogen whether liquid or vapor phase. This will ensure a temperature well below −150° and prevent ice crystal growth and enzyme activity. A few investigators have described changes in cells stored in liquid nitrogen,[20–22] but most investigators have not been able to observe detectable loss or gain of properties when cells are properly stored in it. Twenty-two cell cultures stored in liquid nitrogen in our laboratory have been recovered periodically and observed for viability as determined by trypan blue exclusion, plating efficiency, and cell yield in milk dilution bottles and roller tubes after 7 days of incubation.[23] No significant change in these parameters has been observed after storage in liquid nitrogen for up to 12 years.[23,24]

[18] V. Perry, C. Kraemer, and J. L. Martin, *Tissue Cult. Assoc. Man.* **1,** 119 (1975).
[19] R. Moklebust, N. Diez, and I. Goetz, *Tissue Cult. Assoc. Man.* **3,** 671 (1977).
[20] W. P. Peterson and C. Stulberg, *Cryobiology* **1,** 80 (1965).
[21] H. T. Meryman, "Cryobiology," p. 65. Academic Press, New York, 1966.
[22] L. Berman, M. P. McLeod, and E. P. Powsner, *Lab. Invest.* **14,** 231 (1965).
[23] A. E. Greene, B. H. Athreya, H. B. Lehr, and L. L. Coriell, *Proc. Soc. Exp. Biol. Med.* **124,** 1302 (1967).
[24] A. E. Greene, M. Manduka, and L. L. Coriell, *Cryobiology* **12,** 583 (1975).

## Shipment

In design of a standard procedure for shipment of cell cultures consideration must be given to (1) safe delivery of viable cells, (2) cost, and (3) regulations imposed by government and carriers.

### *Shipment of Frozen Ampules*

Frozen ampules of cell cultures may be shipped in small liquid nitrogen thermos refrigerators, or they may be packed in Dry Ice. Such packages are heavy, costly to ship, and must be returned. Furthermore, they require special handling, and if delayed in route there is danger of evaporation of the refrigerant. After safe arrival at the destination in a frozen state, there are still hazards. After delivery the specimen must be placed in fresh Dry Ice or liquid nitrogen until it is recovered. If stored in a refrigerator (4°) or at −20°, or even at −50°, cells will be damaged. When recovered the ampule must be warmed in a water bath at 37° and immediately cultured in an ideal medium. Because of the above problems routine shipment of frozen ampules is impractical, except for special situations or when a courier can accompany a shipment.

### *Shipment of Monolayer Cell Cultures*

It is recommended that frozen cell cultures be thawed, subcultured, and shipped as an actively growing culture. The procedure is as follows: After rapid thawing of the frozen ampule place the contents in a T25 plastic tissue culture flask containing 5 ml of culture medium with 20% fetal bovine serum. Next day remove the medium and feed. In 4–5 days there should be a confluent sheet of cells. Remove the medium, fill the flask with fresh medium, tape the screw cap in place, package, and ship by air mail special delivery. This procedure also gives the sending laboratory the opportunity to verify, examine, and/or test each culture before it is shipped, and it has resulted in better than 95% success in delivery of viable cell cultures in over 5000 shipments to domestic and foreign laboratories. Upon receipt at the destination the flask should be placed in an incubator at 37° overnight to permit recovery from trauma, exposure, and shaking which may have dislodged some cells during transit. Next day, open the flask and remove and save the excess culture fluid for use when the flask is subcultured.

### *Regulations*

Most cell cultures are free of microbial contaminants and etiologic agents and therefore can be shipped without meeting any special stan-

ETIOLOGIC AGENTS

**BIOMEDICAL MATERIAL**

IN CASE OF DAMAGE
OR LEAKAGE
NOTIFY DIRECTOR CDC
ATLANTA, GEORGIA
404-633-5313

H-388

NOTICE TO CARRIER

This package contains LESS THAN 50 ml OF AN ETIOLOGIC AGENT, N.O.S., is packaged and labeled in accordance with the U.S. Public Health Service Interstate Quarantine Regulations (42 CFR, Section 72.25 (c) (1) and (4)), and MEETS ALL REQUIREMENTS FOR SHIPMENT BY MAIL AND ON PASSENGER AIRCRAFT.

This shipment is EXEMPTED FROM ATA RESTRICTED ARTICLES TARIFF 6-D (see General Requirements 386 (d(1)) and from DOT HAZARDOUS MATERIALS REGULATIONS (see 49 CFR, Section 173.386(d)(3)). SHIPPER'S CERTIFICATES, SHIPPING PAPERS, AND OTHER DOCUMENTATION OR LABELING ARE NOT REQUIRED.

Date ______ Signature of Shipper ______

Institute for Medical Research
Copewood Street

Address

Camden, New Jersey 08103

FIG. 1. These labels may be obtained from a number of companies including Marion Manufacturing Co., Atlanta, Georgia, and Labelmaster, 6001 N. Clark St., Chicago, Illinois.

dards as to volume, labels, and packaging except that they arrive safely and in an undamaged condition. However, some cell cultures may be contaminated and some do contain etiologic agents, e.g., lymphocyte cell lines transformed with Epstein-Barr virus. If known etiologic agents are in the cell culture a package and labeling procedure for shipment of cell cultures that will meet all requirements for etiologic agents should be followed.[25] The packaging requirements for etiologic agents are as follows:

They shall be placed in securely closed, watertight primary receptacles which shall be enclosed in a durable, watertight secondary packaging. Several primary receptacles may be enclosed in a single secondary packaging, providing that the total volume of all the primary receptacles so enclosed does not exceed 50 ml. The space at the top, bottom and sides between the primary receptacles and secondary packagings shall contain sufficient non-particulate absorbent material such as cotton wool, to absorb completely the contents of the

[25] U. S. Public Health Service, "Interstate Quarantine Regulations" (42 CFR, Sect. 72.25). USPHS, Washington, D.C., 1972.

primary receptacle in case of breakage or leakage. Each set of primary receptacle and secondary packaging shall then be enclosed in an outer packaging constructed of corrugated fiberboard, wood, or other material of equivalent strength. The maximum amount of etiologic agents that may be carried in any one package is 50 milliliters.

Packages containing an etiologic agent must resist breaking or leakage of the contents including: (1) a water spray test for 30 min, (2) a free drop test through a distance of 30 feet, and (3) a puncture test of a 3.2-cm steel cylinder weighing 7 kg when dropped 1 m onto the exposed surface of the package. The outside of the package should display the two labels shown in Fig. 1 if it contains etiologic agents.

### Packaging Recommendations

The T25 plastic flask is filled to the top with complete culture medium, and the screw cap is tightened firmly and anchored in place with masking tape. The flask is placed in a liquid-tight polyethylene envelope #101.[26] The loose space around the flask is filled with absorbent cotton, and this envelope is placed in a larger liquid-tight polyethylene envelope #302 and sealed; this is then placed in a tongue and grooved Styrofoam utility mailer #3731–32[27] and sealed with Fiberglas tape. The outer carton is of heavy corrugated cardboard sealed with vinyl sealing tape. The completed package, ready for mailing, weighs approximately 5 ounces and has sufficient insulation to withstand subzero temperatures without freezing the contents.

[26] Briton liquid tight bags distributed by Medical Associates Int., Inc., P.O. Box 123, Topeka, Kansas 66601.

[27] Utility Mailer #3731–32 obtained from Cole Plamer, 7425 Noah Park Ave., Chicago, Illinois 60648.

# [4] Safety Considerations in the Cell Culture Laboratory

*By* W. EMMETT BARKLEY

The basic tenet of safety in research that involves potentially hazardous organisms is strict adherence to good laboratory practice. This tenet demands an awareness of the possible risks associated with the research materials that are handled, knowledge of mechanisms by which exposure may occur, use of procedures and techniques that reduce the potential for exposure, and continuous vigilance to guard against compromise and error. The cell culture worker is not unfamiliar with these basic principles, although the primary intent of their application has been the protection of



ISBN 0-12-181958-2

the cell culture rather than the cell culture investigator. Indeed, the experienced cell culture worker is aware that cell cultures are susceptible to contamination; knows the potential sources of contamination and the means by which contaminants can be introduced into cell cultures; is proficient in the use of procedures and techniques for preventing cell culture contamination; and well understands that a relaxation in good laboratory practice often results in cell culture contamination. Fortunately, the methods of the successful investigator are also applicable to protecting the individual from potential hazards that may be associated with research involving cell cultures.

### Biohazards and Cell Cultures

Cell cultures represent a potential biohazard because of their capacity to infect the laboratory worker with unrecognized viruses. The significance of this potential biohazard is emphasized by the few but serious cases of laboratory-acquired infections associated with the preparation and handling of primary monkey cell cultures. The most notable examples involve infections with *Herpesvirus simiae* (B virus) and Marburg virus.[1]

No laboratory-acquired infections have been associated with the use of continuous cell culture lines that are assumed to be free of infectious virus. These cell lines, however, should not be considered free from hazard since they may harbor latent viruses. For example, transformed cell lines may spontaneously produce viruses with oncogenic potential in animals, and human lymphoid cell lines may harbour Epstein-Barr virus.[2,3] Lymphoid cell lines may also represent a unique potential hazard if they are obtained from persons who plan to maintain them in culture. In the event of accidental ingestion or injection, these cells may not be rejected since they would possess common histocompatibility antigens. If the cells were transformed in culture, they may have the capacity to create transformed foci in the cell culture worker which may subsequently progress to true malignancy.[3]

Cell cultures used in virological studies should be assessed to present the same degree of hazard as the infectious virus under study. In such studies, actual hazards can be assessed easily and appropriate safeguards

[1] R. N. Hull, *in* "Biohazards in Biological Research" (A. Hellman, M. N. Oxman, and R. Pollack, eds.), p. 3. Cold Spring Harbor Lab., Cold Spring Harbor, New York, 1973.

[2] G. J. Todaro, *in* "Biohazards in Biological Research" (A. Hellman, M. N. Oxman, and R. Pollack, eds.), p. 114. Cold Spring Harbor Lab., Cold Spring Harbor, New York, 1973.

[3] J. A. Schneider, *in* "Biohazards in Biological Research" (A. Hellman, M. N. Oxman, and R. Pollack, eds.), p. 191. Cold Spring Harbor Lab., Cold Spring Harbor, New York, 1973.

established. Where infectious viruses are not intentionally handled, an assumption of hazard should be the basis for establishing safeguards.

### Potential for Exposure

Analysis of comprehensive surveys of laboratory-acquired infections reveal that recognized accidents are not the cause for most infections among laboratory workers.[4–6] Indeed, fewer than 20% of documented infections can be attributed to accidental contact, ingestion, or injection with infectious materials.[6] The actual causes for most laboratory-acquired infections are not known. However, the knowledge that most microbiological practices create aerosols suggests that inhalation of undetected aerosols may contribute significantly to occupational illness among laboratory workers who handle infectious materials.[7] The cell culture worker should be aware of the possibility of inadvertent inhalation exposures during the conduct of cell culture work. This potential for exposure should be considered in the selection of cell culture procedures.

### Recommended Safety Practices

Cell culture workers are encouraged to become familiar with the safe practices used in infectious disease research. Excellent review articles on good laboratory practices in infectious disease laboratories have been published.[8–12] This information will be of great value to those planning to conduct research with potentially hazardous viruses. Here, emphasis will be placed on four topic areas of good laboratory practice that pertain specifically to the cell culture technique.

#### *Good Laboratory Practice*

*Pipetting.* Good pipetting practice requires the use of pipetting aids to prevent exposures by ingestion and careful technique to reduce the poten-

[4] S. E. Sulkin and R. M. Pike, *Am. J. Public Health* **41,** 719 (1951).
[5] S. E. Sulkin, *Bacteriol. Rev.* **25,** 203 (1961).
[6] R. M. Pike, *Health Lab. Sci.* **13,** 105 (1976).
[7] R. L. Dimmick, W. Vogl, and M. A. Chatigny, *in* "Biohazards in Biological Research" (A. Hellman, M. N. Oxman, and R. Pollack, eds.), p. 246. Cold Spring Harbor Lab., Cold Spring Harbor, New York, 1973.
[8] A. G. Wedum, *Public Health Rep.* **79,** 619 (1964).
[9] G. B. Phillips, *J. Chem. Educ.* **42,** Part One, A43; *ibid.* Part Two, A117 (1965).
[10] H. M. Darlow, *Methods Microbiol.*" **1,** 169 (1969).
[11] M. A. Chatigny and D. I. Clinger, *in* "An Introduction to Experimental Aerobiology" (L. Dimmick and A. B. kers, eds.), p. 194. Wiley, New York, 1969.
[12] M. A. Chatigny, *Adv. Appl. Microbiol.* **3,** 131 (1961).

tial for creating aerosols. There are a variety of pipetting aids available that range from simple bulb- and piston-actuated systems to more sophisticated devices which contain their own vacuum pumps. In selecting a pipetting aid, consideration should be given to the kind of pipetting procedure that is to be performed, the ease of operation, and the accuracy of delivery that is needed. It is likely that several pipetting aids will be necessary; no universal aid is available that is appropriate for all pipetting procedures.

Pipetting technique that reduces the potential for creating aerosols should be developed. Pipettes should be plugged with cotton. Rapid mixing of liquids by alternate suction and expulsion through a pipette should be avoided. No material should be forcibly expelled out of a pipette, and air should not be bubbled through liquids with a pipette. Pipettes that do not require expulsion of the last drop are preferable to other types. Care should be taken to avoid accidentally dropping cultures from the pipette. When pipetting infectious materials, place a disinfectant-soaked towel on the working surface. Liquids from pipettes should be discharged as close as possible to the fluid or agar level of the receiving vessel, or the contents should be allowed to run down the wall of the receiving vessel. Dropping of the contents from a height into vessels should be avoided. Contaminated pipettes should be carefully placed in a pan containing enough suitable disinfectant to allow complete immersion.

*Hygienic Practice.* The cell culture worker should avoid eating, drinking, and smoking in the laboratory. Hands should be washed after handling cell cultures and before leaving the cell culture laboratory. To avoid contact contamination, the investigator should develop the habit of keeping hands away from mouth, nose, eyes, and face. Laboratory gowns or other suitable clothing should be worn when handling cell cultures; the same clothing should not be worn to eating areas or outside of the laboratory.

*Decontamination.* Work surfaces should be disinfected with a suitable chemical disinfectant following the completion of procedures involving cell cultures. Cell culture wastes which may be contaminated with virus should be disinfected with a suitable chemical disinfectant or sterilized by autoclaving before disposal. Contaminated items such as glassware and laboratory equipment should be disinfected or autoclaved before washing, reuse, or disposal. Contaminated materials that are to be autoclaved or incinerated at a site away from the laboratory should be placed in a durable, leak-proof container that is closed before removal from the laboratory.

*Control of Aerosols.* Most techniques used in a cell culture laboratory have the capacity to create aerosols. Therefore careful attention should be given to measures that can reduce the potential for inhalation exposures to aerosols that are produced by laboratory operations. Although the extent of aerosol production can often be controlled by good technique, efforts to eliminate exposure require the use of containment equipment.

Such special containment equipment for use in preventing the creation of aerosols is available. For example, capped safety centrifuge cups and sealed centrifuge heads can prevent the production of aerosols while centrifuging.

The most important measure for controlling aerosols is the use of ventilated safety cabinets for containment of operations that produce considerable aerosols. The cell culture worker is encouraged to use this containment equipment for operations that involve grinding, blending, homogenizing, and resuspending tissues or cell cultures. This should be standard practice if the cell culture materials are suspected of being contaminated with pathogenic microorganisms or viruses. Specific recommendations for the effective use of this equipment are presented in the following section.

*Ventilated Safety Cabinets*

The most versatile ventilated safety cabinet available to the cell culture worker is the Class II biological safety cabinet. This cabinet is commonly known as a laminar flow safety cabinet. Its versatility is derived from the protection afforded to both the cell culture and the investigator. Cell cultures within this cabinet are protected from airborne contamination by an airflow of uniform velocity that has been filtered by high-efficiency particulate air (HEPA) filters before reaching the work space. The HEPA filters used are virtually impenetrable by microorganisms and viruses. The cabinet provides protection to the cell culture worker by preventing, under normal operating conditions, airborne cells or viruses released during experimental activity from escaping the cabinet and entering the laboratory. This protection is achieved by an air barrier at the front opening of the cabinet and the filtration capability of the exhaust air HEPA filters. It must be emphasized that the protection features of the Class II cabinet are only applicable to airborne contamination; protection against exposures by ingestion, injection, and direct contact must be achieved by standard good laboratory practice.

The effectiveness of laminar flow safety cabinets in providing protection for the cell culture and those working with them, will depend on the performance capability of the cabinet, the location of the cabinet within the laboratory, and certain operational practices. The National Sanitation Foundation (NSF), Ann Arbor, Michigan, has developed a national standard for Class II cabinets.[13] The selection of a Class II cabinet that has been certified by NSF provides an assurance of performance capability. The performance capability, however, should be validated by measuring airflow and filter efficiency once the cabinet has been installed in the laboratory.

Cabinets should be located in an area of the laboratory away from doorways, supply air diffusers, and spaces of high activity. This is done to reduce the adverse effects of room air currents that could compromise the operation of the cabinet. Generally, the best location for the cabinet is along a side wall at a position farthest from the laboratory door.

The operational practices listed below are recommended when using laminar flow safety cabinets. Their use will provide excellent protection for cell cultures against contamination. These practices are most important, however, when the cell cultures are experimentally infected or contaminated with viruses.

1. Cabinet users should be encouraged to wear long-sleeved gowns with knit cuffs and gloves. This reduces the shedding of skin flora into the work area of the cabinet and protects the hands and arms from contact contamination.

2. The work surface of the cabinet should be decontaminated with a chemical disinfectant before placing equipment and material into the cabinet.

3. The cabinet should not be overloaded. Ideally, everything needed for the complete cell culture procedure should be placed in the cabinet before starting work so that nothing passes in or out through the air barrier until the procedure is completed.

4. Nothing should be placed over the exhaust grills located at the front and rear of the work surface. This practice will help maintain uniform airflow within the cabinet.

5. The arrangement of equipment and material on the work surface is an important consideration. Contaminated items should be segregated from clean ones and should be located so that they never have to be passed over clean items. Discard trays should be located to the rear of the cabinet.

6. After all equipment and material has been placed in the cabinet, the cabinet user should wait a few minutes before beginning work. This time period allows the cabinet air to remove any contaminants that may have been introduced into the cabinet during the loading process.

7. Always use aseptic techniques when performing cell culture work within the cabinet. This is important in preventing contact contamination.

[13] "National Sanitation Foundation Standard No. 49 for Class II Biohazard Cabinetry." Natl. Sanit. Found., Ann Arbor, Michigan, 1976.

8. When working with pathogenic viruses, all contaminated items should be contained or disinfected before they are removed from the cabinet. Trays of discarded pipettes and glassware should be covered before they are removed from the cabinet.

9. After all items have been removed from the cabinet, the cabinet should be allowed to operate for a few minutes to assure the removal of airborne contaminants. The work surface should then be decontaminated with a chemical disinfectant.

## Special Considerations and Recommendations

### *Lymphoid and Virus-Transformed Human Cell Lines*

The Human Genetic Mutant Cell Repository operated by the Institute for Medical Research, Camden, New Jersey, under sponsorship of the National Institute of General Medical Sciences has recommended minimum safety guidelines for working with lymphoid and virus-transformed human cell lines.[14] In addition to emphasizing the importance of good laboratory practice, the guidelines encourage the creation of biosafety committees to oversee work involving human cell lines, assign responsibilities to the laboratory supervisor for preparing safety protocols, recommend certain medical surveillance and screening procedures, and limit access to the cell culture laboratory to authorized individuals. The concern addressed by these guidelines is the potential hazard that may be associated with the release of Epstein-Barr virus or a virus whose genome may have been integrated in the DNA of the transformed cell.

Cell culture workers who handle long-term lymphoid lines or their derivatives are advised to have serum samples collected and stored annually. Anti-EB virus titers should be obtained on these samples. Serum samples should also be obtained and evaluated following accidental exposure to these cell cultures.

The guidelines endorse the use of ventilated safety cabinets in all cell culture work. Individuals who obtain long-term lymphoid lines from the repository are required to use ventilated safety cabinets as well as to adhere to the principles of the recommended guidelines.

### *Recommendations of the Medical Research Council, Canada, for Special Facility Safeguards in Cell Culture Work*

The Medical Research Council, Canada, published in February 1977, "Guidelines for the Handling of Recombinant DNA Molecules and Ani-

[14] "The Human Genetic Mutant Cell Repository." Inst. Med. Res., Camden, New Jersey, 1977.

mal Viruses and Cells."[15] These guidelines recommend that good laboratory practice be complemented by special facility safeguards for certain cell culture work. The guidelines require that cell culture work involving (1) the co-cultivation or fusion of human cells with cells of nonhuman mammals and (2) short-term nonhuman primate cells be confined to segregated rooms or cubicles that have exhaust air ventilation. The exhaust air from these laboratories is to be discharged outdoors. However, recirculation of air is allowed providing that the air is treated by filtration with HEPA filters. These requirements have been established to prevent the dissemination of cells and viruses, which the exhaust air may contain, to uncontrolled areas of the laboratory facility in the event that they are accidentally released within the segregated laboratory area.

The facility precautions are based on a possibility of hazard that is considered to be greater than that associated with most cell culture work. For co-cultivation and fusion studies with human cell lines, the principal concern is the possibility of unmasking a virus with oncogenic potential for man. The hazard concern for short-term nonhuman primate cells is the evidence that these cells may contain *Herpesvirus simiae* (B virus).

*Cell Cultures and Infectious Viruses*

When cell cultures are used with, or are known to contain, an infectious virus, the safety precautions selected for the cell culture work should be based on recommended safety guidelines for the infectious virus. Safety guidelines have been developed for viruses pathogenic to man and for viruses known to cause cancer in animals. The Center for Disease Control has classified pathogenic viruses on the basis of hazard and has recommended minimal safety conditions for their control.[16] The National Cancer Institute has published safety standards for research involving oncogenic viruses.[17] The investigator will find these documents useful in establishing specific safety protocols for work with cell cultures and infectious viruses.

[15] "Guidelines for the Handling of Recombinant DNA Molecules and Animal Viruses and Cells." Med. Res. Counc., Ottawa, Canada, 1977.

[16] "Classification of Etiological Agents." DHEW, PHS, CDC, Office of Biosafety, Washington, D.C., 1974.

[17] "National Cancer Institute Safety Standards for Research Involving Oncogenic Viruses." DHEW Publ. No. (NIH) 75-900. DHEW, PHS, NIH, Washington, D.C., 1974.

# [5] Media and Growth Requirements

By RICHARD G. HAM and WALLACE L. MCKEEHAN

## Introduction

This brief review of culture media and their applications is addressed primarily to the use of cell culture as an experimental tool in fields of research other than analysis of cellular growth requirements. Our emphasis is on the types of media available, and on principles involved in selection of the medium that is best suited for the experiment that is to be done. Environmental requirements for cellular multiplication and theoretical aspects of medium development are reviewed in greater detail elsewhere.[1-4]

## Types of Cultured Cells and Culture Systems

The choice of which culture medium to use and what supplement (if any) to add to it is strongly influenced by the type of cell that is to be grown and by the way that cell is to be grown. Thus, it will be helpful to begin with a brief outline of the types of cultured cells and culture systems that can be employed in various kinds of research.

### *Classification of Cells*

The following criteria are often employed to describe cells:

*Karyotype:* Are the chromosomes fully normal for the species both in number and morphology? If so, the cell is considered diploid (or in special circumstances haploid or tetraploid). If not, it is considered to be aneuploid or heteroploid.

*Multiplication potential:* Does the cell cease to multiply after a limited number of divisions in culture (finite life span), or is it a permanent line capable of continuous long-term multiplication?

*Anchorage dependence:* Does the cell require attachment to a substrate for multiplication, or will it also multiply in suspension culture? This is

[1] R. G. Ham and W. L. McKeehan, *In Vitro* **14,** 11 (1978).

[2] R. G. Ham and W. L. McKeehan, *in* "Nutritional Requirements of Cultured Cells" (H. Katsuta, ed.). Jpn. Sci. Soc. Press, Tokyo (in press).

[3] R. G. Ham, C. Waymouth, and P. J. Chapple, eds., "The Growth Requirements of Vertebrate Cells in Vitro." Cambridge Univ. Press, London and New York (in press).

[4] G. H. Rothblat and V. J. Cristofalo, eds., "Growth, Nutrition and Metabolism of Cells in Culture," 3 vols. Academic Press, New York, 1972-1977.





ISBN 0-12-181958-2

often tested by suspending individual cells in soft agar and determining whether they will form colonies.

*Density-dependent inhibition of multiplication:* Do the cells arrest their cell cycles in $G_1$ (or $G_0$) when the culture becomes crowded, or will they continue to multiply and pile up on one another?

*Malignancy:* Will the cells form an invasive malignant tumor when injected into appropriate test animals (e.g., "nude" mice; this volume [31])?

*Differentiation:* Does the cell exhibit tissue-specific differentiated properties in culture?

Exact classification of cultured cells is difficult because many of them tend to blend into a continous spectrum, rather than fitting into sharply defined classes. At the extremes of the spectrum, the fully normal cell is usually thought of as being diploid, finite in life span, anchorage dependent, density inhibited, and not malignant, while the classical permanent cell line is aneuploid, capable of indefinite multiplication, not anchorage dependent, not inhibited by density, and malignant. Unfortunately, these differences do not always occur as a set. There are apparently "normal" cells that are not anchorage dependent, e.g., chondrocytes and various blood-related cells, and others that do not exhibit density-dependent inhibition, e.g., keratinocytes. There are also permanent lines, such as 3T3, that are aneuploid and capable of indefinite multiplication, but are nevertheless anchorage dependent, density inhibited, and not malignant by most tests. Such cells can be "transformed" by oncogenic viruses to yield malignant lines that are no longer anchorage dependent or density inhibited (this volume [30]). Because of this, they are sometimes referred to as "normal," but it is better, in view of their aneuploidy and absence of a finite life span in culture, to refer to them as "nontransformed" permanent lines. Ideally, the term "normal" should only be used to describe cells that do not differ in any major way from the cells that occur in intact animals. Note that the term "nontransformed" is used to refer to normal cells with finite life spans as well as to the permanent lines that lack transformed properties.

### *Relationship between Cell Type and Growth Requirements*

Different types of cells have different growth requirements. Although fully normal cells and nontransformed lines are beginning to receive more attention, fully transformed permanent lines have dominated previous studies of cellular growth requirements and, at present, are the only types that can be grown in "defined" media consisting of low-molecular-weight nutrients without protein supplements. (Quotation marks are used about the term "defined," since in most cases growth in such media appears to be dependent on trace impurities such as inorganic trace elements that are not included in their formulas, as will be discussed in greater detail below.)

The growth requirements of normal cells and nontransformed lines are not as well characterized, and at the present time, it is necessary to use poorly characterized supplements of serum proteins or other biological materials to obtain satisfactory multiplication of such cells. The data are still incomplete, but it currently appears highly probable that most such cells have specific requirements for protein growth factors that cannot easily be replaced by improvements in the low-molecular-weight portion of the culture medium or in other aspects of the culture system (this volume [6]).

## Types of Culture Systems

The major distinctions among culture systems are based on cellular population density and on whether the cells are grown attached to a substrate or in suspension. When the cell density is so low that the descendants of single cells form discrete colonies (clones), the procedures are described as "clonal." The term "cloning" should be used only for procedures in which a new culture is established from the progeny of a single cell. All other experiments involving formation of colonies from single cells are referred to as "clonal growth." Experiments involving cloning or clonal growth are normally done either with the cells attached to a solid substrate under a liquid medium, or else with the cells suspended in a semisolid medium, e.g., in soft agar.

Dense cultures of cells attached to a substrate are often called "monolayer" cultures. Cellular yield in monolayer cultures is limited by the surface available on which cells can grow. This can be increased by use of roller bottles, which are rotated in such a way that their entire inner surfaces are bathed with medium (this volume [16] and [17]), or by use of capillary perfusion systems (this volume [14]) in which medium is circulated through tightly packed artificial capillaries, or by use of suspensions of particulate microcarriers on which the cells can attach and multiply (this volume [15] and [16]).

True suspension cultures, in which the cells are not attached to a substrate, are generally done with continually stirred or shaken culture vessels of various sizes, ranging from small spinner flasks with magnetic stirrers to huge industrial-scale fermentors (this volume [16] and [17]). Suspension cultures offer the special advantages of ease of handling of large numbers of cells and ability to harvest cells without the use of enzymes. For certain types of cells that do not readily attach to each other or to substrates, suspension cultures without agitation are also possible. Both the cellular population density and whether the cells are attached or suspended have significant effects on cellular growth requirements, as is discussed below.

*Cellular Population Density.* Early in the history of cell culture, it was observed that large populations of cells would often multiply under conditions where single isolated cells would not. Sanford and co-workers[5] isolated single cells in capillary tubes with small volumes of medium and demonstrated that the critical variable was the volume of medium per cell and that growth was due to "conditioning" of the medium. Subsequently, Puck and Marcus[6] used a feeder layer of nonmultiplying heavily irradiated cells to condition the medium in a petri dish so that it would support multiplication of single isolated nonirradiated cells to form macroscopically visible colonies. (See this volume [21]).

At least three separate mechanisms are involved in "conditioning" of culture media: (1) Inhibitory materials that bind to the cells are neutralized without reaching a dose per cell that is large enough to block growth. (2) Low-molecular-weight metabolic intermediates that diffuse readily out of cells accumulate in the extracellular medium in amounts which, by equilibration, maintain adequate intracellular levels for biosynthesis and metabolism. Such substances, which must be for biosynthesis and metabolism. Such substances, which must be supplied from an exogenous source when the volume of medium per cell is too large for effective conditioning, are referred to as "population-dependent" growth requirements. The most common examples are carbon dioxide, pyruvate, and the so-called "nonessential" amino acids, particularly serine. (3) Macromolecules synthesized by the cells sometimes must build up to a critical level before growth will occur. In addition to cases where a particular type of cell produces the macromolecules needed for its own multiplication, there are numerous cases in which macromolecules released by one type of cell are beneficial for another.

*Density-Dependent Inhibition.* Most types of cells stop or slow their multiplication at a characteristic population density in any particular medium. The effect tends to be more pronounced with nontransformed cells than with those that have undergone a malignant transformation. The exact mechanisms involved are not fully worked out and remain quite controversial. However, it is likely that for most types of cells, some combination of cell contact, localized or generalized depletion of nutrients

[5] K. K. Sanford, W. R. Earle, and G. D. Likely, *J. Natl. Cancer Inst.* **9,** 229 (1948).
[6] T. T. Puck and P. I. Marcus, *Proc. Natl Acad. Sci. U.S.A.* **41,** 432 (1955).

or macromolecular growth factors, and the production of cell-generated inhibitors (chalones) is involved. The composition of the basal medium and the amount of serum added to it both affect the cellular population density at which density-dependent inhibition (sometimes called "contact inhibition") occurs.

One popular assay for multiplication-promoting factors from serum measures stimulation of DNA synthesis in cultures of nontransformed cells (either normal diploid fibroblasts or permanent lines similar to 3T3) that have ceased to grow in a low serum medium due to density-dependent inhibition. Caution must be exercised in interpreting data from such an assay, however, since it measures only the first factor to become limiting in crowded cultures, and usually does not detect all of the requirements of the cells for sustained multiplication.

*Exhaustion of Media.* Two other phenomena related to cellular population density are depletion of nutrients and accumulation of waste products. For clonal growth, these are normally not serious problems, but in more crowded monolayer and suspension cultures they must be considered. In such cultures, measurements should be made to determine that heavily metabolized nutrients such as glucose and some of the amino acids are not seriously depleted, and that products of cellular metabolism, particularly acids, have not accumulated to a level that is inhibitory to further growth. The simplest test for both is generally the addition of fresh medium. Depletion of oxygen can also be a serious problem, particularly in large-scale suspension cultures where the surface-to-volume ratio is not favorable (see also this volume [16]). Continuous perfusion with fresh medium and controlled gassing with oxygen are desirable for optimum multiplication of crowded cultures, but caution must be exercised with the oxygen, since too much oxygen is often highly inhibitory to cultured cells.

*Suspension Cultures.* Apart from cellular population density, the second major difference in culture systems is whether the cells are grown in suspension or attached to a substrate. Among cells that are not anchorage dependent and can be grown either attached or suspended, the culture method that is utilized affects at least two classes of growth requirements. The first set is related to specific requirements for attachment. Media for suspension culture generally have very low levels of calcium, to minimize attachment of cells to the culture vessel and to each other. Media for substrate-attached growth generally contain substantially more divalent cations, and many nontransformed cells require quite high concentrations of calcium for optimal growth. The second effect of substrate attachment is that it tends to improve cellular efficiency of nutrient uptake, possibly because of the larger exposed surface of the attached and flattened cells. Cells that can be grown either anchored or suspended tend to require higher nutrient concentrations and/or more serum protein when grown in suspension than when grown on a substrate. In addition, suspension culture media generally have enhanced buffering to minimize acidification by the dense cultures that they must accommodate (see also this volume [16]).

*Modifications of the Culture System that Minimize the Requirement for Serum Protein*

*Treatment of the Culture Surface with Basic Polymers.* For some types of anchorage-dependent cells, the nature of the culture surface has a major effect on cellular growth and the requirement for serum proteins. Glass has a negative surface charge, as does typical cell culture plastic, which is produced by selective displacement of protons from polystyrene to generate a net negative surface charge. Recent experience in this laboratory has shown that growth of normal and nontransformed cells with minimal amounts of serum protein can be improved substantially by coating the culture surface with a positively charged polymer.[7] Essentially any positively charged polymer will work. The following procedure uses poly-D-lysine to avoid possible interference with amino acid balance studies and possible introduction of impurities carried by natural polymers such as histone or protamine, which are also active.

Dissolve 0.10 mg/ml poly-D-lysine · HBr (mol wt 30,000–70,000) in distilled water and sterilize by passage through Millipore type GS (0.22 $\mu$m) filter membrane. Pipette 0.5 ml of this solution into each 60 × 15 mm plastic tissue culture petri dish. Rock the dish gently to be certain that the solution spreads uniformly over the entire surface. After 5 min, remove the solution as completely as possible with a Pasteur pipette attached to an aspirator. Add 1.5 ml of sterile water to the dish, rock it gently, and completely remove the water by aspiration. Care must be exercised to remove all solutions completely, since free polylysine tends to inhibit growth when added directly to the culture medium. The dishes can be used immediately, or allowed to dry and used at a later time. It is also possible to do the coating without sterile precautions and then sterilize the coated dishes with ultraviolet (UV) irradiation.

*Minimizing Damage During Subculturing.* In cases where stock cultures of cells are maintained as attached monolayers, the procedures used to remove the cells from that substrate and dilute them into new subcultures or into clonal growth experiments have a profound effect not only on viability, but also on growth requirements of the surviving cells. The

[7] W. L. McKeehan and R. G. Ham, *J. Cell Biol.* **71,** 727 (1976).

requirement for serum protein of human diploid fibroblasts and chicken embryo fibroblasts, as well as other nontransformed anchorage-dependent cells, can be reduced significantly by use of a subculturing procedure designed to minimize mechanical damage to cells and the internalization of trypsin by pinocytosis.[8]

Essential features of that procedure include use of the lowest possible concentration of trypsin, use of the smallest possible volume of trypsin solution, exposure of the cells to the trypsin solution for the shortest possible time, and performance of all operations at ice-bath temperatures. The minimum concentration of trypsin and the minimum time of exposure that can be used depend on the type of cell, the cellular population density, the amount of serum in the medium the monolayer was grown in, and the length of time since the last subculturing. Since trypsin solutions are subject to self-digestion, it is desirable to divide a freshly prepared trypsin solution into small aliquots and store each frozen until it is to be used. In addition, it is desirable to measure enzymic activity of each new trypsin solution and to dilute it such that the cells are always exposed to the same number of trypsin activity units.[8] The following procedure is recommended for human fetal lung fibroblasts (WI-38, MRC-5, etc.) grown to a confluent monolayer in medium MCDB 104[9] supplemented with 1.0 mg/ml dialyzed fetal bovine serum protein (equivalent to about 2% serum). The volumes that are given are for a 25 $cm^2$ culture flask, and should be increased or decreased accordingly for culture vessels with different surface areas.

Chill all solutions and media that are to be used in an ice bath. Place the culture flask containing cells to be subcultured on ice for 15 min. Remove the medium and rinse the cellular monolayer twice with 0.5 ml aliquots of solution I (4.0 m*M* glucose; 3.0 m*M* KCl; 122 m*M* NaCl; 1.0 m*M* $Na_2HPO_4$; 3.3 $\mu M$ phenol red; 30 m*M* HEPES–NaOH, final pH of solution 7.6). Cover the cells with 2.0 ml of twice-crystallized trypsin at a concentration of 100 $\mu$g/ml in solution I. (This solution should have an activity of about 3.6 units/ml measured under the actual trypsinization conditions.[8]) Immediately remove as much as possible of the trypsin solution. Close the flask tightly to prevent evaporation and leave it on ice for about 5 min with only a thin film of trypsin solution covering the cells. Progress of the digestion may be checked by brief examination with an inverted microscope, but the flask should not be removed from the ice long enough to warm appreciably. When the cells are partially loosened, add 5 ml of chilled serum-free culture medium to the flask and shake

[8] W. L. McKeehan, *Cell Biol. Int. Rep.* **1,** 335 (1977).

[9] W. L. McKeehan, K. A. McKeehan, S. L. Hammond, and R. G. Ham, *In Vitro* **13,** 399 (1977).

gently to suspend the cells. Break up clumps by gentle pipetting and dilute the cells as needed for subculture. If a hemocytometer count is made for clonal growth, the suspended cells should be kept on ice during that procedure. The cells are not warmed above ice-bath temperature until they are placed in their final culture medium and returned to the cell culture incubator. As long as they are maintained in the cold, the cells do not round up, but rather remain elongated and irregular in shape. All mechanical procedures involving the cells are done as gently as possible.

For other types of cells, and for cells with different culture histories, it is necessary to vary the amount of trypsin and time of exposure. For example, for human foreskin fibroblasts, previously grown under the same conditions as the human fetal lung cells described above, about 500 μg/ml crystalline trypsin will be needed, and the digestion time must be increased to 15 min.

## Cellular Growth Requirements

Many different variables combine to determine whether or not cells multiply *in vitro*. We have already considered the fact that different types of cells have different growth requirements, and the fact that many aspects of the culture system that have nothing directly to do with medium composition affect growth requirements, including cellular population density, substrate attachment or its absence, the nature of the substrate, and the way the cells are handled during subculturing. In this section, we will consider in sequence the following aspects of cellular growth requirements:

1. The medium must supply all essential nutrients. These include all raw materials needed for the synthesis of new cells, substrates for energy metabolism, vitamins and trace minerals whose function is primarily catalytic, and bulk inorganic ions whose functions are both catalytic and physiological.
2. Physiological parameters, such as temperature, pH, osmolality, and redox potential, must be kept within acceptable limits.
3. The culture system must be free from toxic or inhibitory effects, including those due to excess amounts of essential components.
4. Precise quantitative adjustments, including balance relationships among the components of the culture system, are very important.
5. Serum, which is frequently added to defined basal media to stimulate multiplication, interacts with virtually every other variable in the culture system, and it also serves as a source of macromolecular growth factors that are essential for multiplication of many, but not all, types of cells in currently available basal media.

6. Cells multiply only when *all* of their requirements have been satisfied. A holistic approach is very helpful in dealing with the many different variables that influence multiplication and the complex interactions that occur among them.

### *Qualitative Requirements for Low-Molecular-Weight Nutrients*

Typical cell culture media contain a mixture of defined low-molecular-weight nutrients dissolved in a buffered physiological saline solution. The term "nutrient" is generally restricted to substances that enter cells and are utilized, either as substrates for biosynthesis or metabolism, or else as catalysts in those processes. Most media also contain a few nonnutrient components, such as phenol red as a pH indicator, and, sometimes, HEPES as a buffer. The following classes of nutrients are found in typical culture media.

*Bulk Ions.* The bulk ions may be required primarily for physiological roles, such as maintenance of membrane potentials and osmotic balance, rather than as nutrients in the narrow sense of the term, but they are included here in view of their universal requirement and the fact that most, if not all, also have co-factor roles in various enzymic reactions. Cultured cells require sodium, potassium, calcium, magnesium, chloride, and phosphate for multiplication. Bicarbonate (or carbon dioxide) is also required for a number of biosynthetic reactions, and an exogenous source must be provided when the cellular population is sparse. However, when the volume of medium per cell is small, e.g. in crowded cultures, or when loss of carbon dioxide by diffusion is inhibited, the cellular requirement can be satisfied entirely by metabolically generated carbon dioxide.

*Trace Elements.* Inorganic trace elements tend to be ubiquitous contaminants of the chemicals used in preparation of media, and most so-called "defined" media list in their formulas few, if any, of the trace elements known to be essential for whole animal nutrition. Limited data suggest that growth in such media is dependent on trace elements that are present as contaminants.[10–12] Cultured cells clearly require iron, zinc, and selenium, and there are data suggesting requirements for copper, manganese, molybdenum, and vanadium, although considerable background growth occurs without deliberate addition of those elements, presumably because of contamination.[9,13]

[10] W. G. Hamilton and R. G. Ham, *In Vitro* **13,** 537 (1977).

[11] J. A. Thomas and M. J. Johnson, *J. Natl. Cancer Inst.* **39,** 337 (1967).

[12] T. Takaoka and H. Katsuta, *Exp. Cell Res.* **67,** 295 (1971).

[13] W. L. McKeehan, W. G. Hamilton, and R. G. Ham, *Proc. Natl. Acad. Sci. U.S.A.* **73,** 2023 (1976).

It is likely that additional elements whose requirements are currently masked by contaminants will prove to be essential for cellular multiplication in the future. Until all such requirements are known, there is a constant danger of growth failure in previously adequate media due to increased purity of chemicals. This has already happened due to selenium deficiency in the case of medium F12.[10] Cultured cells grown in protein-free media or in media supplemented with minimal amounts of exhaustively dialyzed serum hold great promise for future studies of trace element requirements and metabolism, particularly in the case of humans, where direct studies involving long-term dietary depletion are not a realistic alternative.

*Sugars.* All commonly used cell culture media contain a six-carbon sugar as a source of metabolic energy and reduced carbon for biosynthesis. Glucose is the most common form, although in some media it is partially or fully replaced by galactose, which is metabolized more slowly and leads to less buildup of acid in the culture medium. There are reports that a mixture of pyruvate and ribose will replace glucose in some systems, but this mixture is not widely used.

*Amino Acids.* Thirteen amino acids, arginine, cyst(e)ine, glutamine, histidine, isoleucine, leucine, lysine, methionine, phenylalanine, threonine, tryptophan, tyrosine, and valine, are generally considered essential for cultured cells. Under specialized conditions, some types of cells can synthesize enough glutamine to satisfy their needs, and in rare cases, sufficient cystine can be made from methionine to support multiplication. Among the "nonessential" amino acids, serine is frequently required on a population-dependent basis for growth of sparse cultures. Asparagine is required by certain malignant cells, including a number of lines of leukemia. Proline is required by the widely used mutant Chinese hamster ovary line, CHO, and may be beneficial on a population-dependent basis for other types of cells. Glycine is sometimes essential in cases of borderline folic acid deficiency, or if folic acid uptake or metabolism is inefficient, as has been reported for several types of primary cultures. All 20 of the amino acids that are involved in protein synthesis are often included in culture media to reduce cellular biosynthetic loads, and it is sometimes possible to manipulate experimental conditions to make growth dependent on normally nonessential amino acids such as glutamic acid, aspartic acid, and alanine. The quantitative requirement for essential amino acids becomes larger when nonessential amino acids are not provided, since in such cases a portion of the essential amino acids must be degraded to provide a nitrogen source for synthesis of the nonessential amino acids. Hydroxyproline is included in the formulas of a number of older media, but there is no convincing evidence that it is actually required. The tripeptide glutathione is also sometimes included, but again evidence for its requirement is not convincing.

*Vitamins.* Cultured cells generally require the B vitamin: biotin, folic acid, niacinamide, pantothenic acid, pyridoxine, riboflavin, and thiamin. Vitamin $B_{12}$ is probably also required, but it is often difficult to achieve a clean background. $\alpha$-Lipoic acid (thioctic acid) is sometimes added to culture media, but there is little evidence that it is actually needed. Coenzymes can generally be substituted for the parent vitamins, but, with the possible exception of folinic acid, they appear to offer no special advantage. Ascorbic acid may be beneficial, particularly to cells that synthesize collagen in culture, but its instability to oxidation makes it very difficult to work with, and there continue to be conflicting reports about its effectiveness. The fat-soluble vitamins appear not to be essential, at least for the types of cells that have been carefully studied in serum-free media. The possible role of fat-soluble vitamins in multiplication of normal cells is still clouded by the requirement of such cells for serum proteins, which could be serving as carriers of fat-soluble vitamins.

*Choline and Inositol.* Both choline and meso-inositol are required by most types of cultured cells, although some Chinese hamster ovary lines show no response to inositol, even at clonal density in completely synthetic media. These compounds are often listed as "vitamins" in the formulas of culture media, but their metabolic roles are clearly as substrates, rather than as catalysts.

*Other Organic Nutrients.* The organic nutrients listed above (glucose, 13 "essential" amino acids, 7 or 8 B vitamins, choline, and meso-inositol) constitute the apparent minimal qualitative requirements for organic nutrients for multiplication of a number of highly adapted permanent cell lines. These are the only nutrients in certain "minimal" media, such as the widely used Minimum Essential Medium of Eagle (Tables I and II, see text p. 84 for main description of Tables I and II). However, a number of additional requirements can be demonstrated for various other cultures. A purine source (adenine or hypoxanthine) together with thymidine is often beneficial, particularly in cases where folic acid is in short supply or used inefficiently. Under some conditions uridine may also be beneficial.

Some types of cells appear to have specific requirements for polyunsaturated fatty acids, and in other cases, mixtures of mono- and polyunsaturated fatty acids, e.g., oleic plus linoleic, are reported to be beneficial. Cholesterol is also reported to be beneficial in certain cases. Free lipids present two major problems in aqueous culture media—they are rather insoluble and they are quite toxic. It currently appears that the use of membranous vesicles (liposomes) generated by sonication of phosphatidylcholine (lecithin) suspensions[14] may provide a soluble and relatively nontoxic source of lipids for cells grown in adqueous media.[15,16] Human diploid fibroblasts and chicken embryo fibroblasts both appear to have definite requirements for lipids.[1,2]

Putrescine or other polyamines are required for growth of Chinese hamster ovary cells in protein-free media and are beneficial to certain other cells. Many types of cultured cells exhibit a population-dependent requirement for pyruvate or other small $\alpha$-oxo acids. The requirement is relatively nonspecific and may reflect the need for a substrate that can be used to oxidize NADH, rather than the need for a specific metabolic intermediate. A variety of other organic compounds have been incorporated into some of the more complex media on the basis that they might be beneficial, and it probably does no harm to include them. However, this philosophy can be a serious barrier to the development of serum-free media since substances that do not appear to be harmful in the presence of large amounts of serum sometimes prove to be quite inhibitory in media with lower concentrations of protein.

### *Physiological Requirements*

In addition to satisfying nutrient requirements, the culture environment must have physicochemical properties that fall within physiologically acceptable limits for cellular survival and multiplication. Important parameters include the following:

*Temperature.* Cellular multiplication rate typically increases with temperature until a limiting value is reached, above which further increases in temperature rapidly become inhibitory. Most laboratories do not have enough incubators available to perform properly controlled experiments on the effects of temperature on cellular multiplication, and except for studies involving temperature-sensitive mutations of cells or viruses, little attention is usually given to this variable. However, different species of warm-blooded animals do have different body temperatures, and certain tissues and organs such as skin and testes are normally maintained at temperatures below the overall body temperature. Therefore, studies of temperature optima could prove to be very profitable in certain cases. Typically, cultures of cells from warm-blooded animals are grown at 37° and cultures from cold-blooded animals at temperatures near the upper limit of optimum body temperature for the intact animal.

[14] G. Poste, D. Papahadjopoulos, and W. J. Vail, *Methods Cell Biol.* **14,** 33 (1976).
[15] N. N. Iscove and F. Melchers, *J. Exp. Med.* **147,** 923 (1978).
[16] W. L. McKeehan and R. G. Ham, Abstracts 1978 TCA Meeting, *In Vitro* **14,** 353 (1978).

TABLE I*

PARTIAL LISTING OF CELL CULTURE MEDIA[a]

| No.[b] | Name of medium and developer(s)[c] | | Original assay system[d] | | | | Com. av.[e] | Table II[f] | Comments |
|---|---|---|---|---|---|---|---|---|---|
| | | | Type | Species | Assay | Supplements | | | |
| A. Early media for primary cultures | | | | | | | | | |
| 1. | V-605 | A. Fischer *et al.* (1948) | Pri | CkE | Out | dE, dS | N | – | Primarily of historical interest. |
| 2. | — | White (1949) | Pri | CkE | Sur | None | N | – | Primarily of historical interest. |
| 3. | 199 | Morgan *et al.* (1950) | Pri | CkE | Sur | None | G | + | Often used for growth of nontransformed cells (human, chicken, mouse, etc.) with serum. |
| B. Media for growth of permanent lines with serum supplementation | | | | | | | | | |
| 4. | Basal (BME) | Eagle | Trn | Hum, Mse | Mlr | dS | G | + | Widely used, including monolayer growth of human diploid cells with serum. |
| 5. | Minimum Essential (MEM) | Eagle | Trn | Hum, Mse | Mlr | dS | G | + | Currently the most widely used medium. Has been modified for suspension and clonal cultures. Used with many different species and kinds of cells. |
| 6. | Alpha-MEM | Stanners *et al.* | Trn | Mse, SyH | Mlr | wS | L | – | Enriched MEM; used for various hard-to-grow cell types. |
| 7. | 5A | McCoy *et al.* | Trn | Rat | Clo | dS | G | + | Used for clonal and monolayer growth of diploid human and rat cells. |
| 8. | RPMI 1640 | Moore *et al.* | Trn | Hum | Sus | wS | G | + | Modification of McCoy 5A for human lymphoblastoid and leukemic lines in suspension culture. There are many closely related RPMI and GEM media. A few cases of serum-free growth have been reported. |
| 9. | N16 | Puck *et al.* | Trn | Hum | Clo | wS | L | – | Used at 40% concentration; supports clonal growth of HeLa with fetuin and serum albumin; lacks population-dependent factors needed for clonal growth of some cells. |
| 10. | "Fischer's" | G. A. Fischer *et al.* | Trn | Mse | Sus | wS | L | – | Used for mouse leukemia in suspension culture. |
| 11. | L-15 | Leibovitz | Trn | Hum, Mky | Mlr | wS | G | + | Designed for use without $CO_2$ incubation; used for human diploid fibroblasts with serum. |
| C. Media for growth of permanent lines with purified protein or hormone supplements | | | | | | | | | |
| 12. | F10 | Ham | Trn | ChH | Clo | FA | G | + | Used for clonal growth of differentiated cells from chicken embryos, often with double amino acids and pyruvate (H. Coon). |

| | | | | | | | | |
|---|---|---|---|---|---|---|---|---|
| 13. "Serumless" | Neumann and Tytell | Trn | Rat, Hum | Mlr | Salmine, insulin | L | + | Supports growth of several lines that normally require serum protein. |
| 14. $WO_5$ | Jenkin *et al.* | Trn | Mky | Mlr | Albumin | L | – | Waymouth MB 752/1 with serum albumin and fatty acids; also used with other species. |
| 15. — | Higuchi | Trn | various | Mlr | Insulin, protamine | N | + | Medium also contains lecithin and methyl-cellulose. Established lines from several species grow well. |
| 16. — | Yamane *et al.* | Trn | various | Sus | Insulin, albumin | N | + | Modified MEM; basic polymers reduce the requirement for albumin; fatty acids are essential. |
| 17. F12 + hormones | Sato (Hayashi *et al.*) | Trn | various | Mlr | Hormones | N | – | Complex mixtures of hormones and transferrin support growth of several lines without serum in medium F12 (cf. No. 34 below). |
| D. Media for monolayer growth of permanent lines without protein supplementation | | | | | | | | |
| 18. NCTC-109 | Evans, McQuilkin *et al.* | Trn | Mse | Mlr | None | L | – | Cells must be adapted; high cysteine may be inhibitory; supplanted by NCTC 135 (No. 19). |
| 19. NCTC 135 | Evans *et al.* | Trn | Mse | Mlr | None | G | + | NCTC 109 without high cysteine; cells must be adapted to serum-free growth. |
| 20. CMRL 1066 | Parker | Trn | Mse | Mlr | None | G | + | Extensively modified medium 199; cells must be adapted to serum-free growth. |
| 21. MB 752/1 | Waymouth | Trn | Mse | Mlr | None | G | + | Less complex than NCTC and CMRL media; cells must be adapted; used from many types of cells, often with serum supplementation. |
| 22. MD 705/1 | Waymouth (Kitos *et al.*) | Trn | Mse | Mlr | None | L | – | Modified MB 752/1 with trace elements. |
| 23. MAB 87/3 | Gorham and Waymouth | Pri | Mse CkE | Mlr | None | L | + | Further modification of MB 752/1; used for direct establishment of cultures in serum-free medium. |
| 24. MAP 954/1 | Waymouth (Donta) | Trn | Rat | Mlr | None | N | – | Supports serum-free growth of rat glioma cells that synthesize tissue-specific protein. |
| 25. — | Mohberg and Johnson | Trn | Mse | Mlr | None | N | – | Modified MB 752/1 of Waymouth with amino acids adjusted to reflect utilization. |
| 26. DM 120, 145 | Katsuta, Takaoka *et al.* | Trn | various | Mlr | None | T | + | Serum-free growth of various cell lines in relatively simple media; DM-145 contains inositol, which DM-120 lacks. Cells must be adapted to serum-free growth. |
| 27. DM-160 | Katsuta and Takaoka | Trn | various | Mlr | None | T | – | Modified DM145; also used with serum for glandular epithelium from rat stomach. |
| 28. A2 + APG | Holmes and Wolfe | Trn | Hum | Mlr | None | L | – | Extremely complex medium designed for growth of permanent human lines without serum supplementation. Cells must be adapted. |
| 29. MCDB 411 | Agy *et al.* | Trn | Mse | Mlr | None | N | + | Supports monolayer growth of mouse Cl300 neuroblastoma serum-free; clonal growth with insulin. No adaptation required. |

* See text p. 84 for main description of Table I.

TABLE I (*continued*)

| No.[b] | Name of medium and developer(s)[c] | | Original assay system[d] | | | | Com. av.[e] | Table II[f] | Comments |
|---|---|---|---|---|---|---|---|---|---|
| | | | Type | Species | Assay | Supplements | | | |
| E. Media for suspension culture of permanent lines without protein supplementation | | | | | | | | | |
| 30. | — | Sinclair | Trn | Mse | Sus | None | N | – | Modification of Mohberg and Johnson modification of MB 752/1 for suspension culture. |
| 31. | — | Birch and Pirt | Trn | Mse | Sus | MC | N | – | Adjusted for efficient nutrient utilization; cell yield comparable to media with serum. |
| 32. | 7C's | Ling *et al.* | Trn | Mse, Rat | Sus | None | N | – | Unusually high nutrient concentrations; medium designed for high cell yield. |
| 33. | — | Nagle and Brown | Trn | various | Sus | MC | N | – | Contains no glutamine; autoclavable; cat kidney line will grow in Nagle series of media. |
| F. Media for clonal growth of permanent lines without protein supplementation | | | | | | | | | |
| 34. | F12 | Ham | Trn | ChH | Clo | None | G | + | Originally supported clonal growth of Chinese hamster ovary and lung lines. Growth was dependent on selenium present as a contaminant. Supplanted by MCDB 301 for protein-free growth. F12 is widely used with serum for differentiated cells from rat, rabbit, chicken embryos. Also used with hormones (Cf. No. 17). |
| 35. | MCDB 301 | Hamilton and Ham | Trn | ChH | Clo | None | N | + | Modified F12 with complex trace element supplement, improved balance. Supports clonal and monolayer growth of Chinese hamster ovary (CHO) lines without added protein. |
| G. Modern media for nontransformed cells | | | | | | | | | |
| 36. | Dulbecco's Modified Eagle's Medium (DME) | Dulbecco | Pri, Fin | Mse | Mlr | wS | G | + | Modification of BME (not of MEM); widely used for untransformed mouse and chicken cultures. |
| 37. | IMEM-ZO | Richter *et al.* | Pri | Rat | Mlr | wS | L | + | Enriched MEM; supports growth with small amounts of serum; also used for clonal growth of human diploid fibroblasts. |
| 38. | CMRL-1415 CMRL 1415-ATM | Healy and Parker | Pri | Mse | Sur | None | C | + | Survival of mouse embryo fibroblasts without supplement and growth with protein fractions; CMRL 1415-ATM for use without carbon dioxide. |

| | | | | | | | | |
|---|---|---|---|---|---|---|---|---|
| 39. CMRL 1969 | Healy and Parker | Pri, Fin | Hum, Mky | Mlr | wS | C | + | Used with serum for diploid human and monkey cultures. |
| 40. E, G | Williams *et al.* | Pri | Rat | Sur | None | L | + | Maintenance of normal liver cells with high rate of survival. |
| 41. DM | Parsa *et al.* | Pri | Rat | OC | None | N | – | Maintenance of organ culture of rat pancreas with differentiation and some multiplication. |
| 42. SM-20<br>SM-20I | Halle and Wollenberger | Pri | CkE, Rat | Sur | None | N | – | Survival of heart tissue with maintenance of beating. |
| 43. MPNL65/C | Waymouth (Leiter *et al.*) | Pri | Mse | Mlr | None | N | – | For mouse pancreas; contains dexamethasone; has high osmolarity. |
| 44. AMBD 647/3 | Waymouth | Pri | Mse | Sur | Insulin | N | – | Autoclavable medium with sodium betaglycerophosphate buffer system. For maintenance of dense populations of differentiated mouse cells (e.g., liver, kidney). Modified MAB 87/3. |
| 45. F12K | Kaighn | Pri | Rat | Mlr | wS | N | – | Modification of F12; incorporates twice the amino acids and pyruvate of H. Coon and other adjustments; used for differentiated rat and chicken cultures. |
| 46. MCDB 104<br>MCDB 105 | McKeehan *et al.* | Fin | Hum | Clo | dS | L<br>N | –<br>+ | Clonal growth of diploid human fibroblasts with minimal protein supplements. MCDB 104 is for 5% $CO_2$, 105 for 2%. Long-term survival without protein supplementation. |
| 47. MCDB 202 | Ham and McKeehan | Pri, Fin | CkE | Clo | dS | N | + | Clonal growth of chicken embryo fibroblasts with minimal protein supplementation; can also be used with cells from other species. |
| 48. MCDB 401 | Shipley and Ham | Pri, Non | Mse | Clo | dS | N | + | Clonal growth of mouse embryo fibroblasts and 3T3 with minimal protein supplement. |
| 49. MCDB 501 | McKeehan *et al.* | Pri | Duck | Clo | dS | N | + | Clonal growth of duck embryo fibroblasts with minimal protein supplementation. |

[a] This table lists the widely used commercially available media plus a number of newer or less known media of potential future usefulness.

[b] At least one literature reference is given below for each medium listed. Note that the reference given is not always the original publication of the medium, and that, in some cases, additional references are given to make it easier to follow the development of the medium in the literature.


1. A. Fischer, T. Astrup, G. Ehrensvard, and V. Oehlenschlager, *Proc. Soc. Exp. Biol. Med.* **67,** 40–46 (1948).
2. P. R. White, *J. Cell. Comp. Physiol.* **34,** 221–241 (1949).
3. J. F. Morgan, H. J. Morton, and R. C. Parker, *Proc. Soc. Exp. Biol. Med.* **73,** 1–8 (1950).
4. H. Eagle, *Science* **122,** 501–504 (1955).
5. H. Eagle, *Science* **130,** 432–437 (1959).
6. C. P. Stanners, G. L. Eliceiri, and H. Green, *Nature (London)*, New Biol. **230,** 52–54 (1971).
7. T. A. McCoy, M. Maxwell, and P. F. Kruse, Jr., *Proc. Soc. Exp. Biol. Med.* **100,** 115–118 (1959).
8. G. E. Moore, R. E. Gerner, and H. A. Franklin, *J. Am. Med. Assoc.* **199,** 519–524 (1967).
9. T. T. Puck, S. J. Cieciura, and A. Robinson, *J. Exp. Med.* **108,** 945–955 (1958).
10. G. A. Fischer and A. C. Sartorelli, *Methods Med. Res.* **10,** 247–262 (1964).
11. A. Leibovitz, *Am. J. Hyg.* **78,** 173–180 (1963).
12. R. G. Ham, *Exp. Cell Res.* **29,** 515–526 (1963).

TABLE I (*continued*)

13. R. E. Neuman and A. A. Tytell, *Proc. Soc. Exp. Biol. Med.* **104,** 252–256 (1960); R. E. Neuman and A. A. Tytell, *ibid.* **107,** 876–880 (1961).
14. H. M. Jenkin and L. E. Anderson, *Exp. Cell Res.* **59,** 6–10 (1970); S. J. Morrison and H. M. Jenkin, *In Vitro* **8,** 94–100 (1972).
15. K. Higuchi, *Methods Cell Biol.* **14,** 131–144 (1976).
16. I. Yamane, O. Murakami, and M. Kato, *Cell Struc. Func.* **1,** 279–284 (1976).
17. I. Hayashi, J. Larner, and G. Sato, *In Vitro* **14,** 23–30 (1978).
18. V. J. Evans, J. C. Bryant, M. C. Fioramonti, W. T. McQuilkin, K. K. Sanford, and W. R. Earle, *Cancer Res.* **16,** 77–86 (1956); W. T. McQuilkin, V. J. Evans, and W. R. Earle, *J. Natl. Cancer Inst.* **19,** 885–907 (1957).
19. V. J. Evans, J. C. Bryant, H. A. Kerr, and E. L. Schilling, *Exp. Cell Res.* **36,** 439–474 (1964).
20. R. C. Parker, "Methods of Tissue Culture," p. 77. Harper (Haeber), New York, 1961.
21. C. Waymouth, *J. Natl. Cancer Inst.* **22,** 1003–1017 (1959).
22. P. A. Kitos, R. Sinclair, and C. Waymouth, *Exp. Cell Res.* **27,** 307–316 (1962).
23. L. W. Gorham and C. Waymouth, *Proc. Soc. Exp. Biol. Med.* **119,** 287–290 (1965); C. Waymouth, *TCA Manual* **3,** 521 (1977).
24. S. T. Donta, *Methods Cell Biol.* **9,** 123–137 (1975).
25. J. Mohberg and M. J. Johnson, *J. Natl. Cancer Inst.* **31,** 611–625 (1963).
26. H. Katsuta, T. Takaoka, and K. Kikuchi, *Jpn. J. Exp. Med.* **31,** 125–136 (1961); T. Takaoka and H. Katsuta, *Exp. Cell Res.* **67,** 295–304 (1971).
27. H. Katsuta and T. Takaoka, *Methods Cell Biol.* **14,** 145–158 (1976).
28. R. Holmes and S. W. Wolfe, *J. Biochem. Biophys. Cytol.* **10,** 389–401 (1961); R. Holmes, *J. Cell Biol.* **32,** 297–308 (1967).
29. P. Agy, G. Shipley, and R. G. Ham, unpublished results; Composition is summarized in Table II, also in Ham and McKeehan [47] below.
30. R. Sinclair, *Exp. Cell Res.* **41,** 20–33 (1966).
31. J. R. Birch and S. J. Pirt, *J. Cell Sci.* **7,** 661–670 (1970).
32. C. T. Ling, G. O. Ley, and V. Richters, *Exp. Cell Res.* **52,** 469–489 (1968).
33. S. C. Nagle Jr. and B. L. Brown, *J. Cell. Physiol.* **77,** 259–263 (1971).
34. R. G. Ham, *Proc. Natl. Acad. Sci. U.S.A.* **53,** 288–293 (1965).
35. W. G. Hamilton and R. G. Ham, *In Vitro* **13,** 537–547 (1977).
36. H. J. Morton, *In Vitro* **6,** 89–108 (1970). (Official Tissue Culture Association version—many other variants have also been published.)
37. A. Richter, K. K. Sanford, and V. J. Evans, *J. Natl. Cancer Inst.* **49,** 1705–1712 (1972).
38. G. M. Healy and R. C. Parker, *J. Cell. Biol.* **30,** 531–538 and 539–553 (1966).
39. G. M. Healy, S. Teleki, A. V. Seefried, M. J. Walton, and H. G. Macmorine, *Appl. Microbiol.* **21,** 1–5 (1971).
40. G. M. Williams and J. M. Gunn, *Exp. Cell Res.* **89,** 139–142 (1974); G. M. Williams, E. Bermudez, and D. Scaramuzzino, *In Vitro* **13,** 809–817 (1977).
41. I. Parsa, W. H. Marsh, and P. J. Fitzgerald, *Exp. Cell. Res.* **59,** 171–175 (1970).
42. W. Halle and A. Wollenberger, *Am. J. Cardiol.* **25,** 292–299 (1970).
43. E. H. Leiter, D. L. Coleman, and C. Waymouth, *In Vitro* **9,** 421–433 (1974).
44. C. Waymouth, *in* "Growth Requirements of Vertebrate Cells in Culture" (R. G. Ham, C. Waymouth, and P. C. Chapple, eds.). Cambridge Univ. Press, London and New York, 1978 (in press).
45. M. E. Kaighn, *J. Natl. Cancer Inst.* **53,** 1437–1442 (1974).
46. W. L. McKeehan, K. A. McKeehan, S. L. Hammond, and R. G. Ham, *In Vitro* **13,** 399–416 (1977); W. L. McKeehan, D. Genereux, and R. G. Ham *Biochem. Biophys. Res. Commun.* **80,** 1013 (1978).
47. R. G. Ham, and W. L. McKeehan, *in* "Nutritional Requirements of Cultured Cells" (H. Katsuta, ed.). Jpn. Sci. Soc. Press, Tokyo, 1978 (in press).

48. G. D. Shipley and R. G. Ham, unpublished results; composition is summarized in Table II, also in Ham and McKeehan [47] above.
49. W. L. McKeehan, K. A. McKeehan, and R. G. Ham, unpublished results; composition is summarized in Ham and McKeehan [47] above.

[c] Related groups of media developed for similar purposes are listed together. The list of "developers" of media emphasizes leaders of the research groups and does not include all authors of the publications listed in footnote *b*.

[d] The original assay system used to develop the medium is described in abbreviated form, as follows:

"Type" refers to type of cell: Fin = diploid culture with finite life span *in vitro*; Non = permanent line of "nontransformed" cells (e.g., 3T3); Pri = primary culture initiated with cells removed from an experimental animal; Trn = permanent line of transformed cells.

Species: ChH = Chinese hamster; CkE = Chicken embryo; Duck = duck embryo; Hum = human; Mky = monkey; Mse = mouse; Rat = rat; SyH = Syrian hamster.

Assay: Clo = clonal growth (colony formation on a solid substrate); Mlr = monolayer growth on a solid substrate; Out = outgrowth from a tissue explant; Sur = survival either attached or in suspension without significant cellular multiplication; Sus = multiplication of cells in fluid suspension culture; OC = organ culture.

Supplements: AF = serum albumin plus fetuin; dE = dialyzed embryo extract; dS = dialyzed serum (from various species); MC = methylcellulose; wS = whole serum (from various species); other supplements are written out; None = no macromolecular supplement.

[e] Commercial availability (Com. Av.) is indicated as follows: C = not available from companies in the United States, but can be obtained from Connaught Laboratories, Toronto, Ontario, Canada. G = generally available from most suppliers in the United States; L = available from a limited number of suppliers in the United States; N = to the best of our knowledge not available as a regular catalog item from any commercial suppliers in the United States (companies should be checked frequently for new additions to their listings); T = not available from companies in the United States, but can be obtained from Kyokutō Pharmaceutical Industries, Ltd., 3–9 Nihonbashi Honcho, Chuo-ku, Tokyo, Japan.

[f] A plus (+) indicates that the composition of the medium is summarized in Table II.

TABLE II*

SUMMARY OF COMPOSITIONS OF SELECTED CULTURE MEDIA[a]

| Name of medium and number from Table I | | Essential amino acids | | | | | | | | | | | | | |
|---|---|---|---|---|---|---|---|---|---|---|---|---|---|---|---|
| | | Arginine | Cysteine | Half-cystine[b] | Gluta-mine | Histi-dine | Isoleu-cine | Leucine | Lysine | Methio-nine | Phenyl-alanine | Threo-nine | Trypto-phan | Tyro-sine | Valine |
| MEM | 5 | 6.0E-4 | — | 2.0E-4 | 2.0E-3 | 2.0E-4 | 4.0E-4 | 4.0E-4 | 4.0E-4 | 1.0E-4 | 2.0E-4 | 4.0E-4 | 5.0E-5 | 2.0E-4 | 4.0E-4 |
| BME | 4 | 1.0E-4 | — | 1.0E-4 | 2.0E-3 | 5.0E-5 | 2.0E-4 | 2.0E-4 | 2.0E-4 | 5.0E-5 | 1.0E-4 | 2.0E-4 | 2.0E-5 | 1.0E-4 | 2.0E-4 |
| DME | 36 | 4.0E-4 | — | 4.0E-4 | 4.0E-3 | 2.0E-4 | 8.0E-4 | 8.0E-4 | 8.0E-4 | 2.0E-4 | 4.0E-4 | 8.0E-4 | 8.0E-5 | 4.0E-4 | 8.0E-4 |
| Yamane | 16 | 6.0E-4 | — | 2.0E-4 | 4.0E-3 | 2.0E-4 | 4.0E-4 | 4.0E-4 | 4.0E-4 | 1.0E-4 | 2.0E-4 | 4.0E-4 | 5.0E-5 | 2.0E-4 | 4.0E-4 |
| IMEM-Z0 | 37 | 6.0E-4 | — | 2.0E-4 | 2.0E-3 | 2.0E-4 | 4.0E-4 | 1.0E-3 | 4.0E-4 | 1.0E-4 | 1.9E-4 | 4.0E-4 | 4.9E-5 | 2.0E-4 | 3.9E-4 |
| L15 | 11 | 2.9E-3 | 9.9E-4 | — | 2.1E-3 | 1.6E-3 | 1.9E-3[c] | 9.5E-4 | 5.1E-4 | 1.0E-3[c] | 1.5E-3[c] | 5.0E-3[c] | 9.8E-5 | 1.7E-3 | 1.7E-3[c] |
| 199 | 3 | 3.3E-4 | 6.3E-7 | 1.7E-4 | 6.8E-4 | 1.0E-4 | 3.0E-4[c] | 9.1E-4[c] | 3.8E-4 | 2.0E-4[c] | 3.0E-4[c] | 5.0E-4[c] | 9.8E-5[c] | 2.2E-4 | 4.3E-4[c] |
| CMRL 1415 | 38 | 2.9E-3 | 1.5E-3 | 2.0E-4 | 2.0E-3 | 2.0E-4 | 4.0E-4 | 4.0E-4 | 3.2E-4 | 1.0E-4 | 1.9E-4 | 4.0E-4 | 4.9E-5 | 2.0E-4 | 3.9E-4 |
| CMRL 1969 | 39 | 3.3E-4 | 6.3E-7 | 1.7E-4 | 1.4E-3 | 1.0E-4 | 1.5E-4 | 4.6E-4 | 4.8E-4 | 1.0E-4 | 1.5E-4 | 2.5E-4 | 4.9E-5 | 2.2E-4 | 2.1E-4 |
| CMRL 1066 | 20 | 3.3E-4 | 1.5E-3 | 1.7E-4 | 6.8E-4 | 9.5E-5 | 1.5E-4 | 4.6E-4 | 3.8E-4 | 1.0E-4 | 1.5E-4 | 2.5E-4 | 4.9E-5 | 2.2E-4 | 2.1E-4 |
| NCTC 135 | 19 | 1.5E-4 | — | 8.7E-5 | 9.3E-4 | 1.3E-4 | 1.4E-4 | 1.6E-4 | 2.1E-4 | 3.0E-5 | 1.0E-4 | 1.6E-4 | 8.6E-5 | 9.1E-5 | 2.1E-4 |
| MB 752/1 | 21 | 3.6E-4 | 5.7E-4 | 1.2E-4 | 2.4E-3 | 7.8E-4 | 1.9E-4 | 3.8E-4 | 1.3E-3 | 3.4E-4 | 3.0E-4 | 6.3E-4 | 2.0E-4 | 2.2E-4 | 5.5E-4 |
| MAB 87/3 | 23[c] | 3.6E-4 | 5.7E-4 | 1.2E-4 | 2.4E-3 | 7.2E-4 | 1.9E-4 | 3.8E-4 | 1.3E-3 | 3.4E-4 | 3.0E-4 | 6.3E-4 | 2.0E-4 | 2.2E-4 | 5.5E-4 |
| DM 145 | 26 | 5.7E-4 | 5.1E-4 | — | 6.8E-4 | 1.9E-4 | 1.1E-3 | 3.0E-3 | 6.8E-4 | 5.4E-4 | 4.8E-4 | 8.4E-4 | 2.0E-4 | 2.8E-4 | 7.3E-4 |
| Williams' G | 40 | 2.9E-4 | 3.3E-4 | 3.3E-4 | 2.0E-3 | 1.3E-4 | 3.8E-4 | 4.6E-4 | 5.8E-4 | 1.3E-4 | 2.4E-4 | 5.0E-4 | 1.7E-4 | 2.2E-4 | 5.1E-4 |
| McCoy's 5a | 7 | 2.0E-4 | 2.0E-4 | — | 1.5E-3 | 1.0E-4 | 3.0E-4 | 3.0E-4 | 2.0E-4 | 1.0E-4 | 1.0E-4 | 1.5E-4 | 1.5E-5 | 1.0E-4 | 1.5E-4 |
| RPMI 1640 | 8 | 1.1E-3 | — | 4.2E-4 | 2.1E-3 | 9.7E-5 | 3.8E-4 | 3.8E-4 | 2.2E-4 | 1.0E-4 | 9.1E-5 | 1.7E-4 | 2.4E-5 | 1.1E-4 | 1.7E-4 |
| Neuman & Tytell | 13 | 2.0E-4 | 2.0E-4 | — | 1.5E-3 | 1.0E-4 | 3.0E-4 | 3.0E-4 | 2.0E-4 | 1.0E-4 | 1.0E-4 | 1.0E-4 | 3.0E-5 | 1.0E-4 | 1.0E-4 |
| Higuchi | 15 | 1.5E-4 | 1.3E-4 | — | 1.4E-3 | 3.0E-4 | 2.5E-4 | 2.0E-4 | 1.5E-4 | 1.0E-4 | 2.0E-4 | 1.0E-4 | 2.9E-5 | 2.5E-4 | 3.0E-4 |
| F10 | 12 | 1.0E-3 | 2.0E-4 | — | 1.0E-3 | 1.0E-4 | 2.0E-5 | 1.0E-4 | 1.0E-4 | 3.0E-5 | 3.0E-5 | 3.0E-5 | 3.0E-6 | 1.0E-5 | 3.0E-5 |
| F12 | 34 | 1.0E-3 | 2.0E-4 | — | 1.0E-3 | 1.0E-4 | 3.0E-5 | 1.0E-4 | 2.0E-4 | 3.0E-5 | 3.0E-5 | 1.0E-4 | 1.0E-5 | 3.0E-5 | 1.0E-4 |
| MCDB 301 | 35 | 1.0E-3 | 1.0E-4 | — | 3.0E-3 | 1.0E-4 | 3.0E-5 | 1.0E-4 | 2.0E-4 | 3.0E-5 | 3.0E-5 | 1.0E-4 | 1.0E-5 | 3.0E-5 | 1.0E-4 |
| MCDB 105 | 46 | 1.0E-3 | 5.0E-5 | — | 2.5E-3 | 1.0E-4 | 3.0E-5 | 1.0E-4 | 2.0E-4 | 3.0E-5 | 3.0E-5 | 1.0E-4 | 1.0E-5 | 3.0E-5 | 1.0E-4 |
| MCDB 202 | 47 | 3.0E-4 | 2.0E-4 | — | 1.0E-3 | 1.0E-4 | 1.0E-4 | 3.0E-4 | 2.0E-4 | 3.0E-5 | 3.0E-5 | 3.0E-4 | 3.0E-5 | 5.0E-5 | 3.0E-4 |
| MCDB 501 | 49 | 1.0E-3 | 2.0E-4 | — | 1.0E-3 | 1.0E-5 | 1.0E-4 | 5.0E-5 | 2.0E-4 | 1.0E-5 | 5.0E-5 | 3.0E-4 | 1.0E-5 | 5.0E-5 | 3.0E-4 |
| MCDB 401 | 48 | 3.0E-4 | — | 4.0E-4 | 5.0E-3 | 2.0E-3 | 1.0E-3 | 2.0E-3 | 8.0E-4 | 2.0E-4 | 3.0E-4 | 5.0E-4 | 1.0E-5 | 2.0E-4 | 2.0E-3 |
| MCDB 411 | 29 | 3.0E-4 | 1.0E-4 | — | 1.0E-3 | 3.0E-5 | 3.0E-4 | 3.0E-4 | 5.0E-4 | 5.0E-5 | 1.0E-4 | 1.0E-4 | 1.0E-5 | 3.0E-5 | 1.0E-4 |

| Name of medium and number from Table I | | Nonessential amino acids | | | | | | | Amino acid derivatives[c] | | | | | | |
|---|---|---|---|---|---|---|---|---|---|---|---|---|---|---|---|
| | | Alanine | Asparagine | Aspartate | Glutamate | Glycine | Proline | Serine | α-Amino butyrate | Glutathione | Hydroxyproline | Ornithine | Putrescine | Taurine | Thyroxine |
| MEM | 5 | — | — | — | — | — | — | — | — | — | — | — | — | — | — |
| BME | 4 | — | — | — | — | — | — | — | — | — | — | — | — | — | — |
| DME | 36 | — | — | — | — | 4.0E-4 | — | 4.0E-4 | — | — | — | — | — | — | — |
| Yamane | 16 | — | — | — | — | 1.0E-4 | — | 1.0E-4 | — | — | — | — | 6.2E-6 | — | — |
| IMEM-Z0 | 37 | — | 4.0E-4 | — | — | — | — | 4.0E-4 | — | — | — | — | 9.9E-7 | — | — |
| L15 | 11 | 5.1E-3[e] | 1.9E-3 | — | — | 2.7E-3 | — | 1.9E-3 | — | — | — | — | — | — | — |
| 199 | 3 | 5.6E-4[e] | — | 4.5E-4[e] | 9.1E-4[e] | 6.7E-4 | 3.5E-4 | 4.8E-4[e] | 3.6E-7 | 1.6E-7 | 7.6E-5 | — | — | — | — |
| CMRL 1415 | 38 | 3.4E-4 | — | 7.5E-5 | 6.8E-5 | 2.3E-4 | 2.6E-4 | 1.1E-4 | — | 3.3E-5 | 7.6E-5 | — | — | — | — |
| CMRL 1969 | 39 | 2.8E-4 | — | 2.3E-4 | 4.1E-4 | 6.7E-4 | 3.5E-4 | 2.4E-4 | — | 1.6E-7 | 7.6E-5 | — | — | — | — |
| CMRL 1066 | 20 | 2.8E-4 | — | 2.3E-4 | 5.1E-4 | 6.7E-4 | 3.5E-4 | 2.4E-4 | — | 3.3E-5 | 7.6E-5 | — | — | — | — |
| NCTC 135 | 19 | 3.5E-4 | 6.1E-5 | 7.4E-5 | 5.6E-5 | 1.8E-4 | 5.3E-5 | 1.0E-4 | 5.3E-5 | 3.0E-5 | 3.1E-5 | 5.6E-5 | — | 3.3E-5 | — |
| MB 752/1 | 21 | — | — | 4.5E-4 | 1.0E-3 | 6.7E-4 | 4.3E-4 | — | — | 4.9E-5 | — | — | — | — | — |
| MAB 87/3 | 23[e] | 1.3E-4 | 1.8E-4 | 4.5E-4 | 1.0E-3 | 6.7E-4 | 4.3E-4 | 1.3E-4 | — | 4.9E-5 | — | — | — | — | — |
| DM 145 | 26 | 4.5E-3 | — | 1.9E-4 | 1.0E-3 | 2.0E-4 | 1.0E-4 | 7.6E-4 | — | — | — | — | — | — | — |
| Williams' G | 40 | 1.0E-3 | 2.3E-4 | 3.8E-5 | 3.9E-4 | 6.7E-4 | 5.2E-4 | 2.4E-4 | — | 1.6E-7 | — | — | — | — | — |
| McCoy's 5a | 7 | 1.5E-4 | 3.0E-4 | 1.5E-4 | 1.5E-4 | 1.0E-4 | 1.5E-4 | 2.5E-4 | — | 1.6E-6 | 1.5E-4 | — | — | — | — |
| RPMI 1640 | 8 | — | 3.3E-4 | 1.5E-4 | 1.4E-4 | 1.3E-4 | 1.7E-4 | 2.9E-4 | — | 3.3E-6 | 1.5E-4 | — | — | — | — |
| Neuman & Tytell | 13 | 1.0E-4 | 3.0E-4 | 1.0E-4 | 1.0E-4 | 1.0E-4 | 1.0E-4 | 2.5E-4 | — | 1.6E-6 | 1.0E-4 | — | — | — | — |
| Higuchi | 15 | — | 1.0E-3 | — | — | — | 1.0E-3 | 1.0E-3 | — | — | — | — | — | — | 2.5E-8 |
| F10 | 12 | 1.0E-4 | 1.0E-4 | 1.0E-4 | 1.0E-4 | 1.0E-4 | 1.0E-4 | 1.0E-4 | — | — | — | — | — | — | — |
| F12 | 34 | 1.0E-4 | 1.0E-4 | 1.0E-4 | 1.0E-4 | 1.0E-4 | 3.0E-4 | 1.0E-4 | — | — | — | — | 1.0E-6 | — | — |
| MCDB 301 | 35 | 1.0E-4 | 1.0E-4 | 1.0E-4 | 1.0E-4 | 1.0E-4 | 3.0E-4 | 1.0E-4 | — | — | — | — | 1.0E-6 | — | — |
| MCDB 105 | 46 | 1.0E-4 | 1.0E-4 | 1.0E-4 | 1.0E-4 | 1.0E-4 | 3.0E-4 | 1.0E-4 | — | — | — | — | 1.0E-9 | — | — |
| MCDB 202 | 47 | 1.0E-4 | 1.0E-3 | 1.0E-4 | 1.0E-4 | 1.0E-4 | 5.0E-5 | 3.0E-4 | — | — | — | — | 1.0E-9 | — | — |
| MCDB 501 | 49 | 1.0E-4 | 1.0E-3 | 1.0E-4 | 1.0E-4 | 1.0E-4 | 5.0E-5 | 3.0E-4 | — | — | — | — | 1.0E-9 | — | — |
| MCDB 401 | 48 | — | 1.0E-4 | — | — | 1.0E-4 | — | 1.0E-4 | — | — | — | — | — | — | — |
| MCDB 411 | 29 | — | 3.0E-5 | 1.0E-5 | — | — | — | 1.0E-4 | — | — | — | — | 3.0E-6 | — | — |

* See text p. 84 for main description of Table II.

TABLE II (*continued*)

| Name of medium and number from Table I | | Water-soluble vitamins and coenzymes[d] | | | | | | | | | | | |
|---|---|---|---|---|---|---|---|---|---|---|---|---|---|
| | | Ascorbic acid | Biotin | Folic acid | Folinic acid | p-Aminobenzoic acid | α-Lipoic acid | Nicotinic acid | Nicotinamide | NAD | NADP | Pantothenic acid | Coenzyme A |
| MEM | 5 | — | — | 2.3E-6 | — | — | — | — | 8.2E-6 | — | — | 4.6E-6 | — |
| BME | 4 | — | 1.0E-6 | 1.0E-6 | — | — | — | — | 1.0E-6 | — | — | 1.0E-6 | — |
| DME | 36 | — | — | 9.1E-6 | — | — | — | — | 3.3E-5 | — | — | 1.7E-5 | — |
| Yamane | 16 | — | — | 2.3E-6 | 1.7E-6 | — | — | — | 8.2E-6 | — | — | 4.6E-6 | — |
| IMEM-Z0 | 37 | — | 4.1E-7 | — | 1.6E-6 | — | 9.7E-7 | — | 8.2E-6 | — | — | 4.2E-6 | — |
| L15 | 11 | — | — | 2.3E-6 | — | — | — | — | 8.2E-6 | — | — | 4.2E-6 | — |
| 199 | 3 | 2.8E-7 | 4.1E-8 | 2.3E-8 | — | 3.6E-7 | — | 2.0E-7 | 2.0E-7 | — | — | 4.2E-8 | — |
| CMRL 1415 | 38 | 2.8E-4 | 4.1E-6 | 2.3E-6 | — | — | — | — | — | 1.1E-5 | 1.3E-6 | 2.1E-6 | — |
| CMRL 1969 | 39 | 2.8E-7 | 4.1E-6 | 2.3E-6 | — | 3.6E-7 | — | — | 8.2E-6 | — | — | 4.2E-6 | — |
| CMRL 1066 | 20 | 2.8E-4 | 4.1E-8 | 2.3E-8 | — | 3.6E-7 | — | 2.0E-7 | 2.0E-7 | 1.1E-5 | 1.3E-6 | 4.2E-8 | 3.3E-6 |
| NCTC 135 | 19 | 2.8E-4 | 1.0E-7 | 5.7E-8 | — | 9.1E-7 | — | 5.1E-7 | 5.1E-7 | 1.1E-5 | 1.3E-6 | 1.0E-7 | 3.3E-6 |
| MB 752/1 | 21 | 9.9E-5 | 8.2E-8 | 9.1E-7 | — | — | — | — | 8.2E-6 | — | — | 4.2E-6 | — |
| MAB 87/3 | 23 | 9.9E-5 | 8.2E-8 | 1.1E-6 | — | — | — | — | 8.2E-6 | — | — | 4.2E-6 | — |
| DM 145 | 26 | 2.3E-4 | 8.2E-9 | 2.3E-8 | — | — | — | — | 4.1E-5 | — | — | 4.6E-6 | — |
| Williams' G | 40 | 1.1E-5 | 2.0E-6 | 2.3E-6 | — | — | — | — | 8.2E-6 | — | — | 4.2E-6 | — |
| McCoy's 5a | 7 | 2.8E-6 | 8.2E-7 | 4.5E-7 | — | 7.3E-6 | — | 4.1E-6 | 4.1E-6 | — | — | 8.4E-7 | — |
| RPMI 1640 | 8 | — | 8.2E-7 | 2.3E-6 | — | 7.3E-6 | — | — | 8.2E-6 | — | — | 1.0E-6 | — |
| Neuman & Tytell | 13 | 2.8E-6 | 8.2E-7 | 4.5E-6 | 2.1E-7 | — | — | — | 4.1E-6 | — | — | 8.4E-7 | — |
| Higuchi | 15 | — | 8.2E-9 | 2.3E-6 | — | — | — | — | 8.2E-7 | — | — | 8.4E-7 | — |
| F10 | 12 | — | 1.0E-7 | 3.0E-6 | — | — | 1.0E-6 | — | 5.0E-6 | — | — | 3.0E-6 | — |
| F12 | 34 | — | 3.0E-8 | 3.0E-6 | — | — | 1.0E-6 | — | 3.0E-7 | — | — | 1.0E-6 | — |
| MCDB 301 | 35 | — | 3.0E-8 | 3.0E-6 | — | — | 1.0E-6 | — | 3.0E-7 | — | — | 1.0E-6 | — |
| MCDB 105 | 46 | — | 3.0E-8 | — | 1.0E-9 | — | 1.0E-8 | — | 5.0E-5 | — | — | 1.0E-6 | — |
| MCDB 202 | 47 | — | 3.0E-8 | — | 1.0E-8 | — | 1.0E-8 | — | 5.0E-5 | — | — | 1.0E-6 | — |
| MCDB 501 | 49 | — | 3.0E-8 | — | 1.0E-6 | — | 1.0E-8 | — | 5.0E-5 | — | — | 1.0E-6 | — |
| MCDB 401 | 48 | — | — | — | 1.0E-6 | — | — | — | 5.0E-5 | — | — | 5.0E-5 | — |
| MCDB 411 | 29 | — | 3.0E-8 | — | 3.0E-9 | — | 1.0E-6 | — | 3.0E-7 | — | — | 1.0E-6 | — |

| Name of medium and number from Table I | | Water-soluble vitamins and coenzymes (continued) | | | | | | | | | | Fat-soluble vitamins | | | |
|---|---|---|---|---|---|---|---|---|---|---|---|---|---|---|---|
| | | Pyri-doxine | Pyri-doxal | Pyri-doxal-5-Phos-phate | Ribo-flavin | Ribo-flavin-5-Phos-phate | FAD | Thia-mine | Thia-mine Mono-phos-phate | Thia-mine Pyro-phos-phate | Vitamin $B_{12}$ | Calci-ferol (Vitamin D) | Mena-dione (Vitamin K) | α-Toco-pherol Phos-phate (Vitamin E) | Vitamin A |
| MEM | 5 | — | 6.0E-6 | — | 2.7E-7 | — | — | 3.0E-6 | — | — | — | — | — | — | — |
| BME | 4 | — | 1.0E-6 | — | 1.0E-7 | — | — | 1.0E-6 | — | — | — | — | — | — | — |
| DME | 36 | — | 2.0E-5 | — | 1.1E-6 | — | — | 1.2E-5 | — | — | — | — | — | — | — |
| Yamane | 16 | — | 6.0E-6 | — | 2.7E-7 | — | — | 3.0E-6 | — | — | — | — | — | — | — |
| IMEM-Z0 | 37 | — | 4.9E-6 | — | 2.7E-7 | — | — | 3.0E-6 | — | — | 1.0E-6 | — | — | — | — |
| L15 | 11 | 4.9E-6 | — | — | — | 1.9E-7 | — | — | 2.8E-6 | — | — | — | — | — | — |
| 199 | 3 | 1.2E-7 | 1.2E-7 | — | 2.7E-8 | — | — | 3.0E-8 | — | — | — | 2.5E-7 | 5.8E-8 | 2.0E-8 | 3.5E-7 |
| CMRL 1415 | 38 | — | — | 4.0E-6 | — | — | 1.3E-6 | — | — | 2.1E-6 | — | — | — | — | — |
| CMRL 1969 | 39 | — | 4.9E-6 | — | — | 1.9E-7 | — | 3.0E-6 | — | — | — | — | — | — | — |
| CMRL 1066 | 20 | 1.2E-7 | 1.2E-7 | — | 2.7E-8 | — | 1.3E-6 | 3.0E-8 | — | 2.1E-6 | — | — | — | — | — |
| NCTC 135 | 19 | 3.0E-7 | 3.1E-7 | — | 6.6E-8 | — | 1.3E-6 | 7.4E-8 | — | 2.1E-6 | 7.4E-6 | 6.3E-7 | 1.5E-7 | 4.5E-8 | 8.7E-7 |
| MB 752/1 | 21 | 4.9E-6 | — | — | 2.7E-6 | — | — | 3.0E-5 | — | — | 1.5E-7 | — | — | — | — |
| MAB 87/3 | 23 | 4.9E-6 | — | — | 2.7E-6 | — | — | 3.0E-5 | — | — | 1.5E-7 | — | — | — | — |
| DM 145 | 26 | 4.9E-6 | — | — | 2.7E-6 | — | — | 3.0E-5 | — | — | 3.7E-9 | — | — | — | — |
| Williams' G | 40 | — | 4.9E-6 | — | 2.7E-7 | — | — | 3.0E-6 | — | — | 1.5E-7 | 2.5E-6 | 3.0E-8 | 1.8E-8 | 3.0E-7[u] |
| McCoy's 5a | 7 | 2.4E-6 | 2.5E-6 | — | 5.3E-7 | — | — | 5.9E-7 | — | — | 5.5E-10 | — | — | — | — |
| RPMI 1640 | 8 | 4.9E-6 | — | — | 5.3E-7 | — | — | 3.0E-6 | — | — | 3.7E-9 | — | — | — | — |
| Neuman & Tytell | 13 | 2.4E-6 | 2.5E-6 | — | 5.3E-7 | — | — | — | — | — | 5.5E-10 | — | — | — | — |
| Higuchi | 15 | — | 2.5E-7 | — | 2.7E-7 | — | — | 5.9E-7 | — | — | 1.5E-9 | — | — | — | — |
| F10 | 12 | 1.0E-6 | — | — | 1.0E-6 | — | — | 3.0E-6 | — | — | 1.0E-6 | — | — | — | — |
| F12 | 34 | 3.0E-7 | — | — | 1.0E-7 | — | — | 1.0E-6 | — | — | 1.0E-6 | — | — | — | — |
| MCDB 301 | 35 | 3.0E-7 | — | — | 1.0E-7 | — | — | 1.0E-6 | — | — | 1.0E-6 | — | — | — | — |
| MCDB 105 | 46 | 3.0E-7 | — | — | 3.0E-7 | — | — | 1.0E-6 | — | — | 1.0E-7 | — | — | — | — |
| MCDB 202 | 47 | 3.0E-7 | — | — | 3.0E-7 | — | — | 1.0E-6 | — | — | 1.0E-7 | — | — | — | — |
| MCDB 501 | 49 | 3.0E-7 | — | — | 3.0E-7 | — | — | 1.0E-6 | — | — | 1.0E-7 | — | — | — | — |
| MCDB 401 | 48 | 1.0E-4 | — | — | 1.0E-6 | — | — | 1.0E-4 | — | — | — | — | — | — | — |
| MCDB 411 | 29 | 3.0E-7 | — | — | 1.0E-7 | — | — | 1.0E-6 | — | — | 1.0E-6 | — | — | — | — |

*(continued)*

TABLE II (*continued*)

| Name of medium and number from Table I | | Carbohydrates and their derivatives | | | | | | | | | | Nucleic acid derivatives (Purine) | | | |
|---|---|---|---|---|---|---|---|---|---|---|---|---|---|---|---|
| | | Acetate | Deoxy-ribose | Galac-tose | Glucos-amine | Glucose | Glucono-lactone | Glu-curo-nate | Glucu-rono-lactone | Pyru-vate | Ribose | Adenine | AMP | ATP | Deoxy-adeno-sine |
| MEM | 5 | — | — | — | — | 5.6E-3 | — | — | — | — | — | — | — | — | — |
| BME | 4 | — | — | — | — | 5.0E-3 | — | — | — | — | — | — | — | — | — |
| DME | 36 | — | — | — | — | 5.6E-3 | — | — | — | 1.0E-3 | — | — | — | — | — |
| Yamane | 16 | — | — | — | — | 1.7E-2 | — | — | — | 1.0E-3 | — | — | — | — | 4.0E-6 |
| IMEM-Z0 | 37 | — | — | — | — | 1.1E-2 | — | — | — | 1.0E-3 | — | — | — | — | — |
| L15 | 11 | — | — | 5.0E-3 | — | — | — | — | — | 5.0E-3 | — | — | — | — | — |
| 199 | 3 | 6.1E-4 | 3.7E-6 | — | — | 5.6E-3 | — | — | — | — | 3.3E-6 | 4.9E-5 | 5.8E-7 | 1.8E-6 | — |
| CMRL 1415 | 38 | — | — | 2.8E-3 | — | 2.8E-3 | — | — | — | 2.0E-3 | — | — | — | — | 4.0E-5 |
| CMRL 1969 | 39 | — | — | — | — | 5.6E-3 | — | — | — | — | — | — | — | — | — |
| CMRL 1066 | 20 | 6.1E-4 | — | — | — | 5.6E-3 | — | 1.8E-5 | — | — | — | — | — | — | 4.0E-5 |
| NCTC 135 | 19 | 3.7E-4 | — | — | 1.8E-5 | 5.6E-3 | — | 7.7E-6 | 1.0E-5 | — | — | — | — | — | 4.0E-5 |
| MB 752/1 | 21 | — | — | — | — | 2.8E-2 | — | — | — | — | — | — | — | — | — |
| MAB 87/3 | 23 | — | — | — | — | 2.8E-2 | — | — | — | — | — | — | — | — | — |
| DM 145 | 26 | — | — | — | — | 5.6E-3 | — | — | — | — | — | — | — | — | — |
| Williams' G | 40 | — | — | 5.6E-3 | — | 5.6E-3 | — | — | — | 9.1E-4 | — | — | — | — | — |
| McCoy's 5a | 7 | — | — | — | — | 1.7E-2 | — | — | — | — | — | — | — | — | — |
| RPMI 1640 | 8 | — | — | — | — | 1.1E-2 | — | — | — | — | — | — | — | — | — |
| Neuman & Tytell | 13 | — | — | — | — | 1.7E-2 | — | — | — | 1.0E-3 | — | — | — | — | — |
| Higuchi | 15 | — | — | — | — | 1.0E-2 | 5.0E-4 | — | — | 1.0E-3 | — | — | — | — | — |
| F10 | 12 | — | — | — | — | 6.1E-3 | — | — | — | 1.0E-3 | — | — | — | — | — |
| F12 | 34 | — | — | — | — | 1.0E-2 | — | — | — | 1.0E-3 | — | — | — | — | — |
| MCDB 301 | 35 | — | — | — | — | 1.0E-2 | — | — | — | 1.0E-3 | — | — | — | — | — |
| MCDB 105 | 46 | — | — | — | — | 4.0E-3 | — | — | — | 1.0E-3 | — | 1.0E-5 | — | — | — |
| MCDB 202 | 47 | — | — | — | — | 8.0E-3 | — | — | — | 5.0E-4 | — | 1.0E-6 | — | — | — |
| MCDB 501 | 49 | — | — | — | — | 1.0E-3 | — | — | — | 5.0E-4 | — | 1.0E-7 | — | — | — |
| MCDB 401 | 48 | — | — | — | — | 5.5E-3 | — | — | — | 1.0E-3 | — | — | — | — | — |
| MCDB 411 | 29 | — | — | — | — | 5.0E-3 | — | — | — | 3.0E-4 | — | 3.0E-8 | — | — | — |

| Name of medium and number from Table I | | Nucleic acid derivatives (purine) (continued) | | | | | Nucleic acid derivatives (pyrimidine) | | | | | | | |
|---|---|---|---|---|---|---|---|---|---|---|---|---|---|---|
| | | 6,8-Dihydroxypurine | Guanine | Deoxyguanosine | Hypoxanthine | Xanthine | Deoxycytidine | 5-Methylcytosine | 5-Methyldeoxycytidine | Thymine | Thymidine | Uracil | Uridine | UTP |
| MEM | 5 | — | — | — | — | — | — | — | — | — | — | — | — | — |
| BME | 4 | — | — | — | — | — | — | — | — | — | — | — | — | — |
| DME | 36 | — | — | — | — | — | — | — | — | — | — | — | — | — |
| Yamane | 16 | 2.2E-6 | — | — | — | — | 1.3E-7 | — | — | — | 8.3E-6 | — | — | — |
| IMEM-Z0 | 37 | — | — | — | — | — | — | — | — | — | — | — | — | — |
| L15 | 11 | — | — | — | — | — | — | — | — | — | — | — | — | — |
| 199 | 3 | — | 1.6E-6 | — | 2.2E-6 | 2.0E-6 | — | — | — | 2.4E-6 | — | 2.7E-6 | — | — |
| CMRL 1415 | 38 | — | — | 3.7E-5 | — | — | 4.4E-5 | — | 4.1E-7 | — | 4.1E-5 | — | — | 2.1E-6 |
| CMRL 1969 | 39 | — | — | — | — | — | — | — | — | — | — | — | — | — |
| CMRL 1066 | 20 | — | — | 3.7E-5 | — | — | 4.4E-5 | — | 4.1E-7 | — | 4.1E-5 | — | — | 2.1E-6 |
| NCTC 135 | 19 | — | — | 3.7E-5 | — | — | 3.8E-5 | 8.0E-7 | — | — | 4.1E-5 | — | — | 1.8E-6 |
| MB 752/1 | 21 | — | — | — | 1.8E-4 | — | — | — | — | — | — | — | — | — |
| MAB 87/3 | 23[c] | — | — | — | 1.8E-4 | — | — | — | — | — | 3.3E-5 | — | — | — |
| DM 145 | 26 | — | — | — | — | — | — | — | — | — | — | — | — | — |
| Williams' E | 40 | — | — | — | — | — | — | — | — | — | — | — | — | — |
| McCoy's 5a | 7 | — | — | — | — | — | — | — | — | — | — | — | — | — |
| RPMI 1640 | 8 | — | — | — | — | — | — | — | — | — | — | — | — | — |
| Neumann & Tytell | 13 | — | — | — | — | — | — | — | — | — | — | — | — | — |
| Higuchi | 15 | — | — | — | — | — | — | — | — | — | — | — | — | — |
| F10 | 12 | — | — | — | 3.0E-5 | — | — | — | — | — | 3.0E-6 | — | — | — |
| F12 | 34 | — | — | — | 3.0E-5 | — | — | — | — | — | 3.0E-6 | — | — | — |
| MCDB 301 | 35 | — | — | — | 3.0E-5 | — | — | — | — | — | 3.0E-6 | — | — | — |
| MCDB 105 | 46 | — | — | — | — | — | — | — | — | — | 3.0E-7 | — | — | — |
| MCDB 202 | 47 | — | — | — | — | — | — | — | — | — | 3.0E-7 | — | — | — |
| MCDB 501 | 49 | — | — | — | — | — | — | — | — | — | 1.0E-8 | — | — | — |
| MCDB 401 | 48 | — | — | — | — | — | — | — | — | — | 1.0E-6 | — | — | — |
| MCDB 411 | 29 | — | — | — | — | — | — | — | — | — | 3.0E-6 | — | — | — |

TABLE II (*continued*)

| Name of medium and number from Table I | | Lipids and their derivatives: Choles-terol | Choline | *i*-Inosi-tol | Lecithin (mg/1000 ml)[e] | Linole-ate | Oleate | Bulk inorganic ions[f]: Calcium | Mag-nesium | Potas-sium | Sodium | Chloride | Nitrate | Phos-phate | Sulfate |
|---|---|---|---|---|---|---|---|---|---|---|---|---|---|---|---|
| MEM | 5 | — | 8.3E-6 | 1.1E-5 | — | — | — | 1.8E-3 | 9.8E-4 | 5.4E-3 | 1.4E-1 | 1.3E-1 | — | 9.7E-4 | — |
| BME | 4 | — | 1.0E-6 | — | — | — | — | 1.0E-3 | 5.0E-4 | 5.0E-3 | 1.2E-1 | 1.1E-1 | — | 1.0E-3 | — |
| DME | 36 | — | 2.9E-5 | 3.9E-5 | — | — | — | 1.8E-3 | 8.1E-4 | 5.4E-3 | 1.6E-1 | 1.2E-1 | 7.5E-7 | 9.1E-4 | 8.1E-4 |
| Yamane | 16 | — | 1.2E-4 | 7.8E-5 | — | — | — | 1.8E-3 | 9.8E-4 | 5.4E-3 | 1.5E-1 | 1.3E-1 | — | 9.7E-4 | 3.0E-6 |
| IMEM-Z0 | 37 | — | 4.0E-4 | 2.0E-4 | — | 3.0E-7 | — | 1.8E-3 | 1.0E-3 | 5.4E-3 | 1.5E-1 | 1.3E-1 | — | 1.1E-3 | 1.0E-4 |
| L15 | 11 | — | 8.3E-6 | 1.1E-5 | — | — | — | 1.3E-3 | 1.8E-3 | 5.8E-3 | 1.5E-1 | 1.5E-1 | — | 7.9E-4 | 8.1E-4 |
| 199 | 3 | 5.2E-7 | 3.6E-6 | 2.8E-7 | — | — | — | 1.8E-3 | 8.1E-4 | 5.4E-3 | 1.5E-1 | 1.3E-1 | 1.2E-6 | 1.0E-3 | 8.6E-4 |
| CMRL 1415 | 38 | — | 7.2E-6 | 1.1E-5 | — | — | — | 1.3E-3 | 9.7E-4 | 5.4E-3 | 1.4E-1 | 1.3E-1 | — | 1.8E-3 | 9.7E-4 |
| CMRL 1969 | 39 | — | 7.2E-6 | 1.1E-5 | — | — | — | 1.3E-3 | 8.1E-4 | 5.4E-3 | 1.5E-1 | 1.5E-1 | — | 1.8E-3 | 8.1E-4 |
| CMRL 1066 | 20 | 5.2E-7 | 3.6E-6 | 2.8E-7 | — | — | — | 1.8E-3 | 8.1E-4 | 5.4E-3 | 1.5E-1 | 1.3E-1 | — | 1.0E-3 | 8.1E-4 |
| NCTC 135 | 19 | — | 9.0E-6 | 6.9E-7 | — | — | — | 1.8E-3 | 8.3E-4 | 5.4E-3 | 1.5E-1 | 1.3E-1 | — | 1.0E-3 | 8.3E-4 |
| MB 752/1 | 21 | — | 1.8E-3 | 5.6E-6 | — | — | — | 8.2E-4 | 2.0E-3 | 2.6E-3 | 1.3E-1 | 1.1E-1 | — | 2.7E-3 | 8.1E-4 |
| MAB 87/3 | 23 | — | 1.8E-3 | 4.6E-6 | — | — | — | 8.2E-4 | 1.6E-3 | 3.5E-3 | 1.3E-1 | 1.1E-1 | — | 3.6E-3 | 4.1E-4 |
| DM 145 | 26 | — | 1.8E-3 | 1.1E-5 | — | — | — | 2.4E-3 | 4.9E-4 | 4.0E-3 | 1.5E-1 | 1.5E-1 | — | 1.4E-3 | — |
| Williams' G | 40 | — | 1.1E-5 | 1.1E-5 | — | 1.0E-7[g] | — | 1.8E-3 | 8.1E-4 | 5.4E-3 | 1.5E-1 | 1.2E-1 | — | 1.0E-3 | 8.1E-4 |
| McCoy's 5a | 7 | — | 3.6E-5 | 2.0E-4 | — | — | — | 1.8E-3 | 8.1E-4 | 5.4E-3 | 1.4E-1 | 1.2E-1 | — | 1.0E-3 | 8.1E-4 |
| RPMI 1640 | 8 | — | 2.1E-5 | 1.9E-4 | — | — | — | 4.2E-4 | 4.1E-4 | 5.4E-3 | 1.4E-1 | 1.1E-1 | 8.4E-4 | 5.6E-3 | 4.2E-4 |
| Neuman & Tytell | 13 | — | 3.6E-5 | 5.0E-5 | — | — | 3.4E-5[g] | 1.8E-3 | 8.1E-4 | 5.4E-3 | 1.3E-1 | 1.2E-1 | — | 9.9E-4 | 8.1E-4 |
| Higuchi | 15 | 5.0E-6 | 7.2E-5 | 1.1E-5 | 1.0 | — | 1.8E-6 | 1.8E-3 | 1.4E-3 | 5.4E-3 | 1.5E-1 | 1.4E-1 | — | 7.2E-4 | 5.5E-6 |
| F10 | 12 | — | 5.0E-6 | 3.0E-6 | — | — | — | 3.0E-4 | 6.2E-4 | 4.4E-3 | 1.4E-1 | 1.3E-1 | — | 1.7E-3 | 6.2E-4 |
| F12 | 34 | — | 1.0E-4 | 1.0E-4 | — | 3.0E-7 | — | 3.0E-4 | 6.0E-4 | 3.0E-3 | 1.5E-1 | 1.4E-1 | — | 1.0E-3 | 6.0E-6 |
| MCDB 301 | 35 | — | 1.0E-4 | 1.0E-4 | — | 3.0E-7 | — | 6.0E-4 | 6.0E-4 | 3.0E-3 | 1.5E-1 | 1.4E-1 | — | 1.0E-3 | 6.0E-6 |
| MCDB 105 | 46 | — | 1.0E-4 | 1.0E-4 | — | 1.0E-8 | — | 1.0E-3 | 1.0E-3 | 3.0E-3 | 1.3E-1 | 1.2E-1 | — | 3.0E-3 | 1.0E-3 |
| MCDB 202 | 47 | — | 1.0E-4 | 1.0E-4 | — | 2.0E-7 | — | 2.0E-3 | 1.5E-3 | 3.0E-3 | 1.4E-1 | 1.3E-1 | — | 5.0E-4 | 1.5E-3 |
| MCDB 501 | 49 | — | 1.0E-4 | 1.0E-4 | — | 2.0E-7 | — | 1.0E-3 | 5.0E-3 | 3.1E-3 | 1.4E-1 | 1.3E-1 | — | 1.0E-4 | 1.5E-3 |
| MCDB 401 | 48 | — | 1.0E-4 | 4.0E-5 | — | — | — | 1.6E-3 | 8.0E-4 | 4.0E-3 | 1.3E-1 | 1.2E-1 | — | 5.0E-4 | 8.0E-4 |
| MCDB 411 | 29 | — | 1.0E-5 | 1.0E-4 | — | 3.0E-7 | — | 1.0E-3 | 1.0E-3 | 1.0E-3 | 1.3E-1 | 1.1E-1 | — | 1.0E-3 | 4.0E-6 |

| Name of medium and number from Table I | | Inorganic trace elements[h] | | | | | | | | | | | | | |
|---|---|---|---|---|---|---|---|---|---|---|---|---|---|---|---|
| | | Chromium | Cobalt[i] | Copper | Fluorine | Iodine | Iron | Manganese | Molybdenum | Nickel | Selenium | Silicon | Tin | Vanadium | Zinc |
| MEM | 5 | — | — | — | — | — | — | — | — | — | — | — | — | — | — |
| BME | 4 | — | — | — | — | — | — | — | — | — | — | — | — | — | — |
| DME | 36 | — | — | — | — | — | 2.5E-7 | — | — | — | — | — | — | — | — |
| Yamane | 16 | — | — | — | — | — | 2.9E-6 | — | — | — | — | — | — | — | 1.0E-7 |
| IMEM-Z0 | 37 | — | 1.0E-6 | — | — | — | 2.0E-6 | — | — | — | — | — | — | — | 4.9E-7 |
| L15 | 11 | — | — | — | — | — | — | — | — | — | — | — | — | — | — |
| 199 | 3 | — | — | — | — | — | 4.1E-7 | — | — | — | — | — | — | — | — |
| CMRL 1415 | 38 | — | — | — | — | — | — | — | — | — | — | — | — | — | — |
| CMRL 1969 | 39 | — | — | — | — | — | — | — | — | — | — | — | — | — | — |
| CMRL 1066 | 20 | — | — | — | — | — | — | — | — | — | — | — | — | — | — |
| NCTC 135 | 19 | — | 7.4E-6 | — | — | — | — | — | — | — | — | — | — | — | — |
| MB 752/1 | 21 | — | 1.5E-7 | — | — | — | — | — | — | — | — | — | — | — | — |
| MAB 87/3 | 23 | — | 2.4E-7[j] | 2.0E-7 | — | — | 3.0E-6 | 9.5E-8 | 1.4E-7 | — | — | — | — | — | 1.0E-7 |
| DM 145 | 26 | — | 3.7E-9 | [k] | — | — | [k] | [k] | — | — | — | — | — | — | — |
| Williams' G | 40 | — | 1.5E-7 | 3.7E-9 | — | — | 3.1E-9 | 4.0E-10 | 2.8E-9 | — | 1.4E-10 | — | — | — | 1.1E-8 |
| McCoy's 5a | 7 | — | 5.5E-10 | — | — | — | — | — | — | — | — | — | — | — | — |
| RPMI 1640 | 8 | — | 3.7E-9 | — | — | — | — | — | — | — | — | — | — | — | — |
| Neuman & Tytell | 13 | — | 5.5E-10 | — | — | — | 2.0E-5 | — | — | — | — | — | — | — | — |
| Higuchi | 15 | — | 1.5E-9 | — | — | 1.0E-7[m] | 5.0E-6 | — | — | — | — | — | — | — | 4.9E-7 |
| F10 | 12 | — | 1.0E-6 | 1.0E-8 | — | — | 3.0E-6 | — | — | — | — | — | — | — | 1.0E-7 |
| F12 | 34 | — | 1.0E-6 | 1.0E-8 | — | — | 3.0E-6 | — | — | — | — | — | — | — | 3.0E-6 |
| MCDB 301[l] | 35 | 1.0E-9 | 1.0E-6[j] | 1.0E-8 | 1.0E-8 | 1.0E-9 | 3.0E-6 | 1.0E-9 | 7.0E-8 | 1.0E-9 | 1.0E-8 | 1.0E-9 | 1.0E-8 | 1.0E-8 | 3.0E-6 |
| MCDB 105 | 46 | — | 1.0E-7 | 1.0E-9 | — | — | 5.0E-6 | 1.0E-9 | 7.0E-10 | 5.0E-10 | 3.0E-8 | 5.0E-7 | 5.0E-10 | 5.0E-9 | 5.0E-7 |
| MCDB 202 | 47 | — | 1.0E-7 | 1.0E-9 | — | — | 5.0E-6 | 5.0E-10 | 7.0E-9 | 5.0E-12 | 3.0E-8 | 5.0E-7 | 5.0E-12 | 5.0E-9 | 1.0E-7 |
| MCDB 501 | 49 | — | 1.0E-7 | 1.0E-9 | — | — | 5.0E-6 | 5.0E-10 | 7.0E-9 | 5.0E-12 | 3.0E-8 | 5.0E-7 | 5.0E-12 | 5.0E-9 | 1.0E-7 |
| MCDB 401 | 48 | — | — | 5.0E-9 | — | — | 1.0E-6 | 1.0E-9 | 2.1E-8 | 3.0E-10 | 1.0E-8 | 1.0E-5 | — | 5.0E-9 | 1.0E-6 |
| MCDB 411 | 29 | — | 1.0E-6 | 1.0E-8 | — | — | 3.0E-6 | 5.0E-9 | 2.1E-8 | 3.0E-10 | 3.0E-8 | 1.0E-5 | — | 5.0E-9 | 1.0E-6 |

*(continued)*

TABLE II (*continued*)

| Name of medium and number from Table I | | Buffers and indicators | | | | Chelating agents, detergents, and solvents | | | | Proteins and polymers | | | |
|---|---|---|---|---|---|---|---|---|---|---|---|---|---|
| | | Bicarbonate[n] | Carbon dioxide (gas phase, %) | HEPES | Phenol red | EDTA | Ethanol | Tween 20 (mg/1000 ml) | Tween 80 (mg/1000 ml) | Insulin (mg/1000 ml) | Methylcellulose (mg/1000 ml) | Protamine (mg/1000 ml) | Salmine (mg/1000 ml) |
| MEM | 5 | 2.4E-2 | 5[o] | — | 1.4E-5[r] | — | — | — | — | | — | — | — |
| BME | 4 | 2.0E-2 | 5[o] | — | 1.4E-5 | — | — | — | — | — | — | — | — |
| DME | 36 | 4.4E-2 | 10 | — | 4.2E-5 | — | — | — | — | — | — | — | — |
| Yamane | 16 | 2.4E-2 | 5[o] | — | 1.4E-5 | — | — | — | — | 1.0 | — | — | — |
| IMEM-Z0 | 37 | 2.6E-2 | 5 | — | 2.8E-5 | — | — | — | — | — | — | — | — |
| L15 | 11 | — | air[p] | — | 2.8E-5 | — | — | — | — | — | — | — | — |
| 199 | 3 | 2.6E-2 | 8 | — | 4.2E-5 | — | — | — | 20.0 | — | — | — | — |
| CMRL 1415 | 38 | 1.2E-2 | 5 | — | 5.6E-5 | — | — | — | — | — | — | — | — |
| CMRL 1969 | 39 | 6.7E-3 | air[p] | — | 5.6E-5 | — | — | — | — | — | — | — | — |
| CMRL1066 | 20 | 2.6E-2 | 5[o] | — | 5.6E-5 | — | 3.5E-4 | — | 5.0 | — | — | — | — |
| NCTC 135 | 19 | 2.6E-2 | 10 | — | 5.5E-5 | — | 8.7E-4 | — | 12.5 | — | — | — | — |
| MB 752/1 | 21 | 2.7E-2 | 5[o] | — | 2.8E-5 | — | — | — | — | — | — | — | — |
| MAB 87/3 | 23 | 2.7E-2 | 5[o] | — | 4.2E-6 | — | — | — | — | 8.0 | — | — | — |
| DM 145 | 26 | 1.2E-2 | 5[o] | — | [s] | — | — | — | — | — | — | — | — |
| Williams' G | 40 | 2.6E-2 | 5 | — | 2.8E-5 | — | — | — | — | — | — | — | — |
| McCoy's 5a | 7 | 2.6E-2 | 8 | — | 7.1E-6 | — | — | — | — | — | — | — | — |
| RPMI 1640 | 8 | 2.4E-2 | 5[o] | — | 1.4E-5 | — | — | — | — | — | — | — | — |
| Neuman & Tytell | 13 | 2.6E-2 | 8 | — | 7.1E-6 | — | 1.3E-2 | 15 | — | 1.0 | — | — | 5 |
| Higuchi | 15 | 1.4E-2 | [q] | 1.2E-2 | 2.8E-5 | 5.0E-6 | 7.8E-3 | — | — | — | 600 | 2 | — |
| F10 | 12 | 1.4E-2 | 5 | — | 3.3E-6 | — | — | — | — | — | — | — | — |
| F12 | 34 | 1.4E-2 | 5 | — | 3.3E-6 | — | — | — | — | — | — | — | — |
| MCDB 301 | 35 | 1.4E-2 | 5 | — | 3.3E-6 | — | — | — | — | — | — | — | — |
| MCDB 105 | 46 | — | 2 | 3.0E-2 | 3.3E-6 | — | — | — | — | — | — | — | — |
| MCDB 202 | 47 | — | 2 | 3.0E-2 | 3.3E-6 | — | — | — | — | — | — | — | — |
| MCDB 501 | 49 | — | 2 | 3.0E-2 | 3.3E-6 | — | — | — | — | — | — | — | — |
| MCDB 401 | 48 | 1.4E-2 | 5 | — | 3.3E-5 | — | — | — | — | — | — | — | — |
| MCDB 411 | 29 | 1.4E-2 | 5 | 2.8E-2 | 3.3E-6 | — | — | — | — | — | — | — | — |

[a] This table summarizes the compositions of 27 of the culture media described in Table I. Literature references are listed in footnote *b* of Table I. Concentrations are expressed in moles per liter except where indicated otherwise. An abbreviated exponential notation is used in which the letter "E" signifies "10 to the power." Thus, 6.0E-4 means $6.0 \times 10^{-4}$ moles/liter. This summary does not provide sufficient information for preparation of media. Details such as which salt is used and the sequence for adding components to avoid precipitation can be very important. Anyone seeking to prepare a medium should consult the original literature or a standard technique manual such as the "TCA Manual" published by the Tissue Culture Association, 12111 Parklawn Drive, Rockville, Md. 20852.

[b] The value given is for half-cystine residues (two times the moles per liter of cystine) so that a direct comparison of molar concentrations can be made with cysteine and other amino acids.

[c] This heading includes amino acids not directly involved in protein synthesis, the small peptide glutathione, and various products of amino acid metabolism.

[d] Coenzymes and derivatives of vitamins are listed immeidately after the parent vitamin. *p*-Aminobenzoic acid is treated as a derivative of folic acid, although it is actually a precursor in bacterial systems and presumably cannot be converted to folic acid in vertebrate systems. Acidic vitamins are named as acids, although they are actually in the ionic form at physiological pH.

[e] Concentrations are given in weight units for substances of inexact composition, molecular weight, or purity, including lecithin, various polymers, and purified proteins.

[f] The total amount of each ion in the medium is given without regard to the type of salt or organic molecule that supplies it in the medium.

[g] Supplied as methyl ester rather than free acid or salt.

[h] Only those trace elements deliberately added as part of the medium formulation are listed. Substantial amounts of trace elements are frequently present as contaminants (cf. footnote *k* below). All trace elements listed on this table have been reported to be required by intact animals.

[i] Except where indicated cobalt is added in the form of Vitamin $B_{12}$.

[j] Free cobalt is added in addition to Vitamin $B_{12}$. The amount shown is the total from both sources.

[k] DM 145 has no deliberately added copper, iron, or manganese in its formula. However, analysis of typical batches has shown that these elements are present as contaminants as follows: copper, 6.3E-7 to 9.4E-7; iron, 1.9E-6; manganese, 5.5E-7 [T. Takaoka and H. Katsuta, *Exp. Cell Res.* **67**, 295–304 (1971)].

[l] In addition to the trace elements listed in the table, MCDB 301 also contains the following elements which are known to accumulate in biological systems and which may in some cases have beneficial effects: A1, 5.0E-9; Ag, 1.0E-9; Ba, 1.0E-8; Br, 1.0E-9; Cd, 1.0E-8; Ge, 5.0E-9; Rb, 1.0E-8; Ti, 5.0E-9; Zr, 1.0E-8.

[m] Iodine is supplied in the form of thyroxine.

[n] Only added bicarbonate is listed. Substantial amounts of bicarbonate may also be formed when buffered media equilibrate with an atmosphere enriched in carbon dioxide.

[o] The original paper does not indicate the amount of carbon dioxide to be used with the medium. These media are commonly used either with 5% carbon dioxide in air or with metabolically generated carbon dioxide from dense cultures in a closed culture vessel.

[p] These media are designed for use in unsealed vessels in air, with no enrichment of carbon dioxide other than that due to cellular metabolism. CMRL 1415 without bicarbonate (CMRL 1415-ATM) can also be used in this manner.

[q] Carbon dioxide is used to adjust the pH of the bicarbonate stock solution during preparation of the medium. All additional requirements are satisfied by metabolic carbon dioxide.

[r] The original paper does not indicate the amount of phenol red in MEM. However, it is commonly used in the same amount as in BME.

[s] The amount of phenol red (if any) in DM 145 is not stated in the original paper.

[t] The indicated amino acids are in the DL form in the original formulations. All others are in the L form.

[u] Vitamin A acetate.

[v] The two references cited in Table I for MAB 87/3 contain discrepancies in the formula. The concentrations in this table are from C. Waymouth, *TCA Manual* **3**, 521 (1977), with the addition of arginine HCl, 0.36 m*M*, which is included in the instructions but omitted from the summary table in that paper.

*pH*. The pH range that supports optimum cellular multiplication is generally quite narrow and varies somewhat with the type of cell. For example, the optimum range for clonal growth of human diploid fetal lung fibroblasts (WI-38) with minimal amounts of serum protein is pH 7.15–7.45, with some growth from pH 6.8–7.8. The optimum for chicken embryo fibroblasts under similar conditions is pH 6.95–7.30, with some growth from pH 6.4–7.7. Permanent lines of transformed cells are often much more tolerant of acidic conditions than diploid cells.

*Carbon Dioxide Tension*. The population-dependent requirement of cultured cells for carbon dioxide and/or bicarbonate ion as substrates for biosynthesis has already been described above. Since bicarbonate and carbonic acid are in a rapid equilibrium relationship determined by the pH, and since carbonic acid equilibrates with water and gaseous-phase carbon dioxide, it is impossible to maintain an adequate concentration of carbonic acid and bicarbonate in the culture medium without also maintaining an increased partial pressure of carbon dioxide in the gaseous phase. For monolayer or suspension cultures this can be accomplished by limiting diffusion of metabolic carbon dioxide out of the culture vessel, although there is danger of over-acidification due to excess accumulation of carbon dioxide. For clonal cultures, and for crowded cultures where better pH control is desired, the culture chamber is gassed with a mixture of humidified carbon dioxide in air, and the culture flasks or dishes are not tightly closed. Most culture media are designed for use with 5% carbon dioxide, which is close to the value found in body fluids, and some are used with as much as 10%. Recent studies in our laboratory indicate that a concentration of 2% is sufficient, and possibly even superior, for clonal growth of sensitive diploid cells with low levels of serum protein.

*Buffering*. The necessity of using carbon dioxide in the incubation chamber creates additional problems of pH control. A bicarbonate–carbon dioxide buffer will provide excellent pH control as long as the cultures are left in the incubator, but in the absence of supplementary buffering, media that contain bicarbonate become alkaline very rapidly due to loss of $CO_2$ when removed from the incubator. Current practice is moving toward the use of organic buffers, such as $\alpha$-glycerolphosphate or HEPES, at concentrations sufficient to reduce the pH change on removal of cultures from the incubator to a few tenths of a pH unit. We currently add 30 m*M* HEPES to media that are to be used with 2% carbon dioxide. Sufficient sodium hydroxide is added to bring the pH to the desired value under incubation conditions.

Since both temperature and $CO_2$ tension affect the final pH of the medium, it is important to measure pH under actual incubation conditions. However, after determination of the amount of pH change that occurs when a particular medium is placed in the incubator, it is possible to adjust the pH at room temperature and in air so that it will shift to the correct value in the incubator. It is unnecessary to include bicarbonate in the formulation of a medium whose pH is adjusted in this manner, since bicarbonate is formed from dissolved carbon dioxide as the buffered medium equilibrates in the incubator. One other potential area of concern is possible inhibitory effects of large quantities of artificial buffers. In the case of HEPES, there does not appear to be a problem up to at least 50 m*M*, provided that appropriate osmotic corrections are made.

*Osmolality.* The optimum range of osmotic pressures for cellular growth is quite narrow and varies with the type of cell and the species. For human fetal lung fibroblasts, optimum clonal growth occurs between about 250 and 325 mosmol/kg, with some growth from about 200–375 mosmol/kg. For chicken embryo fibroblasts, the optimum is about 275–325 mosmol/kg, and the range of suboptimum growth is about the same as for the human cells. Cells from amphibia require significantly lower osmolality than avian or mammalian cells, and this is often achieved by diluting standard culture media with distilled water. Cells from parts of the body such as the kidneys that are exposed to differing osmolalities may also have different requirements.

The narrowness of the optimum range of osmolality makes it necessary to adjust the concentration of sodium chloride when major additions are made to a culture medium, such as the use of 30–50 m*M* HEPES buffer (plus enough NaOH to bring it to a physiological pH). Osmotic considerations may also be important when media are used with high concentrations of carbon dioxide. With an atmosphere of 10% carbon dioxide and a pH of 7.4, the combined concentrations of carbonic acid and bicarbonate ion generated by carbon dioxide dissolving in a buffered medium add up to about 30 m*M*. (See also this volume [16] for a discussion of osmolarity.)

*Humidification of Incubators.* Since rapid equilibration between the culture medium and the gaseous carbon dioxide–air mixture in the culture chamber is frequently important, culture dishes usually cannot be sealed tightly. The narrow limit of acceptable osmolality requires that essentially all evaporation be prevented during the time that the dishes are in the incubator, which may be 2 weeks or longer for some clonal growth experiments. Limitation of evaporation is accomplished by maintaining the relative humidity very close to saturation in the incubation chamber. This can be a difficult task in a dry climate. One of the specific requirements is that the water that is used for humidification be kept at the same temperature as the rest of the incubator. Each time that the incubator door is opened, dry air enters and cools the water by evaporation. If there is inadequate heat transfer, the water may remain cool for a prolonged period and fail to humidify the incubation chamber adequately to prevent evaporation and maintain the osmolality of the culture medium. Similarly, if there is a cool spot anywhere in the incubator, such as a poorly insulated door, it will serve as a site of condensation and prevent the partial pressure of water vapor from reaching the needed level. To avoid these problems, there must not be any surface in the incubator, including that of the water used for humidification, that is more than a few tenths of a degree centrigrade below the temperature of the medium in the culture dishes.

One of the best ways to monitor humidification is to measure the osmolality of control dishes of medium at the beginning and the end of the incubation period for the experiments being performed. If there is a significant increase in osmolality, the humidity is not high enough and evaporation is occurring. Excessive humidification that leads to condensation on the culture dishes is also undesirable, since it increases the risk of microbial contamination.

*Oxygen Tension.* A minimal amount of oxygen is essential for the multiplication of most types of cells in culture. However, the partial pressure of oxygen in normal body fluids is significantly less than that of air. Systematic studies of the effects of oxygen tension on cultured cells are technically difficult due to diffusion gradients, leakage, and release of dissolved oxygen from plastic culture vessels. However, there are numerous reports that growth of cultured cells can be improved by reducing the percentage of oxygen in the gaseous phase to between 1 and 10%. The trace element, selenium, is a component of the enzyme glutathione peroxidase, which destroys metabolically generated peroxides. The inhibitory effects of excess oxygen are much more severe when cells are marginally deficient in selenium than when adequate amounts are supplied.

There have been a number of attempts to correlate cellular growth with redox potential measured by inserting electrodes into the culture medium, but the results are difficult to interpret because a cell growth medium is not an equilibrium system. In fact, for cellular growth to occur at all, the cell must be able to catalyze an energy-yielding reaction among components of the culture system (normally glucose and oxygen). Cysteine and various other reducing agents are present in large amounts in some of the older culture media, but there is little evidence that the reducing conditions that they were intended to generate are actually needed, and in at least some cases, very high concentrations of cysteine are clearly inhibitory.

### *Freedom from Toxicity*

The requirement that the culture system must be free from toxic or inhibitory effects might seem so obvious that it would need no discussion.

However, there are many subtle types of toxic and inhibitory effects that are not immediately obvious. A sufficiently large excess of any component of the culture system is inhibitory, even if only for osmotic reasons, and many essential nutrients become inhibitory at surprisingly low levels. Contaminants in the chemicals or water used in medium preparation can be toxic. Membrane-type sterilizing filters may introduce substantial amounts of detergents into culture media. Because of this, detergent-free filters should always be specified. Short-wavelength visible light interacts with certain components of culture media (riboflavin, tyrosine, tryptophan) to generate toxic photoproducts.[17] Therefore, it is necessary to keep exposure of media to ordinary laboratory fluorescent lights at a minimum. Toxicity can sometimes be generated by interactions among components of the culture medium that are not individually toxic. Thus, for example, bovine serum cannot be used in media that contain spermine or spermidine because its high content of spermine oxidase will generate toxic oxidation products from these polyamines. Putrescine is safe to use with bovine serum, however, as it is not affected by the enzyme. In certain cases, inhibitory contaminants may alter responses to other components of the culture system. For example, the presence of traces of cadmium may increase substantially the amount of zinc needed for optimal multiplication. Similarly, in certain cases, a substantial portion of the requirement for serum protein may be to overcome inhibitory effects.

*Quantitative Requirements*

The full significance of quantitative requirements for cellular multiplication is just beginning to be realized. For classically studied permanent cell lines, precise quantitative adjustments do not seem to be very important, although there is some evidence that balance relationships affect growth.[18] For normal cells that are now receiving increasing study, however, such adjustments are proving to be critically important.[1,2,9]

For every required component in a culture medium, a three-part growth response curve, similar in principle to the idealized example shown in Fig. 1, can be obtained. In the first region on the left, multiplication is roughly proportional to the amount of the component added. In the second region, in the center, the growth response is "saturated," and the response curve exhibits a "plateau," within which growth is virtually independent of concentration of the component. Finally, on the right, above a critical threshold, the component becomes inhibitory and the growth response is negative.

[17] R. J. Wang, *In Vitro* **12,** 19 (1976).
[18] R. G. Ham, *In Vitro* **10,** 119 (1974).

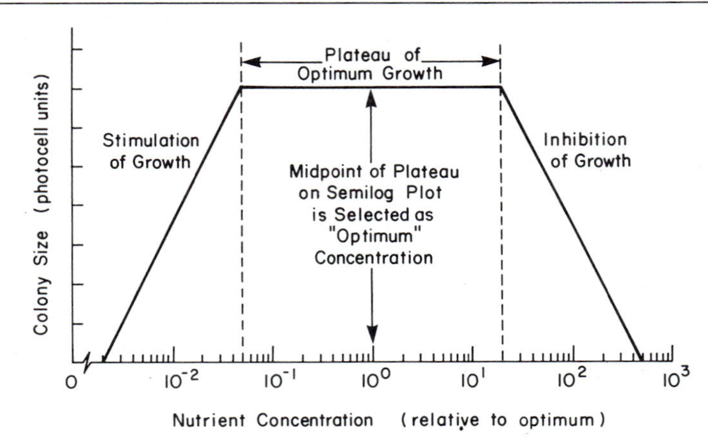


FIG. 1. Idealized clonal growth response illustrating the method used to select the "optimum" concentration of a nutrient for inclusion in a culture medium. From R. G. Ham and W. L. McKeehan, *In Vitro* **14,** 11 (1978).

Response curves of this type can be used to establish an "optimum" concentration for each component of the medium. Generally, it is desirable to select a concentration that is as far removed as possible both from deficiency and from inhibitory effects. This is accomplished by selecting the midpoint of the plateau of the growth response on a semilog plot, as illustrated in Fig. 1.

We have recently systematically adjusted the concentrations of all components of the medium to optimum concentrations for clonal growth of several different types of normal cells with minimal amounts of serum protein. This was done sequentially at progressively lower concentrations of serum protein until no further reduction could be achieved.[1,2,9] The result was a very substantial decrease in the amount of serum protein needed,[19] as well as a significant improvement in growth. Parallel studies with normal human, mouse, chicken, and duck fibroblast-like cells have shown that each has a characteristic pattern of quantitative requirements, and each grows best in a medium that has been optimized specifically for its own needs.[1,2,9,20] For example, medium MCDB 105, which was optimized for human diploid fibroblasts and medium MCDB 202, which was optimized for chicken embryo fibroblasts, both contain the same 56 components, but they differ in the quantities that were found to be optimal for

[19] W. L. McKeehan, D. P. Genereux, and R. G. Ham, *Biochem. Biophys. Res. Commun.* **80,** 1013 (1978).

[20] W. L. McKeehan, K. A. McKeehan, and R. G. Ham, *in* "The Growth Requirements of Vertebrate Cells in Vitro" (R. G. Ham, C. Waymouth, and P. J. Chapple, eds.). Cambridge Univ. Press, London and New York (in press).

clonal growth of 26 of those 56 components (Table II). None of the "standard" media developed over the years for permanent established lines will support clonal growth of these cells with comparably small amounts of serum protein.

Another aspect of quantitative requirements that must be considered is the balance among interrelated components of the culture system. For example, in addition to osmotic requirements and absolute requirements for sodium and potassium, the $Na^+/K^+$ ratio appears to be important, with different optimum values for different types of cells. Another set of balance relationships that appears to be very important occurs among the amino acids.[18] The ratios of amino acids in media optimized for various normal and nontransformed cells differ significantly. (Compare MCDB 105, 202, and 401 in Table II.)

### *Serum and Its Relationship to Other Growth Requirements*

The reason that serum is such an effective supplement for promoting cellular multiplication appears to be that it contains a large number of different growth-promoting activities in a physiologically balanced blend. The multitude of different growth-promoting roles that serum is capable of playing can perhaps best be appreciated in terms of the many different ways in which the requirement for serum can be partially replaced. In a more detailed review, we have listed and discussed 24 different ways in which serum could conceivably contribute to cellular multiplication.[2] Only a few of the more significant ways are described here.

Whole serum contains most of the low-molecular-weight nutrients needed for cellular multiplication, and at one time the culture of permanent cell lines such as Mouse L and HeLa was routinely done in mixtures of serum, embryo extract, and saline, with no defined organic nutrients added to the medium other than glucose. Dialyzed serum retains many small molecular nutrients in bound form, and it also possesses many other growth-promoting activities. Large amounts of serum protein allow cells to multiply with concentrations of essential nutrients that are inadequate for multiplication with smaller amounts of serum protein. Dialyzed serum is also able to reduce the inhibitory effects both of contaminants and of essential nutrients that are present in excess amounts. In some cases, serum macromolecules may buffer toxic nutrients by binding them and releasing them in small amounts as their free concentration in the medium is reduced by cellular metabolism. Serum protein neutralizes trypsin and other proteases and also makes cells better able to overcome the damage done to them by harsh subculturing procedures. Serum has an undefined effect on the interaction between cells and their substrate, which can be replaced by coating the substrate with polylysine. Serum provides carrier proteins for water-insoluble substances, such as lipids, that appear to be required by normal cells. Serum may also provide peptide hormones or hormone-like growth factors, or other proteins that are specifically required for cellular multiplication (see also this volume [6]).

The preceding is only a minimal outline of the very large number of different roles that serum appears to perform in supporting the multiplication of various types of cells in various types of basal media. The number of such roles that are required appears to vary greatly from one type of cell to another. As mentioned earlier, there are some types of very highly adapted cells that can be grown without serum protein in a number of different defined media. There are also a few other cell types that can be grown without protein supplementation, but only in media that have been tailored precisely to satisfy their individual needs.[10,21]

Appropriate adjustments in the many different variables that serum interacts with greatly reduce the amount of serum protein that is needed for growth of a variety of "normal" cells, but it currently appears unlikely that this approach will make it possible to eliminate completely the requirement of such cells for macromolecular growth factors. Requirements in the milligram per milliliter range have proven easy to replace, but for all of the diploid cells that we have studied, requirements in the microgram per milliliter range are not replaceable by any combination of polylysine-coated dishes, improved subculturing, and qualitative and quantitative optimization of the basal medium used to assay for the multiplication-promoting activity. Preliminary fractionation studies have reduced the total amount of active material needed to a few micrograms per milliliter in the case of human diploid fibroblasts, and the number of components that can be detected in the active fraction suggests that the ultimate requirements for individual active components will be in the nanogram per milliliter range when purification is completed.[19]

Although the complete set of growth requirements has not been fully worked out for most types of cells, there are numerous reports in the literature that peptide hormones or hormone-like growth factors are needed in the nanogram or microgram per milliliter range for multiplication of a variety of types of cells. A detailed discussion of these factors is beyond the scope of this article, but they have recently been reviewed by Gospodarowicz and Moran,[22] and a number of them are discussed in a recent symposium volume.[3] They range from well-characterized peptide hormones like insulin to specialized growth factors, such as epidermal growth factor, fibroblast growth factor, and the somatomedins, and on to

[21] R. G. Ham, *Proc. Natl. Acad. Sci. U.S.A.* **53,** 288 (1965).

[22] D. Gospodarowicz and J. S. Moran, *Annu. Rev. Biochem.* **45,** 531 (1976).

factors that have not yet been fully characterized chemically, such as "platelet" factor and "multiplicaton stimulating activity" from conditioned medium. Gordon Sato's research group has recently reported complete replacement of the serum requirements of several permanent cell lines with mixtures of hormones and peptide growth factors[23] (see also this volume [6]).

Some of the peptide growth factors, such as epidermal growth factor and fibroblast growth factor, are beginning to be produced commercially. Such factors will undoubtedly play increasingly important roles in the study of cellular growth requirements during the next few years, as will numerous factors that are currently under study, but not yet fully purified or characterized. It currently seems reasonable to anticipate that growth of certain normal diploid cells in "defined" media supplemented only with nanogram per milliliter quantities of purified macromolecular factors will become possible within a few years.

### *Holistic Approach to Cellular Growth Requirements*

There are so many individual requirements for multiplication of cells in culture that it is difficult to gain a clear perspective of the total set of conditions that must be satisfied before cells will multiply. The problem is further complicated by a multitude of complex interactions that occur among the individual variables, including quantitative balance relationships, the existence of alternate methods for satisfying many of the individual requirements, and the ability of serum to interact with virtually every other variable in the entire set. Because of these interactions, we have found it desirable to adopt a holistic view of cellular growth requirements, in which all variables that affect cellular multiplication in any way are viewed as members of a single complex interacting set, which is so constituted that a change in any one of its variables can potentially have an effect on the growth response to any other variable in the entire set.[1] Systematic study of the effects of individual variables requires specifically that all interactions be recognized and that all other significant variables be held constant while the effects of each single variable or interacting group on cellular multiplication are being studied.

## Preliminary Decisions to Be Made before Choosing a Medium

The choice of which medium to use will normally be one of the last decisions made before an experiment is undertaken. Preliminary decisions

[23] I. Hayashi, J. Larner, and G. Sato, *In Vitro* **14**, 23 (1978).

related to the experimental objective that first must be made include: (1) what kind of cell to use; (2) what type of culture (monolayer, suspension, clonal, etc.) to use; and (3) what degree of chemical definition is needed.

*Choosing Which Cell to Use*

In most cases the first decision to be made will be what cell type to use in the experiment. That choice is ultimately determined by the goals of the experiment. Important considerations include: (1) What differentiated properties are needed in the cell that is selected? (2) Does it matter which species is used? (3) Does the experiment require the use of normal diploid cells or can it be done just as well with permanent lines of transformed cells whose growth requirements are simpler and better understood?

*Differentiated Properties.* For some types of research that directly involve differentiated functions, highly specialized differentiated cells must be used. In such cases, the primary consideration is whether to use a normal diploid culture or a permanent line of differentiated tumor cells. The tumor lines will generally grow in simpler media and be easier to handle, but are also less likely to show fully normal regulation of expression of their differentiated properties. For experiments that do not directly involve differentiation, there is usually a considerable latitude in the type of cell that will allow the experimental objective. In such cases, it is desirable to select a cell whose growth requirements are relatively well understood and that will grow well in the type of culture and with the degree of chemical definition needed for the experiment.

*Species.* In certain cases the experimental objectives will dictate cells from a particular species. Examples include the use of human cells for medically related studies and the use in culture of the same species that was used in a previous study that did not involve culture. Since species has a significant effect on quantitative growth requirements and balance relationships, particularly for normal cells, it is important to employ a medium that will work well with the specific species. In cases where species is relatively unimportant to the experimental objectives, it is desirable to select a species whose growth requirements are relatively well understood and whose growth properties in culture are compatible with the experiment to be performed.

*Normal versus Transformed.* Strong arguments can often be mustered on both sides of the question of whether to use a normal diploid cell or a transformed permanent cell line. The permanent lines, and particularly those that can be grown without serum, offer obvious advantages in terms of ease of handling and more precisely defined culture systems. However, at the same time, they are grossly abnormal cells that have undergone extensive adaptation and chromosomal evolution in culture. They have generally lost essentially all growth regulation, and except for differentiated tumor lines, they exhibit few, if any, differentiated properties. If the experimental variable that is being studied can be shown not to differ significantly between permanent cell lines and normal cells in the intact animal, it may be worthwhile to employ the permanent lines and the more highly defined media that accompany them. However, such a choice should be made cautiously, and the results that are obtained from such a study should ultimately be verified with unaltered normal cells.

*Choosing a Culture System*

The choice of a culture system is influenced both by the type of cell that has been selected and by the experimental objectives. The major considerations involved are:

1. Is the actual experiment to be done in cell culture, or is cell culture simply being used to grow cells for use in other types of experimentation?
2. How many cells are needed?
3. Is it important to be able to evaluate the responses of individual cells?
4. Are the cells to be grown attached or suspended?

The first three questions are all interrelated. If cell culture is to provide cells that are then harvested and used for other experiments, it is likely that reasonably large numbers of cells will be needed. In such cases, it generally will be desirable to use monolayer or suspension cultures to obtain the required number of cells with a minimum of effort. For very large yields, specialized systems such as roller bottles and capillary perfusion are useful. In all cases where high cellular yield is necessary, it is important to select media that will support dense growth and to renew the media either by periodic replacement or by continuous perfusion as needed for optimal growth. In cases where cellular responses are being measured in culture, the number of cells needed depends on the sensitivity of the assay procedure. Large populations may be needed to obtain enough material for chemical analysis. However, it may be more important to look at the responses of individual cells in clonal growth assays. In the latter instance, it is important to select a medium that satisfies all of the population-dependent requirements of isolated cells. Such media generally have lower concentrations of nutrients to avoid inhibitory effects, and are not particularly well suited for dense cultures.

The option of using either attached or suspended cultures is available only when a cell type that is not anchorage-dependent is selected. In most cases, suspension culture is used only as a convenient means of growing large numbers of cells, rather than for performing actual experiments. If suspension culture is selected, it is usually necessary either to select a medium designed for suspension culture or to modify a conventional medium by reducing its calcium content. Many of the media designed specifically for suspension culture have particularly high concentrations of nutrients to avoid possible depletion by the dense cellular populations that are involved.

*How Much Chemical Definition Is Needed*

Because all chemicals contain some amount of impurity, there is no such thing as a medium that is truly chemically defined. However, the options available for cell culture range from "defined" media prepared from the best available grades of chemicals and water to media that are supplemented with large amounts of grossly undefined additives such as whole serum and embryo extract. The decision of what degree of chemical definition to seek for a particular experiment is influenced both by the type of cell that is used and by the experimental objectives.

At present, all *normal* cells require at least some serum protein for sustained multiplication, and many whose requirements are not yet well characterized require large amounts of whole serum. Most *permanent* lines can be grown with small amounts of serum protein, and, at least in dense cultures, many can be grown in synthetic media without any supplementation.[24,25] In a few cases, clonal growth without serum supplementation is also possible.[10]

When serum does not interfere in any way with the objectives of the experiments that are to be performed, it is often simpler to use serum-containing media than to try to replace the serum. Such an approach will usually work when the only objective is to grow large numbers of a particular cell. However, when cellular metabolism or interactions between the cells and their environment are being studied, experimental results are likely to be influenced by components of the undefined and not strictly reproducible supplements.

Where the macromolecular components of serum do not interfere, but a reasonably defined small molecular environment is needed, it is often convenient to use dialyzed serum as a supplement. Although dialyzed serum is frequently referred to as "serum protein" in this article, it is important to remember that it is actually a complex mixture that also contains substantial amounts of lipids and carbohydrates in the form of

[24] K. Higuchi, *Adv. Appl. Microbiol.* **16,** 111 (1973).
[25] H. Katsuta and T. Takaoka, *Methods in Cell Biol.* **6,** 1 (1973).

lipoproteins and glycoproteins, as well as an assortment of bound small molecules, all of which can interfere in some types of experiments. For example, with cells that can be grown without serum protein, small amounts of serum protein effectively replace the requirement for biotin. It is not clear at this time whether the effect is due to bound biotin, or to end-product substitution of fatty acids from lipoproteins, whose synthesis would otherwise require biotin. Similar results are also observed for most trace elements and for certain amino acids. For example, tryptophan binds strongly to serum albumin, and cysteine reversibly forms —S—S— linkages with serum proteins, which can sometimes be reduced to release enough cysteine for cellular multiplication. If dialyzed serum is employed, it is often convenient to prepare large standardized batches by dialysis against distilled water, followed by lyophilization and storage as a dry powder at $-20°$.[13]

As has been described above, it is frequently possible to reduce the amount of serum protein needed for cellular multiplication by employing media and culture conditions that have been optimized for the particular type of cell that is being grown. In some instances, still further reduction of the amounts of undefined additives can be achieved by use of hormones[23] or partially purified growth factors[19] in place of dialyzed serum.

Whenever the experimental objectives are compatible with the use of permanent cell lines, there are definite advantages, such as freedom from interfering substances, in using a defined medium that does not require protein supplementation. However, if such a system is employed, it is important to avoid the introduction of contaminants from all other sources including the cellular inoculum. Cells that have been previously grown as stock monolayer or suspension cultures in a richly supplemented medium are likely to introduce large amounts of undefined substances into the "defined" experimental medium. Even if the cells are washed thoroughly, such substances may be carried over intracellularly and perhaps even released into the medium by death of a portion of the inoculum in amounts sufficient to influence experimental results. One of the special advantages of the clonal growth technique is that it minimizes the effects of such carry-over by using only a very small inoculum (typically 100 cells in 5 ml of medium) and by requiring that each cell undergo repeated cycles of multiplication, which rapidly dilute out any intracellular carry-over effects.

Finally, as alluded to earlier, serious questions remain about the purity of "defined" media. In cases where contaminants are a serious problem, such as trace element research, it may prove necessary to prepare the "defined" media from specially purified chemicals and water in a carefully decontaminated laboratory to achieve the required freedom from contamination. In such cases, it may also prove necessary to add extra nutrients to achieve growth comparable to that in the less-pure medium.

## Culture Media and Which One to Use

The research literature describes hundreds of different media for growth of vertebrate cells in culture. Not all are widely used, but there is a sufficient number to cause confusion to the newcomer to cell culture. About a dozen are so generally used that they are available from virtually every commercial supplier, and many others are available from some of the larger companies. In addition, there are a number of newer and less well known media that are not currently available commercially, but which may be very useful in certain types of research. All of the most widely used media plus a number of the less common ones that are of potential interest are listed in Table I. The chemical composition of a number of these media is summarized in Table II. [See pp. 56–71 for Tables I and II.]

### *Brief History of Commonly Used Media*

The development of modern culture media can be divided into three somewhat overlapping eras, which are reflected in the organization of Table I.

1. The first era (part A of Table I) occurred prior to the general availability of permanent cell lines, and employed cells that had only been in culture a short time to measure growth (or survival) responses. Of this group the only medium still used to a significant extent is medium 199 of Morgan, Morton and Parker, which was originally developed for prolonged survival of chicken embryo heart fibroblasts without serum. When supplemented with serum, it supports quite good growth of many types of cells, and despite its age, it remains the medium of choice in some cases.

2. The second era, which began in the early 1950s (and is still continuing to some extent), was based on the use of permanent cell lines to measure growth responses (parts B, C, D, E, and F of Table I). Two lines, mouse L and HeLa, dominated the rapid progress in medium design made in the late 1950s. With only a few exceptions, the media that currently enjoy the most general usage were either developed through use of mouse L or HeLa to measure growth responses, or else through use of other permanent lines that had first been adapted to grow in media originally developed for mouse L or HeLa. Media from that era include formulas such as the widely used Eagle's basal and minimum essential media that

were designed for use with dialyzed serum, as well as a number of media that support growth of permanent lines without serum supplementation. The latter have been reviewed in detail by Higuchi[24] and by Katsuta and Takaoka.[25]

3. The third era (part G of Table I) has been under way for many years, but is just now beginning to pick up momentum. It represents a return to cells that are far more "normal" in their properties for the measurement of growth responses. The classical permanent lines generally underwent extensive nutritional adaptation during the process of becoming established in culture, and media developed to satisfy their needs generally do not support growth of normal cells without heavy serum supplementation. Current research on cell growth requirements is focusing increasingly on normal cells and on nontransformed lines whose growth requirements are more like those of normal cells. At the present time, the growth requirements of normal cells are not yet fully understood, and the limited number of improved media for such cells that have been described in the literature are just beginning to become available commercially.

*Media Designed for Use with Serum or Serum Protein*

It is difficult to organize a discussion of media and their applications, since many are currently used for purposes very different from the one for which they were originally developed, and since certain popular media are used for widely divergent purposes. The organization of Table I is based on the original purpose for which each medium was developed, and an attempt has been made to indicate other uses in the "comments" column. The discussion that follows is somewhat loosely organized, but progresses in general from serum-supplemented media to "defined" media, and within each type of medium, from transformed permanent lines to normal cells.

*Eagle's Media.* The most widely used of all cell culture media, Eagle's minimum essential medium (MEM), was designed specifically for monolayer growth of mouse L and HeLa cells with modest amounts of dialyzed serum, as was its predecessor, Eagle's basal medium (BME).[26] Both of these media are simple and easy to prepare (Table II), and both support good monolayer growth of many different types of cells when

[26] There are actually several "basal" media described by Eagle that differ from one another in minor details. The Tissue Culture Association recommends use of the term "Eagle's Basal Medium" only to describe the formula for HeLa cells given in H. Eagle, *Science* **122,** 501 (1955). That formula has been summarized in Table II.

supplemented with whole or dialyzed serum. MEM is the result of careful quantitative balancing and can be used with an optional supplement of nonessential amino acids and pyruvate for clonal growth.[27] Modified versions for suspension culture are also available from commercial suppliers.

*Dulbecco's Medium.* There are innumerable modifications of Eagle's media. Perhaps the best known is that of Dulbecco. This medium has evolved over the years, and there is a good deal of confusion about various versions of it. In an attempt to clarify the situation, the Tissue Culture Association has designated the formula summarized in Table II as the official version,[28] and has requested that all others be designated "modified." DME contains 4 times the amino acids and vitamins of one of the early versions of BME, and it also has a few added components (Table II). The official version contains 1 mg/ml of glucose, but for some types of cells, a variant with 4.5 mg/ml of glucose is superior. Dulbecco's modified Eagle's medium should be abbreviated "DME" and not "DMEM," since the latter leads to confusion with Eagle's MEM; DME was actually developed from BME rather than from MEM. DME is widely used with whole or dialyzed serum for nontransformed cultures, particularly of mouse and chicken cells.

*Clonal Growth.* A number of media were developed specifically for clonal growth of transformed cells, including McCoy's medium 5A for Walker carcinosarcoma 256 cells, and Ham's F10 and F12 for Chinese hamster lines. All three media have been used widely for primary cultures of differentiated cells from various species including rats, rabbits, and chickens. In many cases, the concentrations of amino acids and pyruvate in F10 and F12 are doubled, as first proposed by Hayden Coon. Medium F12 was originally developed for protein-free growth, but it no longer works well without serum protein unless selenium is added (see below).

*Fischer's Medium.* References to "Fischer's" medium are potentially ambiguous. Albert Fischer was one of the pioneers in medium development (Table IA), but his media are now primarily of historical interest. More recently, Glenn Fischer and co-workers developed a medium for growth of mouse leukemic cells in suspension culture that is rather widely used (No. 10, Table IB). That medium is currently offered by several commercial suppliers simply as "Fischer's medium."

*Media Used without Carbon Dioxide.* Several media have been developed with special buffering for use without carbon dioxide gassing.

[27] H. Eagle, *Science* **130,** 432 (1959).

[28] H. Morton, *In Vitro* **6,** 89 (1970). This article, available as a monograph from the Tissue Culture Association, contains recommended formulations for a variety of commercially prepared culture media.

The most widely used of these is medium L-15 of Leibovitz, which substitutes galactose for glucose and contains high concentrations of the basic amino acids. This formula has been used with serum for a variety of cells, both permanent lines and diploid primary cultures. A second such medium is the CMRL 1415-ATM modification of CMRL 1415, which was designed for primary cultures of mouse embryo cells. It is also useful for a variety of other types of cells, and growth can sometimes be improved by reducing its unusually high concentration of cysteine.

*Media for Normal Cells.* As described above and in Table I, many of the media developed originally for established lines are widely used with serum for growth of normal diploid cells and permanent lines of nontransformed cells. In addition, there are a number of media that have been developed specifically for such cells (parts A and G of Table I). With the exception of Medium 199 and similar media from the era of development that preceded the availability of permanent cell lines, all of these formulas are modifications of previous media for permanent lines. However, some, such as the MCDB series, represent complete reworkings of the formulas to satisfy the needs of the nontransformed cells.

A number of the formulas listed in Table IG were developed primarily for survival and maintenance of function of normal cells without serum supplements, and will be discussed below under the defined medium heading. Dulbecco's DME and Richter's IMEM-ZO are both modifications of original formulations by Eagle. As already discussed, DME is one of the most widely used media for primary cultures of mouse and chicken cells. IMEM-ZO has been reported to support normal rat and human cell growth, including clonal growth with relatively low concentrations of serum proteins. CMRL 1415 is used with serum for growth of mouse cells, and CMRL 1969 for human and monkey cells. F12K is a further modification of Coon's modification of F12, and it has been reported to work well for a number of types of differentiated rat cells.

The MCDB series of media is the result of a process of qualitative and quantitative optimization for clonal growth with minimal amounts of serum protein. In each case, they will support clonal growth of diploid fibroblast-like cells from the species they were designed for (MCDB 104, 105, human; MCDB 202, chicken; MCDB 401, mouse; MCDB 501, duck) with far less dialyzed serum than conventional media. The composition of MCDB 202 is such that it will also work reasonably well with cells from a variety of other species. These media are specifically recommended for clonal growth of their respective species, although they will not necessarily work for all types of cells from a particular species. They also can be used for monolayer growth with modest amounts of dialyzed serum. At the time of this writing, only MCDB 104 is available commercially.

*Media Used with Purified Proteins.* Table IC lists a number of cases in which proteins of varying purity have been used in place of serum. These range from rather crude fetuin and serum albumin used in the development of medium F10 to mixtures of hormones and hormone-like factors used by Gordon Sato and his co-workers to replace serum protein.[23] Recent media from Waymouth's laboratory for differentiated cells (Table IG) are also supplemented with hormones. Several protein-supplemented "serumless" media are listed in Table IC, including formulas by Neumann and Tytell, by Jenkin and co-workers, by Higuchi, and by Yamane and co-workers. These appear to support good growth of a number of types of cells and may be of value when whole serum cannot be employed. However, none have gained widespread usage, and they are either not available commercially or else are listed only by a few companies.

### *"Defined" Media*

A number of media have been developed that support growth of permanent lines without the use of serum or other undefined supplements. The probable role of contaminants in such growth makes it questionable to call those media "defined," but in the absence of a suitable alternative, we will employ that term with implied reservations. The following sections consider in order defined media for monolayer growth, suspension culture, and clonal growth of permanent lines, and then maintenance without growth of normal cells.

*Defined Media for Monolayer Cultures.* During the late 1950s and early 1960s several series of defined media for monolayer cultures of mouse L cells were developed (Table ID). Media NCTC 135, CMRL 1066, and MB 752/1 are all widely used and readily available from commercial suppliers. As can be seen from Table II, Waymouth's MB 752/1 is somewhat simpler than the other two, and on that basis it is probably preferable for use in situations where all three support growth equally well. Newer formulations from Waymouth's laboratory with trace element supplements (MD 705/1 and MAB 87/3) are available from some suppliers. All have been used with serum for growth of a variety of types of cells that do not grow in them in the absence of serum.

A fourth series of defined media has been used for many years by Katsuta and Takaoka and their co-workers in Japan, although it has never achieved generalized usage in other parts of the world. DM 120, DM 145, and the more recent DM 160 are all simple and seemingly reliable formulas that appear to deserve more widespread usage than they have received.

There are relatively few reports of the growth of differentiated tumor lines in defined media. One such report is growth of the $C_6$ glial tumor line by Donta in medium 924 MAP/1, developed originally by Waymouth. A second is the growth of monolayers of C1300 mouse neuroblastoma cells in medium MCDB 411, which has been optimized qualitatively and quantitatively for growth of the C1300 neuroblastoma, and will support multiplication without delay or apparent adaptation of cells previously grown exclusively in serum-containing media. It can also be used without supplementation to initiate cultures directly from transplantable C1300 tumors that have not previously been grown in culture, again without delay or the need for adaptation. Another set of media developed in the same manner are F12 and its more recent replacement, MCDB 301, which are discussed below under clonal growth, although they also support monolayer growth of Chinese hamster ovary lines. F12 and MCDB 301 have an unusually high concentration of zinc, which should be reduced if they are used for other types of cells, such as mouse L.[21]

*Defined Media for Suspension Culture.* A number of completely synthetic media for suspension culture have been described in the literature, including several designed specifically for high cellular yield, but none have become widely used. Suspension cultures in defined media require very close attention to details, and it appears that in most cases, the addition of serum does not interfere sufficiently with experimental objectives to justify the use of defined media. Media that are available for defined suspension cultures include minor modifications of NCTC 109 and MB 752/1, and various special-purpose formulas (Table IE). Birch and Pirt adjusted concentrations of nutrients for efficient utilization and high yields in suspension cultures of LS cells. Ling *et al.* used unusually high concentrations of nutrients to develop a medium (7C's) that would yield a maximum number of cells. Nagle and Brown developed a suspension culture medium that is free of glutamine, autoclavable, and effective for cells from many different species. However, none are widely used or generally available commercially. RPMI 1640 and several other formulas developed by Moore and his co-workers for lymphoblastic cells in suspension culture are commercially available. They are usually used with serum, but will also support growth of some lines without supplementation.

*Defined Media for Clonal Growth.* Clonal growth in defined media has not been widely pursued, but there are a few cases in which it is possible. Medium F12 was developed for clonal growth of Chinese hamster ovary and lung lines without protein supplementation, but later failed to support such growth. Recent studies have shown that the main problem was a requirement for selenium, which was apparently provided as a contaminant in the original studies.[10] The current medium of choice for such growth is MCDB 301, which corrects for the selenium deficiency and also incorporates other improvements. However, it is not yet commercially available. Commercially prepared F12 can be used with a selenium supplement, but in many cases it will also be necessary to add linoleic acid, which is extremely labile to oxidation and is often not present in an active form in commercial F12, including dry powder formulations.

*Defined Media for Normal Cells.* A number of defined media have been developed specifically for maintenance of normal cells. Although little or no growth occurs without serum supplementation, the cells remain fully viable, and in some cases they continue along developmental pathways during prolonged periods in culture. When supplemented with serum, these media generally support excellent growth of normal cells. The original assay used for development of medium 199 was long-term survival of chicken embryo heart fibroblasts in the absence of serum supplementation. When supplemented with serum, medium 199 supports good growth of many types of cells, and it is still widely used. In our own laboratory, we have recently found it to be the best starting point among commercially available media for the development of a medium that is specifically optimized for human epidermal keratinocytes.[29]

Medium CMRL 1415 was developed for survival of mouse embryo fibroblasts and for their growth when supplemented with partially purified serum fractions. Our experience suggests that clonal growth of mouse embryo fibroblasts with dialyzed serum can be improved by omitting the very high level of cysteine in CMRL 1415 and using only the lower level of cystine.[30] However, even better growth can be obtained with medium MCDB 401 plus dialyzed serum. Medium MCDB 401 and most of the other MCDB media, which were developed for growth of normal or nontransformed cells with small amounts of dialyzed serum, will also support prolonged survival of their respective cells, even at clonal density, without protein supplementation.[1]

Several special-purpose defined media for differentiated cells have appeared in the recent literature, including media E and G of Williams for rat liver cells; the very complex defined medium of Parsa for maintenance and differentiation of pancreatic rudiments; medium SM-20 of Halle and Wollenberger for beating heart cells; and modifications of Waymouth's media for liver and pancreatic cells (Table IG).

### *Making the Choice among Media*

The final choice of a medium for a specific series of experiments is usually based on two pragmatic considerations: (1) Will it yield the data that

[29] D. M. Peehl and R. G. Ham, Abstracts 1978 TCA Meeting, *In Vitro* **14**, 352 (1978).

[30] R. G. Ham and P. M. Sullivan, *in* "Cell Culture and its Application" (R. T. Acton and J. D. Lynn, eds.), p. 533. Academic Press, New York, 1977.

are needed? (2) Is its use feasible on the scale that is needed? The former is determined by trying the medium, often in competition with other media, and the latter by practical considerations such as how hard the medium is to prepare and whether or not it is commercially available.

For laboratories not equipped to prepare their own media, lack of commercial availability can be a serious obstacle to trying new or unusual media. It is possible to obtain custom-made batches of media, but the companies generally require a rather large minimum order to recover their expenses, and they do not have testing programs for determining whether the batch that they prepare actually measures up to the claims made about it in the literature. (Lack of adequate testing can also be a problem with standard media. Often media are tested only for sterility and ability to support monolayer growth with whole serum. Special-purpose media, such as those designed for clonal growth with dialyzed serum, are frequently not tested to see if they will actually do what they are supposed to. If problems arise, it is desirable to find out from the supplier exactly what testing has been done.)

For the novice, it is generally desirable to begin with media and culture conditions that previously have been reported to support growth of the cell in question. Cell culture is a complex art with many different things that can go wrong. It is therefore highly desirable to begin with conditions that are known to work for the cell, and to wait until the initial problems of getting it to grow have been resolved before attempting innovations. In many instances, if the original conditions work well and are compatible with the planned experiment, no further experimentation with media may be necessary.

If previously reported culture systems are not adequate, or if the cell type has not previously been cultured from the species that is to be used, it will be necessary to begin a program of systematic testing of media. Generally it is desirable to begin by attempting to obtain good growth with whole serum, and then to proceed as far as needed toward chemical definition. A generalized outline for that process is given in Table III, and the details are presented in several other papers.[1,2,9,20]

Species is a major factor in determining which medium to use, but tissue specificity can also have a major effect. Media that tend to have rather broad species applicability include 199, MEM, DME, F12K, and MCDB 202. For human and monkey cells, 199, BME, MEM, L15, 5a, RPMI 1640, CMRL 1969, MCDB 104, and MCDB 202 can be tried. For rat and rabbit, MEM, 5a, F12, F12K, and MCDB 104 may work. For non-transformed mouse cells, DME, CMRL 1415, MCDB 202, MCDB 401, and some of the newer Waymouth formulations can be tried. For chicken cells, 199, DME, F12K, and MCDB 202 are possibilities. Media for certain invertebrate cells are discussed in this volume [39]. Other formulations

TABLE III

GENERALIZED SEQUENCE OF STEPS FOR MINIMIZING SERUM REQUIREMENTS AND FOR DEVELOPING DEFINED MEDIA[a]

1. Obtain growth *in vitro* of the type of cell of interest by any means necessary.
2. Modify media, supplements, and culture techniques as needed to obtain clonal growth.
3. Survey readily available media and supplements and select the combination that supports the best clonal growth.
4. Replace all undefined supplements, e.g., serum, with dialyzed supplements plus ultrafiltrates, hydrolysates, etc., if they are needed.
5. Identify all low-molecular-weight nutrients needed for clonal growth in the presence of dialyzed supplements and add them to the synthetic portion of the medium.
6. Reduce the amount of the dialyzed supplement(s) to a level that supports less than optimum growth.
7. a. Adjust all components of the synthetic portion of the medium sequentially to experimentally determined optimum concentrations for clonal growth with minimal amounts of serum protein, beginning with the one found to be most limiting in a preliminary survey.
   b. Systematically test low-molecular-weight compounds considered likely to have growth-promoting activity, e.g., trace elements and nutrients known to be required by other types of cells or organisms.
   c. Test complex low-molecular-weight mixtures such as hydrolysates, extracts, and ultrafiltrates for growth-promoting activity.
   d. Attempt to separate low-molecular-weight growth-promoting activity from the dialyzed supplements through the use of dissociating conditions. If activity is detected in steps 7c or 7d, return to step 5.
   e. Attempt to improve growth through refinement of the culture techniques (e.g., use of treated dishes, trypsinization at low temperature, etc.).
8. Whenever growth is improved significantly, reduce the concentration of the dialyzed supplement until it is again growth-limiting, and repeat all parts of step 7.
9. Keep repeating steps 7 and 8 until no further benefits can be obtained.
10. Isolate and characterize all macromolecular growth-promoting activities that continue to be required under the "optimum" conditions developed in step 9.
11. Identify all "contaminants" both in the synthetic medium and in the "purified" macromolecular factors, and if any affect growth, add them to the synthetic medium at their experimentally determined optimum concentrations.
12. Verify experimentally that each component of the "final" medium is at its optimum concentration, and make any changes that may be needed as a result of adjustments made in steps 10 and 11.

[a] After all requirements for multiplication of isolated cells have been identified and optimized, it may be necessary to make additional adjustments for monolayer growth, e.g., by adjusting nutrients to the upper end of their optimum plateaus to avoid depletion and adding factors needed to reduce the effects of density-dependent inhibition.

may work in many instances. In cases of extreme difficulty, it is probably desirable to test virtually every medium that can be obtained easily. In addition, it is often helpful to test combinations of media. For example, a mixture of 199 and F12 is superior to either alone for clonal growth of chicken embryo fibroblasts, and that combination formed the starting point for the development of MCDB 202.

## Future Prospects

Looking ahead, the prospects are that overall progress in the understanding of cellular growth requirements and the development of new media may be rather slow. We are currently at the brink of a complete understanding of the growth requirements of a very limited number of normal diploid cells. However, the number of laboratories actively enganged in these studies is sufficiently small so that, at our present pace, many years will be required to gain a broad general understanding of cellular growth requirements.

Hopefully, an expanding awareness of how much we still need to learn about the growth requirements of normal cells, together with a growing realization of the urgency for gaining such knowledge, will provide the stimulus to draw many more qualified investigators into the field and accelerate our progress. However, until then, cell culture will remain in many respects an art rather than an exact science. Any investigator who is unwilling to work with an undefined system, or with one of the very limited number of cell types whose requirements are beginning to be understood, will find it necessary to do pioneering work on the growth requirements of his cell before he can undertake the study that he ultimately plans to do with that cell.

## Acknowledgments

Special thanks are due to Susan Jennings for assembling the data for the tables and to Karen Brown for typing the manuscript and tables; also to Dr. Raymond Kahn for use of a computer printout that provided much of the raw data for Table II. Development of the MCDB series of media was supported by contract 223-74-1156 from the Bureau of Biologics, U.S. Food and Drug Administration, and by grants AG00310 and CA15305 from the National Institutes of Health.

# [6] The Growth of Cells in Serum-Free Hormone-Supplemented Media

*By* JANE BOTTENSTEIN, IZUMI HAYASHI, SHARON HUTCHINGS, HIDEO MASUI, JENNIE MATHER, DON B. MCCLURE, SUGAYUKI OHASA, ANGIE RIZZINO, GORDON SATO, GINETTE SERRERO, RICHARD WOLFE, and REEN WU

## Introduction (G. Sato)

A series of experiments from our laboratory indicated that hormonal depletion of the serum component of the medium is necessary to demonstrate hormone-dependent growth of cells in culture.[1-3] From these observations, it was proposed that the main role of serum in cell culture is to provide complexes of hormones and that it should be possible to substitute hormones for serum in cell culture media.[4] Experimental support for this hypothesis was first provided in the case of $GH_3$, HeLa, and BHK cells and extended to other cell lines.[5-7] In this article we present the practical details for growing a number of cell lines in serum-free hormone-supplemented medium.

In the absence of serum, greater than usual care must be taken in preparation of the synthetic portion of the medium. The water used in preparation of our medium is distilled in glass stills in three steps. In the first, deionized water is distilled and collected. It is placed in the boiler of a second still to which is added approximately 10 g of potassium permanganate and 1 ml of concentrated sulfuric acid per liter. The distillate from the second still is freshly distilled from a third still just prior to the preparation of the medium. Careful preparation of water is essential for consistent results with serum-free medium.

Serum and even dialyzed serum can mask nutritional requirements of

[1] J. L. Clark, K. L. Jones, D. Gospodarowicz, and G. H. Sato, *Nature (London)*, New Biol. **236,** 180 (1972).

[2] H. A. Armelin, K. Nishikawa, and G. H. Sato, *in* "Control of Proliferation and Animal Cells" (B. Clarkson and R. Baserga, eds.), Cold Spring Harbor Lab., Cold Spring Harbor, New York, 1974.

[3] K. Nishikawa, H. A. Armelin, and G. Sato, *Proc. Natl. Acad. Sci. U.S.A.* **72,** 483 (1975).

[4] G. H. Sato, *Biochem. Actions Horm.* **3,** 391 (1975).

[5] I. Hayashi and G. Sato, *Nature (London)* **259,** 132 (1976).

[6] J. Mather, R. Wu, and G. Sato, *in* "Growth Requirements of Vertebrate Cells in Vitro" (R. G. Ham and C. Waymouth, eds.). Cambridge Univ. Press, London and New York. (in press).

[7] S. E. Hutchings and G. H. Sato, *Proc. Natl. Acad. Sci. U.S.A.* **75,** 901 (1978).



ISBN 0-12-181958-2

cells in culture. For this reason we use one of the more complex media, Ham's F12,[8] as the basic synthetic medium. We also supplement the medium with the trace elements recommended by Ham.[9] Ham's F12 medium, supplemented with trace elements, is sufficiently complex to satisfy the nutritional requirements of cells in the absence of serum. However, the concentration of the components is often not optimal. This is remedied by using mixtures of F12 and DME.[10,11]

Commercially available powdered media are adequate for serum-free work although each batch must be checked for suitability.

Serum also serves the function of a trypsin inhibitor in conventional tissue culture procedures. In the absence of serum, we use soybean trypsin inhibitor to neutralize the trypsin after subculture. After trypsinization, the cells are suspended in serum-free F12 medium containing soybean trypsin inhibitor (1 mg/ml), centrifuged, and resuspended in serum-free F12 for plating.

Hormones, growth factors, and transferrin are maintained as ×100 stocks and added to the medium just prior to plating the cells. Stock cultures are usually kept in a 1:1 mixture of Ham's F12 and Dulbecco's modified Eagle's medium (DME), supplemented with sodium bicarbonate (1.2 g/liter), *N*-2-hydroxyethylpiperazine-*n*′-2-ethanesulfonic acid, pH 7.2 (HEPES) (15 m*M*), penicillin (192 units/ml), streptomycin (200 μg/ml), ampicillin (25 μg/ml), 5% (v/v) horse serum, and 2.5% (v/v) fetal calf serum. The cells are grown on plastic tissue culture dishes in a humidified atmosphere of air containing 5% $CO_2$.

In those cases in which subculture involves trypsinization, two techniques for growing cells in serum-free medium are used. The first involves the transfer of cells by trypsinization from stock medium containing serum, directly to serum-free medium. In the second procedure, cells are first plated into serum-containing medium and, after a day, the medium is replaced with serum-free medium after washing the cells with serum-free medium. In those cases in which this procedure is followed, the initial incubation in serum-containing medium may be essential. There are two likely explanations for the requirement for serum pretreatment: (1) The serum may be necessary for the repair of trypsinization damage after subculture; or (2) the residual serum left after its removal, even after one or more washes, may be furnishing unknown growth factors. The latter possibility seems unlikely in view of the extensive growth which is possi-

[8] R. G. Ham, *Proc. Natl. Acad. Sci. U.S.A.* **58,** 288 (1965).

[9] W. L. McKeehan, W. G. Hamilton, and R. G. Ham, *Proc. Natl. Acad. Sci. U.S.A.* **72,** 2023 (1976).

[10] H. Eagle, *Science* **122,** 501 (1955).

[11] J. D. Smith, G. Freeman, M. Vogt, and R. Dulbecco, *Virology* **12,** 185 (1960).

ble after transfer to serum-free medium. In any case, we intend to settle this question by growing these cells on beads[12] (see also this volume [15]). This system of culture allows the subculture of cells without trypsinization.

In the course of defining the requirements of $C_6$ glial cells, an active factor was found in the rat submaxillary salivary gland (see section on $C_6$ cells). The crude extract of salivary gland can be maximally active at concentrations as low as 2 $\mu$g protein/ml of medium; the purification of this material is still in progress in our laboratory. In order to avoid any implication of its possible physiological role, we have named this factor rat submaxillary gimmel factor or gimmel. Gimmel is of great importance because it has been found to be required for the growth in serum-free medium of Balb 3T3 and Swiss 3T3 cells (see sections on Balb and Swiss 3T3 cells). In addition, it is possible that a factor other than MSA is in conditioned Buffalo rat liver cultures and is responsible for the growth stimulation of rabbit aortic intimal cells (see section on intimal cells). The preliminary evidence supporting this notion is that the growth activity for intimal cells and the MSA activity (stimulation of DNA synthesis in serum-deprived 3T3 cells) do not copurify on Dowex columns. These two examples illustrate how the serum-free approach readily lends itself to the discovery of novel factors of possible physiological significance.

We have also included an example of hormone binding to cells grown in serum-free, hormone-supplemented media (see section on [$^{125}$I]EGF binding to HeLa cells). The significance of this section is that the phenomenon of hormone internalization, degradation, and "down" regulation of receptors does not seem to occur in defined media. Since EGF is required for growth of these cells, one can tentatively conclude that internalization and "down" regulation of receptors is not necessary for the mitogenic activity. If this generalization holds up in future experimentation, it represents an unexpected technical advantage in doing experiments in serum-free hormone-supplemented media.

The cell lines used in the study are rat neuroblastoma,[13] $GH_3$,[14] HeLa,[15] Balb 3T3,[16] rat ovarian cells (RF),[17] B16 melanoma M2R,[18] mouse testicu-

[12] D. W. Levine, J. S. Wong, D. I. C. Wang, and W. G. Thilly, *Somatic Cell Genet.* **3,** 149 (1977).

[13] D. Schubert, S. Heinemann, W. Carlisle, H. Tarikas, B. Kimes, J. Patrick, J. Steinbach, W. Culp, and B. Brandt, *Nature (London)* **249,** 224 (1974).

[14] Y. Yasumura, A. H. Tashjian, Jr., and G. H. Sato, *Science* **154,** 1186 (1966).

[15] G. O. Gey, W. D. Coffman, and M. T. Kubicek, *Cancer Res.* **12,** 264 (1952).

[16] S. A. Aaronson and G. J. Todaro, *J. Cell. Physiol.* **72,** 141 (1968).

[17] H. Masui, S. Ohasa, and S. Mackensen, in preparation.

[18] J. Mather and G. H. Sato, (S. Federoff, ed.). *in* "Cell Tissue and Organ Cultures in Neurobiology," p. 619. Academic Press, New York, 1978.

lar cells TM4,[19] rat glioma $C_6$,[20] mouse embryonal carcinoma,[21,22] Swiss 3T3,[23] and a line of cells isolated from rabbit aortic intima.[24]

*Note*

Bovine crystalline insulin, human transferrin, and steroid hormones are purchased from either Sigma or Calbiochem. Epidermal growth factor (EGF) and fibroblast growth factor (FGF) are purchased from Collaborative Research, Waltham, Massachusetts. A gift of a somatomedin preparation from Dr. Knut Uthne, Kabi Co., Stockholm, Sweden made it possible to start this work. Gifts of EGF from Dr. Stanley Cohen, Vanderbilt University, and FGF from Dr. Denis Gospodarowicz, University of California Medical School, San Francisco, were helpful in the early stages of this work. Highly purified pituitary hormones and somatomedins were generously provided by Dr. Harold Papkoff of the University of California, San Francisco, and Dr. Judson Van Wyk, University of North Carolina, Chapel Hill, respectively. Partially purified pituitary hormones are obtained from the National Institutes of Health Hormone Distribution Program. We are very grateful to Dr. Brian Kimes of the NCI. Without his timely assistance, this work would not have been possible. This work was supported by grants from the National Science Foundation, NIH GM 17019, NIH CA 19731, NIH GM 17702, and NCI CA 18885.

## Neuroblastoma Cells in Serum-Free Medium (J. Bottenstein)

### *Cell Culture in the Presence of Serum*

The B104 rat neuroblastoma[13] stock cultures are maintained in a 1:1 mixture of Ham's F12 (F12) and Dulbecco's modified Eagle's medium (DME) supplemented with 5% (v/v) horse serum, 2.5% (v/v) fetal calf serum, $NaHCO_3$ (1.2 g/liter), HEPES buffer (15 m$M$), penicillin (40 mg/liter), ampicillin (8 mg/liter) and streptomycin (90 mg/liter). Stocks are grown in 25 cm² tissue culture flasks at 37° in a humidified atmosphere containing 5% $CO_2$ and are subcultured every 4–6 days. After an initial 3-ml wash with PBS, 2 ml of 0.1% trypsin–0.9 m$M$ EDTA in PBS for 3 min

[19] J. Mather, unpublished results.

[20] P. Benda, J. Lightbody, G. Sato, L. Levine, and W. Sweet, *Science* **161,** 370 (1968).

[21] H. Jakob, T. Boon, J. Gaillard, J.-F. Nicolas, and F. Jacob, *Ann. Microbiol.* (*Paris*) **124b,** 269 (1973).

[22] K. Artzt, P. Dubois, D. Bennett, H. Condamine, C. Babinet, and F. Jacob, *Proc. Natl. Acad. Sci. U.S.A.* **70,** 2988 (1973).

[23] H. Green and O. Kehinde, *Cell* **1,** 113 (1974).

[24] V. Buonassisi and L. Ozzello, *Cancer Res.* **33,** 874 (1973).

TABLE I
EFFECTS OF GROWTH FACTORS ON B104 CELLS[a]

| Addition | Cell number × $10^3$ | | |
|---|---|---|---|
| | Day 3 | Day 5 | Growth rate (hr) |
| None | 179 ± 19 | 78 ± 18 | — |
| 4F | 493 ± 11 | 24 ± 5 | — |
| 4F + $Na_2SeO_3$ | 573 ± 14 | 1856 ± 97 | 28 |
| 10% fetal calf serum | 1307 ± 30 | 6395 ± 186 | 21 |

[a] Data are expressed as the mean ± SD of triplicate samples. Initial inoculum is $7 \times 10^4$.

at 37° is used to detach the cells. To inactivate trypsin, 2 ml of medium (*vide supra*) are added and the cell pellet obtained by centrifugation is diluted with medium to the appropriate plating density, usually 7–14 × $10^4$ cells/flask.

### *Cell Culture in the Absence of Serum*

Cells are subcultured as above and plated in medium (*vide supra*) at a density of 7 × $10^4$ cells/60-mm dish. After 18 hr this medium is removed, cells are washed twice with 3 ml of serum-free medium, and 3 ml of serum-free experimental medium are added. The four factors (4F) are: bovine insulin (5 μg/ml), human transferrin (5 μg/ml), putrescine dihydrochloride (100 μ*M*), and progesterone (20 n*M*). The media are changed on day 3 only. Table I shows that with 4F alone, cell number increases about 3-fold at day 3, but is markedly reduced by day 5. The addition of 30 n*M* selenium ($Na_2SeO_3$) to 4F has a slight stimulatory effect at day 3 and allows the cells to continue growing at the same rate. Thus, in a serum-free condition with 4F and $Na_2SeO_2$, the growth is 75% of that when 10% fetal calf serum is present.

## **$GH_3$ Cells in a Defined Medium (I. Hayashi)**

$GH_3$ is a clonal line derived from a spontaneous, transplantable rat pituitary tumor.[14] The cells produce growth hormone and prolactin and respond to thyroid hormones for growth.[25] Stock cells are maintained in DME supplemented with 12.5% fetal calf serum. In our experience, it is important that cells be kept in medium containing at least 10% serum to

[25] H. H. Samuels, J. S. Tsai, and R. Cintron, *Science* **181**, 1253 (1973).

get good hormone response. The stocks are subcultured about once a week.

For growing $GH_3$ cells in serum-free condition, exponentially growing stock cells are trypsinized (0.1% trypsin in PBS) and then treated with soybean trypsin inhibitor (0.1% of the inhibitor in PBS). The treated cells are washed by centrifugation, resuspended in serum-free F12 medium, and inoculated into hormone-containing, serum-free F12 medium at $0.5 \times 10^5$ (35-mm dishes) or $1.0 \times 10^5$ (60-mm dishes) per plate. Unlike the other lines described here, we have found that $GH_3$ cells prefer F12 medium to any other media, or the combination of F12 and DME. These cells have not responded to hormones under serum-free conditions unless F12 was used. The hormones and their final concentration in the medium that are required for growth of $GH_3$ cells in the serum-free condition are as follows: 3,3′,5-triiodothyronine ($3 \times 10^{-11}$ $M$), thyrotropin-releasing hormone (1 ng/ml), parathyroid hormone (0.5 ng/ml), transferrin (5 $\mu$g/ml), insulin (5 $\mu$g/ml), fibroblast growth factor (1 ng/ml), and somatomedin C (1 ng/ml).[26] When these hormones are added to the serum-free F12 medium, the cells not only grow as well as in serum-supplemented medium, but also can be maintained for a long period by serial propagation. Requirement for parathyroid hormone, which has the least effect on short-term growth of these cells, will become obvious in such long-term culture in serum-free condition. For long-term growth in serum-free, hormone-supplemented medium, trace metals such as selenium and cadium are added at the final concentration of 50n$M$ and 0.5$\mu M$, respectively. Some of the other trace metals, especially manganese, are strongly inhibitory to $GH_3$ cells and, therefore, are not used for these cells.

One of the immediate interests in defining hormone requirements of a cell line and growing it in a serum-free condition is the question of the universality of the requirement. For instance, will the hormone combination that allows the growth of $GH_3$ cells also allow the growth of normal rat pituitary cells which produce growth hormone and/or prolactin? To examine these possibilities, we made primary cultures from young normal rat pituitaries. The pituitaries were washed several times in serum-free medium, minced in the presence of trypsin inhibitor, and plated in serum-free F12 medium supplemented with the hormones mentioned above. Comparison of such cultures with those in serum-supplemented (10% fetal calf serum) F12 medium showed that there is at least a significant suppression of fibroblast growth in serum-free, hormone supplemented media. At present, it is too early to determine whether the cells in serum-free, hormone-supplemented medium are the normal equivalent of $GH_3$ cells which produce growth hormone and/or prolactin, but the results indicate

[26] I. Hayashi, J. Larner, and G. Sato, *In Vitro* **14**, 23 (1978).

that, in the future, hormone-supplemented serum-free medium could be used to select specific cell types from mixed populations of cells in primary cultures.

### Single-Cell Plating of HeLa Cells in Serum-free, Hormone-Supplemented Medium (S. Hutchings)

The cell strain, designated HeLa-S, was obtained from Dr. John Holland of the University of California, San Diego, La Jolla, California. Stock cultures are maintained in a 1:1 mixture of F12 and DME supplemented with serum as described in the introduction.

Stock cultures in the exponential phase of growth are trypsinized (0.1% trypsin with 0.03% EDTA in PBS), suspended in serum-free F12, SFF12, containing 0.1% soybean trypsin inhibitor, and centrifuged. The cell pellet is resuspended in SFF12 and plated at a low density (about 1,000 cells) in 100 mm plastic dishes containing 10 ml of medium. The medium is always warmed and equilibrated with the atmosphere in a $CO_2$ incubator before use for either the primary plating or medium changes. The defined medium consists of F12 medium containing trace elements,[9] to which is added insulin (5 $\mu$g/ml), transferrin (0.5 $\mu$g/ml), EGF (50 ng/ml), FGF (100 ng/ml), and hydrocortisone (100 m*M*). At day 5, the medium is changed, and at day 9 the cells are fixed in methanol and stained with 0.1% crystal violet. The plating efficiency in the defined media is approximately 50% of that in serum-containing media. Omission of any one of the above-mentioned components results in reduced plating efficiency.

The most stringent test of the adequacy of a medium is its ability to support colonial growth from single cells at low plating density. Serum-free, hormone-supplemented media support such growth.

### Normal Rat Follicular Cells in Serum-Free Medium (H. Masui)

A follicular cell line (RF-1) has been established in culture from the ovaries of normal Fisher rats. The medium used for establishing these cells was a mixture of DME and F12 medium (1:1) supplemented with 2.5% fetal calf serum and 12.5% horse serum. Stock cultures of RF-1 cells are maintained in the medium. RF-1 cells have characteristics of normal cells, such as diploid chromosome number ($2n = 42$), density-dependent inhibition of growth, and no tumorigenicity in Fisher rats or nude mice. RF-1 cells produce several steroids from exogenous cholesterol; the main product seems to be testosterone.[17]

When RF-1 cells are grown in a mixture of DME and F12 medium (1:1) supplemented with various concentrations of fetal calf serum, the rate of cell growth increases with the serum concentration and reaches a maximum between 3 and 6%. When either DME or F12 medium is used instead of the mixture, cells do not grow even in the presence of serum.

To grow RF-1 in serum-free media, cells are trypsinized and plated in the DME/F12 mixture supplemented with 2.5% fetal calf serum and 12.5% horse serum. The cells are kept in this serum-containing medium for at least 6 hr. Thereafter, they are washed with serum-free medium and the medium is replaced with DME/F12 (1:1) supplemented with bovine insulin (2 $\mu$g/ml), hydrocortisone (4 ng/ml), triiodothyronine ($3 \times 10^{-10}$ $M$), and transferrin (5 $\mu$g/ml). Usually $9 \times 10^4$ cells are plated per 60-mm dish and increase to 3–4 $\times 10^6$ cells after 5 days of growth. During the exponential phase of growth, the doubling time is about 26 hr which is comparable to the doubling time in serum-supplemented media.

## Growth of Balb/c Mouse Fibroblasts in Serum-Free Medium (D. McClure)

Serum-free medium supplemented with insulin, fibroblast growth factor, transferrin, and a crude extract prepared from female rat submaxillary gland, rat submaxillary gimmel factor, supports continuous division cycles of Balb/c 3T3 mouse fibroblasts washed free of serum. Under these serum-free conditions, the growth rate and saturation density are similar to those observed for the same cells cultured in DME containing 10% fetal calf serum. Morphologically, these cells appear more epithelioid than their serum-grown counterparts.

Trypsinized stock cultures of logarithically growing Balb/c 3T3 cells, maintained at an early passage number, are subcultured at $5 \times 10^4$ cells per dish in DME plus 10% fetal calf serum and incubated overnight at 37° in a humidified atmosphere containing 5% $CO_2$. On day 1, after washing each culture twice with serum-free DME, the medium is changed to serum-free medium, consisting of three parts DME and one part F12, plus trace elements. Hormone additions are made directly into each culture dish from freshly prepared concentrated stock solutions. The final concentrations of insulin, transferrin, hydrocortisone, EGF, and NIH-LH ( a source of FGF) are 2 $\mu$g/ml, 5 $\mu$g/ml, $10^{-7}$ $M$, 50 ng/ml, and 1 $\mu$g/ml, respectively. Rat submaxillary gimmel factor is added at a concentration of 2 $\mu$g/ml. The preparation of gimmel factor is described in the section on $C_6$ glial cells.

Balb/c 3T3 fibroblasts, washed free of serum and incubated in serum-free media supplemented with insulin, transferrin, hydrocortisone, NIH-LH (FGF), EGF, and a crude preparation of rat submaxillary gland, grow from a low cell density ($3 \times 10^3$ cells/cm$^2$) to form a confluent monolayer ($7 \times 10^4/cm^2$) with an approximate generation time of 20 hr. Hormone-supplemented medium without the addition of rat submaxillary gimmel factor is sufficient to support only two rounds of replication after which the cells cease to divide at an appreciable rate. Addition of as little as 2 $\mu$g/ml of crude gimmel factor is sufficient to support continuing cell growth. By omitting each individual component from the above mixture of hormones and growth factors it can be further shown that only insulin, transferrin, NIH-LH, and rat submaxillary extract are necessary for Balb/c 3T3 cells growth in the complete absence of serum.

Others have succeeded in growing Balb/c 3T3 cells, maintained in low serum, by the addition of hormones and polypeptide growth factors to the culture medium.[27,28] Here, we demonstrate that Balb/c 3T3 fibroblasts exhibit continuing growth in hormone-supplemented medium in the complete absence of serum. We suggest that rat submaxillary gimmel factor replaces the residual serum requirement hitherto found necessary for the growth of 3T3 cells in totally defined medium.

### Melanoma Cells in Defined Medium (J. Mather)

A clonal line of melanoma, designated M2, was isolated from the B16 transplantable tumor[18] obtained from the Jackson Laboratory, Bar Harbor, Maine. This line metastasizes in 50% of the animals in which cells have been injected subcutaneously. A group of hormones has been shown to replace the serum requirement for growth of these cells in culture. Cells are grown in F12/DME (1:1) supplemented with sodium selenite ($10^{-7}$ *M*) and the following hormones: insulin (5 $\mu$g/ml), transferrin (5 $\mu$g/ml), ovine follicle-stimulating hormone (purified FSH) (40 $\mu$g/ml), luteinizing hormone-releasing hormone (LRH) (10 ng/ml), nerve growth factor (NGF) (10 ng/ml), and either testosterone (5 n*M*) or progesterone (1 n*M*).[6,18] The cells can be transferred by trypsinization directly into the serum-free medium.

Transferrin seems to be acting primarily, but not exclusively, as an iron transport protein in culture since 75% of the growth stimulation seen with transferrin can be obtained by increasing the ferrous iron concentration in the medium. However, the remainder of the stimulation cannot be accounted for solely be increased iron transport. Insulin has been shown to stimulate glucose incorporation into both glycogen and fatty acids in these cells and appears to be acting in a manner consistent with our understanding of insulin action in classic target tissues. Although the cells

[27] D. Gospodarowicz and J. Moran, *Proc. Natl. Acad. Sci. U.S.A.* **71**, 4584 (1974).
[28] H. Armelin, *Proc. Natl. Acad. Sci. U.S.A.* **70**, 2702 (1973).

are stimulated by either progesterone or testosterone, the increase in cell number with testosterone is greater than that seen with equivalent amounts of progesterone and the effects do not seem to be additive. When radioactive hormone was given and the products analyzed by thin-layer chromatography greater than 99% of the testosterone migrated as testosterone after 24 hr in culture. In a similar experiment three peaks were obtained from [$^3$H]progesterone, none of which comigrated with testosterone; the products have not been identified. The results indicate that progesterone does not affect the cells by being metabolized to testosterone nor do the two hormones act via a common steroid metabolic product.

The metastatic cells have been injected subcutaneously into C57 BL/6 mice. After 2 weeks, when no metastases are visible and only a small tumor is formed, the mouse was sacrificed and the lungs, lymph nodes, and spleen removed, minced, and plated into medium supplemented with the hormones which will support the growth of melanoma cells. After 10 days, the only cells surviving were small colonies of melanoma cells, presumably arising from microscopic metastatic foci in the organs plated. Thus, the hormone-supplemented medium can be used to select for a desired cell type in primary cultures.

### Testicular Cells in Defined Medium (J. Mather)

An epithelial cell line, TM4,[19] was isolated from normal immature (11-day) Balb/c mouse testes. This line has been cloned 3 times and has been in culture for over 2 years. The doubling time is 16 hr. The cells do not form tumors in Balb/c–nu/nu mice.

The cell line can be grown in F12/DME (1:1) supplemented with a mixture of hormones and growth factors. These cells are growth stimulated by insulin (5 $\mu$g/ml), transferrin (5 $\mu$g/ml), epidermal growth factor (3 ng/ml), FSH (0.5 $\mu$g/ml), somatomedin C (1 ng/ml), ovine growth hormone (6.5 $\mu$U/ml), and retinoic acid (50 ng/ml). Under hormone-supplemented conditions, growth rates are obtained equivalent to those seen in the same medium supplemented with 5% fetal calf serum.

Primary cultures of Sertoli cells from immature testes can be grown in the above hormone-supplemented medium. Survival times are increased significantly over those of primary cultures plated in serum-free medium (6 weeks as opposed to 5–10 days) with no fibroblast overgrowth.

### Rat Glial $C_6$ Cells in Serum-free Medium (S. Ohasa)

The maintenance of stock cultures is carried out as follows: rat glial $C_6$ cells in stock culture bottles (100 $\times$ 20 mm) are incubated with 0.1%

trypsin solution (in PBS containing 0.9 m$M$ EDTA) at room temperature for 2 min and suspended in 10 ml of a mixture of DME/F12 (1:1) supplemented with 6% fetal calf serum. The medium is pipetted up and down with a 10-ml sterile pipette in order to detach the cells from the culture dish. Cells are diluted to the desired cell density and plated. Usually, $5 \times 10^5$ cells per 100-mm plate are inoculated and cultured in DME/F12 with 6% fetal calf serum at 37° under a mixture of air containing 5% $CO_2$. This process is repeated every 5 days.

For growth in serum-free medium, the cells in stock cultures are detached by trypsinization, suspended in 10 ml of 1:1 DME/F12 with 6% serum, and then inoculated into a 60-mm culture dish. After incubating the cells in 5 ml of the serum-supplemented medium for 1 day the cells are washed 3 times with 3 ml DME/F12 in order to remove the added serum and then cultured in 5 ml of serum-free DME/F12 supplemented with transferrin (5 $\mu$g/ml), insulin (2 $\mu$g/ml), NIH-LH (0.4 $\mu$g/ml), and an extract of rat submaxillary gland (5–10 $\mu$g/ml).

The activity of rat submaxillary gland extract was discovered in experiments with $C_6$ glial cells. The $C_6$ system is currently being used as the assay in the purification of the active factor, gimmel factor. Gimmel factor is prepared as follows: Submaxillary glands (1 g) from a young, male or female Fisher rat are minced, suspended in 5 ml of PBS containing 0.2 $M$ sucrose, and homogenized with a glass/Teflon homogenizer. The homogenate is centrifuged at 27,000 $g$ for 30 min, and the clear supernatant solution is collected to give a submaxillary gland extract. The sample is diluted with PBS to yield a 1 mg/ml protein solution, sterilized by filtration, and then used for cell culture.

TABLE II

EFFECTS OF HORMONES ON THE $C_6$ CELL GROWTH[a]

| Addition | Cell no. $\times 10^{-5}$ | Relative cell no. |
|---|---|---|
| None (control) | 4.7 ± 0.9 | 1.0 |
| Serum (6%) | 48.2 ± 0.5 | 10.3 |
| NIH-LH (0.4 $\mu$g/ml) | 10.3 ± 0.1 | 2.2 |
| FGF (20 ng/ml) | 8.6 ± 0.4 | 1.8 |
| Ins (2 $\mu$g/ml) | 12.2 ± 0.7 | 2.6 |
| Trans (5 $\mu$g/ml) | 8.4 ± 0.1 | 1.8 |
| Gimmel (10 $\mu$g/ml) | 15.0 ± 0.6 | 3.2 |
| Ins + trans + FGF* | 36.8 ± 0.3 | 7.8 |
| Ins + trans + FGF* + gimmel | 53 8 ± 0.7 | 11.4 |

[a] $C_6$ cells ($8 \times 10^4$) were plated into a 60-mm culture dish, incubated in 5 ml DME/F12 with 6% FCS, washed 3 times with 3 ml DME/F12, and then cultured for 4 days in 5 ml DME/F12 with the indicated factors. The added factors were serum (FCS), FGF or NIH-LH (indicated as FGF*), Ins (insulin), Trans (transferrin), and gimmel (submaxillary gland extract) which was prepared from Fisher rat submaxillary gland. The indicated cell number per 60-mm culture dish is an average of 2 determinations ± standard error.

$C_6$ cell growth is stimulated by insulin, FGF or NIH-LH, transferrin, and rat submaxillary gimmel factor when they are added individually. The growth rate is the greatest when all four factors are present resulting in an approximate doubling time of 15 hr. Rat submaxillary gimmel factor cannot be replaced by purified EGF or NGF which are present in the submaxillary gland, suggesting that gimmel factor is distinct from them.

### Embryonal Carcinoma Cells in Serum-Free Medium (A. Rizzino)

Two embryonal carcinoma cell lines, PCC.4 aza-1[21] and $F_9$,[22] have been cultured in the absence of serum for over 70 and 20 generations, respectively.[29] The cultures were kept at 37° under a mixture of air containing 5% $CO_2$. The cells are subcultured (when grown in serum or in serum-free media) by removing the spent medium and replacing it with fresh serum-free medium (see below). This is repeated once more for the purpose of washing the cells. The medium is then pipetted up and down over the cells with a sterile Pasteur pipette in order to detach the cells from the culture dish. The cells are diluted to the desired cell density and plated. Usually $3–6 \times 10^5$ cells are plated per 25 $cm^2$ "T" flask. In the case of $F_9$ cells, trypsin (0.1%)–EDTA (0.9 m*M*) in PBS can also be used to detach the cells from the culture dish; this is unnecessary unless a single cell suspension is desired. When trypsin–EDTA is used, 3 ml are added to a 25 $cm^2$ "T" flask long enough to detach the cells after which an additional 3 ml of serum-free medium are added. The cells are collected by centrifugation at room temperature; trypsin inhibitor is unnecessary. Once the cells reach high density, about $2 \times 10^6$ cells per flask, they may be subcultured.

The serum-free medium used for embryonal carcinoma cells is referred to as EM-1. This medium is composed of F12 supplemented with fetuin (500 $\mu$g/ml, Calbiochem-Pedersen type), bovine insulin (10 $\mu$g/ml), human transferrin (5 $\mu$g/ml), and 10 $\mu M$ $\beta$-mercaptoethanol. For optimal growth the medium should be changed daily; this is more important for PCC.4 aza-1 than for $F_9$ cells. The medium (EM-1) is prepared every day from the appropriate stock solutions. DME cannot be used to replace F12.

### Mouse Swiss 3T3 Fibroblasts, Clone 3T3-L1 in a Serum-Free Medium (G. Serrero)

Clone 3T3-L1 of mouse Swiss 3T3 fibroblasts, isolated in Dr. Howard Green's laboratory, has the property of differentiating into adipocytes when the cultures are confluent. Attempts have been made to grow these cells in a serum-deprived medium supplemented with hormones and other growth factors in order to study differentiation into adipocytes.

The cells are routinely cultured in DME supplemented with 10% fetal calf serum. Cells are plated at a density of 2000 cells/$cm^2$ and grown to

[29] A. Rizzino and G. Sato, *Proc. Natl. Acad. Sci. U.S.A.*, **75**, 1844 (1978).

confluency. The medium is changed every other day with cultures maintained at 37° in a humid atmosphere containing 5% $CO_2$. The cells can be grown in a serum-free medium consisting of 75% DME and 25% F12 enriched with a mixture of trace elements;[9] insulin (0.5 μg/ml); NIH-LH-B9, a biological standard of bovine luteinizing hormone (2 μg/ml); and rat submaxillary gimmel factor prepared as described in the section on $C_6$ cells.

In this medium, the cells have the same morphology as those cultured in the presence of 10% fetal calf serum. In serum-rich medium, they divide with a generation time of 20 hr and reach a saturation density of $75 \times 10^3$ cells/cm$^2$; in hormone-supplemented medium, their generation time is 28 hr and they reach a saturation density of approximately $36 \times 10^3$ cells/cm$^2$. The efficiency of this serum-free medium for supporting the growth and viability of 3T3-L1 depends on the presence of gimmel; its omission results in an 80% inhibition of growth.

The nature of the factor responsible for the multiplication of 3T3 is under study. Experiments show that purified EGF is unable to replace gimmel in supporting the growth of 3T3. The omission of NIH-LH-B9, insulin, or transferrin results in growth inhibition of 26%, 40%, and 72%, respectively. In the work described above, the cells were plated and incubated with DME/10% fetal calf serum for 15 hr after trypsinization before being cultured in serum-free medium. This "serum-incubation" appears to be necessary for the cells to recover from the trypsinization and to spread on the dishes, a process that seems important in the response of the cells to the hormones and growth factors. Attempts to eliminate the step of serum incubation are in progress. Addition of serum fractions such as fetuin, and Cohn fractions I and IV, seem to improve greatly the plating efficiency and the spreading of 3T3-L1 in a serum-free medium just after trypsinization.

Another clone of mouse Swiss 3T3, 3T3-C2, that does not differentiate into adipocytes also can be grown in the absence of serum in a medium similar to that established for 3T3-L1.

### Rabbit Intimal Cell Line in a Serum-Free Medium (G. Serrero)

The cell line discussed in this section was established by Dr. Vincenzo Buonassisi from cells isolated by trypsinization of the subendothelial portion of the intima of rabbit aorta. Cells are routinely cultured in F12 medium supplemented with 10% fetal calf serum and kept at 37° in a humidified atmosphere containing 5% $CO_2$. The cells have a fibroblast-like appearance in a subconfluent stage and become enlarged and oriented in a parallel fashion in the culture dish when they reach confluency. The cells grow in a serum-free medium that consists of F12 enriched with trace elements,[9] insulin (10 μg/ml), transferrin (5 μg/ml), NIH-LH-B9 (2 μg/ml), progesterone (1 n*M*), growth hormone (50 ng/ml), and 5% of a serum-free medium conditioned by growing cells. The medium is conditioned by Buffalo rat liver cells (BRL clone 3A, kindly supplied by Dr. Peter Nissley). BRL cells produce and release into their culture medium a growth factor known as "multiplication stimulating activity" (MSA)[30–32] that promotes DNA synthesis in chick embryo fibroblasts and 3T3 cells. However, preliminary experiments suggest that the factor stimulating the multiplication of the intimal cell line in serum-free medium is not MSA. Characterization of the responsible factor is in progress.

In the medium described above, the intimal cells have similar doubling times (24 hr) and the same morphology (the cells become elongated and oriented when they reach confluency) as do cells cultivated in the presence of 10% fetal calf serum. The cells have been grown continuously and transferred in the defined medium. They have been kept for 2 months in the absence of serum and have been passed 10 times in the synthetic medium without loss of their growth properties.

Subconfluent or confluent cells are washed free of medium with PBS and detached from the plates with 1 ml of trypsin (1 mg/ml) per 100-mm dish. The excess is immediately removed by aspiration so that only a thin film remains on the plate. After 2 min at 37°, the reaction is stopped by adding 1 ml soybean trypsin inhibitor (1 mg/ml). Cells are then detached and washed free of trypsin and trypsin inhibitor by suspension into serum-free F12. Cells are inoculated at a density of $2 \times 10^3/cm^2$. Incubation with serum is not necessary for spreading the cells on the dishes and the start of optimal growth.

In the absence of BRL conditioned medium, the cells cannot be transferred continuously in the defined medium; their growth drops after the second subculture. If insulin and transferrin are also omitted, growth is reduced by 70–90% and there are drastic modifications of cell morphology.

## [$^{125}$I]-Labeled EGF Binding to HeLa-S in a Defined System (R. A. Wolfe and R. Wu)

Many cell lines exhibit marked growth responses to EGF.[33,34] Furthermore, [$^{125}$I]EGF retains this activity in a bioassay, and its binding to

[30] N. C. Dulak and H. M. Temin, *J. Cell. Physiol.* **81**, 161 (1973).
[31] N. C. Dulak and H. M. Temin, *J. Cell. Physiol.* **81**, 153 (1973).
[32] G. L. Smith and H. M. Temin, *J. Cell. Physiol.* **84**, 181 (1974).
[33] D. Gospodarowicz and J. S. Moran, *Am. Rev. Biochem.* **45**, 531 (1976).
[34] G. Sato and L. Reid, *in* "Biochemistry and Mode of Action of Hormones" (H. V. Rickenberg, ed.). Univ. Park Press, Baltimore, Maryland (1978).

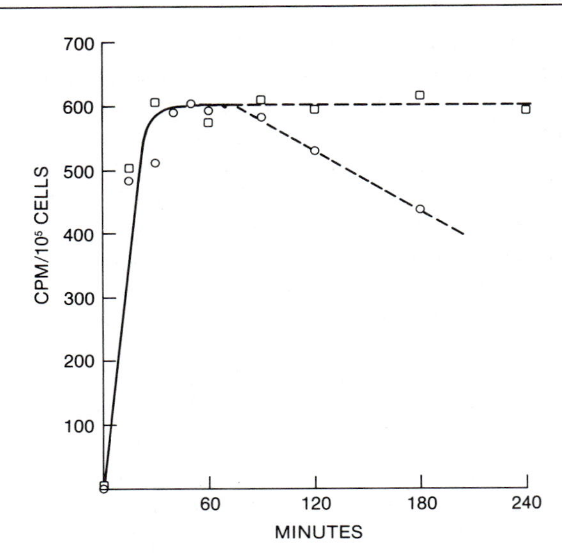


FIG. 1 Time course of $^{125}$I–EGF binding. Six hundred microliters of standard binding mixture [SF-4X(-EGF), 8 ng/ml $^{125}$I–EGF] are applied to the dishes at 37° for various times. □, cells grown in the hormone-supplemented medium; ○, cells grown in 10% FCS.

human fibroblasts has been shown to be both specific and saturable.[35] We have labeled EGF via the chloramine T technique[36] and developed a miniaturized binding assay that does not require the presençe of serum, albumin, or other undefined components.

[$^{125}$I]EGF binds to HeLa-S cells that are grown in 10% fetal calf serum in a manner consistent with other reports. There are approximately 140,000 specific binding sites per cell, and the apparent dissociation constants are in the range of $1–7 \times 10^{-10}$, depending upon growth and assay conditions. At 37°, binding is maximal after 45–60 min and, upon continued incubation, decreases with an apparent half-life of 1–1.5 hr. This must be due to a decrease in the number of binding sites available since the ligand remains capable of binding quantitatively to fresh cells after a 9-hr incubation period. We have suggested several hypotheses: the [$^{125}$I]EGF–receptor complex may be internalized and degraded with the labeled fragments released into the media; the receptor could be masked by a membrane component or by a conformational change; or the iodinated component of the EGF could be removed by a protease leaving the receptor occupied by the unlabeled residue.

[35] G. Carpenter and S. Cohen, *J. Cell Biol.* **71,** 159 (1976).

[36] W. M. Hunter and F. C. Greenwood, *Nature (London)* **194,** 495 (1962).

Cells grown in the serum-free system described below do *not* exhibit this decrease in apparent binding capacity upon exposure to the ligand. Furthermore, the addition of a very small amount of serum to the assay mixture, comparable to the amount of serum resisting the prewash, does not alter this phenomenon. Thus, this decrease in apparent binding capacity is due to a residual effect of culturing the cells in the presence of large amounts of many undefined materials; i.e., serum. It is quite possible that serum proteases adsorbed by the cells are responsible for this effect. It must also be stressed that EGF is a potent mitogen in this system, decreasing the doubling time from 40 to 20 hr.[6] Therefore, the decrease in apparent binding capacity, whatever the mechanism, is not involved with the growth-promoting activity of EGF in this system.

The serum requirement of HeLa-S cells can be replaced by the addition of several hormones and transport proteins to the media.[7] The cells grow in F12 supplemented with insulin (5 μg/ml), transferrin (5 μg/ml), FGF (50 ng/ml), hydrocortisone (50 n*M*), and EGF (50 ng/ml). Stock cultures are maintained in this "completely supplemented" media and in media supplemented with all of the above factors except EGF, subsequently referred to as SF-4X (minus EGF). Cells are plated at a density such that they will be nearly confluent 5 days after plating (250–300,000/35 mm dish). The media is changed after 2 and 4 days of growth and the binding assay performed on the fifth day.

All washes include a second aspiration of medium from a tilted dish. The cells are washed twice with 2 ml of F12 before the assay is performed in SF-4X (−EGF). After adding 0.6 ml of the assay mixture (containing approximately 4 ng labeled [$^{125}$I]EGF), the dishes are incubated at 37°. The assay mixture is then removed, and the plates are washed 4 times with 2 ml of cold PBS. After trypsinization in 1 ml of trypsin solution (1 mg/ml of trypsin and 0.9 m*M* EDTA) is complete, 0.6 ml of the solution is used for determination of radioactivity and 0.2 ml is used to determine the cell number. Nonspecific binding is assayed in the presence of 2.5 μg unlabeled EGF.

# [7] Use of Antibiotics in Cell Culture Media

*By* D. PERLMAN

Like Death and Taxes, contamination or the threat of contamination is always with us and we need all the weapons to combat them.[1]

## Introduction

Bacterial, yeast, and fungal contamination are hazards to those studying various phases of metabolism and growth of mammalian cells *in vitro*. Although we have used antibiotics for more than 30 years to eliminate or suppress unwanted microbial contaminants, many who add them to tissue cultures for this purpose are unfamiliar with the products used, the limitations for their use, and the practical value of this strategy.

The basic requirements are:

1. The antibiotic *must eliminate* the microbial contaminant. (Bactericidal compounds are preferred over bacteriostatic.)
2. The antibiotic must not inhibit growth and metabolism of the mammalian cells in tissue culture.
3. The antibiotic must provide "protection" for the complete experimental period.
4. The antibiotic should not affect any ultimate use intended for the mammalian cells, e.g., virus production or the preparation of antigens.
5. The antibiotic should be nontoxic and safe as far as handling by laboratory personnel is concerned.
6. The antibiotic should be compatible with other components of the culture media.
7. The antibiotic should be inexpensive and should not contain excipients, e.g., buffers, which affect cell growth or metabolism adversely.

## Materials

Most of the antibiotics considered useful in coping with microbial contamination in mammalian cells in tissue culture may be obtained by: (1) direct purchase from a local pharmacy (clinical, community, or hospital); (2) purchase from a biological supply house; or (3) purchase from the manufacturer.

[1] L. L. Coriell, *Natl. Cancer Inst., Monog.* **7**, 33 (1962).

METHODS IN ENZYMOLOGY, VOL. LVIII

ISBN 0-12-181958-2

Those available from the local pharmacy are products that are usually intended for parenteral administration in the therapy of acute infectious diseases. They frequently contain buffers, organic acids, and other excipients which aid their usefulness as therapeutic dosage forms, but might lead to problems in mammalian tissue culture, e.g., the presence of ascorbic acid in the tetracycline formulation. The purity of the antibiotic in these preparations is usually very high, and the manufacturer frequently can supply information on the quantity and identity of the impurities. However, the presence of buffers may result in the potency of the material on a weight basis being as low as 10% that of the pure material. The parenteral dosage form usually is sterile and the container fabricated of materials such that there is minimal carry-over of plasticizers and other minor chemical contaminants. The capsular dosage form usually is "semisterile" and the filler and other excipients are likely to be reasonably water soluble. This formulation should be dissolved in the basal medium and then sterilized by filtration through Millipore or equivalent filters; asbestos filters should not be used since the antibiotic may react with the filter.

The antibiotics available from biological and chemical supply houses include some primarily intended for use in mammalian culture media and others which are sold on a "user's risk basis." The former should have been tested for cytotoxicity in mammalian cell cultures, and the vendor should have data to support claims of low or minimal cytotoxicity; each manufacturing batch should have its own documentation. The latter materials are sometimes of uncertain origin and not infrequently contain cytotoxic contaminants and/or microbial contaminants. The user should be warned to evaluate these materials for cytotoxicity over a period of time before adding them to a critical experiment. This evaluation might include study of the interaction of the antibiotic (or mixture of antibiotics) with constituents of the medium and other factors.

Several antibiotics may be obtained directly from the manufacturer. These are usually in the form of nonsterile powders, and the administrative difficulties in obtaining them frequently outweigh the advantages of obtaining them through this route. Most of the manufacturers are reluctant to sell a few grams of material, and most laboratories are not interested in purchasing kilogram amounts. The result is often a gift of a few grams, and no guarantees are involved with regard to potency, presence of microbial and/or chemical contaminants, or cytotoxicity.

Some of the antibiotics which may be useful in the elimination or at least suppression of contaminating microorganisms from mammalian cell cultures are listed in Table I together with information on the useful concentration and the length of time protection might be expected under

TABLE I
SOME ANTIBIOTICS USEFUL FOR ELIMINATION OR SUPPRESSION OF CONTAMINATING MICROORGANISMS FROM MAMMALIAN CELL CULTURES

| Antibiotic | Antimicrobial spectrum | Recommended concentration in μg/ml media[a] | Stability in media at 37° days[b] |
|---|---|---|---|
| Amoxicillin | Gram-positive and gram-negative bacteria | 100 | 3 |
| Amphotericin B (as deoxycholate complex) | Fungi and yeasts | 2.5 | 3 |
| Amphotericin B methyl ester | Fungi and yeasts | 1.0 | 3 |
| Ampicillin | Gram-positive and gram-negative bacteria | 100 | 3 |
| Carbenicillin | Gram-positive and gram-negative bacteria, especially pseudomonads | 100 | 3 |
| Cephalothin | Gram-positive and gram-negative bacteria | 100 | 3 |
| Chloramphenicol | Gram-negative bacteria | 5 | 5 |
| 7-Chlortetracycline | Gram-positive and gram-negative bacteria | 10 | 1 |
| 6-Demethyl-7-chlortetracycline | Gram-positive and gram-negative bacteria | 5 | 5 |
| Dihydrostreptomycin | Gram-positive and gram-negative bacteria | 100 | 5 |
| Doxycycline | Gram-positive and gram-negative bacteria | 20 | 5 |
| Erythromycin | Gram-positive bacteria and mycoplasma | 100 | 3 |
| Gentamycin | Gram-positive and gram-negative bacteria and mycoplasma | 50 | 5 |
| 5-Hydroxytetracycline | Gram-positive and gram-negative bacteria | 5 | 3 |
| Kanamycin | Gram-positive and gram-negative bacteria and mycoplasma | 100 | 5 |
| Lincomycin | Gram-positive bacteria | 100 | 4 |
| Neomycin | Gram-positive and gram-negative bacteria | 50 | 5 |
| Nystatin | Fungi and yeasts | 50 | 3 |

TABLE I (*continued*)

| Antibiotic | Antimicrobial spectrum | Recommended concentration in μg/ml media[a] | Stability in media at 37° days[b] |
|---|---|---|---|
| Paromomycin | Gram-positive and gram-negative bacteria | 100 | 5 |
| Polymyxin B | Gram-negative bacteria | 50 | 5 |
| Penicillin G | Gram-positive bacteria | 100 | 3 |
| Penicillin V | Gram-positive bacteria | 100 | 3 |
| Tetracycline | Gram-positive and gram-negative bacteria and mycoplasma | 10 | 4 |
| Streptomycin | Gram-positive and gram-negative bacteria | 100 | 3 |
| Tylosin | Gram-positive bacteria and mycoplasma | 100 | 3 |
| Viomycin | Gram-positive and gram-negative bacteria | 50 | 5 |

[a] Concentration effective in controlling "light" infection and at the same time not cytotoxic to $L_{929}$ and KB cells in serum-containing media.

[b] Length of time in which at least 10% of initial activity could be demonstrated in incubation at 37° up to 5 days in serum-containing media.

normal circumstances. Since most are more cytotoxic in the absence of serum, the recommended concentrations should be reduced by $\frac{1}{3}$ (or more) if serum-free media are used.

Commercial sources for some of the antibiotics listed in Table I are mentioned in Table II. All of these preparations are likely to contain some excipients, and the user is advised to check the label on the vial or the package insert to learn the nature of these materials (as some of them may have some effect on the mammalian cell growth and/or metabolism). Unfortunately, the nature of these added excipients that will vary from one manufacturer to the next, and each source should be treated as unique.

Although a few antibiotics are heat or light stable, most should be considered as labile, and all preparations should be refrigerated until needed. Once the dry powder has been dissolved in water it may be frozen until used. The exceptions to this latter procedure are nystatin and amphotericin B. The former is water insoluble and once suspended in a buffer will form aggregates of considerable size. Amphotericin B is available as a colloidal dispersion (containing sodium deoxycholate); upon refrigeration the colloid may form an amorphous gel.

TABLE II

SOME COMMERCIALLY AVAILABLE ANTIBIOTIC PREPARATIONS WHICH MAY BE USED AS TISSUE CULTURE MEDIA SUPPLEMENTS

| Antibiotic | Commercial source | Comments |
|---|---|---|
| Amoxicillin | Beecham Laboratories<br>Roche | Take capsules and dissolve contents and sterile filter |
| Amphotericin B | E. R. Squibb and Sons | Available as power for reconstitution; contains deoxycholate and buffer |
| Ampicillin | Bristol Laboratories<br>Ayerst Laboratories<br>Pfizer Laboratories<br>E. R. Squibb and Sons<br>Wyeth Laboratories | Use injectable formulation |
| Carbenicillin | Beecham Laboratories<br>Pfizer Laboratories | Take capsules and dissolve contents and sterile filter |
| Cephalothin | Eli Lilly Company | Use injectable formulation |
| Chloramphenicol | Park, Davis and Company | Use injectable formulation |
| 7-Chlotetracycline | Lederle Laboratories | Use injectable formulation note buffer composition |
| 6-Demethyl-7-chlor-tetracycline | Lederle Laboratories | Use injectable formulation |
| Dihydrostreptomycin | Eli Lilly Company<br>Pfizer Laboratories | Use injectable formulation |
| Doxycycline | Pfizer Laboratories | Use injectable formulation |
| Erythromycin | Abbott Laboratories<br>Bristol Laboratories<br>Eli Lilly Company<br>Upjohn Company | Use injectable formulation; note composition of buffer |
| Gentamycin | Schering Corporation | Distributed as sterile solution |
| 5-Hydroxytetra-cycline | Pfizer Laboratories | Use injectable formulation |
| Kanamycin | Bristol Laboratories | Use injectable formulation |
| Lincomycin | Upjohn Company | Use injectable formulation |
| Neomycin | S.B. Penick Company<br>Pfizer Laboratories<br>E.R. Squibb and Sons<br>Upjohn Company | Distributed as sterile solution |
| Nystatin | E.R. Squibb and Sons | Distributed as sterile powder for resuspension |
| Penicillin G potassium | Bristol Laboratories<br>E.R. Squibb and Sons<br>Pfizer Laboratories<br>Wyeth Laboratories | Use injectable formulation and note composition of buffer |
| Penicillin V potassium | Bristol Laboratories<br>Abbott Laboratories<br>Eli Lilly Company<br>Wyeth Laboratories | Use injectable formulation and note composition of buffer |

TABLE II (*Continued*)

| Antibiotic | Commercial source | Comments |
|---|---|---|
| Paromomycin | Parke, Davis and Company | Take capsules and dissolve contents and sterile filter |
| Polymyxin B | Burroughs Wellcome Company | Use injectable formulation |
| Streptomycin | Eli Lilly Company<br>Merck Sharp and Dohme<br>Pfizer Laboratories | Use injectable formulation |
| Tetracycline | Lederle Laboratories<br>Pfizer Laboratories<br>Rachelle Laboratories<br>Bristol Laboratories<br>E.R. Squibb and Sons | Use injectable formulation and note composition of buffer |
| Tylosin | Eli Lilly Company | Distributed as sterile solution |
| Viomycin | Pfizer Laboratories | Use injectable formulation |

## Some Strategic Considerations

Any procedure evaluating the use of antibiotics in preventing or eliminating microbial contaminants from mammalian cell cultures must include:

1. A study of the cytotoxicity of the candidate antibiotic to a series of mammalian cell lines growing in a series of different media;
2. A determination of the stability of the antibiotic in the inoculated and uninoculated tissue culture media incubated at 37° in tissue culture vessels for up to 5 days; and
3. An experimental series where mammalian cells are deliberately contaminated with bacteria, yeasts, fungi, and/or mycoplasma and there is a determination of how effective the antibiotic is in *eliminating* (not suppressing) the contaminating organisms. These ''purified'' mammalian cell cultures have to be tested for microbial contamination after four to six transfers in antibiotic-free media.

Since many of the contaminating microorganisms likely to be a problem in mammalian cell cultures are also likely to be antibiotic resistant, any antibiotic evaluated in step 2 above should not decompose very rapidly in tissue culture media when incubated at 37°. (Some of the practical parameters are summarized in our earlier publications.[2–4])

[2] D. Perlman and S. A. Brindle, *Antimicrob. Agents Chemother.* p. 458 (1964).

[3] D. Perlman, N. A. Giuffre, and S. A. Brindle, *Proc. Soc. Exp. Biol. Med.* **105,** 1015 (1961).

[4] D. Perlman, S. B. Rahman, and J. B. Semar, *Appl. Microbiol.* **15,** 82 (1967).

Our studies have shown that many antibiotics may be fairly well tolerated by mammalian cells and chick fibroblasts growing in various media. For practical purposes we recommend benzylpenicillin (penicillin G) and dihydrostreptomycin for elimination or suppression of bacteria, and gentamicin or tylosin to prevent growth of mycoplasmas. Amphotericin B is fairly useful in controlling fungal or yeast contamination but we question whether it should be used on a continuous basis since it is somewhat toxic to most cells.

If the bacterial contaminants turn out to be resistant to the combination of penicillin with dihydrostreptomycin, treatment of the contaminated mammalian cell cultures with tetracycline or chloramphenicol may be one approach to rescuing the cell line. Cross-resistance among antibiotics should be given some consideration when resistant contaminating organisms are the enemy: organisms resistant to penicillin G are likely to be resistant to ampicillin, penicillin V, and perhaps the cephalosporins, cephalothin and cephalexin. Organisms resistant to dihydrostreptomycin may be resistant to kanamycin, gentamycin, neomycin, paromomycin, and ribostamycin.

# [8] Laboratory Management of Cell Cultures

*By* KEN WOLF

## Introduction

Animal cell or tissue cultures—widely used in different scientific disciplines—are biological systems that are seldom completely predictable or amenable to absolute control. Accordingly, cell culture has some attributes of a science while other characteristics resemble those of an art. As such, some practitioners are consistently more adept or successful than others. However, anyone who has had more than passing experience with cell cultures soon learns that things can and do go awry; in fact, catastrophes sometimes happen.

Contamination is an ever-present threat; bacteria, yeasts, and molds can flourish in tissue culture media. The presence of such microbial contaminants, fortunately, is usually quickly and readily visible. Moreover, in many situations, antibiotics are used to minimize the risk of contamination. Cell cultures may also become contaminated but show little or no visible signs of a problem; virus is one example of nonapparent infection, but mycoplasmas are by far the most frequent culprits. Detection of mycoplasma contamination with any degree of confidence is a challenging feat. A variety of methods have been developed for this purpose and for



ISBN 0-12-181958-2

the control of mycoplasma and other microbiological contamination, and they are reviewed in this volume [2].

In addition to microbial contamination, cultures of cell lines can become adulterated, i.e., cell lines can become mixed. Different human and animal cell lines, for example, have been found to be partly or even wholly HeLa cells. One wonders about the validity of the conclusions when the cells were not what they were thought to be. This subject is treated, in part, in this volume [13].

Culture media, serum, and components of media can be toxic and jeopardize the cells' performance and, at the extreme, even their continued existence. Precautions to be taken in assuring quality control of serum for cell culture have been covered in this volume [2]. Additional problems are encountered when laboratory glassware has harmful residues or when plastic tissue culture ware leaks or has not been uniformly coated to permit cell attachment. Major disasters or catastrophes happen when there is critical equipment failure, e.g., when freezers malfunction or incubators overheat and cultures are killed.

Disciplined and systematic management of cell culture activities will minimize the number of things that can go wrong and, more importantly, will greatly reduce the impact on operations when problems do occur. There are ways by which the real catastrophe can be avoided or completely negated.

The following system of management employs a dichotomous approach, redundancy wherever possible, quality control, safekeeping measures, and careful record-keeping. The system is probably of greatest advantage to continuing or long-term operations, but some of its features lend themselves to short-term work as well. It is especially appropriate for modest laboratories—those with limited budgets and personnel—and in the long run it will actually save time.

## Sequence of Steps in Managing Cell Cultures

1. Determine whether or not there is actually a need to adopt a system of management. If, for instance, short-term work is planned and quality cells of the kind to be used are readily available—such as from the American Type Culture Collection—it may be most practical and economical to purchase certified cells, use them for the particular application, and simply discard them when work has been finished.

2. If it is determined that management method is needed, the quality of starting stock should be appropriate, i.e., American Type Culture Collection Certified Cell Lines or the equivalent. Mongrel cultures should be avoided. Propagation of the cells should be divided into *stock cultures* that

are grown to maintain the lineage under the best possible conditions and *working cultures* that are to be used for the actual work applications. Propagation of *stock cultures* must be accorded the highest work priority and be carried out with strict aseptic technique and without antibiotics.

3. The second element of dichotomy of the system is in carrying two sets of *stock cultures*. To minimize the risk of culture loss due to equipment failure, the two sets should be housed in different incubators. Each set of cultures should have its own pretested, reserved, single-use portions of frozen, antibiotic-free media; separate records; and separate schedules for handling. Ideally, the division of risk and responsibility should be between two persons, but the principle can be employed by an individual who does the work as two independent activities.

4. Each *stock culture* should be maintained on two different lots of appropriate medium. Stock culture medium stored at $-20°$ or lower remains useful for several years, and single-use portions minimize risk of contamination and obviate formation of precipitates from repeated freeze–thaw cycles. In practice, stock cultures can be kept at reduced temperature and may require handling only once a month and, in some cases, only several times a year.

5. Serum should be obtained from a supplier who certifies it to be free of mycoplasma. Several different samples are obtained prior to purchase and tested to determine which gives the best plating efficiency for the cells to be used. For this test, a suspension containing about 200 cells is added to duplicate or triplicate 25 $cm^2$ flasks containing 5 ml of medium with at least a 20% level of the several samples of serum to be tested. Incubation is allowed to proceed for 2–3 weeks, after which the flasks are drained, rinsed with balanced salt solution, stained with 1% aqueous or alcoholic crystal violet, washed, and the cell colonies (clones) counted. That serum giving the most clones should be purchased. Additional measures for quality control of serum are considered in this volume [2]. Serum should be stored at $-20°$ or lower and may safely be kept for at least a year without noticeable loss of potency.

6. Medium and other solutions are prepared only with water of tissue-culture quality. A freshly prepared borosilicate glass distillate of deionized water is usually satisfactory. Preferred storage is under sterile conditions in borosilicate glass. For the small user, USP water for injection is suitable; it is available from pharmacies and hospital supply houses in liter bottles.

7. Locally prepared media or solutions must be thoroughly tested for sterility before use. Samples of at least 5% are suggested as is incubation for 7–10 days at the usual incubation temperature and at 37°. Details are presented in this volume [2].

8. To minimize adulteration of one cell line with another, only one cell line should be opened and handled at any given time. All cultures must be identified at the time of handling.

9. Stock cultures should be propagated in multiples. Because plastic vessels permit slow gas exchange, 16 × 125 glass tubes with rubber-lined screw caps, small flasks, or bottles are advantageous since they assure gas retention, provide multiple cultures in reduced space, and are easily examined microscopically.

10. Stock cultures are to be handled with minimal frequency. To accomplish this use reduced seeding levels and low-temperature incubation.

11. Maintain detailed records of medium and solutions that are prepared. Include relevant data on all components and notations of when, how, and by whom cells are handled.

12. Stock cultures should be tested at least once a year and preferably twice for mycoplasma and other possible contaminants.

13. Replace regular working cultures annually with newly expanded stock cultures.

14. Attempts should be made to disperse the cells from stock cultures with an autoclaved solution of EDTA that does not contain trypsin. Trypsin solutions are not sterilized, and they represent a possible source of culture contamination. Accordingly, for greatest security, stock cultures should be dispersed with an autoclaved solution of EDTA.

15. Frozen stocks of unique cell populations or those that are epecially valuable should be maintained in liquid nitrogen and at more than one facility. The American Type Culture Collection, Rockville, Maryland 20852 offers such a safe-deposit service.

# [9] Dispersion and Disruption of Tissues

*By* MARK M. BASHOR

Initial studies in tissue culture carried out near the beginning of this century involved the explantation of fragments of tissue from an organism into an artificial *in vitro* environment. Propagation of such explants usually involved aseptically cutting or trimming the original explant to reduce its size so that appropriate gas and nutrient diffusion could occur, and so that multiple cultures could be obtained. From these pioneering efforts have developed the closely related disciplines of cell, tissue, and organ culture.

[1] The opinions or assertions contained herein are the private views of the author and are not to be construed as official or as reflecting the views of the Department of the Army or the Department of Defense. Citation of trade names in this report does not constitute an official endorsement or approval of the use of such items.




ISBN 0-12-181958-2

This article will present materials and methods generally applicable to the preparation of primary cell cultures from organs or tissues of warm-blooded animals.

Methods used for preparing cell suspensions from intact tissue may be broadly classified as (1) mechanical (cutting, mincing, shearing, sieving); (2) chemical (usually omission of divalent cations, with or without the addition of chelating agents); and (3) enzymic (digestion with trypsin, collagenase, pronase, hyaluronidase). Most techniques are, however, combinations of two or all three categories. The choice of procedures depends upon the nature and amount of available tissue, the intended use of the cultures, and perhaps precedents described in the literature. However, in all instances the methods should be designed to minimize physical and chemical trauma to the isolated cells in every way possible.

## Balanced Salt Solutions and Growth Media

The concept of a balanced salt solution (BSS) is generally credited to the elegant studies of Sydney Ringer[1a] published between 1880 and 1895. The first such solution to be designed specifically for mammalian cells was by Tyrode[2] in 1910, and many others have since been developed. These solutions serve several important functions in cell culture. They are useful for washing tissues and cells, as vehicles for treating cells and tissues with various agents, and they provide the cells with water and inorganic ions, while maintaining a physiological pH (7.2–7.6) and osmotic pressure. Several BSS formulations also include glucose as a source of energy. Some of the more popular formulations are presented in Table I. Earle's[3] and Hank's[4] BSS and Dulbecco and Vogt's[5] phosphate-buffered saline (PBS) are the most commonly used. These are all available commercially as sterile solutions and in some instances as sterile concentrates or dry powders. If they are to be formulated in the laboratory it is advised that the solutions which contain bicarbonate be filter-sterilized by pressure rather than vacuum filtration; the use of 95% air/5% $CO_2$ as pressurant is also advised. Due to $CO_2$ volatility it is sometimes useful to prepare the bicarbonate as a separate sterile solution (7.5%) and add it to the BSS just prior to use.

The choice of BSS to be used during the preparation of cell suspensions depends to a certain extent upon the procedures which will be used. If all steps are to be performed under an air atmosphere, then a BSS

[1a] S. Ringer, *J. Physiol. (London)* **18,** 425 (1895).

[2] M. V. Tyrode, *Arch. Int. Pharmacodyn. Ther.* **20,** 205 (1910).

[3] W. R. Earle, *J. Natl. Cancer Inst.* **4,** 165 (1943).

[4] J. W. Hanks and R. E. Wallace, *Proc. Soc. Exp. Biol. Med.* **71,** 196 (1949).

[5] R. Dulbecco and M. Vogt, *J. Exp. Med.* **99,** 167 (1954).

TABLE I

COMPOSITION OF SELECTED BALANCED SALT SOLUTIONS[a,b]

| | Ringer[c] | Tyrode[d,e] | Gey[f] | Earle[g] | Puck[h] | Hanks[i] | Dulbecco (PBS)[j,k] |
|---|---|---|---|---|---|---|---|
| NaCl | 9.00 | 8.00 | 7.00 | 6.80 | 8.00 | 8.00 | 8.00 |
| KCl | 0.42 | 0.20 | 0.37 | 0.40 | 0.40 | 0.40 | 0.20 |
| $CaCl_2$ | 0.25 | 0.20 | 0.17 | 0.20 | 0.012 | 0.14 | 0.10 |
| $MgCl_2 \cdot 6\ H_2O$ | | 0.10 | 0.21 | | | 0.10 | 0.10 |
| $MgSO_4 \cdot 7\ H_2O$ | | | 0.07 | 0.10 | 0.154 | 0.10 | |
| $Na_2HPO_4 \cdot 12\ H_2O$ | | | 0.30 | | 0.39 | 0.12 | 2.31 |
| $NaH_2PO_4 \cdot H_2O$ | | 0.05 | | 0.125 | | | |
| $KH_2PO_4$ | | | 0.03 | | 0.15 | 0.06 | 0.20 |
| $NaHCO_3$ | | 1.00 | 2.27 | 2.20 | | 0.35 | |
| Glucose | | 1.00 | 1.00 | 1.00 | 1.10 | 1.00 | |
| Phenol red | | | | 0.05 | 0.005 | 0.02 | |
| Atmosphere | air | air | 95% air/5% $CO_2$ | 95% air/5% $CO_2$ | air | air | air |

[a] Amounts are given as grams per liter of solution.

[b] In some instances the values given represent calculations from data presented by the authors to account for the use of hydrated or anhydrous salts.

[c] S. Ringer, *J. Physiol. (London)* **18,** 425 (1895).

[d] M. V. Tyrode, *Arch. Int. Pharmacodyn. Ther.* **20,** 205 (1910).

[e] R. C. Parker, "Methods of Tissue Culture," 3rd ed., p. 57. Harper (Hoeber), New York, 1961.

[f] G. O. Gey and M. K. Gey, *Am. J. Cancer* **27,** 55 (1936).

[g] W. R. Earle, *J. Natl. Cancer Inst.* **4,** 165 (1943).

[h] T. T. Puck, S. J. Cieciura, and A. Robinson, *J. Exp. Med.* **108,** 945 (1958).

[i] J. H. Hanks and R. E. Wallace, *Proc. Soc. Exp. Biol. Med.* **71,** 196 (1949).

[j] PBS, phosphate-buffered saline.

[k] R. Dulbecco and M. Vogt, *J. Exp. Med.* **99,** 167 (1954).

which is not buffered with bicarbonate should be used, since loss of $CO_2$ will cause the solution to become too alkaline. On the other hand, if some or all steps will be done under an air/$CO_2$ atmosphere, then a BSS using a bicarbonate buffer will have to be employed to prevent acidification of the medium by dissolved $CO_2$. Phenol red may be included, usually at a concentration of 20–50 mg/liter, as a visual indicator of pH. Glucose may be included or omitted, depending upon the time elapsing between extirpation of the tissue and the final dispersion of the cells in complete growth medium. The final consideration in the choice of BSS composition is the inclusion or exclusion of the divalent cations calcium (Ca) and magnesium (Mg). As early as 1900, it was recognized that these ions increased the stability of the intracellular matrix, and many investigators have confirmed this in tissue culture experiments. For this reason several of the BSS are also commercially available (or are prepared in the laboratory) as calcium-and magnesium-free solutions (CMF–BSS). The further addition of chelators, or the readdition of Ca or Mg when certain enzymes are used as dispersing agents, is discussed below.

It is not the intent of this section to present an extensive discussion of growth media, as this is covered elsewhere in this volume [5]. Many formulations are available commercially as sterile solutions and in some instances as sterile concentrates or dry powders. Most such media were originally formulated for a particular cell type, such as RPMI-1640 for lymphocytes or Dulbecco's modified Eagle's medium (DMEM) for fibroblasts. In almost all instances these media are further supplemented with an animal serum or sera (horse, calf, fetal calf, human; 2–20%). Regardless of the type or composition of the growth medium, or the nature of the supplements which are added to it, all represent no more than an elaboration of nutrients, hormones, or "growth factors" to a BSS. Occasionally these media, usually without a serum supplementation, are used in place of BSS during the intermediate stages of tissue dispersal. More frequently, however, they are used only as a final rinse, immediately prior to suspension of the cells in the complete (serum-supplemented) growth medium and plating in the culture vessels as described in a later section.

The use and choice of antibiotics is discussed seperately from several points of view in this volume [2] [3] and [7].

## Procurement and Preparation of the Tissue

The source or sources of tissue to be cultured will vary with the design and needs of a particular laboratory. Two precautions should be observed in all cases insofar as is possible. The tissue must be kept moist, and a minimal amount of time should elapse between extirpation of the tissue and initiation of culture. Under ideal circumstances the culture laboratory will be situated adjacent to the surgical rooms. Recognizing that this ideal is seldom realized, the following suggestions are made. Samples obtained from a hospital operating room may be placed by the surgeon in sterile dishes containing a BSS that one supplied. Once this is accomplished, an assistant can move them to the culture laboratory. When a considerable time may elapse between excision and culture, it may be advisable to place the tissue in a BSS containing glucose or in a complete culture medium. If extensive transportation is involved, as between a slaughterhouse and the laboratory, chilling with ice may be required. For small samples a 100–150 mm sterile, disposable tissue culture dish may be employed. These are conveniently sealed for transportation by stretching a length of Parafilm "M" (American Can Co., Neenah, Wisconsin) around the circumference two or three times. The closest situation to ideal is when tissues are removed from laboratory animals, in which case the excision of tissue can be accomplished readily in or very near the culture laboratory. The animal is anesthetized or sacrificed, the area of incision shaved (if necessary), sterilized by liberal application of 70% ethanol, and the sterile field isolated by draping the animal with sterile cloth or gauze. The incision is then made, the tissue located, removed aseptically, and placed immediately in sterile BSS or complete medium. The instruments should be chosen and the surgery accomplished in such a way as to minimize mechanical trauma to the tissue.

When the tissue is of fetal origin, the uterus is first removed *in toto* to a suitable sterile dish, the fetuses dissected out, decapitated, transferred to clean sterile dishes, and the tissue(s) then removed. Fresh sterile BSS is used whenever the material needs to be moistened or rinsed free of blood or small pieces of debris; excess BSS can be aspirated using sterile Pasteur pipettes connected by tubing to a vacuum flask. Small surgical instruments which are used repeatedly may be cleaned and sterilized readily by keeping at hand a beaker of detergent solution and a small brush (a toothbrush is adequate), one or two beakers of distilled water for rinsing, and a container in which the instruments can be immersed in 70% ethanol. Before use, instruments are removed and dried by evaporation or flaming (carefully).

## Dispersion of the Tissue

It is only necessary to consider the extreme differences in tissue architecture exhibited by tissues such as brain, heart, skeletal muscle, liver, lung, and kidney to appreciate that there is no "standard" procedure for preparing cell suspensions from these or any other tissues. Perhaps as we learn more about the nature of the cell surface and cell–cell interactions we will be able to devise better, more uniform methods for disrupting the intracellular matrix. In spite of the diversity of these structures, there are certain steps in the dispersion process which are common to virtually all procedures. After discussing these, specific examples and exceptions will be considered.

*Mincing the Tissue*

It is necessary to render the tissue into pieces of sufficiently small size so that the solution or solutions used for dispersion can easily penetrate the matrix. The size of the tissue fragments is generally of the order of 1–3 mm cubes. Except where very large pieces or amounts of tissue must be processed, the use of scissors should be avoided due to compression injury to the cells adjacent to the cut. Tissue samples that are obtained from surgery or biopsy, from adult laboratory animals, and particularly tissue of fetal origin are obtained in amounts of only a few grams, and frequently much less. These amounts may be minced conveniently in sterile 100-mm dishes containing medium or BSS. This may be accomplished by passing two very sharp cataract knives or scalpels[6] smoothly in opposite directions through the tissue. Care should be taken to accomplish clean cuts through the tissue and to avoid tearing the material. Once the tissue is minced, the fragments can be transferred with a large-bore pipette to the sterile vessels in which the cells will be dispersed.

*Selecting and Preparing Dispersant Solutions and Treating the Tissue*

As mentioned, mechanical, chemical, and enzymic methods are used to prepare cell suspensions from tissue. Mechanical methods include homogenizing (very harsh and probably should be avoided), shaking on a gyrorotary shaker, vortexing, triturating with fine-bore or wide-bore pipettes, and forcing the minced tissue through a nylon or stainless-steel mesh; the last is frequently used for preparing lymphocytes from spleen, thymus, and lymph nodes.[7–9] Typically, however, a procedure will use an enzymic and/or chemical treatment to weaken the intracellular matrix and a mild mechanical treatment to release the cells. The use of trypsin for preparing cell suspensions was first reported by Rous and Jones in 1916.[10] Since then the use of hog or bovine pancreatic trypsin preparations has

[6] A useful scalpel for this is the Bard-Parker No. 11 blade (disposable) in the No. 7 handle (Becton, Dickinson and Co., Waltham, Mass.). These are sold by most general laboratory supply companies.

[7] H. R. Hendricks and I. L. Eestermans, *J. Immunol. Methods* **12,** 345 (1976).

[8] O. Braendstrup, V. Andersen, and O. Werdelin, *Cell. Immunol.* **25,** 207 (1976).

[9] M. Ohishi and K. Onoue, *Cell. Immunol.* **18,** 220 (1975).

[10] P. Rous and F. A. Jones, *J. Exp. Med.* **23,** 555 (1916).

become very popular. Several other enzymes have also been used, notably pronase,[11-13] collagenase,[14,15] hyaluronidase,[14-17] elastase,[15,17] pancreatin,[17,18] papain,[17] and deoxyribonuclease.[17,18] It should be mentioned that virtually all of these enzymes normally contain several activities. Crude trypsin may contain, in addition to trypsin, chymotrypsin, elastase, ribonuclease, deoxyribonuclease, and amylase. This may be of considerable importance in designing a protocol for preparing cell suspensions, since in some instances the efficacy of a trypsin-induced dissociation has been attributed to the presence of contaminating activities and substitution of more highly purified crystalline trypsin has proven less effective or even harmful. On the other hand, some investigators have found that crude trypsin contains toxic or harmful substances, whereas the highly purified enzyme is suitable. A statement of the rationale for including deoxyribonuclease in some dispersing media may be instructive at this point with regard to "contaminating" enzyme activities. Essner *et al.*[19] found that clumps of a rat ascites hepatoma could be dispersed by several treatments. Disaggregation by trypsin or chymotrypsin proceeded in two phases, the first being dissolution of the clumps into single cells, the second phase being the agglutination of freed cells by a mucous-like coagulum which formed. Auerbach and Grobstein,[20] studying dissociation and reaggregation of embryonic mouse tissues, found that certain lots of trypsin produced a "gummy" material which enmeshed the cells and cell clumps, preventing complete disaggregation. This material was not formed when crude pancreatin was included in their trypsin solutions. Steinberg[18] subsequently observed the appearance of a "slimy" material which trapped cells that had been obtained by trypsin dissociation of heart and liver cells from 5-day chick embryos. This material was shown to be DNA which was released from damaged cells, hydrated, and uncoiled to form a viscous material which readily enmeshed the cells. Steinberg found that inclusion of crystalline deoxyribonuclease in the dissociation medium would completely prevent the appearance of the gel.

In recent years better preparations of several enzymes have become available for tissue culture work. As an example, Worthington Biochemi-

[11] D. Weinstein, *Exp. Cell Res.* **43,** 234 (1966).

[12] J. C. Sullivan and I. A. Schafer, *Exp. Cell Res.* **43,** 676 (1966).

[13] J. F. Foley and B. Aftonomos, *J. Cell. Physiol.* **75,** 159 (1970).

[14] P. O. Seglen, *Exp. Cell Res.* **76,** 25 (1973).

[15] P. O. Seglen, *Exp. Cell Res.* **82,** 391 (1973).

[16] L. M. J. Rinaldini, *Exp. Cell Res.* **16,** 477 (1959).

[17] L. L. Wiseman and W. R. Hammond, *J. Exp. Zool.* **197,** 429 (1976).

[18] M. S. Steinberg, *Exp. Cell Res.* **30,** 257 (1963).

[19] E. Essner, H. Sato, and M. Belkin, *Exp. Cell Res.* **7,** 430 (1954).

[20] R. Auerbach and C. Grobstein, *Exp. Cell Res.* **15,** 384 (1958).

cal Corporation (Freehold, New Jersey) produces four types of crude collagenase. Type I, having a "normal balance" of enzyme activities, is recommended for fat cells and adrenal tissue; Type II, high in clostripain activity, is recommended for liver, bone, thyroid, heart, and salivary tissue; Type III, low in proteolytic activity, is recommended for mammary tissue; Type IV has low tryptic activity and is recommended for pancreatic islets.

Just as there is no "standard" procedure for tissue dissociation, so too is there no "standard" buffer for dispersing tissues. Generally, the solutions are prepared in BSS or CMF–BSS rather than a culture medium. The choice between bicarbonate and nonbicarbonate buffered BSS has been discussed. The concentrations of trypsin most frequently used are 125–250 $\mu$g/ml. When highly purified crystalline trypsin preparations are used, these concentrations may be reduced to 10–50 $\mu$g/ml. If, in addition to CMF–BSS, a chelator is desired, EDTA[21] or EGTA[22] have been used at concentrations of approximately 20 $\mu$g/ml. Rinaldini,[23] in an extensive survey of enzymes suitable for preparing dispersions of embryonic chick cells, used Tyrode's CMF–BSS with 100 $\mu$g/ml sodium citrate as a chelator. Use of the potassium chelator, tetraphenylboron, should probably be avoided.[24,25]

Collagenase is usually used at a concentration of 0.1–0.3 mg/ml, and sometimes in conjunction with trypsin (0.25 mg collagenase and 0.1 mg trypsin per milliliter). Pronase has been used at concentrations of 1–2 mg/ml. Deoxyribonuclese is sometimes added, as discussed above, in concentrations ranging from 1–50 $\mu$g/ml. It should be remembered that collagenase requires the presence of Ca ions whereas DNase requires Mg ions.

With the exception of pronase and some collagenase preparations, all of these enzymes are readily dissolved in BSS or CMF–BSS and may be sterilized by filtration through 0.2 $\mu$m sterile filters. Pronase and collagenase may require centrifugation to remove insoluble particles prior to filtration. The solutions may be stored at $-10°$ for 3–6 months.

Once a solution or solutions have been chosen for dispersing the tissue, there remains the choice of method(s) for treating the minced tissue. Again there is an endless list of variables and variations, but fortunately only a few basic principles. The procedure of Dulbecco and Vogt[5] for

[21] EDTA, ethylenediaminetetraacetic acid.

[22] EGTA, ethyleneglycol bis($\beta$-aminoethyl ether) *N,N'*-tetraacetic acid.

[23] L. M. Rinaldini, *Exp. Cell Res.* **16,** 477 (1959).

[24] S. R. Hilfer and J. M. Brown, *Exp. Cell Res.* **65,** 246 (1971).

[25] P. R. Kerkof, S. Smith, H. T. Gagné, D. R. Pitelka, and S. Abraham, *Exp. Cell Res.* **58,** 445 (1969).

preparing cell suspensions of monkey kidney (used in the production of polio vaccines) still serves as a model of general steps for dispersing tissue. Tissue from the kidneys is minced (as described above) and transferred to a 250-ml flask or centrifuge bottle, washed several times in PBS, and suspended in PBS containing 0.25% trypsin prewarmed to 37°. After 10-min incubation at 37°, this solution is removed and discarded.[26] This is replaced with 20 ml of fresh, prewarmed trypsin and the fragments agitated by pipetting back and forth 10–15 times with an automatic pipette.[27] After allowing the undigested fragments to settle, the turbid supernatant is transferred to a centrifuge bottle. Warm trypsin is again added, the fragments agitated as before, and the released cells collected. This is repeated several times until the fragments lose their brownish color and become clumps of whitish connective tissue. The combined supernatant fluids are washed 3 times by centrifugation (3 min at 300 g) and resuspension in PBS. After the last centrifugation, the packed cells are suspended in 10–20 times their volume of complete culture medium, the cells are enumerated by hemocytometer count, and diluted to contain $3 \times 10^5$ cells/ml before plating in culture vessels.

Although agitation of the tissue fragments in dispersant solution is still frequently accomplished by trituration with large-bore and small-bore pipettes, Rappaport[28] modified the procedure of Dulbecco and Vogt and introduced special "trypsinizing flasks." These are Erlenmeyer flasks modified for use with a magnetic stirrer by indenting the sides at right angles to the bottom at four equally spaced points around the circumference to produce maximum turbulence and mixing.[28] A constriction in the neck just below a side arm permits trapping of fragments in the bottom of the flask while pouring the suspended cells from the side arm. Recognizing that trypsin-treated cells are unusually fragile, Rappaport also designed a similar apparatus that allowed continuous addition of fresh trypsin while removing cells as they were suspended. Several other designs are commercially available[29] such as that devised by Shipman and Smith[30] for continuous flow of dispersant. Alternatives to both repetitive pipetting and magnetic stirring are swirling on a gyrorotary shaker or gentle vortexing.

[26] This initial treatment is used in many other procedures and serves to remove further blood cells and cells injured during the preparation of tissue fragments.

[27] Pipette fillers such as the Bel-Art Pi-Pumps (A. H. Thomas Co., Philadelphia, Pennsylvania), operated by a thumbwheel, used with sterile cotton-plugged pipettes, is a convenient aid. An automatic pipette filler such as the Drummond Pipet-Aid (Bellco Glass Co., Inc., Vineland, New Jersey) is also useful.

[28] C. Rappaport, *Bull. W. H. O.* **14,** 147 (1956).

[29] Bellco Glass Co., Inc., Vineland, New Jersey 08360.

[30] C. Shipman, Jr. and D. F. Smith, *Appl. Microbiol.* **23,** 188 (1972).

After suspending the cells with an enzyme solution it may be necessary to reduce or stop the enzyme action by immediate dilution into serum-containing BSS or growth medium prior to centrifugation and washing procedures; serum contains potent trypsin inhibitors.

Separation of released cells from undigested fragments, connective tissue, and other debris can be accomplished by allowing the fragments to settle (as in the procedure of Dulbecco and Vogt), by very brief centrifugation (15–30 sec at 300 g), by filtration through fine mesh nylon netting,[31] stainless-steel screen,[32] or several layers of washed cheesecloth or gauze. The filtered cells are transferred to sterile centrifuge tubes[33] and pelleted by centrifugation for 5–10 min at 300–600 g. The supernatant fluid can be drawn off without disturbing the cell pellet by placing a Pasteur pipette attached to a vacuum flask at the top of the tube and slowly inverting the tube as the medium is drawn off. Cells are resuspended in fresh BBS or medium by very gentle swirling or vortexing, and the centrifugation is repeated. Washing may be repeated once or twice more before suspending the cells in complete growth medium, with the cell density determined by hemocytometer count or with an electronic particle counter. The cells are diluted as necessary to a final density of $10^5$–$10^6$ cells/ml, inoculated into the culture vessels, and placed in the incubator.

## Alternative Methodologies

While the sequence that has been described is applicable to most tissues, there are interesting and very useful deviations from the general outline. Some of these may be used for dispersing tissues other than those described below.

### *Organ Perfusion: Adult Rat Liver*

Adult rat liver is an organ which is particularly amenable to mechanical perfusion, both *in situ* and *in vitro*. Seglen[14,15] has described a method for preparing hepatocyte suspensions from adult rat liver which has been widely used and occasionally modified. The procedure is unique in that the dispersant solutions are perfused through the liver rather than treating minced tissue. The method also involves sequential treatment of the tissue with different solutions. The method as used by Williams *et al.*[34] is briefly

[31] Available from Tetko, Monterey Park, California 91754.

[32] Available from Perforated Products Inc., Brookline, Massachusetts 02146.

[33] Disposable, sterile plastic centrifuge tubes, with caps, are available in 15- and 40-ml sizes from several manufacturers of tissue culture plasticware.

[34] G. M. Williams, E. Bermudez, and D. Scaramuzzino, *In Vitro* **13,** 890 (1977).

stated as follows. The liver of an adult rat is perfused for 4 min with Hanks's CMF–BSS buffered with 0.05 *M* HEPES[35] and containing 0.5 m*M* EGTA.[22] After this treatment the liver is perfused for 10–12 min with a collagenase solution (100 units/ml) in Williams's Medium E[36] buffered with 0.05 *M* HEPES. The cells are combed into fresh collagenase solution, centrifuged (50 g for 4 min), and resuspended in complete growth medium before inoculating into culture vessels.

### *Sequential Enzyme Treatment: Mouse Mammary Gland*

It is sometimes possible and advantageous to use a sequential enzyme-treatment procedure for preparing cell suspensions. A good example is a method described by Wiepjes and Prop[37] for preparing epithelial cells from mouse mammary glands. Virgin mice, 9–10 weeks old, are sacrificed and the skin and mammary glands are removed. The glands are dissected free, minced, and the fragments treated for 15 min at 36° with a CMF–BSS containing glucose. Gentle agitation is provided by a gyrorotary shaker. The floating tissue fragments are transferred to a solution of 0.125% collagenase, 0.1% hyaluronidase, 0.4% demineralized bovine serum abumin in CMF–BSS and incubated for 45 min. A subsequent incubation on the gyrorotary shaker for 60 min in 1.25% pronase dissolved in Medium 199 completely disrupts the tissue. Twice the volume of serum is added, and the cells are sedimented by centrifugation (2 min at 300 g). The supernatant liquid and floating fat cells are removed and discarded. The pellet is gently suspended in complete culture medium containing approximately 1 μg/ml DNase, filtered, washed twice with complete culture medium, and plated in culture dishes.

### *Enzymic Dispersion after Culture: Rat Pineal Gland*

It is not always necessary to prepare cell dispersions from minced tissue immediately after excision and fragmentation. The pieces of tissue can be rinsed with several changes of BSS, transferred directly to culture dishes containing complete growth medium, and placed in the incubator, changing the medium as necessary. During this time of culture the explants will attach to the bottom of the dish and cells may begin to grow out of the explant and onto the surface of the dish surrounding the explant. Should the explants tend to float, a smaller volume of medium may be used, or the explants may be held against the bottom of the dish by

[35] HEPES, *N*-2-Hydroxyethylpiperazine-*N'*-2-ethanesulfonic acid.

[36] G. M. Williams and J. M. Gunn, *Exp. Cell Res.* **89**, 139 (1974).

[37] G. J. Wiepjes and F. J. A. Prop, *Exp. Cell Res.* **61**, 451 (1970).

overlaying a piece of washed, sterilized dialysis tubing or a sterile filter soaked with the medium. At the discretion of the investigator the medium is drawn off, the attached fragments and cells rinsed with BSS, treated with a dispersing agent, and suspended in fresh medium. At this stage, cells can be separated from remaining fragments by centrifugation or filtration, or they may again be cultured together.

Nathanson *et al.*[38] have used such a method for preparing cell cultures from adult rat and avian pineal glands. Essentially their method involves removing and mincing the isolated glands into 10–15 fragments. The explants from 2–4 glands are then transferred to 100-mm tissue culture dishes and incubated at 37° in a modified Ham's F-12 medium containing 10% fetal calf serum and antibiotics. After 2–4 weeks of culture, the dishes are rinsed with Hanks's BSS and incubated for 10 min at 37° with 10 ml of 0.25% trypsin. Gentle trituration with a pipette is used to suspend the cells and fragments which are then transferred to Erlenmeyer flasks and incubated an additional 10 min on an incubator gyrorotary shaker at 37°. This suspension is again gently pipetted, transferred to centrifuge tubes containing an equal volume of ice-cold complete culture medium, and centrifuged at 200 g for 3 min. The pellet is then resuspended in fresh cold complete medium, and the remaining clumps of cells are removed by centrifugation (50 g for 1 min). The supernatant is centrifuged again (200 g for 3 min) to pellet the single cells which are then suspended in fresh medium and plated at a density of $2 \times 10^4$ to $10^5$ cells per 100-mm dish.

*Mechanical (Nonenzymic) Dispersal of Tissue: Chick Embryo Skeletal Muscle*

The use of mechanical forces to prepare suspensions of lymphocytes from spleen and thymus[7–9] has been mentioned. The procedures cited rely on forcing the tissue through a fine-mesh nylon or stainless-steel screen. An entirely different mechanical procedure has been successful for preparing chick embryo skeletal muscle cultures.[39,40] The procedure relies on the dispersion of cells from the tissue fragments by the shearing forces of circularly flowing medium. The swirling action is provided by suspending the fragments in medium in a test tube which is "vortexed" on a suitable laboratory mixer[41] for a minute or less. The cells are then filtered through nylon cloth, diluted, and placed in culture vessels. The authors state that

[38] M. A. Nathanson, S. Brinkley, and S. R. Hilfer, *In Vitro* **13,** 843 (1977).

[39] A. I. Caplan, *J. Embryol. Exp. Morphol.* **36,** 175 (1976).

[40] J. C. Bullaro and D. H. Brookman, *In Vitro* **12,** 564 (1976).

[41] Vortex-Genie, Scientific Industries, Inc., Bohemia, New York 11716; Super-Mixer, Lab-Line Instruments, Inc., Melrose Park, Illinois 60160.

the advantages are that the method is more rapid than conventional trypsinization; that the cells are released directly into the growth medium, myoblasts having a much shorter recovery period before fusion occurs; and that a lower level of fibroblast contamination is achieved.

Similar procedures have been used for other types of muscle[42] and for neural tissue,[43] and they may prove useful for other tissues as well.

## Determining Cell Yield and Viability

Judging a procedure for the preparation of cell suspensions from living tissues as successful or unsuccessful depends to a certain extent upon the needs and requirements of the investigator and the purpose for which the cell cultures are being established. Important considerations in making such a judgement are: the adequacy of the cell yield, i.e., the number of cells obtained per gram or milligram of tissue; the viability of the cells; and the ability of the cells to retain in culture the properties and functions expected of them.

The cell yield is determined by counting the number of cells in a known volume of the final suspension and calculating the total number of cells obtained from the original mass of tissue (See this volume [11]).

The most commonly used criterion for cell viability is based on the assumption that viable cells will exclude certain dyes such as trypan blue, whereas nonviable cells will take up the dye (See this volume [13]). Other criteria for viability include measures of the ability of the cells to attach to the surface of the culture vessel, i.e., the fraction of cells inoculated into a vessel that will attach to the surface. Occasionally, the ability of the cells to take up or incorporate a radioactive compound may be useful. The ability of cells to survive normal culture conditions or clonal growth conditions is a further criteria of viability. The integrity of the cell membrane and the general appearance of the suspended cells should be monitored by phase-contrast microscopy. If possible, the ultrastructure of the cells should be examined by scanning and transmission electron microscopy.

[42] S. M. Heywood, A. S. Havaranis, and H. Herrmann, *J. Cell. Physiol.* **82**, 319 (1973).
[43] R. E. Mains and P. H. Patterson, *J. Cell Biol.* **59**, 329 (1973).

# [10] Monolayer Culture Techniques

*By* JAMES A. MCATEER and WILLIAM H. J. DOUGLAS

## Introduction

Monolayer cell cultures are frequently established from single cell suspensions prepared by the enzymic dissociation of organ fragments. The cell preparation is inoculated into a culture vessel containing fluid medium and incubated in a controlled atmosphere. Viable cells settle and attach to the substrate within several hours. The resultant primary culture is a mixed cell population which contains many of the cell types present in the tissue of origin. In order to obtain monolayer cultures consisting of a single cell type, it is necessary to isolate that cell from the primary preparation. Cell separation and isolation methods such as density gradient centrifugation,[1] electrophoresis,[2] or affinity column separation[3] can be applied to the initial cell suspension prior to culture. Alternatively, the cell population can be enriched using *in vitro* methods that select cell types based on their attachment or growth characteristics. For example, fibroblasts and macrophages attach to culture surfaces more rapidly than other cells and thus they can be removed by selective adherence techniques.[4] Fibroblasts replicate faster than most cell types. In mixed cell cultures they tend to overgrow the population and obscure other cells of interest. Dilution plating at clonal density[5] is often used to separate cell types so that colonies which develop may be physically isolated and then subcultured as pure populations.

The cells routinely studied in monolayer culture can be derived in the laboratory from primary cultures, or obtained from commercial suppliers and nonprofit organizations such as the American Type Culture Collection[6] and Institute for Medial Research.[7] The most commonly studied, well-characterized cell lines include the WI-38 and IMR-90 (human diploid, fibroblastic), 3T3 (mouse embryonic, fibroblastic), and HeLa (hu-

[1] T. G. Pretlow II, *Int. Pathol.* **16,** 42 (1975).

[2] R. C. Boltz, Jr., P. Todd, M. J. Streibel, and M. K. Louie, *Prep. Biochem.* **3,** 383 (1973).

[3] S. F. Schlossman and L. Hudson, *J. Immunol.* **110,** 313 (1973).

[4] F. H. Kasten, *in* "Tissue Culture: Methods and Applications" (P. F. Kruse, Jr. and M. K. Patterson, Jr., eds.), p. 72. Academic Press, New York, 1973.

[5] R. D. Cahn, H. G. Coon, and M. B. Cahn, *in* "Methods in Developmental Biology" (F. H. Wilt and N. K. Wessels, eds.), p. 493. Crowell (Collier), New York, 1967.

[6] The American Type Culture Collection. Rockville, Maryland, 20852.

[7] The Human Genetic Mutant Cell Repository. Inst. Med. Res., Camden, New Jersey, 08103.





ISBN 0-12-181958-2

man cervical adenocarcinoma, epithelioid) lines. Numerous other cell lines from many vertebrate, insect, and plant species have been established, characterized, and catalogued, and specific techniques for them are presented throughout this volume.

Basic monolayer cell culture techniques are well established and accessible. Publications are available that detail culture procedures[8] and techniques courses are sponsored by organizations such as the Tissue Culture Association.[9]

The information presented in this article provides background, resources, and practical procedures for the investigator interested in establishing a research program involving monolayer cell culture.

## Properties of Monolayer Cultures

Most cell populations maintained in monolayer culture form multiple cell layers and in the case of smooth muscle, form large, grossly visible colonies. Only a few cell types, e.g., vascular endothelial and 3T3 cells, grow as true monolayers.

Cell types which require attachment to a rigid substrate in order to replicate *in vitro* are classified as anchorage-dependent cells. These include most, but not all, cells maintained in monolayer culture. Many abnormal cell types, e.g., transformed and neoplastic cells, may grow well in fluid suspension culture or in a soft agar matrix.

The growth kinetics of cells in monolayer culture follow a characteristic pattern. Following seeding the cells undergo a quiescent period (lag phase) during which there is no cell division. The duration of the lag phase is dependent, in part, upon the cell type, seeding density, media composition, and previous handling of the cells. The cells then enter a log phase of growth in which there is an exponential increase in cell number. During the log phase the cells exhibit their highest metabolic activity. When culture conditions will no longer support cell division the population enters a stationary phase, during which the cell number remains constant. The attainment of stationary phase is largely dependent upon nutritional factors within the system and can occur before or after confluency is reached.

The term "confluent" describes a cultured population which occupies all available growth surface. If nutritional conditions are adequate, a confluent cell population may continue to replicate in log phase. In practice, however, it is not wise to maintain cells beyond the point of confluency

[8] Tissue Culture Association Manual: Techniques, Methods and Procedures for Cell, Tissue and Organ Culture. Tissue Cult. Assoc., Rockville, Maryland, 20852.

[9] Course Secretary, W. Alton Jones Cell Science Center, Lake Placid, New York.

since essential metabolites are rapidly depleted at high cell density. Some cell types will not replicate beyond confluency and are described as being contact inhibited. "Contact inhibition" is a widely misused term which encompasses the phenomena of contact inhibition of movement[10] and contact inhibition of replication.[11] These are separate, probably unrelated, aspects of cell behavior. Contact inhibition of movement describes the well-documented condition that occurs when the directional movement of a cell changes upon contact with another cell. Contact inhibition of replication is the cessation of growth due to physical crowding of cells at confluency. This occurrence is not well documented and in most cases may be a misinterpretation of a well-known cellular response to nutrient depletion more correctly described as a density-dependent inhibition of growth.[12]

Cultured cells can become "transformed." Cellular transformation is a spontaneous or induced change, e.g., viral transformation, in the characteristics of a cell type expressed as an abnormal karyotype, a change in morphology, or alteration in cell behavior or growth potential. Spontaneous cellular transformation often occurs with extended time in culture. Since a transformed cell population will no longer respond characteristically to experimental conditions, the age of the cells at the onset of transformation is an important parameter. The age of a culture is usually expressed as either its passage number or number of population doublings. Passage number refers to the number of times the cells have been subcultured since the time of explantation. This is not a precise estimate of cell age for it is affected by numerous variables including: the number of cells replanted at each passage, cell survival after inoculation, and the cell density (i.e., confluency, semiconfluency) at subcultivation. The number of population doublings is a more critical estimate of cell age since it accounts for the time required for the cells to double in number. The number of population doublings a culture has undergone is calculated by knowing the number of cells originally planted and the cell number at the next subcultivation.

Normal diploid cells *in vitro* have a finite lifespan, while transformed cells apparently possess unlimited proliferative potential. The process of *in vitro* cellular aging is considered to be a deterioration of the cellular processes that are necessary to support continued replication. As a cultured cell population ages its population doubling time progressively increases until proliferation eventually ceases. During the process of sene-

[10] M. Abercrombie and J. E. M. Heaysman, *Exp. Cell Res.* **6,** 293 (1954).

[11] L. N. Castor, *J. Cell. Physiol.* **72,** 161 (1968).

[12] M. G. P. Stoker and H. Rubin, *Nature (London)* **215,** 171 (1967).

sence, cells undergo alterations such as chromosomal aberrations and changes in membrane structure and permeability.

**Environmental Factors in Cell Culture**

The physical culture environment should be well defined and precisely controlled. Numerous factors influence the growth of cultured cells including temperature, pH, osmolality, humidity, and the composition of the overlying gas phase. The control of these variables is essential for interpretation of cellular response and requires an understanding of not only the cellular requirement for each factor, but the interactions between factors. The overall culture milieu must be constructed with the knowledge that cells *in vitro* continuously alter their environment. Cells utilize nutrients and thus deplete the media of certain constituents. They also produce metabolites and "waste products," which can interact with medium components to degrade growth-promoting factors or alter pH. The range of proper incubation temperature is usually quite narrow, and for most cells from homeothermic animals it is between 35–37°. Cells can survive for extended periods at a lower than optimal temperature but do not tolerate high temperatures. Lower temperatures depress cellular metabolism, usually without deleterious effect. Cultures remain viable at room temperature for periods of up to several days, long enough to allow shipment through the postal system of living cells that have been properly packaged. Cultures derived from poikilothermic animals are normally incubated at lower temperatures. Cells from cold-water fishes[13] are cultured at 15–20°, warm-water fishes at 22–27°, and insect cells[14] have been cultured in a wide temperature range, 23–37°. (See this volume [39] and [40]).

Optimum pH of the medium is usually within the range 7.2–7.5, though some cells will survive at the extremes, pH 6.6–7.8. The conditions which support maximum growth in turn range ±0.5 units. The determination of proper pH within a culture system is dependent upon both the cell type and the specific physiological parameter being assessed. For example, a certain cell type may exhibit its greatest secretory activity at a pH different from that which supports the most rapid growth.

The maintenance of pH involves the interaction of several factors, most importantly the buffer system and the overlying gas phase. Cells in culture have a metabolic requirement for both bicarbonate ion and $CO_2$. Some cells produce sufficient $CO_2$ as a metabolic by-product to satisfy this

[13] K. Wolf and M. C. Quimby, *Tissue Cult. Assoc. Man.* **2,** 445 (1967).

[14] J. L. Vaughn, *in* "Invertebrate Tissue Culture" (C. Vago, ed.), Vol. 1, p. 3. Academic Press, New York, 1971.

requirement, however, bicarbonate must be added to the medium. Many media are buffered with some type of bicarbonate buffer system. These contain bicarbonate in excess of that needed by the cells, and require $CO_2$ from the gas phase to activate the buffer and establish the bicarbonate $\leftrightarrows$ $CO_2$ equilibrium. The $pCO_2$ required to equilibrate the buffer is dependent upon the bicarbonate concentration, but usually it falls in the range of 12–38 mm Hg. With most commercially available tissue culture media this requires an atmosphere of approximately 5% $CO_2$. Media can be constructed with organic buffers which do not require equilibration with $CO_2$.[15] These buffers make it possible to culture within closed systems in which the $O_2$ requirement is easily met by the ambient air (gas phase) within the vessel. Recently, however, the use of organic buffers has been criticized. Reports indicate that certain organic buffers may be toxic to cultured cells.[16]

Cultured cells require oxygen. A $pO_2$ of 15 to 75 mm Hg is usually adequate for monolayer culture. Open culture systems (Petri dishes, or flasks with loosened caps) are usually incubated in a 95% air–5% $CO_2$ mixture. Higher oxygen concentrations can be detrimental, and $O_2$ levels only slightly higher than ambient have been shown to be toxic to some cultured cells.[17]

A very high relative humidity (>98%) must be maintained when working with open culture systems. Since cells in monolayer are submerged in fluid medium there is no danger of drying at the cell surface, although evaporation from the medium must be avoided. The humidity of the incubator should be in equilibrium with the gas phase immediately overlying the media.

Other environmental factors are known to influence monolayer culture. Visible light has been shown to affect cells in culture. Direct metabolic effects are known, as well as light-induced production of toxic compounds in some media.[18] Cells should be cultured in the dark and should be exposed to room lighting as little as possible. Vibration in the laboratory may also be detrimental to certain systems. Cells in clonal growth are particularly sensitive to vibration.

The chemical composition of the culture medium is of obvious importance, not only from a nutritional standpoint (detailed in this volume [5] but with regard to osmolality. Media osmolality for mammalian cell

[15] N. E. Good, G. D. Winget, W. Winter, T. N. Connolly, S. Izawa, and R. M. M. Singh, *Biochemistry* **5,** 467 (1966).

[16] H. Eagle, *Science* **174,** 500 (1971).

[17] A. Mizrahi, G. V. Vosseller, Y. Vagi, and G. E. Moore, *Proc. Soc. Exp. Biol. Med.* **139,** 118 (1972).

[18] R. J. Wang, *In Vitro* **12,** 19 (1976).

culture is usually held in the range of 280–300 mosmol/kg. The optimal value may depend upon the specific cell type.

### Substrates for Monolayer Culture

Cells in monolayer culture require a substrate for attachment. Ideally this surface should be nontoxic, biologically inert, and optically transparent. It must allow cell attachment and cell movement required for replication and migration during growth.

The interaction between cell and substrate is an important factor in the status of the culture system. If the two are incompatible, the culture will not survive. Cells may not adhere at all to some surfaces while other substrates may allow attachment but inhibit growth.

A variety of materials have been used successfully as culture substrates. Cells can be grown on cellophane, carbon or collagen-coated glass and plastic, Teflon, silicone rubber, polycarbonate and cellulose ester (Millipore) filters, and many other surfaces. Routinely most laboratories use commercially available borosilicate glass or specially treated polystyrene plastic culture ware. Plastic tissue culture products are more popular because of the problems encountered with washing and rinsing glassware. The plastics used for monolayer culture are specially treated during manufacture to provide a hydrophilic surface which promotes both cell attachment and growth. These products are designated "culture grade" as opposed to "bacteriological grade" vessels which will not support the growth of cultured cells. Under certain culture conditions even culture grade plastics are not suitable for cell growth. When diploid cells are seeded at clonal density in medium containing a low serum concentration plating efficiency is poor. Pretreatment of the plastic culture vessel with polylysine permits efficient clonal growth.[19] (See also [5]).

Various types of culture vessels are available for specific applications. Those designed for routine monolayer culture are made with a uniform, transparent growth surface to facilitate their use with an inverted microscope. These include the petri dish, multi-well plate, screw-cap culture flask (T-flask), and Leighton tube.

Monolayer culture systems used primarily in industry for the mass propagation of anchorage-dependent cells are designed to provide the maximum surface area for growth. The roller bottle[20] is used for this purpose. The system consists of a cylindrical borosilicate glass or polycarbonate plastic bottle which rests horizontally on a roller device

[19] W. L. McKeehan and R. G. Ham, *Cell Biol.* **71,** 727 (1976).

[20] W. L. Whittle and P. F. Kruse, Jr., *in* "Tissue Culture: Methods and Applications" (P. F. Kruse, Jr. and M. K. Patterson, Jr., eds.), p. 327. Academic Press, New York, 1973.

within an incubator or controlled-temperature room. The bottle is partially filled with medium, and as it turns the cells adherent to its walls are alternately bathed in the fluid medium and exposed to the gas phase.

## Routine Cell Culture Procedures

Subculturing or passaging cells involves detaching the cells from their substrate and transferring them to new culture vessels. This is done when cells have utilized the growth surface available to them or have reached a population density which suppresses their growth. Since the replicative capacity of a cultured cell population declines when confluency is reached, cells are often passaged at "semiconfluency," when they are still in log-phase growth.

Various methods can be used to remove cells from the culture surface. Mechanical means such as scraping with a silicone rubber spatula removes the cells in clumps or irregular patches. This is somewhat disruptive to the cells and usually results in a poor recovery. Dissociation with a proteolytic enzyme such as trypsin,[21] collagenase,[22] or pronase[23] gives a better cell yield and much higher plating efficiency. Enzymes are usually prepared in a $Ca^{2+}$–$Mg^{2+}$ free balanced salt solution. They are used separately, in combination, or with the addition of a divalent cation chelating agent such as EDTA. Enzyme concentrations of 0.05–0.5% are used, and crude enzymes are usually preferred. (See also this volume [9]).

Enzymic dissociation damages cells, especially the cell surface. Thus it is desirable to minimize the duration of contact between the cells and enzyme solution. It is also wise to avoid harsh physical manipulation of the cells while in the presence of enzymes, although gentle pipetting is usually necessary to prepare a monodisperse suspension.

The following protocol outlines one procedure for passaging cells in monolayer culture.

### *Dissociation*

1. Remove and discard the culture medium.
2. Wash the cell sheet thoroughly with warm Moscona's saline[24]; discard the fluid.

[21] C. Shipman, Jr., *in* "Tissue Culture: Methods and Applications" (P. F. Kruse, Jr. and M. K. Patterson, Jr., eds.), p. 5. Academic Press, New York, 1973.

[22] S. R. Hilfer, *in* "Tissue Culture: Methods and Applications" (P. F. Kruse, Jr. and M. K. Patterson, Jr., eds.), p. 16. Academic Press, New York, 1973.

[23] R. B. L. Gwatkin, *in* "Tissue Culture: Methods and Applications" (P. F. Kruse, Jr. and M. K. Patterson, Jr., eds.), p. 3. Academic Press, New York, 1973.

[24] A. A. Moscona, *Exp. Cell Res.* **3,** 535 (1952).

3. Add warm enzyme solution (collagenase 0.1%, trypsin 0.1%, chicken serum 1% in Moscona's saline) (1 ml/5 $cm^2$ culture surface). After 1 min withdraw the solution, leaving only a very thin layer of fluid covering the cells.

4. Observe the cells with an inverted phase-contrast microscope; note distribution and predominant cell morphology.

5. Cap the flask tightly and incubate at 37° for 3–5 min.

6. Again observe the cells with a phase microscope; look for free-floating cells and attached cells which have pulled in their processes and are rounding up. If the majority of cells do not appear to be detached return the flask to the incubator.

7. After several minutes of incubation hold the flask horizontally and give its side a sharp rap with the hand. This will help detach loosely adherent cells. Often the cells loosen in sheets and these can be seen to drift across the substrate surface when the flask is tipped at an angle. Observe the cells by phase microscopy. Repeat if necessary.

8. Add several milliliters of culture medium to the flask. Use a pipette to gently triturate the cells.

*Seeding*

9. A viable cell count[25] (dye exclusion) is then performed on the suspension so that other flasks can be seeded at a known cell density.

10. Cell dilutions are made and aliquots seeded to new culture vessels containing warm (equilibrated) medium.

11. The newly seeded culture vessels are incubated undisturbed to facilitate cell attachment.

## Morphological Evaluation of Monolayer Cultures

Many cultured cells possess distinctive morphological characteristics observable by phase-contrast microscopy. This enables the investigator to identify specific cell types in mixed cultures and to carry out procedures such as clonal isolations. Routine observations by phase microscopy are essential for proper assessment of a cell population. The morphology of cultured cells may change in response to slight alterations in culture conditions. An abnormal morphological change in the cell population is often the first indication that a problem exists within the system. Therefore, it is useful to frequently monitor culture morphology and to keep a record of the observations. A Polaroid-type camera back affixed to an inverted microscope is very useful for this purpose.

[25] H. J. Phillips, *in* "Tissue Culture: Methods and Applications" (P. F. Kruse, Jr. and M. K. Patterson, Jr., eds.), p. 406. Academic Press, New York, 1973.

Fixed and stained cell preparations of various types are used for morphological analysis of monolayer cultures. Clonal plates are prepared for gross visual examination by fixing the cultures *in situ* (buffered 2.5% glutaraldehyde) and then staining directly with 0.5% crystal violet. For light microscopical analysis of monolayer cultures *in situ* cells can be grown on coverslips, fixed with Bouin's solution, stained with hematoxylin and eosin, and mounted on microscope slides.[26] This provides good resolution, and cell structure is well preserved.

Critical morphological analysis of cultured cells by high-resolution light microscopy and transmission electron microscopy necessitates embedding and sectioning of the cultures. The cells need not be removed from their substrate for processing, and all steps through embedding can be carried out within the culture vessel. Since monolayer cultures are only one to several cell layers in thickness, fixation and dehydration are rapid, and infiltration with resin is uniform. Conventional preparation methods can be used with the precaution of avoiding solvents that might dissolve the tissue culture vessel. Processing cells *in situ* does not hinder sectioning since resin will infiltrate between the cells and substrate allowing them to be separated following polymerization.[27] The tissue block can then be oriented for enface or transverse sectioning. Certain types of resins may work better with specific brands of plastic culture ware. One reliable combination is the use of an Epon mixture[28] with Lux or Corning plastic surfaces. Processing cells grown in glass dishes is more difficult because polymerized resin does not readily separate from the dish. Cells grown on glass coverslips are easier to work with. If the coverslip is embedded so that resin contacts only the side which bears cells the coverslip can be separated from the polymerized plastic by immersing them repeatedly in liquid nitrogen and then boiling water.

The preparation of cultured cells for scanning electron microscopy presents no unusual difficulties. Processing by the critical-point drying method is recommended.[29] One necessary precaution is to assure that the cells are grown on a substrate that can be conveniently placed in the chamber of the critical-point dryer. Alternatively, pieces of the culture surface must be cut from the culture vessel without damaging the cells. The cells may then be fixed, critically point dried, and coated by routine methods.

[26] J. Fogh and J. A. Sykes, *In Vitro* **7,** 206 (1972).

[27] W. H. J. Douglas, E. P. Dougherty, and G. W. Phillips, *Tissue Cult. Assoc. Man.* **3,** 581 (1977).

[28] C. C. Haudenschild, R. S. Cotran, M. A. Gimbrone, Jr., and J. Folkman, *J. Ultrastruct. Res.* **50,** 22 (1975).

[29] K. R. Porter, D. Kelley, and P. M. Andrews, *Proc. Annu. Stereoscan Symp., 5th, 1972* p. 1 (1972).

# [11] Measurement of Growth and Viability of Cells in Culture

*By* MANFORD K. PATTERSON, JR.

A variety of techniques has been proposed for the quantitation and measurement of viability of cells grown in tissue culture. Only a few have emerged as being suitable for routine purposes. Since the preparation of tissues for cell culture or cells for subculture is metabolically destructive, the use of cell count alone for quantitation may be misleading. A desired adjunct for growth studies, therefore, is a determination of the viable cell population.

Cell culture measurements can be divided into four major catagories: (1) visual methods which employ the use of light microscopy and devices commonly used in hematology; (2) chemical methods which employ commonly used analytical biochemical procedures adapted to tissue culture; (3) electronic systems using flow-through cells or apertures for measurement of incorporated dyes or cell numbers; and (4) an array of miscellaneous procedures.

## Cell Preparation

The basic adaptations of analytical biochemical procedures for tissue culture material have been for increased sensitivity because of the limited amount of tissue available and to accomplish *in situ* dissolution of the cells. The latter has been necessary to avoid the cellular alterations caused by the hydrolytic enzymes used to dissociate or detach cells although, under certain circumstances, cells can be detached[1] or dissociated[2] by nonenzymic procedures.

In the procedures to be described it should be noted that often more than one measurement can be made on a single preparation of cells. For example, cells prepared for direct nuclei counting or for nucleic acid analysis also can be used for protein determination.

The handling of cells prior to analysis is extremely important for a variety of reasons, namely: (1) The exogenous growth media may contain components that are being assayed in the cell, e.g., proteins that adsorb to cells. (2) Leakage of cellular components may occur if cellular integrity is lost at this point. (3) Loss of cells may result from severe mechanical handling. (4) Use of hydrolytic enzymes may lead to the loss of measura-

[1] M. K. Patterson, Jr., *in* "Tissue Culture: Methods and Applications" (P. F. Kruse, Jr. and M. K. Patterson, Jr., eds.), p. 192. Academic Press, New York, 1973.

[2] C. Waymouth, *In Vitro* **10,** 97 (1974).




ISBN 0-12-181958-2

TABLE I
BALANCED SALT SOLUTIONS[a]

| | Dulbecco[b] | Earle[c] | Gey[d] | Hanks[e] | Puck F[f] | Puck G[f] |
|---|---|---|---|---|---|---|
| $CaCl_2$ | 0.10[g] | 0.20[g] | 0.17[g] | 0.14[g] | | |
| $CaCl_2 \cdot 2\ H_2O$ | | | | | 0.016 | 0.016 |
| KCl | 0.20 | 0.40 | 0.37 | 0.40 | 0.285 | 0.40 |
| $KH_2PO_4$ | 0.20 | | 0.03 | 0.06 | 0.083 | 0.150 |
| $MgCl_2 \cdot 6\ H_2O$ | 0.10[g] | | 0.21 | | | |
| $MgSO_4 \cdot 7\ H_2O$ | | 0.20 | 0.07 | 0.20 | 0.154 | 0.154 |
| NaCl | 8.0 | 6.80 | 7.00 | 8.00 | 7.40 | 8.0 |
| $NaHCO_3$ | | 2.20 | 2.27 | 0.33 | 1.20 | |
| $Na_2HPO_4 \cdot 7\ H_2O$ | 2.16 | | 0.226 | 0.09 | 0.29 | 0.29 |
| $NaH_2PO_4 \cdot H_2O$ | | 0.14 | | | | |
| Glucose | | 1.00 | 1.00 | 1.00 | 1.10 | 1.10 |
| Phenol red | | 0.01[g] | | 0.01[g] | 0.0012 | 0.0012 |

[a] Usually prepared as a ×10 solution and diluted prior to filtration. Cell suspensions are generally washed with calcium- and magnesium-free solutions. Concentrations given are g/liter, ×1 solution.
[b] Sterilize by autoclaving; R. Dulbecco and M. Vogt, *J. Exp. Med.* **99,** 167 (1954).
[c] Sterilize by filtration; J. C. Bryant, *Tissue Cult. Assoc.* **1,** 185 (1975).
[d] Sterilize by filtration; G. O. Gey and M. K. Gey, *Am. J. Cancer* **27,** 45 (1936).
[e] Sterilize by filtration; J. H. Hanks, *Tissue Cult. Assoc.* **1,** 3 (1975).
[f] Sterilize by filtration; T. T. Puck, S. J. Ciecura, and A. Robinson, *J. Exp. Med.* **108,** 945 (1958).
[g] Prepare as separate solution and mix prior to use; may be omitted if solution is to be used for wash.

ble materials. Proper handling of cultures requires careful attention to pH, temperature, and osmotic pressure during washing, and a balance between adequate removal of growth media without loss of cells.

Choice of washing solutions appears to be a simple matter of preference. The basic formulas for the most commonly used are given in Table I. In general the choice should be dictated by the salt solution used as the base for the growth medium. Moreover, calcium-free salt solutions are generally preferable for all cell suspensions since this discourages cell aggregation. A comprehensive listing of balanced salt solutions has been published.[3]

## Growth Measurements

### *Visual Methods*

The most commonly used measurement of growth is direct enumeration of cells employing a hemocytometer. This method is best applied

[3] C. Waymouth, *in* "Cell Biology" (P. L. Altman and D. D. Katz, eds.), Vol. I, p. 61. FASEB, Bethesda, Maryland, 1976.

when only a few samples are to be counted or quantitative information on cell viability is desired. If numerous cultures are to be counted on a routine basis, electronic enumeration should be considered.

### Hemocytometer Counting[4]

*Reagents and Equipment*

*Balanced salt solution–see Table I*
*Hemocytometer–double chamber with Neubauer rulings*
*Microscope–with 16-mm objective and* ×10 ocular
Hand counter—tally register

*Procedure.* Cell suspensions (direct or prepared from tissue or monolayers) are diluted with balanced salt solution to contain approximately 300,000–500,000 cells/ml. Taking care not to overfill, the cell suspension is added to both chamber sides of the hemocytometer using a pipette or dropper. The chamber coverslip should be firmly in place.

Any number of squares in the hemocytometer can be counted. Reproducibility is the key factor so that sufficient squares should be counted to obtain a count statistically representative of the population. Each large square (there are nine on each chamber side) represents an area of 1 $mm^2$ and a depth of 0.1 mm, i.e., a volume of 0.1 $mm^3$. As an example, in 10 1-$mm^3$ squares (the four corner and center squares of each chamber side), 200 total cells are counted. The original cell suspension had been diluted 1:10; therefore, 200 cells × 10 (dilution) × 1000 ($mm^3$ in 1 $cm^3$) equals $2 \times 10^6$ cells/ml in the original cell suspension.

This procedure is also applicable to the direct enumeration of cell nuclei and the dye exclusion test for viable cells. Some cells, e.g., human diploid cells, are difficult to count directly, and a staining procedure is incorporated to enhance visibility of the nuclei.

### Nuclei Counting

*Reagents*

*Citric acid–0.1 M* (1.9212 g/100 ml distilled water) to which a few crystals of thymol are added
Crystal violet—0.2% dissolved in 0.1 *M* citric acid
Balanced salt solution—see Table I

*Procedure.* Monolayer cultures are rinsed 3 times with balanced salt solution (this is optional depending upon media and subsequent use, e.g., if citric acid extract is to be used also for protein determination) and

[4] M. Absher, *in* "Tissue Culture: Methods and Applications" (P. F. Kruse, Jr. and M. K. Patterson, Jr., eds.), p. 395. Academic Press, New York, 1973.

drained. Two milliliters of 0.1 *M* citric acid are added to a T-25[5] flask (5.0 ml for T-75) and allowed to stand at room temperature. The time can vary but a minimum of 10 min has been found adequate for disruption of all cells that have been tested. Frequent shaking of the flask during incubation can aid in disrupting the cells but is not always necessary, since the final suspension is prepared by scraping the flask surface.

Note: Protein analysis may be performed on this extract by adding Lowry's reagent directly to an aliquot (see "Chemical Methods").

A 0.1-ml aliquot is diluted to 1.0 ml with 0.1 *M* citric acid and 1.0 ml of the crystal violet solution added. The resulting suspensions of muclei have been found stable for up to 1 week at 4°.

The suspended nuclei are counted in a hemocytometer; calculations are made as described.

### *Chemical Methods*

The relationship of protein and nucleic acid biosynthesis to the growth of cells has led to the measurement of these components as an expression of cell growth. However, variations in their content occur, and a linear relationship is not always found. The validity of this type of assay should be established by comparison with the more direct visual methods.

#### Protein Determination

Oyama and Eagle[6] were the first to adapt the Lowry[7] assay for protein to measurement of cells in tissue culture. Subsequent studies have shown wide variations in ratio of protein per cell values (Table II).

*Reagents*

- Balanced salt solution—see Table I
- Lowry solution A—20 g $Na_2CO_3$ and 4 g NaOH pellets dissolved to give a final volume of 1 liter (store in plastic bottle)
- Lowry solution B—Equal parts of 1% $CuSO_4 \cdot 5\ H_2O$ and 2.7% sodium potassium tartrate
- Lowry solution C—100 ml solution A, 2 ml solution B
- Phenol reagent (Folin and Ciocalteau)—1 *N* (usually purchased as a 2 *N* solution)
- Human serum albumin, crystallized—Stock solution containing 200 $\mu$g/ml

[5] Tissue culture flasks are available in various sizes constructed of glass or plastic. The "25" represents 25 $cm^2$ surface area.

[6] V. I. Oyama and H. Eagle, *Proc. Soc. Exp. Biol. Med.* **91,** 305 (1956).

[7] O. H. Lowry, N. J. Rosebrough, A. L. Farr, and R. S. Randall, *J. Biol. Chem.* **193,** 265 (1951).

TABLE II
PROTEIN CONTENT OF CELLS IN TISSUE CULTURE

| Cell line | Days after plating 0 | 1 | 2 | 3 | 4 | 5 | References |
|---|---|---|---|---|---|---|---|
| | | | pg/cell | | | | |
| L-5 (monolayer) | *320*[a] | *550* | *480* | *450* | *370* | *350* | b |
| L-5 (suspension) | *350* | *500* | *500* | *500* | *330* | *330* | b |
| HeLa | | 930 | 1010 | 920 | 740 | | c |
| Human leukemia | | 810 | 850 | 780 | 640 | | c |
| Intestinal epithelium | | 890 | 960 | 1010 | 790 | | c |
| WI-38 | *1150* | *940* | *800* | *400* | *390* | *410* | d |
| Jensen sarcoma | *1100* | *700* | *710* | *720* | *400* | *390* | d |
| WISH | *350* | *320* | *300* | *320* | *310* | *320* | d |
| HEp-2 | *390* | *360* | *350* | *410* | *400* | *350* | d |
| Chang (suspension) | *400* | *220* | *180* | *175* | *175* | | e |
| HeLa (suspension) | *230* | *175* | *150* | *150* | *100* | | e |
| CF-3 | | 872 | 787 | 498 | 545 | | f |
| IMR-91 | | 860 | 678 | 588 | 446 | | f |
| Human fetal lung | | | | | | | g |
| 20–29 doublings | | 490 | | | | | |
| 30–39 doublings | | 480 | | | | | |
| 40–49 doublings | | 883 | | | | | |
| 50–59 doublings | | 1287 | | | | | |

[a] Italicized values are extrapolated from curves; data given in picograms per cell.
[b] A. Tsuboi, T. Kurotsu, and T. Terasima, *Exp. Cell Res.* **103**, 257 (1976).
[c] V. I. Oyama and H. Eagle, *Proc. Soc. Exp. Biol. Med.* **91**, 305 (1956).
[d] E. Miedema and P. F. Kruse, Jr., *Biochem. Biophys. Res. Commun.* **20**, 528 (1965).
[e] P. Volpe and T. Eremenko-Volpe, *Eur. J. Biochem.* **12**, 195 (1970).
[f] R. T. Dell'Orco, unpublished.
[g] E. L. Schneider and S. S. Shorr, *Cell,* **6,** 179 (1975); days not specified; cell cultures either in logarithmic growth ($<1.3 \times 10^4$ cells/cm$^2$) or when the monolayer reached light confluency ($1.3–2.6 \times 10^4$ cells/cm$^2$).

*Procedure.* Cell monolayers or suspensions are washed 3 times with balanced salt solutions and allowed to drain by inversion. Sufficient Lowry solution C is added to cover the cell layer, e.g., 5 ml for a T-25 flask, and it is then incubated. Incubation conditions should be sufficient to dissolve the cellular material. Generally, 37° for 1 hr is sufficient or temperatures up to 60° may be used for 5 min. (Following dissolution the solutions may be stored at 4° for several days.) The intensity of the color produced by the reaction of the alkaline copper-tartrate solution C is indicative of the aliquot required to obtain a final readable color. If, for example, no color is detectable in the 5-ml solution in the T-25 flask, then

2- to 3-ml aliquots are required; a deep blue color suggests a 1:2 or 1:10 dilution with solution C. A final volume of 5 ml of sample in solution C plus 1 ml distilled water or standard (0–200 $\mu$g) receives 0.5 ml of phenol reagent, jetted to obtain rapid admixture. A digest from cell-free controls is best used as a blank. Thirty minutes after addition of the phenol reagent, the absorbance of the samples is measured with a spectrophotometer at 660 nm. Color is stable for at least 2 hr. Absorbance is compared with the absorbance values of standards in the range of 0–200 $\mu$g.

*Limitations.* A large number of compounds including organic buffers, sucrose,[8] glycerol,[8,9] Tris,[10] and certain reducing agents[11] interfere with the Lowry protein method, and special sample treatment may be necessary when such reagents are used in cultures.

## DNA Determination

As with protein, the DNA content per cell depends upon the stage of growth and the phase of the cell cycle.[12–15] It is therefore advisable to equate the results of DNA determination with other methods of evaluation. The procedures in general use for tissues are adapted to cells derived from cell culture systems. Several critical steps exist in the estimation of nucleic acid,[16] the majority of which concern extraction and hydrolysis. Since one is dealing with small amounts of tissue, special care must be exercised if quantitation is to be achieved. The disintegration of the cell sample is a critical step in obtaining an accurate determination of nucleic acids because of the widespread presence of nucleases. Two general procedures exist: (1) removal of the cells from the solid matrix, or (2) *in situ* extraction. An example of the former uses a freeze–thaw procedure which separates a particulate fraction containing the cellular DNA and a soluble fraction containing cytoplasmic enzymes[13] prior to removal of interfering substance by cold acid precipitation of the nucleic acids. An example of *in situ* extraction can be found in the procedure to be described (cf. also Setaro and Morley[17]).

[8] A. Bensadoun and D. Weinstein, *Anal. Biochem.* **70,** 241 (1976).
[9] P. Blümel and W. Uecker, *Anal. Biochem.* **76,** 524 (1976).
[10] R. Rej and A. H. Richards, *Anal. Biochem.* **62,** 240 (1974).
[11] F. Higuchi and F. Yoshida, *Anal. Biochem.* **77,** 542 (1977).
[12] E. L. Schneider and S. S. Shorr, *Cell* **6,** 179 (1975).
[13] A. Leyva, Jr. and W. N. Kelley, *Anal. Biochem.* **62,** 173 (1974).
[14] V. J. Cristofalo and D. Kritchevsky, *Med. Exp.* **19,** 313 (1969).
[15] R. Yanishevsky, M. L. Mendelsohn, B. H. Mayall, and V. J. Cristofalo, *J. Cell. Physiol.* **84,** 165 (1974).
[16] H. N. Munro and A. Fleck, *Methods Biochem. Anal.* **14,** 113 (1966).
[17] F. Setaro and C. G. D. Morley, *Anal. Biochem.* **71,** 313 (1976).

Methods of analysis of hydrolyzed extracts for DNA vary widely. Most have used the colorimetric reagent, diphenylamine, which is sensitive to about 10 μg or approximately $10^6$ cells. One method, based on the fact that an antibiotic, mithramycin, binds to DNA and fluoresces at 540 nm in direct proportion to the DNA present, reportedly can detect as little as 0.5 μg or $5 \times 10^4$ cells.[18] Other microfluorometric methods using antibiotics,[19] ethidium bromide,[20-22] and D-aminobenzoic acid[17,23,24] have been reported. Table III gives the literature values for DNA content of cells grown in tissue culture.

*Reagents*

Balanced salt solution—see Table I
Perchloric acid, 1.0 *N*; dilute 8.62 ml 70–72% reagent grade to 100 ml
Sodium hydroxide, 0.3 *N*; 1.2 g/100 ml
Diphenylamine—dissolve 1.5 g in 100 ml glacial acetic acid; add 1.5 ml concentrated $H_2SO_4$; mix fresh and store in dark until used
Acetaldehyde, 16 mg/ml; dilute 0.5 ml to 25 ml with distilled water; store at 4°
Burton reagent—just before use add 0.1 ml of acetaldehyde solution to each 20 ml of diphenylamine reagent required
DNA standard—sodium desoxynucleate, dissolve 0.2 mg/ml of 5 m*M* NaOH; store at 4°; mix equal volume with 0.5 *N* perchloric acid for a final concentration of 100 μg/ml

*Extraction Procedure (for T-25 flask).* Wash cell layer 3 times with balanced salt solution. Wash once with 5 ml cold 0.2 *N* perchloric acid. Add 5 ml cold 0.2 *N* perchloric acid and let stand 10 min at 4°; decant. Add 2 ml of 0.3 *N* NaOH and let stand 1 hr at 37°; transfer to 15-ml centrifuge tube and cool in ice. Add 2 ml of cold 1.0 *N* perchloric acid and let stand 10 min at 4° to precipitate DNA and protein; centrifuge (supernatant contains RNA fraction). Wash the pellet with 2 ml cold 0.2 *N* perchloric acid; centrifuge. Resuspend pellet in 2 ml of 0.5 *N* perchloric acid and heat in water bath for 15 min at 90° to hydrolyze DNA. Centrifuge and retain supernatant liquid. Wash the pellet with 2 ml 0.5 *N* perchloric acid; centrifuge and combine supernatant fluid with that obtained in the immedi-

[18] B. T. Hill and S. Whatley, *FEBS Lett.* **56,** 20 (1975).
[19] B. T. Hill, *Anal. Biochem.* **70,** 635 (1976).
[20] U. Karsten and A. Wollenberger, *Anal. Biochem.* **46,** 135 (1972).
[21] U. Karsten, *Anal. Biochem.* **77,** 464 (1977).
[22] M. J. Blackburn, T. M. Andrews, and R. W. E. Watts, *Anal. Biochem.* **51,** 1 (1973).
[23] W. Y. Fujimoto, J. Teague, and R. H. Williams, *In Vitro* **13,** 237 (1977).
[24] H. L. Hosick, *Tissue Cult. Assoc. Man.* **1,** 45 (1975).

TABLE III
DNA CONTENT OF CELLS IN TISSUE CULTURE

| Cells | DNA (pg/cell) | Method | Reference |
|---|---|---|---|
| WI-38 passage 20–29 | 9.6 ± 3.7 | Colorimetric | a |
| 30–39 | 9.7 ± 2.8 | | |
| 40–49 | 10.6 ± 3.0 | | |
| 50–59 | 8.4 ± 2.5 | | |
| Human fibroblast—day 0 | 7.4 | Colorimetric | b |
| 2 | 9.8 | | |
| 5 | 7.5 | | |
| WI-38 passage 19–54 | 7.94 ± 0.55 | Colorimetric | c |
| Human fibroblast | 9.0 ± 0.4 | Fluorometric | d |
| | 8.3 ± 0.45 | Fluorometric | e |
| | 8.5 | Fluorometric | f |
| Human lymphocytes | 8.0 ± 0.1 | Colorimetric | g |
| | 7.5 ± 0.4 | Absorbance | |
| | 7.8 ± 0.4 | Fluorometric | |
| Granulocytes | 7.5 ± 0.27 | Fluorometric | h |
| Granulocytic leukemic | 8.62 ± 0.67 | Absorbance | i |
| 3T3 mouse | 13.7 | Absorbance | j |

[a] E. L. Schneider and S. S. Shorr, *Cell* **6,** 179 (1975).
[b] A. Leyva, Jr. and W. N. Kelley, *Anal. Biochem.* **62,** 173 (1974).
[c] V. J. Cristofalo and D. Kritchevsky, *Med Exp.* **19,** 313 (1969).
[d] B. T. Hill, *Anal. Biochem.* **70,** 635 (1976).
[e] B. T. Hill and S. Whatley, *FEBS Lett.* **56,** 20 (1975).
[f] U. Karsten, *Anal. Biochem.* **77,** 464 (1977).
[g] U. Karsten and A. Wollenberger, *Anal. Biochem.* **46,** 135 (1972).
[h] W. Y. Fujimoto, J. Teague, and R. H. Williams, *In Vitro* **13,** 237 (1977).
[i] G. T. Rudkin, D. A. Hungerford, and P. C. Nowell, *Science* **144,** 1229 (1964).
[j] H. Morimoto, P. A. Ferchmin, and E. L. Bennett, *Anal. Biochem.* **62,** 436 (1974).

ately preceding step. (Note: The pellet may be dissolved in 0.3 *N* NaOH and used for protein determination.)

*Analytical Procedure (Colorimetric–Burton).* Standard curve: Prepare duplicate standards of 5, 10, 25, and 50 $\mu$g of DNA per tube by adding 0.05, 0.10, 0.25, and 0.50 ml of working standard (100 $\mu$g/ml) to individual tubes and adding 0.5 *N* perchloric acid to a final volume of 2.0 ml. Prepare a standard blank of 2.0 ml 0.5 *N* perchloric acid. To 2 ml of tissue extract and tubes containing the DNA standards, add 4.0 ml of Burton's reagent. Incubate 16–20 hr at room temperature. Read absorbance at 600 nm and calculate by comparing with the absorbance of the DNA standards.

Recent modifications suggest substitution of paraldehyde for acetaldehyde and its inclusion with diphenylamine to produce a more stable reagent[25]; a shortened incubation period (4 hr) reportedly results in an increase in sensitivity when acetaldehyde is omitted and the incubation is conducted with 10,000–15,000 lx of light.[26]

## *Miscellaneous Enumeration Methods*

### ELECTRONIC

For routine purposes, especially when many samples are involved, blood cell counters are used.[27–29] There are several commercial models available that also yield information about the size of cells. The more recently developed flow systems for automated cytology (FMF or flow microfluorometry[30]) have been applied to measurement of cell number and viability.[31] Simultaneous measurement of both DNA and protein has been accomplished.[32]

### HEMATOCRIT

Although a rapid and simple method has been proposed by Waymouth,[33] it is seldom used. The method correlated with hemocytometer counts up to $3 \times 10^6$ cells.

### IN SITU

Efforts to enumerate cells *in situ* have been directed mostly toward the use of individual frames from time-lapse cinematography.[34,35] Others have enumerated cells in a microscopic field.[36–38] These methods are limited to low cell counts.

[25] G. M. Richards, *Anal. Biochem.* **57,** 369 (1974).

[26] J. R. Decallonne and J. C. Weyns, *Anal. Biochem.* **74,** 448 (1976).

[27] M. Harris, *in* "Tissue Culture: Methods and Applications" (P. F. Kruse, Jr. and M. K. Patterson, Jr., eds.), p. 400. Academic Press, New York, 1973.

[28] W. F. Daly, *in* "Tissue Culture: Methods and Applications" (P. F. Kruse, Jr. and M. K. Patterson, Jr., eds.), p. 398. Academic Press, New York, 1973.

[29] M. E. Kaighn and D. J. Merchant, *Tissue Cult. Assoc. Man.* **3,** 649 (1977).

[30] H. A. Crissman, P. F. Mullaney, and J. A. Steinkamp, *Methods Cell Biol.* **9,** 179 (1975).

[31] W. P. Drake, P. C. Ungaro, and M. R. Mardiney, *Transplantation* **14,** 127 (1972).

[32] H. A. Crissman, M. S. Oka, and J. A. Steinkamp, *J. Histochem. Cytochem.* **24,** 64 (1976).

[33] C. Waymouth, *J. Natl. Cancer Inst.* **17,** 305 (1956).

[34] L. N. Castor, *Exp. Cell Res.* **68,** 17 (1971).

[35] G. Froeze, *Exp. Cell Res.* **65,** 297 (1971).

[36] C. W. Jamieson, D. Russin, E. Benes, C. W. DeWitt, and J. H. Wallace, *Nature (London)* **222,** 284 (1969).

[37] G. Martinez-Lopez and L. M. Black, *In Vitro* **9,** 1 (1973).

[38] G. F. -Y. Lee and D. L. Engelhardt, *J. Cell. Physiol.* **92,** 293 (1977).

OTHER

Included here are turbidometry[39] and dry weight[40] measurement.

### *Expression of Growth Data*

Measurement of growth rate implies a change in cell number during a specified period of time. In most instances the data can be expressed simply as cell number; in others it is advantageous to express the data in such a manner that cell doubling time or percent of doubling is easily determined. Expression of cell number as a function of $\log_2$ is used if such is the case. Cell number can easily be converted to $\log_2$ by multiplication of the common logarithm of the cell count by the factor 3.3219. For example:

Cell count at time 0 = $2 \times 10^5$

Cell count at 48 hr = $1.3 \times 10^6$

(a) $\log_2$ at time 0 = (5.3010) (3.3219) = 17.6094

(b) $\log_2$ at 48 hr = (6.1139) (3.3219) = 20.3080

(c) number of doublings in 48 hr (b − a) = 2.6996

Doubling time of culture population, 48 ÷ c = 17.78 hr

Graphical expression of $\log_2$ (cell count) can be made on linear graph paper if the above calculations are used or with log graph paper if direct count is plotted.

## Measurements of Viability

Numerous criteria are used to determine cell viability. A survey by Malinin and Perry[41] listed 40 viability assay methods. These may be grouped into six broad categories: (1) survival and growth in tissue culture, (2) membrane integrity, (3) metabolite incorporation, (4) enzyme spectrum, (5) transplantation potential, and (6) structural alteration, chemical composition, and electroconductivity.

The ability of cells to withstand the rigors of dispensing agents, changes in environment, freezing, and thawing is indicative of cell integrity but is not sufficiently quantitative for description. Beating heart cells

[39] C. Peraino and W. J. Eisler, Jr., *in* "Tissue Culture: Methods and Applications" (P. F. Kruse, Jr. and M. K. Patterson, Jr., eds.), p. 351. Academic Press, New York, 1973.

[40] D. J. Merchant, R. H. Kahn, and W. H. Murphy, *in* "Handbook of Cell and Organ Culture," p. 160. Burgess, Minneapolis, Minnesota, 1964.

[41] T. I. Malinin and V. P. Perry, *Cryobiology* **4,** 104 (1967).

are strong indicators of viability but give no representation of the survival of the cell inoculum.

Although cloning,[42] assaying for "cultivable" cells,[43] trypsin digestion,[44] and uptake of isotope[45] have been used to determine cell viability, all are time-consuming. Most commonly used have been the metabolic stains and dyes whose uptake is dependent upon cellular membrane integrity or whose metabolism is indicative of enzyme activity. Of the latter, the tetrazolium salts are the most common.[44,46] These are water-soluble, colorless substances which, in contact with cellular reductases, are reduced to insoluble chromogens. Following incubation (up to 6 hr) of cells in the presence of tetrazolium salts the appearance of discrete granules in the cytoplasm is indicative of a viable cell.

For routine purposes "dye exclusion" tests are preferred. Schrek[47] was the first to suggest that an intact cell membrane was necessary for the exclusion of certain dyes. Dyes which have been used for this purpose include safranin,[48] eosin,[43,44,47,49] Congo red,[50] erythrocin,[51] nigrosin,[52] trypan blue,[43,44,51,53,54] and alcian blue.[55] The last was compared to the other six stains for use following gluteraldehyde fixation and found to be the only one that was reproducible before and after fixation. Similar results were obtained with cells fixed for up to 1 week. No differences were noted among the dyes when applied to unfixed cells using established procedures. For routine purposes the most commonly used procedure is that with trypan blue and a visual count of the unstained "live" cells. Since trypan blue absorbs effectively at 633 nm, the wavelength of the red Helium–Neon laser, the laser used in certain of the flow-automated cytology systems, electronic enumeration of live–dead cell ratios is possible.[31]

[42] K. B. MacDonald and W. R. Bruce, *Exp. Cell Res.* **50,** 471 (1968).
[43] J. R. Tennant, *Transplantation* **2,** 685 (1964).
[44] J. M. Hoskins, G. G. Meynell, and F. K. Sanders, *Exp. Cell. Res.* **11,** 297 (1956).
[45] B. T. Mossman, *Tissue Cult. Assoc.* **3,** 663 (1977).
[46] F. H. Straus, N. D. Cheronis, and E. Straus, *Science* **108,** 113 (1948).
[47] R. Schrek, *Am. J. Cancer* **28,** 389 (1936).
[48] J. Paul, *in* "Cell and Tissue Culture," p. 284. Williams & Wilkins, Baltimore, Maryland, 1961.
[49] J. H. Hanks and J. H. Wallace, *Proc. Soc. Exp. Biol. Med.* **98,** 188 (1958).
[50] C. F. Geschickter, *Stain Technol.* **5,** 49 (1936).
[51] H. J. Phillips, *in* "Tissue Culture: Methods and Applications" (P. F. Kruse, Jr. and M. K. Patterson, Jr., eds.), p. 406. Academic Press, New York, 1973.
[52] J. P. Kaltenbach, M. H. Kaltenbach, and W. B. Lyons, *Exp. Cell Res.* **15,** 112 (1958).
[53] W. Sawicki, J. Kieler, and P. Briand, *Stain Technol.* **42,** 143 (1967).
[54] S. Tolnai, *Tissue Cult. Assoc. Man.* **1,** 37 (1975).
[55] D. K. Yip and N. Auersperg, *In Vitro* **7,** 323 (1972).

Acridine orange[56] and a fluorescent derivative of disulfonic acid[57] have also been used as a vital stain.

*Reagents and Equipment*

Trypan blue (vital stain, C.I. 23850)–400 mg added to 90 ml water containing 810 mg NaCl, 60 mg $K_2HPO_4$, and 50 mg methyl *p*-hydroxybenzoate. Heat mixture to boiling, cool, and adjust pH to 7.2–7.3 with 1 *N* NaOH (approx. 8 drops). Adjust to final volume of 100 ml.

Hemocytometer

*Procedure.* Prepare cell suspension as described for cell count. In a test tube mix 0.9 ml of cell suspension and 0.1 ml trypan blue. After 5 min, add to hemocytometer. Count blue stained cells as nonviable.

Note: Trypan blue has a greater affinity for serum proteins than for cellular protein.[50] If the background shows heavy staining the concentration of dye should be increased or cells should be centrifuged and resuspended prior to staining. Timing of exposure of cells to dye can be critical in that cells continue to take up stain. If this appears to be the case, cells should be removed from the staining solution prior to counting.[53]

[56] M. R. Melamed, L. R. Adams, A. Zimring, J. G. Murnick, and K. Mayer, *Am. J. Clin. Pathol.* **57,** 95 (1972).

[57] I. Katz, *Tissue Cult. Assoc. Man.* **1,** 41 (1975).

# [12] Cloning

*By* LOLA C. M. REID

## A. Introduction

A cloned cell population is one derived from a single parental cell. Cloned populations provide the experimenter with the distinct advantage of minimal genetic variability in the model system under study. Methods for isolating single cells and propagating them into a population of cells are called cloning, and the majority of them fall into one of three categories of techniques:

*1. Dilution Plating.* Suspensions of cells are diluted with a sufficient volume of medium to permit addition of single cells to dishes.

*2. Cloning of Anchorage-Dependent Cells.* Dilute concentrations of cells are seeded onto one of several possible substrates. The cells are allowed to attach and to grow into colonies of cells. The colonies of cells are then individually subcultured and transferred to other dishes.





ISBN 0-12-181958-2

*3. Cloning of Anchorage-Independent Cells.* Dilute concentrations of cells are seeded into medium solidified by agar or agarose. Cells capable of growing in suspension in the agar-medium form morulae-shaped colonies. These colonies are transferred individually to culture dishes.

All cloning procedures require careful consideration of factors affecting the survival and growth of isolated cells. In the development of cloning techniques it was discovered that cells leak essential nutrients into the medium and secrete chemical messengers that are required for survival and growth.[1-4] In cell culture jargon, the cells are said to "condition the medium," and medium containing these nutrients and chemical messengers is referred to as "conditioned medium." Because an isolated cell in a large volume of medium cannot condition the medium adequately, the ability of the isolated cell to grow in the given medium is a measure of the cell's autonomy to the factors in conditioned medium. A quantitative measure of this is provided by the plating efficiency, the number of colonies of cells (each colony deriving from one cell) that form in plates seeded with dilute concentrations of cells (usually 100–200), calculated as the number of colonies, over the number of cells seeded, multiplied by 100. Many established cell cultures and most primary cell cultures have low plating efficiencies, often less than 1% and as low as 0.001%. Since low plating efficiencies are due to the cells' nutritional needs in low cell densities, special cloning media have been developed that are enriched for amino acids and vitamins.[5-12] The chemical messengers found in conditioned media are provided by feeder layers of cells or by supplementation of the cloning medium with old media from confluent cell populations.

The techniques described include several of the most routinely used methods from each of the three categories of cloning techniques, as well as methods for preparing feeder layers and cloning media supplemented with conditioned media. The capillary technique originated by Sanford[13] and used to produce the first clonal cell cultures is effective but tedious

[1] H. Eagle, *Science* **122,** 501 (1955).
[2] H. Eagle, *Science* **130,** 432 (1959).
[3] K. Sanford, L. Duprée, and A. Covalesky, *Exp. Cell Res.* **31,** 345 (1963).
[4] R. G.Ham, *Exp. Cell Res.* **29,** 515 (1963).
[5] R. G. Ham, *Biochem. Biophys. Res. Commun.* **14,** 34 (1964).
[6] R. G. Ham, *Proc. Natl. Acad. Sci. U.S.A.* **53,** 288 (1965).
[7] R. G. Ham, *In Vitro* **10,** 119 (1974).
[8] W. McKeehan, S. Hamilton, and R. G. Ham, *Proc. Natl. Acad. Sci. U.S.A.* **73,** 2023 (1976).
[9] W. McKeehan, K. McKeehan, S. Hammon, and R. Ham, *In Vitro* **13,** 399 (1977).
[10] S. Iwakata, and J. Grace, *N.Y. State J. Med.* **64,** 2279 (1964).
[11] R. Parker, *Spec. Publ., N.Y. Acad. Sci.* **5,** 503 (1957).
[12] J. Morgan, J. Morton, and R. Parker, *Proc. Soc. Exp. Biol. Med.* **73,** 1 (1950).
[13] K. K. Sanford, W. R. Earle, and G. D. Likely, *J. Natl. Cancer Inst.* **9,** 229 (1948).

even in the simplified form.[14] It has been superceded by easier methods and consequently will not be presented here. Methods for cloning plant cells are essentially analogous to those for mammalian cells (see this volume [41]). Guides to specific nutritional requirements and other modifications necessary for plant cells[15,16] (see this volume [41]) and for cells from invertebrates (see this volume [39]) are available.

## B. Equipment and Supplies Needed for All the Cloning Procedures

*Equipment*

Desk-top centrifuge
Inverted phase microscope with a total magnification of $\times 50$ to $\times 100$
Incubator, maintained at 37° with a water-saturated atmosphere containing 5% $CO_2$ in air
Biogaard or laminar flow hood. The hood minimizes contamination of cultures. If a hood is not available, one can do reasonably well with an inexpensive, plastic-sheltered area in which an ultraviolet light is placed to minimize pathogens.
Hemocytometer for counting cells. Cell counting also can be done with a Coulter counter if it is available.
Bunsen burner

*Supplies*

Plastic culture dishes and flasks can be obtained from a number of suppliers. However, the plating efficiency and growth of cells may vary somewhat on the plastics from different commercial suppliers. Therefore, one should select plastics from one company and use them exclusively for all experiments in which data are to be compared.
Pipettes:
Sterile Pasteur pipettes (7'' and 9''). Some pipettes should be plugged with cotton prior to sterilization.
Sterile 1- and 5-ml volumetric glass or plastic pipettes
Sterile test tubes holding 10 ml
Basal media: See Table I for a listing of possible cloning media.
Serum: Fetal bovine serum is most commonly used in cloning since it has high concentrations of growth factors and pregnancy hor-

[14] K. K. Sanford, *in* "Tissue Culture: Methods and Applications" (P. F. Kruse, Jr. and M. K. Patterson, eds.), pp. 237–241. Academic Press, New York, 1973.

[15] A. C. Hildebrandt, *in* "Tissue Culture: Methods and Applications" (P. F. Kruse, Jr. and M. K. Patterson, eds.), pp. 244–254. Academic Press, New York, 1973.

[16] P. K. Dougall, *in* "Tissue Culture: Methods and Applications" (P. F. Kruse, Jr. and M. K. Patterson, eds.), pp. 261–264. Academic Press, New York, 1973.

TABLE I
CLONING MEDIA[a]

| Name | Use | References |
|---|---|---|
| Eagle's Minimal Essential Medium (MEM) | Established cell lines | 1,2 |
| Ham's Cloning Media | Primary cultures or freshly explanted cells | 5–9 |
| F10 | | 4 |
| F12 | | 6 |
| F12M | | 6 |
| F12K | | 8 |
| MCDB104 | WI-38 and other fibroblast-like cells | 9 |
| RPMI | Lymphocytes and leucocytes | 10 |
| CMRL-1066 | Embryonic cells | 11 |
| Medium 199 | Virus-infected cells | 12 |

[a] The compositions of the media listed can be found in this volume [5] and in the specific references noted. Media should be freshly prepared every 2–3 weeks and stored at 4°. Longer storage results in the loss of labile components such as glutamine. Selection of a medium for a cell population to be cloned should be made after assaying the plating efficiency and growth behavior of the cells in the medium. Ham's F12K medium, in particular, has proven to be an excellent cloning medium for many cell types and is used routinely in many laboratories.

mones. However, if cells grow ideally in medium supplemented with another serum type, it should be used preferentially. For some cell lines serum components may be replaced with specific mixtures of hormones and growth factors. These techniques are presented in this volume [6] and have been summarized elsewhere.[17] Serum used should be heat-inactivated for 30 min at 56°.

Trypan blue solution—0.1% solution in phosphate buffered saline (PBS)

Enzyme solutions—one of the following:

Trypsin—0.1% trypsin in PBS (Gibco or Sigma)

Collagenase—0.1% collagenase in PBS (Sigma). There is usually a significant amount of proteolytic activity present in the collagenase purchased from most commercial suppliers.

Pronase—0.01% in PBS. A 1% stock solution is prepared and stirred for several hours at 4° centrifuged, and the supernatant fluid sterilized by Millipore filtration. The stock solution is diluted to the final concentration prior to use. Particularly effective for fibroblast-like cells; pronase is not as good for epithelial cells.

70% Ethanol solution

[17] G. Sato, and L. Reid, *in* "Biochemistry and Mode of Action of Hormones" (H. V. Rickenberg, ed.), Chapter 7. Academic Press, New York, 1978.

C. Dilution Plating Techniques

In brief, the techniques include the preparation of single cell suspensions from monolayer cultures or from solid tissue pieces in an appropriate cloning media; dilution of the cell suspension to a concentration of 10 cells per milliliter of medium; addition of 0.1 ml of medium to a microtest well; incubation of the cells for several weeks to permit growth of the colonies; and transfer of individual colonies to tissue culture dishes or flasks.

The original dilution plating techniques were aimed at adding a single cell to a tissue culture plate or to a test tube. With the realization of the cells' need to adjust the medium with "conditioning factors," the volume of fluid was reduced to approximately 0.1 ml per cell by adding the cell to a small capillary[13,14] or to a microdrop under paraffin.[18] More recently, the technique has been modified by the introduction of microtest plates (Falcon Plastics) containing 96 wells, each well holding up to 0.4 ml. One can add a single cell in 0.1 ml of medium per well and thereby provide the individual cell with its own culture dish and with a sufficiently small amount of fluid to minimize the conditioning factor problem.[19,20]

*Supplies Necessary for Dilution Plating Techniques*

Falcon Microtest Plates
Tissue culture plates (35 mm diameter)
Cloning medium supplemented with 10% serum (or, where feasible, specific hormones and growth factors)
Enzyme solution (trypsin is adequate for most cell types)

*Procedures*

1. The cell cultures to be cloned should be recently subcultured, be in an active state of growth, and show no signs of ill health.

2. Cloning of freshly explanted normal or neoplastic tissue is difficult and almost always requires use of feeder layers or conditioned media (techniques presented in subsequent sections). Tissue or tumor to be cloned should be sterilely dissected from the animal.

3. Single cell suspensions are prepared from monolayer cultures as follows. Remove the medium from the culture plates and rinse the plates twice with PBS. Add 1 ml of the trypsin solution to a 60-mm plate (2–3 ml for larger plates or flasks) of cells and incubate the plates at 37° for 5–15

[18] A. Lwoff, R. Dulbecco, M. Vogt, and M. Lwoff, *Virology* **1,** 128 (1955).

[19] J. A. Robb, *Science* **170,** 857 (1970).

[20] J. A. Robb, *in* "Tissue Culture: Methods and Applications" (P. F. Kruse, Jr. and M. K. Patterson, eds.), pp. 270–274. Academic Press, New York, 1973.

min. The plates should be watched carefully to ensure that the time of contact with trypsin is minimized. As soon as the majority of cells are rounded and mostly detached from the plates, squirt the cells with the conditioned medium supplemented with serum. Most serum contains a trypsin inhibitor (chicken serum is an exception). By squirting the cells with the medium, trypsin is inactivated and detachment of the cells from the plates is completed. To completely inactivate trypsin requires a 9 : 1 dilution of the trypsin solution with serum-supplemented medium. One can also remove trypsin by centrifuging the cells, removing the supernatant fluid, and resuspending the cells in the cloning medium.

Single cell suspensions are prepared from solid tissues, either normal or neoplastic[42] (see this volume [9]), as follows. The dissected tissue is minced finely with scissors sterilized by dipping them in 95% ethanol. Treat the mince with 0.1% collagenase (Sigma) in PBS at a ratio of 10 ml of enzyme solution to 1 ml of mince. Enzymic digestion should take place in a shaker bath at 37°. After 15 min of enzymic treatment, allow the chunks of tissue to settle to the bottom of the test tube, and pipette the supernatant fluid, containing suspended cells, into another test tube. Centrifuge the cells in a desk-top centrifuge at 900 rpm. Resuspend the pellet in cloning medium. The chunks and tissue mince can be repeatedly treated with collagenase solution until they are totally disaggregated.

4. Count the cells by the trypan blue exclusion assay in order to determine the number of viable cells (see this volume [11]).

5. Dilute the cell suspensions, whether from the monolayer culture or from the solid tissue dispersion, to a concentration of 10 cells/ml.

6. Use 0.1 ml of the medium containing the cells per well. Repeatedly agitate the cell suspension to ensure homogeneous dispersion of the cells throughout the medium. For rapid inoculation of cells into microtiter wells, a Hamilton repeating dispenser (Hamilton Company, Whittier, California, # PB600-1) is useful.[20]

7. Carefully observe each well and score those containing only one cell.

8. Incubate the cultures in 37° incubators with 5% $CO_2$ in air and with 95% relative humidity.

9. During the growth of the colonies, it should be unnecessary to change the medium. However, if colony formation is unusually slow, and a medium change is required, the spent medium can be gently aspirated from the cultures, and fresh medium added. Leave a residue of fluid over the cells while changing medium to prevent desiccation of the cells.

10. When colonies of 500–600 cells have formed, usually in 2–3 weeks, remove the medium from the wells. Rinse twice with phosphate buffered saline. Add 0.1 ml of trypsin solution per well. Allow the cells to

round, and then add cloning medium with 10% serum to suspend the cells. Transfer the cells to an appropriate tissue culture dish (35–60-mm plate) or flask.

### D. Cloning of Anchorage-Dependent Cells

The techniques in this category are slight variations of the ones described above in "Dilution Plating Techniques." The preparation and dilution of single cell suspensions are the same as described there. The cells are seeded onto 60-mm tissue culture dishes or on 60-mm petri dishes coated with either collagen or fibrin. By seeding only 100–200 cells per plate, one can obtain large colonies of cells that are sufficiently separated to permit isolation of individual colonies.

The technique can be used only for cells that require anchorage and spreading on a substrate in order to grow. This includes most normal cells as well as many established cell cultures.

*Supplies Necessary for Cloning of Anchorage-dependent Cells*

Sterilized cloning rings: metal cylinders that are 6 mm in diameter, 12 mm high, and have 1-mm-thick walls

Sterilized Dow-Corning stopcock grease: Layer the grease on a glass petri dish and sterilize. The grease may be kept sterile by taping on the lid and keeping the dish in the hood under a UV lamp.

Forceps

60-mm tissue culture dishes or, for collagen- or fibrin-coated substrates, 60-mm petri dishes.

35-mm tissue culture dishes

*Procedures*

1. Cells vary as to the substrates to which they will anchor. Established cells lines will attach to plastic coated with polycations. However, freshly explanted tissues usually prefer either collagen-coated plates or fibrin-coated plates. Preparation of collagen-coated plates is given elsewhere in this volume [21]. Fibrin-coated plates are prepared as follows: Prepare a solution of cloning medium containing 0.12 units of thrombin per 100 ml of medium. Prepare a solution of 250 mg bovine fibrinogen, 800 mg sodium chloride, and 25 mg sodium citrate per liter of glass-distilled water. Sterilize by filtration. Add 1 ml of the fibrinogen solution and 4 ml of the medium containing thrombin to a tissue culture dish and mix rapidly with a sterile Pasteur pipette. A clear gel will form within several minutes. This technique was first introduced by Schindler[21] who used it to replace

[21] R. Schindler, M. Day, and G. A. Fischer, *Cancer Res.* **19**, 47 (1959).

agar cloning (described in the next section). Schindler suspended the cells in fibrin and permitted them to form colonies shaped like morulae. However, one can also attach the cells to the surface of the fibrin and observe colony formation.

2. Prepare single cell suspensions from monolayer cultures or from tissues by the methods given in "Dilution Plating Techniques."

3. Depending on the plating efficiency of the cells, select an appropriate concentration of them to yield ultimately 1–10 colonies per plate. For example, with a plating efficiency of 5–10%, one should plate 100 cells per plate.

4. Dilute the cell suspensions to an appropriate density so as to add the correct number of cells per 60-mm dish. The 60-mm dishes hold about 5 ml of medium.

5. The plates are incubated at 37° in 5% $CO_2$ in air and with 95% relative humidity.

6. The following day, single cells that are sufficiently isolated to permit easy cloning are marked by encircling them on the plastic with a grease pencil or felt pen.

7. The cells are allowed to grow for several weeks into colonies containing approximately 500–1000 cells.

8. Remove the medium from the plates and rinse them twice with PBS. Leave a residue of the PBS over the cells to prevent their drying.

9. With forceps sterilized by dipping them into 70% ethanol, dip previously sterilized cloning rings into the petri dish containing stopcock grease. Place the grease-coated end down against the plastic so that the ring encircles a colony of cells and so that the greased end forms a seal with the plastic.

10. Add enzyme solution, e.g., 0.1% trypsin, to each of the rings. Usually 1–2 drops of solution will fill the rings.

11. Incubate the cultures for 5–15 min at 37°, observing the cultures periodically with a phase microscope. When the cultures are rounded and ready to detach, use a 9'' Pasteur pipette and pipette the solution gently up and down to complete the detachment process.

12. Pipette the solution of suspended cells into a 35-mm tissue culture dish or 35-mm petri dish coated with collagen or fibrin. Add 2 ml of cloning medium per plate.

### E. Cloning of Cells that Are Anchorage-Independent for Growth

Most cells require anchorage to and spreading on a substrate for growth. However, some cells have acquired the ability to grow in suspension cultures. Factors contributing to this cellular capability include viral transformation (SV40, polyoma), infection of cells with mycoplasma, and malignant transformation.[22–25] Puck *et al.*[26] showed that cells which are anchorage-independent for growth can be cloned in a medium solidified by low concentrations of agar. The method has been expanded into a family of techniques for cloning of cells in semisolid media. The version given below is that of Macpherson[27] who developed methods in which a base layer of agar-medium is overlaid with a thin layer of agar-medium containing the cells to be cloned. By layering the cells at the surface of the base layer, they can be observed readily with an inverted phase microscope.

Agar contains acidic and sulfated polysaccharides that are inhibitory to most cells but not to many virally or malignantly transformed cells. The selective inhibition of the agar-medium is due both to the inability of cells to anchor to the agar matrix and to the inhibitory polyanions within the matrix. Growth of cells in agar-medium has become an assay for viral and/or malignant transformation, since growth of the cells in agar is significantly correlated with tumorigenicity of the cells in immunologically suppressed hosts such as the athymic nude mice.[25,28,29] To reduce the toxicity of the agar, the polyanions were reduced yielding a matrix referred to as agarose. Agarose is used routinely to clone cell types that are anchorage independent but will not grow in agar due to its toxicity. As noted in Section D, one may also use fibrin gels for cloning cells in suspension. However, cells attach to fibrin even though they do not spread on it and so are not strictly anchorage-independent.[30] Normal cells, as well as transformed ones, will form colonies when embedded within a fibrin matrix.

*Supplies Necessary for Cloning Anchorage-Independent Cells*

Agar: prepare a 1.25% agar stock from Difco Bacto-agar. Preparative details are given below.

Cloning medium: prepare a double-strength stock solution. Sterilize by filtration.

Tryptose phosphate broth: Sterilize by autoclaving.

[22] I. Macpherson, and L. Montagnier, *Virology* **23,** 291 (1964).
[23] P. H. Black, *Virology* **28,** 760 (1966).
[24] I. Macpherson, *J. Cell Sci.* **1,** 145 (1966).
[25] V. Freedman, and S. Shin, *Cell* **3,** 355 (1974).
[26] T. Puck, P. I. Marcus, and S. J. Cieciura, *J. Exp. Med.* **103,** 273 (1956).
[27] I. Macpherson, *in* "Tissue Culture: Methods and Applications" (P. F. Kruse, Jr. and M. K. Patterson, eds.), pp. 276–280. Academic Press, New York, 1973.
[28] H. P. Klinger, S. Shin, and V. H. Freedman, *Cytogenet. Cell Genet.* **17,** 185 (1976).
[29] C. D. Stiles, W. Desmond, L. Chuman, G. Sato, and M. Saier, Jr., *Cancer Res.* **36,** 3300 (1976).
[30] L. Reid, C. Stiles, M. Rindler, and M. Saier, submitted for publication.

60-mm petri dishes
35-mm tissue culture dishes

*Procedures*

1. Agar stock. Add 12.5 g of Difco agar to 800 ml of boiling glass-distilled water and heat until the agar dissolves. Make up to 1 liter with water. Sterilize by autoclaving 80-ml portions of the agar solution in screw-capped bottles of 250-ml capacity. Autoclave under low pressure and with the screw caps loosened. Allow the bottled solutions to come to room temperature. Then tighten the bottle caps. Store at room temperature.

2. Melt the agar stock solution by placing the bottle in boiling water until the agar liquefies. Cool for several minutes and place in a water bath at 44° so that the hot water is above the level of the agar.

3. Prepare the complete agar medium by mixing 80 ml of the cloning medium (double strength) with 20 ml tryptose broth and 20 ml serum in the bottle with the agar. Mix by swirling.

4. Pipette 1–2 ml of medium into 60-mm petri dishes or tissue culture dishes and swirl the plate to evenly coat the solution over the bottom of the dish. Allow the base layers to solidify (10–15 min).

5. Prepare a cell suspension as described under "Dilution Plating Techniques."

6. Dilute the cell suspension with cloning medium to a concentration that is 3-fold that of the final, desired concentration. Mix 1 volume of the cell suspension with 2 volumes of the complete agar solution.

7. Spread the cells over the base layer and swirl the plate to assure even spreading.

8. Incubate the cultures at 37° in an atmosphere of 5% $CO_2$ in air and with 95% relative humidity.

9. Allow the cells to form colonies of 500–1000 cells.

10. Colonies are transferred by sucking them from the agar with a Pasteur pipette and transferring to a sterile test tube containing 1 ml of medium. The colony of cells is pipetted to dissociate it from the agar.

11. The cell suspension is transferred to a 35-mm dish to which is added 2 ml of cloning medium.

## F. Preparation of Cloning Media, Feeder Layers, and Conditioned Media

A listing of some of the media used for cloning various cell types is provided in Table I. All of them employ a basal medium enriched in particular nutrients, especially those, such as pyruvate and nonessential amino acids, that are lost from the cells into the medium.

*Feeder Layers*

The low plating efficiency of some cells, especially freshly explanted cells, prevents or seriously complicates the cloning procedure. As noted, the low plating efficiency is due in part to the loss of essential nutrients from the isolated cells. However, in the special cloning media listed in Table I these nutrients are provided. The other factors, available in high-density cultures of cells but not in low density cell cultures, are "chemical messengers" secreted by the cells and regulating survival and/or growth. The chemical messengers playing roles in colony formation are referred to as *Colony Stimulating Activity* factors (CSA). To provide such factors, it is necessary to co-culture the cells to be cloned with a feeder layer of the same or another cell type. Many established cell lines requiring feeder layers under clonal conditions can use feeder layers of the same cell type. Freshly explanted epithelial cells, whether from normal or neoplastic tissue, often do better when cloned over feeder layers of mesenchymal cells (L. C. M. Reid, unpublished data). Many of the CSA factors have been or are being purified in a number of laboratories. Several of the better-characterized CSA are listed in Table II.[31–37] Except for some anaplastic cell lines, supplementation with only the CSA factors is inadequate for colony formation, suggesting that other factors are necessary. Corroborating this interpretation are the findings by Sato and his colleagues that multiple hormones, growth factors, and transfer factors are essential as supplements to basal media to attain growth of many cell types[17] (see this volume [6]).

*Supplies for Preparing Feeder Layers or Conditioned Medium*

Microtiter wells (Falcon)
35-mm dishes
60-mm dishes
Cloning medium

[31] N. Dulak and H. Temin, *J. Cell. Physiol.* **81,** 161 (1973).
[32] G. Smith and H. Temin, *J. Cell. Physiol.* **84,** 181 (1974).
[33] T. Chen, J. Mealey, Jr., and R. L. Campbell, *J. Natl. Cancer Inst.* **55,** 1275 (1975).
[34] T. Bradley and M. Sumner, *Exp. Biol. Med. Sci.* **46,** 607 (1968).
[35] G. Price, E. McCullough, and J. Till, *Blood* **42,** 341 (1973).
[36] T. Landau and L. Sachs, *Proc. Natl. Acad. Sci. U.S.A.* **68,** 2540 (1971).
[37] R. Stanley and P. Heard, *J. Biol. Chem.* **252,** 4305 (1977).

TABLE II
CHEMICAL MESSENGERS AFFECTING COLONY FORMATION

| Name | Tissue source(s) | Active with | Molecular weight | Stability to: Heat | Stability to: Freeze/thaw | Stability to: Proteases | General | References |
|---|---|---|---|---|---|---|---|---|
| Multiplication stimulation activity (MSA) | Liver, serum | Fibroblast cultures | About 10,000 | Yes | Yes | No | A peptide (similar to somatomedins) | [31,32] |
| | Brain | Brain tumor cells | 1,000–10,000 | Yes (up to 66°) | Yes | No | A peptide | [33] |
| | Bone marrow | Haemopoietic precursor cells | 10,000 | Yes | Yes | Partially | A peptide | [34] |
| | Human leukocyte cultures | Human marrow cells; granulocytes | About 1300 | Yes | Yes | No | A hydrophobic peptide | [35] |
| Colony stimulating activity (CSA) | Mouse L cells<br>Serum<br>Urine | Mouse marrow cells; macrophages and granulocytes | 65,000–70,000 | No | No | Partially | Acidic protein | [36] |
| Colony stimulating activity (CSA) | Mouse L cells<br>Serum<br>Urine | Mouse marrow cells; macrophages and granulocytes | 45,000 | No | Yes | No | Glycoprotein | [37] |

*Procedures*

1. Feeder layers may be prepared from any cell population desired. Single cell suspensions are obtained from monolayer cultures or from tissues as described for methods for dilution plating.

2. The suspensions are added to whatever type of dish will be used in the cloning procedure. For microtiter dishes, add 50–100 cells per well; for 35-mm dishes, add $10^4$ cells; for 60-mm dishes, add $10^5$–$10^6$ cells per plate.

3. Incubate the cells at 37° with 5% $CO_2$ in an air atmosphere and with 95% relative humidity. Allow the cells to attach to the plates and grow for 1 or 2 days.

4. Irradiate the plates with 4000–5000 rad (cobalt-60 or cesium-135) to eliminate all mitotic activity. Mitomycin C has been used to mitotically kill the feeder layer cells; it is used less frequently now, since it has been found to adversely affect the cells being cloned.

5. The method of addition of the cells depends on the cloning technique that is used:

a. Dilution technique: The single cells are added directly to wells containing confluent feeder layer.

b. Anchorage-dependent technique: Cells to be cloned may be added directly to the feeder layer. For those cells that grow best on collagen or fibrin substrates, polymerize the collagen or fibrin over the feeder layer. The cells to be cloned are then added to the surface of the substrate.

c. Anchorage-independent technique: As in (b), the agar or agarose matrix is poured as an overlay onto the feeder layer. Cells to be cloned are then added to the surface of the base agar or agarose layer.

6. For conditioned medium, 1–2-day-old medium from confluent feeder layer cultures is removed and mixed at a 3 : 1 ratio with fresh medium. The medium is used as the cloning medium and changed every 4–5 days.

# [13] Cell Culture Characterization: Monitoring for Cell Identification[1]

*By* WARD D. PETERSON, JR., WILLIAM F. SIMPSON, and BHARATI HUKKU

Techniques and cell culture support systems now available enable any bioscience laboratory to successfully propagate animal cells *in vitro*. While the relative ease with which cell cultures can now be started or

[1] Supported in part by National Cancer Institute Contract NOl CP 3-3333.



ISBN 0-12-181958-2

maintained is a great advantage, it frequently happens that a requirement for one cell line or strain in a particular research inquiry is followed by a need for a second, third, and fourth cell line. Soon, either by acquisition or initiation, the number of cell lines escalates. An escalation in the number of cell lines in a laboratory beyond one should be accompanied by the institution of cell identification and monitoring regimens.

The need for cell line monitoring arises from more than just numbers of cell lines. Cells are passaged, cocultivated, transformed, and/or transplanted for numerous investigative purposes. In each of these operations, there is a need to know the status of the cells *in vitro* for at least these reasons: to annotate changes in cells upon passage; to determine the species of the predominant cell population in a cocultivated culture; to confirm the fact of cell transformation; or to establish whether transplantation resulted in a tumor of cell or host origin. Beyond that, cell monitoring apprehends such technical mishaps as mislabeling or inadvertent mixing of cells that may occur in each of the aforementioned operations.

Monitoring also is necessary when cultures are exchanged between laboratories. Cultures often are sent out to another investigator without being newly recharacterized, and even more frequently they are received and worked with by the recipient investigator who also fails to examine them for identity. In this way, subsequent conforming or diverging results may perpetuate misinformation in very costly and time-wasting ways.

Numerous reports during the past 25 years have documented instances of cross-contamination between cell cultures. The subject has been reviewed more than once,[2,3] and it has been recognized by the establishment of repositories for characterized cell populations.[4,5] More recent publications indicate that interspecies and intraspecies contamination of cell cultures may be as high as 20–30%.[6,7] While these recent reports may represent findings from a somewhat biased sample, in that at least a portion of the cell lines were sent for examination because of suspicions by the submitting investigator, it is just as likely that there are many cross-contaminated cultures about which no suspicions are held. In short, the problem of cross-contamination is a substantive and persisting problem that may afflict any laboratory using cell cultures. Monitoring cell cultures for identity is a necessary but often neglected adjunct control in the laboratory.

[2] C. S. Stulberg, *in* "Contamination in Tissue Culture" (J. Fogh, ed.), p. 1. Academic Press, New York, 1973.

[3] P. P. Ludovici and N. R. Holmgren, *Methods Cell Biol.*, **6**, 143 (1973).

[4] American Type Culture Collection, Rockville, Maryland.

[5] Human Genetic Mutant Cell Repository. Institute for Medical Research, Camden, New Jersey.

[6] C. S. Stulberg, W. D. Peterson, Jr., and W. F. Simpson, *Am. J. Hematol.* **1**, 237 (1976).

[7] W. A. Nelson-Rees and R. R. Flandermeyer, *Science* **195**, 1343 (1977).

Selected for description here are three approaches to cell monitoring that utilize genetically stable phenotypic expressions of cells. Cells may be identified by isozymes, chromosomes, or species-specific antigens, and of course by information derived in combination from all three of these different cell markers. Each of the procedures to be described yields rapid results and reasonable certainty of accuracy when properly controlled. As presented, each method has limitations. However, each can be extended to give rather precise intraspecific information about a cell line, as will be briefly indicated in each section. Each method also has certain virtues that will enable a laboratory to quickly adopt one or more of them because of compatibility with the laboratory's research direction and available equipment.

## I. Monitoring of Cells by Isozymes

Extraction of enzymes from cells, and comparison of electrophoretic mobilities of enzymes on zymograms, comprise a convenient method for determining species. By comparing electrophoretic mobilities of only two or three enzymes, species can be readily determined. Further, by choosing isozymes polymorphic in a species, intraspecies characterizations can also be carried out as was demonstrated by Gartler,[8] who chose glucose 6-phosphate dehydrogenase and phosphoglucomutase to examine a series of human cell lines. He demonstrated contamination of many human cell lines by the cell line HeLa,[9] on the evidence that the dehydrogenase Type A and mutase Type$_1{}^1$ and Type$_3{}^1$, characteristic of HeLa cells, were also found in cultures whose presumptive donors could not possess those forms of the two isozymes. For that reason, and since human cells are cultured extensively, a procedure for determining glucose 6-phosphate dehydrogenase mobility is described here.[10,11] The finding of Type A mobility may be helpful in raising suspicions about a human cell line.[12] Also described is the determination of lactate dehydrogenase isozyme mobilities.[13,14] Both isoenzymes are generally abundant in cells, and the

[8] S. M. Gartler, *Natl. Cancer Inst., Monogr.* **36,** 167 (1967).

[9] W. F. Scherer, J. T. Syverton, and G. O. Gey, *J. Exp. Med.* **97,** 695 (1953).

[10] M. C. Rattazzi, L. F. Bernini, G. Fiovelli, and P. M. Mannucci, *Nature (London)* **213,** 79 (1967).

[11] W. D. Peterson, Jr., C. S. Stulberg, N. K. Swanborg, and A. R. Robinson, *Proc. Soc. Exp. Biol. Med.* **128,** 772 (1968).

[12] W. A. Nelson-Rees and R. R. Flandermeyer, *Science* **191,** 96 (1976).

[13] E. S. Vesell, J. Philip, and A. G. Bearn, *J. Exp. Med.* **116,** 797 (1962).

[14] F. Montes De Oca, M. L. Macy, and J. E. Shannon, *Proc. Soc. Exp. Biol. Med.* **132,** 462 (1969).

procedures used are readily carried out. Use of both isozymes give sufficient information to determine most cell species.[15]

## A. Preparation of Isozyme-containing Extracts

*Reagents*

Trypsin–EDTA solution:
1:300 Trypsin, crude, 2.5 g/1
EDTA (ethylenediaminetetraacetic acid), 0.2 g/1
Glucose, 1.0 g/1
NaCl, 1.0 g/1
$NaHCO_3$, 1.89 g/1

*Procedure*

1. Grow 5 × $10^6$ cells in appropriate medium and disperse with trypsin–EDTA if it is an adherent cell line. When cells are monodispersed, stop the action of trypsin–EDTA by adding growth medium and collect the cells in a conical centrifuge tube.
2. Centrifuge the cell suspension at 1500 rpm for 10 min, remove the supernate, and wash the cells 3 times with 10 volumes of 0.9% NaCl.
3. After the last washing, resuspend the cells in an equal volume of 0.9% NaCl. Freeze the cell suspension by placing the centrifuge tube in a methanol—dry ice solution. When frozen remove the tube and thaw the cell suspension slowly at room temperature. Repeat the procedure 3 times. This rapid freeze–thaw procedure will disrupt cells.
4. Centrifuge the suspension at 3000 rpm for 10 min and draw up the supernatant liquid in a Pasteur pipette. Place the sample in a closable 12 × 75 mm tube for storage at −20°. The isozymes are stable for several weeks when stored in this manner.

## B. Electrophoresis of Sample

*Reagents*

Glucose 6-phosphate dehydrogenase buffer, 0.075 *M* Tris-citric acid, pH 7.5:
Tris, 9.08 g/1
EDTA, 1.49 g/1
Citric acid, 15.75 g/1
Add 20 ml of citric acid solution to 1 liter of Tris-EDTA solution.

[15] J. E. Shannon and M. L. Macy, *in* "Tissue Culture: Methods and Application" (P. F. Kruse, Jr. and M. K. Patterson, Jr., eds.), p. 804. Academic Press, New York, 1973.

Lactate dehydrogenase buffer, 0.05 *M* barbital, pH 8.8:
Sodium barbital, 10.31 g/1
EDTA, 0.35 g/1
Adjust to pH 8.8 with 1 *N* NCl.

*Procedure*

1. Pour the buffer solution suitable for electrophoresis of the isozyme into all chambers of the electrophoretic apparatus to appropriate volume. Inexpensive low-voltage electrophoretic equipment[16] is commercially available from a number of manufacturers.

2. Thaw samples and prepare for application to electrophoretic support medium. Inexpensive electrophoretic support media are commercially available as acetate strips, cellulose acetate gel strips, or agarose gel film. Each requires a somewhat different procedure for application. A convenient system is that of the agarose gel film[17] with eight preformed slots to which 1-$\mu$l samples can be added by syringe. The application is simple, and eight samples can be run at a time for comparison of electrophoretic mobility. After application, the film is inverted so that the agarose side is down, and the sample origin is placed at the cathodal side of the box. Wicking of the agarose to the buffer is accomplished by filter paper wicks to complete the circuit across the substrate. Power is turned on, and the sample is electrophoresed at 6 mA (50–150 V) for 90 min at room temperature.

3. After electrophoresis, remove the film preparatory to staining. Empty the electrophoresis chamber of buffer to restore ionic balance throughout the buffer solution before using it in a subsequent electrophoresis run.

## C. Staining of Isoenzymes

*Reagents*

Glucose 6-phosphate dehydrogenase stain:

| | *Amount/ solution* | *Proportion in stain* |
|---|---|---|
| NADP | 10 mg/ml | 0.4 ml |
| Glucose 6-phosphate | 25 mg/ml | 0.6 ml |
| Phenazine methosulfate (PMS) | 2 mg/ml | 1.0 ml |
| Nitro blue tetrazolium (NBT) | 2 mg/ml | 1.0 ml |
| Cobaltous chloride | 0.5 *M* | 0.4 ml |
| Tris · HCl (pH 8.6) | 0.075 *M* | 5.0 ml |

[16] J. K. Turner model 310-211 with electrophoresis chamber 310-120 is used in this laboratory.

[17] Universal electrophoresis film, agarose; Corning ACI, Palo Alto, California.

PMS solution is made fresh each day and should not be exposed to light. NAD, glucose 6-phosphate, and NBT solutions are stable at 4° for 1 week. The stain is bright light yellow and will not react if the color darkens. Prepare stain and incubate for 15 min at 37° immediately before use.

LDH stain:

| | *Amount/ solution* | *Proportion in stain* |
|---|---|---|
| NAD | 10 mg | 10 mg |
| Sodium lactate | 1 *M* | 1.0 ml |
| Nitro blue tetrazolium | 1 mg/ml | 3.0 ml |
| Phenazine methosulfate | 1 mg/ml | 0.3 ml |
| Barbital buffer, pH 8.8 | 0.5 *M* | 1.0 ml |

Stain does not require preincubation before use.

*Procedure*

1. Prepare stain as indicated. Pour into a clean shallow staining dish. Remove electrophoresis film, and place agarose side directly onto stain so that stain comes in contact with area to which isoenzymes have migrated.
2. Incubate the staining dish in a humidified chamber at 37°. Color should be observed within 5 min, and the staining reaction can be developed for 20 min. The staining reaction should be followed every few minutes so that overstaining, with consequent blurring of reaction bands, does not occur.
3. When the staining reaction is satisfactory, remove the electrophoresis film from the staining dish and place it in 1:40 formaldehyde for 10 min to stop the reaction.
4. Comparative electrophoretic mobilities can be determined immediately by removing the film from the formaldehyde solution, shaking the film to remove moisture, and overlaying it on lined or graph paper. By measuring the distance from the origin to the reactive enzyme site, and by noting the patterns, comparison of enzyme mobilities with known and test isoenzyme extracts can be carried out.
5. A permanent record for the electrophoresis run can be made by drying the agarose film at 37° overnight or at 60° for 2 hr. The stained film serves as a record itself, or it may be photographed. To prevent scratching or dislodging of the electrophoresis film with time, an adherent clear plastic sheet may be placed over the reactive side of the film.

*Comment.* The procedure presented above utilizes inexpensive equipment, offers simplicity of operation, and gives rapid results. The entire procedure will take 2–3 hr, and eight different samples can be accommodated. The resolution is not as great as that achieved by starch gel or acrylamide substrates, nor is its application to other enzyme systems as facile when direct visual staining is utilized. If other electrophoretic systems are already in use, it will require very little additional work to develop appropriate cell monitoring regimens.

O'Brien *et al.*[18] have shown that appropriate selection of additional isoenzymes polymorphic in a species can be used to develop a genetic signature for each cell line. Seven isozymes were used to study 30 human cell lines. They showed that the enzyme profiles obtained distinguished between 19 human glucose 6-phosphate dehydrogenase Type B cell lines and 5 dehydrogenase Type A cell lines, demonstrated that two dehydrogenase Type B cell lines thought to be different were actually the same line, and demonstrated that three cell lines suspected of being HeLa cells by other criteria were confirmed as such since all three had a genetic signature identical to that of HeLa. The probability of two different cell lines having the same signature with the seven isozymes by chance alone was calculated to be less than 5%. Extension of the concept seems possible with regard to other species as a means for intraspecies identification of individual cell lines.

## II. Monitoring of Cells by Chromosomal Examination

Examination of chromosomes is the definitive method for determining cell species, and if carried out in a thorough way it is one of the definitive methods for individual cell line identification. The methodology for this is a fairly complicated and time-consuming task—so much so that cytogenetics tends to be completely avoided by those not in that area of study. In fact, however, it is not difficult to quickly obtain basic information on cell cultures by chromosomal examination. Metaphase preparations can be prepared, stained, and examined in a short time. In terms of cell culture monitoring, these procedures are extremely useful for determining the current status of a cell culture. Familiarity with the chromosomal complement of the cells cultured in a laboratory will enable one to quickly denote similarities or differences in a cell line with passage and time in culture. On the basis of such findings, a decision to obtain more expert analysis can be made (see also [27]).

### A. Preparation of Cell Metaphases

*Reagents*

Colcemid (*N*-deacetyl-*N*-methylcolchicine) solution:
Colcemid,[19] 100 mg

[18] S. J. O'Brien, G. Kleiner, R. Olson, and J. E. Shannon, *Science* **195,** 1345 (1977).

[19] CIBA Pharmaceutical Company, Summit, New Jersey. Also available as prepared solution from other companies.

95% ethanol, 5.3 g
Propylene glycol, 10 g
$Na_2HOP_4 \cdot 12\ H_2O$, 6.2 g
$NaH_2PO_4 \cdot 2\ H_2O$, 0.15 g
Distilled water, q.s. 100 ml

Dissolve Colcemid in ethanol and add propylene glycol. Add phosphate buffer to the solution of dissolved Colcemid. Sterilize by autoclaving at 10 lb/in$^2$ for 30 min. Disperse in 1-ml aliquots and store at 4°. For use, dilute 1 ml in 24 ml of Hank's balanced salt solution (HBSS).

Fixative solution:
Absolute methanol, 3 volumes
Glacial acetic acid, 1 volume

*Procedure*

1. Establish culture with actively dividing cells as in logarithmic growth phase. Adherent cells are usually satisfactory 2 days after transfer, when cells cover about $\frac{1}{2}$ of the flask surface. Suspension cultures are satisfactory the day following feeding.
2. Add to the culture one drop of diluted Colcemid solution per 25 ml of growth medium to arrest the dividing cells in metaphase. Following a 1–3 hr incubation period, depending on the rate at which the cells multiply, the Colcemid-containing medium is removed, and 5 ml are placed in a centrifuge tube. Five milliliters of trypsin–EDTA solution are added to the culture flask to disperse an adherent monolayer of cells.
3. When the cells are dispersed, pipette and transfer them to the centrifuge tube with Colcemid medium. Trypsin–EDTA action on the cells is stopped. Centrifuge the cell suspension at 1500 rpm for 10 min. Remove all but 1 ml of supernatant from the cell pellet. Resuspend the cells by gentle repeated pipetting in a manner to avoid foam formation.
4. To 1 ml of cell suspension, add 4 ml of distilled water and suspend the cells in this hypotonic solution for 8–10 min at room temperature. For human lymphoblast cultures, better preparations are obtained by treatment with hypotonic 37.5 m*M* KCl for 18 min. The time in hypotonic solution is critical for obtaining good metaphase preparations, and that given here is only representative. Since each cell culture may require somewhat different hypotonic treatment periods, precise timing can only be determined by trial.
5. Add about 0.7 ml of fixative solution drop by drop, while gently agitating the tube to obtain good mixing. Centrifuge the suspension at 500–800 rpm for 5 min.
6. Remove the supernatant from the cell pellet and add 0.2 ml of fixative solution (approximately 3–4 times the volume of the pellet).

7. Resuspend the cells by gently bubbling air through a Pasteur pipette, and add sufficient fixative solution to bring the cell suspension volume to about 5 ml. Leave the cells in fixative solution for 10 min, and centrifuge the cell suspension as before.

8. Remove the supernatant, and resuspend the cells in 0.5 ml of fixative solution.

9. Take up a small portion of the suspension in a Pasteur pipette. From a distance of 30–40 cm above a horizontally held, cold, wet slide, drop two drops of cell suspension at different locations on the slide. Blow once down the length of the slide which is held 5–10 cm from the mouth, in order to spread the cells on the slide and to remove excess fixative. Then dry the slide on a hot plate at 60° for 2 min, or air dry by placing the slide narrow dimension down against a rack for about 30 min.

10. Examine the slide with a ×10 objective for cell nuclei and for chromosomes spread in metaphase plates. If there are judged to be too many cells for the matephases to be well spread, that is, the metaphases are overlapped, dilute the suspension appropriately with fixative solution and make another trial slide.

11. Once a satisfactory trial slide is obtained, additional slides are made as needed in the same manner. Slides are marked for identity and dried as before.

## B. Staining and Evaluation of Metaphases

*Reagents*

Giemsa stain[20]:
Giemsa powder, 7g
Glycerol, 482 ml
Methanol (absolute), 462 ml

Mix Giemsa powder and glycerol. Place solution in a 60° oven for 2 hr. Allow solution to cool and add methanol. Filter through Whatman #1 paper as used.

Giemsa buffer, pH 7.0:
$Na_2HPO_4$ (anhydrous), 9.5 g/1
$NaH_2PO_4H_2O$, 9.2 g/1

Add 61.1 ml of dibasic phosphate solution and 38.9 ml of monobasic phosphate solution to 900 ml of distilled water. The working solution of

[20] Harleco Company, Philadelphia, Pennsylvania. Also available as prepared working solution from other companies.

Giemsa stain is prepared by adding Giemsa stain to Giemsa buffer in a ratio of 1:25.

*Procedure*

1. Place slides in a glass staining jar containing 1 *N* HCl prewarmed in a 56° water bath and hydrolyze for 11 min.
2. Wash slides in the staining jar with running tap water for 10 min. Then rinse with 1% Giemsa buffer in distilled water to neutralize any remaining acid.
3. Transfer the slides to a staining jar and stain for 11 min with the working dilution of Giemsa stain.
4. Remove the slides and dip them into 1% Giemsa buffer in distilled water to rinse.
5. Shake the rinsed slides to remove excess water, wipe the back of the slide with paper toweling, and firmly blot the metaphase side with bibulous paper. Place the slide on a hot plate at 60° for 2 min, or allow the slide to dry overnight at room temperature.
6. When slides are dry, add one drop of Permount[21] at each end of a 22 × 40 mm #1 coverslip lying on paper toweling. Invert the slide, metaphase side down, and place on the coverslip. Squeeze excess Permount out from between the coverslip and slide by pressing firmly on the slide. Do not change the position of the slide with respect to the coverslip while pressing down. Allow Permount to dry for 15 min prior to examination.
7. Examine slides microscopically by first using a ×16 objective to locate well spread metaphases. Change to a ×100 oil immersion objective to count chromosomes in the metaphase plate. A sample of at least 15 metaphases should be counted to determine the chromosomal modal number. At the same time chromosomes are evaluated morphologically for species characteristics. In the beginning, it will be convenient to make a photographic record of several metaphases of a preparation to compare chromosome morphology with published karyotypes.[22] With experience, counting and sorting out of chromosomes can be done microscopically in a short time. The total amount of actual time in preparation and examination of metaphases approximates 4 hr.

*Comment.* It is certain that chromosomes will be visible if the above method for obtaining stained metaphases is followed. From these, an evaluation of the status of the cell culture can be made to ascertain the species of most cell cultures, degree of ploidy, and notation of abnor-

[21] Fisher Scientific Company, Fairlawn, New Jersey.

[22] T. C. Hsu and K. Benirschke, "An Atlas of Mammalian Chromosomes," Vols. 1–10. Springer-Verlag, Berlin and New York, 1967–1977.

malities. With only a relatively small investment in time and materials much information can be obtained about a cell culture in this way.

However, a chromosome examination in the manner described will not fully define a cell culture to the extent now possible using various chromosome banding techniques[23–25] (see also this volume [27]). Useful intraspecies chromosome markers, such as the human Y chromosome,[26] HeLa or other human tumor cell marker chromosomes,[27–29] or a number of chromosome markers found in cell lines of other species[30] can only be identified with certainty from banded preparations. Even some interspecies identifications require chromosome banding. For example, banding of the number 2 chromosomes distinquishes between the karyotypes of rhesus monkey and baboon.[31] Banding techniques strongly enhance monitoring and cell identification capability.

## III. Monitoring of Cells by Species-Specific Antigen – Antibody Reaction

Membrane antigens that are species specific can be used as a marker of cell identity. Antisera prepared against membranes can be used to determine species of cells in culture by several serologic methods. The method described here employs fluorescence as an indicator of a positive specific reaction. It is a useful method because the proportion of different species of cells in mixed culture populations can be determined.

### A. Preparation of Species-Specific Antiserum

1. Satisfactory antigen sources can be secured by using red blood cells, primary cell strains initiated for the purpose, or well-characterized cell lines as are obtainable from the American Type Culture Collection.[4] Cells can be grown at a rate sufficient to satisfy the inoculation schedule.

[23] T. Casperson, L. Zech, and C. Johansson, *Exp. Cell Res.* **62,** 490 (1970).

[24] M. Seabright, *Lancet* **2,** 971 (1971).

[25] H. A. Lubs, W. H. McKenzie, S. R. Patil, and S. Merrick, *Methods Cell Biol.* **6,** p. 345 (1973).

[26] W. D. Peterson, Jr., W. F. Simpson, P. E. Ecklund, and C. S. Stulberg, *Nature (London), New Biol.* **242,** 22 (1973).

[27] O. J. Miller, D. A. Miller, P. W. Allerdice, V. G. Dev, and M. S. Grewal, *Cytogenetics* **10,** 338 (1971).

[28] W. A. Nelson-Rees, R. R. Flandermeyer, and P. K. Hawthorne, *Science* **184,** 1093 (1974).

[29] W. A. Nelson-Rees, R. R. Flandermeyer, and P. K. Hawthorne, *Int. J. Cancer* **16,** 74 (1975).

[30] W. A. Nelson-Rees, *in* "Proceedings of Workshop on Cell Substrates for Vaccine Production" (J. Petricciani, chm.), p. 33. NIH, Bethesda, Maryland, 1976.

[31] D. S. Markarjan, I. A. Gvaramia, and N. V. Shonia, *Mamm. Chromosomes Newsl.* **17,** 5 (1976).

However, a convenient procedure is to grow the cells to approximately $6 \times 10^7$ cells, disperse the cells by trypsin–EDTA solution, and collect and centrifuge them into a pellet.

2. Resuspend the cells in medium containing 10% serum; count the cells, and adjust the cell concentration of part of the suspension to $3 \times 10^6$ cells/ml, sufficient for five ampules. The remainder, adjusted to $6\times 10^6$ cells/ml, is distributed to at least seven ampules. The cell suspensions are then frozen at −20°.

3. When ready to begin the immunization procedure, rapidly thaw sufficient ampules each day of inoculation to have sufficient cells according to the following schedule: $3 \times 10^5$ cells on day 1; $6 \times 10^5$ cells on day 4; $1.5 \times 10^6$ cells on day 7; $3 \times 10^6$ cells on day 11; and $6 \times 10^6$ cells on days 14, 17, and 21. Cells are centrifuged and washed 3 times with 10 volumes of HBSS and resuspended in 1 ml of HBSS for each dose. Two rabbits are injected in the marginal ear vein with the same cell density.

4. Bleed rabbits by cardiac puncture 7 days following the last inoculation. Forty to eighty milliliters of blood are obtained, which are allowed to clot at room temperature for 2 hr.

5. Separate the serum from the clot, and pool the sera collected from each animal. Store sera in 5-ml portions at −20°.

6. For antiserum of higher specificity, or for antiserum against rabbit cells, inoculation of guinea pigs is satisfactory. Six guinea pigs each are injected intraperitoneally with 1 ml of $6 \times 10^6$ washed cells and subcutaneously in four sites with 0.25 ml of $1.5 \times 10^6$ washed cells suspended in Freund's complete adjuvant.[32] The intraperitoneal injection is repeated 21 days later, and the animals are bled by cardiac puncture on day 28. The serum is processed and stored as indicated above.

### B. Flourescent Labeling of Species-Specific Antiserum[33]

*Reagent*

0.5 *M* carbonate–bicarbonate buffer, pH 9.0:
A. $Na_2CO_3$, 53 g/1
B. $NaHCO_3$, 42 g/1
Mix 1 part A with 4 parts B.

*Procedure*

1. Thaw 10 ml of antiserum and add 20 ml of 27% sodium sulphate. Incubate the solution for 16 hr at 37°. Gamma globulins will precipitate

[32] Difco Laboratories, Detroit, Michigan.

[33] C. S. Stulberg, W. F. Simpson, and L. Berman, *Proc. Soc. Exp. Biol. Med.* **108,** 343 (1961).

during this period. Centrifuge the suspension at 3000 rpm for 30 min and remove the supernatant liquid.

2. Add 3 ml of distilled water to the pellet of gamma globulins to dissolve them and transfer the solution to dialysis tubing. Tightly knot the dialysis tubing and dialyze against two changes of 1 liter of distilled water for 30 min each. Dialyze further against six changes of 0.9% NaCl during an 18-hr period.

3. After removing the protein solution from dialysis tubing, sample 0.1 ml to determine the protein concentration; the Biuret method[34] is satisfactory. On the basis of the protein assay, weigh sufficient fluorescein isothiocyanate (FITC) to have 0.025 mg FITC/mg protein. Mix this with nine parts by weight of Celite 501 powder.[35]

4. The dialyzate is adjusted to pH 9.0 by the addition of 0.5 *M* carbonate–bicarbonate buffer (1 volume to 9 volumes dialyzate), and the FITC–Celite 501 mixture is added slowly with gentle agitation of the suspension. Continue to gently shake the suspension for 10 min during which time the FITC becomes bound to gamma globulin. Centrifuge the suspension at 1500 rpm for 5 min and remove the supernate containing FITC-coupled globulin.

5. Because there is free FITC in the protein solution the preparation is charged onto a Sephadex G-25 (fine) column (15 $\times$ 1 cm). Collect the effluent from the column until it appears colorless. Add an additional 10 ml of saline to the column to remove remaining globulin. Store the coupled globulin in 0.5-ml portions in ampules at $-20°$.

### C. Titering and Use of Coupled Antiserum

*Reagents*

Phosphate buffered saline (PBS) pH 7.5:
$NaH_2PO_4H_2O$, 1.38 g/1
NaCl, 9.0 g/1
Adjust to pH 7.5 with 1 *N* NaOH.

*Procedure*

1. Grow approximately $3 \times 10^6$ cells of the same cell line or a line of the same species as that against which antiserum was prepared. Disperse the cells by trypsin–EDTA, if an adherent cell line, and collect the monodispersed cells. Neutralize the dispersing agent activity by adding 5 ml of growth medium containing serum and centrifuge the cell suspension

[34] J. R. Marrack and H. Hoch, *J. Clin. Pathol.* **2,** 161 (1949).

[35] Johns Manville Co., Waterville, Ohio.

at 1500 rpm for 5 min. Wash the pelleted cells 3 times with 10 volumes of PBS. Adjust the concentration of cells to $3 \times 10^6$ cells/ml.

2. Thaw an ampule of antiserum and prepare a series of 2-fold dilutions, from undiluted to 1:128 inclusive of antiserum in PBS. Place 0.1 ml of each dilution of antiserum in a $12 \times 75$ mm tube and add 0.1 ml of cell suspension to each tube.

3. Place the eight tubes on a low-speed rotary shaker to mix the cell–antiserum suspension thoroughly for 30 min at room temperature.

4. Add 1.0 ml of PBS to each tube, shake, and centrifuge at 3000 rpm for 2 min. Remove the supernatant and wash the cell–antiserum mixture again in the same manner. Remove the supernatant and resuspend the cells in 2 drops of PBS.

5. Place a large drop of the resuspended cells on a glass slide and carefully float a $22 \times 30$ mm #1 coverslip on the drop so as to avoid trapping air bubbles or crushing the cells in suspension.

6. Place the slide on the stage of a fluorescent microscope and observe the preparation with a $\times 16$ or $\times 25$ objective by dark-field illumination. When cells are located, switch to fluorescent illumination. Cells reacting specifically have a halo of peripheral bright green fluorescence delimiting the cell membrane. Nonreactive cells are difficult to discern because of a failure to show reaction. Dead cells have bright green fluorescence across the entire cell. The brightness of a specific reaction will quickly fade (30 sec) if fluorescent illumination continues, so a new microscopic field should be located by changing back to dark-field illumination and repeating the process.

7. Examine each slide in the dilution series as described to qualitatively determine the dilution at which brightness is less intense. The working dilution of antiserum is $\frac{1}{4}$ of that dilution.

8. Test the working dilution of the antiserum on other species of cells maintained in the laboratory and compare its reaction with that against the homologous species of cells. If nonspecific cross-reactions do occur, absorption with 1 volume of the cells of the cross-reacting species and 9 volumes of the working dilution of antiserum will usually eliminate the cross-reaction without significantly diminishing the homologous reaction. Alternatively, use of a higher dilution of antiserum will frequently take care of the problem.

9. For routine monitoring cells should be tested within 4 hr of harvesting. The procedure works only with living cells. It is not well adapted to fixed cell preparations or tissue sections. Excess working dilution antiserum can be refrozen and used one additional time. After that, a fresh working dilution antiserum should be prepared for use. The undiluted

coupled antiserum is stable in frozen storage for several years. After that period of time the antiserum may require titering.

*Comment.* The procedure described is not difficult to perform. Results can be obtained in 2–3 hr, and the test can be activated for use as needed. The only reagent that need be prepared for each day of test is PBS. Antisera are stable for a long period of time, and they can be used for other serologic tests of cell identity, including cytotoxicity, hemagglutination, and mixed agglutination.[36–38] Emphasis is placed on a direct immunofluorescent procedure because, unlike the other methods, it can be used to follow cocultivation experiments involving cells of two different species, as well as detecting early evidence of cross-contamination. As few as one cell of one species can be detected in a mixture of 10,000 cells of another species.[39] The indirect immunofluorescent procedure can also be used.

The limitations of the method derive mostly from the limitations of specificity of antisera that can be prepared. Cross-reactions between closely related species occur. For example, chimpanzee, orangutan, and human cells react to substantially equal degree with human antiserum. Rhesus monkey, African green monkey, and baboon cells react similarly with rhesus antiserum, so one should carefully consider the monitoring capability of this procedure in relation to the species of cells being utilized in the laboratory. However, the method can be adapted for use in intraspecies cell characterization. For example, using reciprocally absorbed antisera, specific human T cell and B cell antisera have been developed to analyze T and B lymphoblast cell lines by indirect immunofluorescence.[40]

[36] A. E. Greene, L. L. Coriell, and J. Charney, *J. Natl. Cancer Inst.* **32,** 779 (1964).
[37] K. G. Brand and J. T. Syverton, *J. Natl. Cancer Inst.* **28,** 147 (1962).
[38] D. Franks, B. W. Gurner, R. R. A. Coombs, and R. Stevenson, *Exp. Cell Res.* 608 (1963).
[39] W. F. Simpson and C. S. Stulberg, *Nature (London)* **189,** 616 (1963).
[40] J. Kaplan and W. D. Peterson, Jr., *Clin. Immunol. Immunopathol.* **8,** 530 (1977).

# [14] Tissue Culture on Artificial Capillaries

*By* P. M. GULLINO and R. A. KNAZEK

Survival and growth of cell populations *in vivo* are supported by a vascular network that delivers nutrients and removes products of cellular metabolism. The capillary culture unit (CCU) described here consists of a network of artificial capillaries that simulates the *in vivo* vascular matrix. Employment of this technique permits cells to grow and to reach a density characteristic of solid tissue. In contrast to standard *in vitro* culture pro-



ISBN 0-12-181958-2

cedures, the CCU preserves the pericellular microenvironment and simulates the physiological conditions in which cells function *in vivo*.

## Principles

Under standard tissue culture procedures, cells are immersed in a pool of medium containing essential nutrients and metabolic products. The concentration of both constituents changes as the cell population grows, thus influencing cell survival and function. In the CCU the culture medium flows within the capillaries and its composition can be kept constant. The cell population grows within the extracapillary space, and nutrients and metabolites are exchanged by diffusion from the perfusate within the capillaries. A more complex form of CCU embodies a double network of capillaries whereby differential pressure between the two capillary sets superimposes extracapillary convection upon diffusion.

In neoplastic tissues, anoxic necrosis was observed *in vivo* when the distrance between the capillary wall and the cells was about 150 μm or more.[1] The CCU was built on the assumption that within a rapidly growing population sufficient oxygen tension could not be maintained at distances greater than 150 μm from the capillary wall. If a bundle of artificial capillaries of about 300 μm in outside diameter are tightly packed in parallel, most of the extracapillary space should be within the oxygen diffusion distance of 150 μm.

## Construction of Culture Units

The standard culture unit consists of polymeric membranes in the shape of tubes bundled together within a transparent cylinder (Fig. 1A,B). We refer to these tube-shaped membranes as "capillaries" although their diameter and wall thickness are much greater than an *in vivo* capillary. Several types of artificial capillaries made of polysulfone, acrylic copolymers, or cellulose acetate are available commercially, as is the complete CCU.[2] Such units, individualized to suit specific needs, can be constructed with a variety of available capillaries, as follows:

A bundle of 100–300 capillaries, having inner and outer diameters of ~200 and ~350 μm, respectively, are arranged in parallel and pulled tightly into a glass or plastic shell about 1 cm in diameter and about 10 cm in length. The bundle is closed with a suture tie at both ends of the shell and immersed in a mixture of 12 g liquid silicone rubber (General Electric

[1] R. H. Tomlinson and L. H. Gray, *Br. J. Cancer* **9**, 539 (1955).

[2] Amicon Corporation, Lexington, Massachusetts; Gulf South Research, New Orleans, Louisiana.

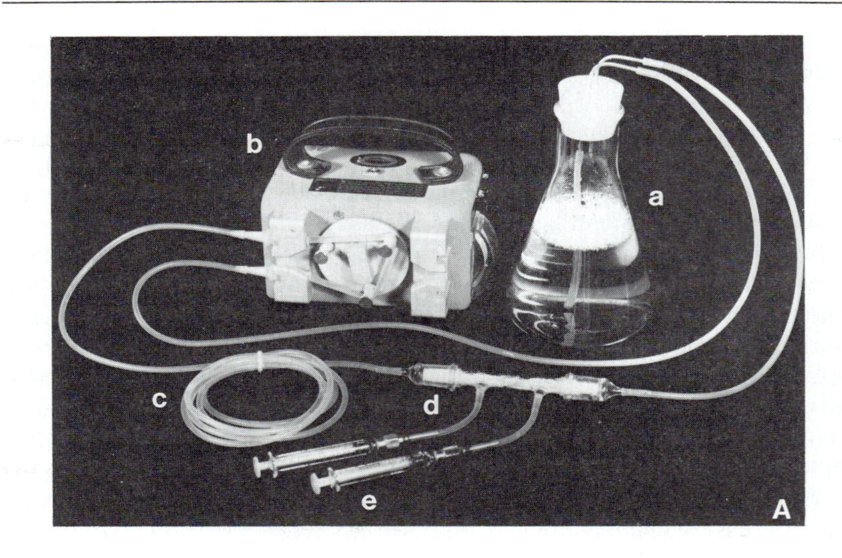


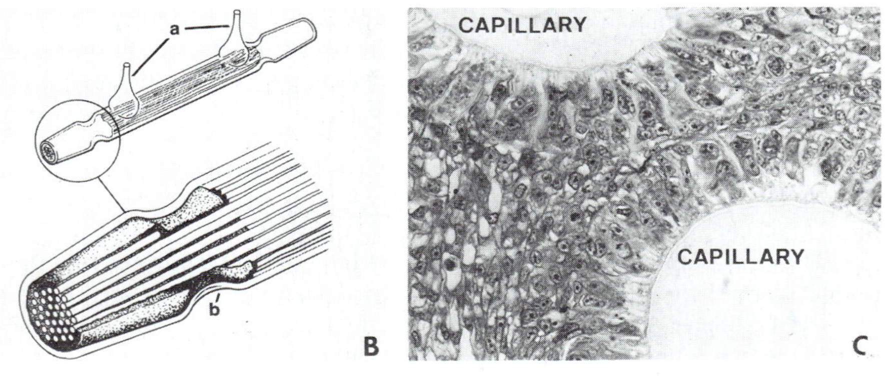


FIG. 1. (A) Capillary culture perfusion circuit: (a) tissue culture medium reservoir; (b) peristaltic pump; (c) silastic tubing coil for oxygenation of perfusate; (d) capillary culture unit; (e) syringes for loading cells onto capillaries. (B) Schematic presentation of capillary culture unit: (a) loading ports; (b) cutaway view shows capillaries encased in polymeric potting material (stippled area). (C) Histologic cross-section of human mammary carcinoma cells (BT20) cultured for 4 weeks within a culture unit. Note the extracapillary space filled with cells that also penetrate the porous outer structure of the capillary walls. Inner diameter of capillary is 200 $\mu$m.

RTV-11) and 8 g Medical Fluid 360 (Dow Corning Corp., Midland, Michigan) that had been formed by catalysis with approximately 40 mg stannous acetate (Tenneco, Nuocure 28). An epoxy resin can be used in place of the silicone rubber mixture. After solidifying overnight, the excess polymer is trimmed flush to the shell to expose the lumen of the capillaries. The process is then repeated for the opposite end. The resultant seal must be tight so that fluid pumped through one end of the CCU will flow only within the capillaries while the extracapillary space, in which the cells are growing, remains undisturbed. With this type of arrangement the exchange between inner and extracapillary spaces occurs mostly by diffusion through the capillary walls.

In the more complex CCU, convective currents are generated within the extracapillary space. The capillaries are intertwined in a braid-like fashion rather than arranged in a parallel array. This unit consists of a central part where the capillaries form a single bundle and two extremities where the capillaries diverge into two bundles, one designated as the "artero-venous" (A-V) circuit and the other as the "lymphatic" circuit. The A-V circuit is filled with medium circulating in a closed recirculating loop to which pressures up to 300 mm Hg can be applied. The lymphatic system is filled with medium circulating at the same rate but at atmospheric pressure. Steady convective currents, therefore, move within the extracapillary space from the artero-venous to the lymphatic networks. The CCU with lymphatic drainage is not available commercially.

### Perfusion Circuit

#### *a. Assembly*

Under standard conditions the CCU is operated at 37° in 95% humidified air + 5% $CO_2$. In this environment the perfusion circuit requires four components: a reservoir for the medium, an oxygenator, a pump, and the CCU interconnected by silicone rubber tubing.

Initial experiments were performed using a flat membrane oxygenator[3] which predisposed the perfusion circuit to leaks and contamination. Subsequent work showed that the oxygenator could be replaced by 2–3 m of silastic tubing (0.125 in. OD, 0.063 in. ID) (Dow Corning Corp.) inserted between the reservoir and the CCU. Since silastic is very permeable to gases, that length of tubing is sufficient to maintain a physiological pH and $pO_2$ in the perfusate flowing at 5 ml/min.

The pump is an extremely important component of the circuit since it must function for several weeks without interruption. The roller-type pump (Holter Co., Extracorporeal Medical Specialties, King of Prussia, Pennsylvania) proved to be the most efficient. It has the advantage of not being in direct contact with the medium, thereby minimizing contamina-

[3] T. Kolobow, W. Zapol, and J. Marcus, *in* "Organ Perfusion and Preservation" (J. C. Norman, ed.), p. 155. Appleton, New York, 1968.

tion, but the disadvantage of wearing out the wall of the pumping chambers. Replacement of this chamber every 10 days is advisable. Another problem is the liberation of microscopic fragments from the inner wall of the pumping chamber into the perfusing medium. After several weeks of uninterrupted activity, occlusion of the capillaries may occur with irreparable damage to the culture.[4]

A CCU with lymphatic drainage requires two circuits, assembled as described above, with an additional device to increase the pressure of the medium circulating within the A-V circuit. This may be attained by pressuring the A-V reservoir with an air–$CO_2$ mixture. The lymphatic circuit is connected to a second reservoir, and the medium is circulated at normal atmospheric pressure. An equal flow rate is usually maintained in both lymphatic and A-V circuits.

It is convenient, although not indispensable, to insert a filter in the perfusion circuit downstream from the pump with an optional bypass. In this manner, sampling sites can be inserted into the circuit.

The recommended sterilization of the CCU is in ethylene oxice for 10 hr followed by extensive aeration; however, certain types of CCUs may be autoclaved, or the unit may be perfused in a 3% formaldehyde solution at 37° overnight.[4] The remainder of the perfusion circuit may be autoclaved. All components are then assembled in an aseptic environment. The assembled circuit must be perfused with sterile, pyrogen-free water for at least 2 and preferably 7 days with at least three changes of water prior to replacing the perfusate with complete medium. The objective of the preliminary flushing is to eliminate residues that are toxic to the cells.

The sterilized circuit should always be handled under rigid aseptic conditions; one should replace the reservoir in a sterile hood while wearing sterile gloves, and the stopper of the reservoir should be covered by aluminum foil to be replaced with each change of the medium.

Seeding of the CCU is done with approximately $5 \times 10^6$ cells obtained either by dissociating tissues or from an established culture. The cell suspension is injected through one shell port while a slight vacuum is drawn by a syringe on the second port (Fig. 1A). The fluid must be transferred slowly to deposit, rather than channel, cells or cell clumps between the capillaries.

A perfusion flow rate of 0.5–1.0 ml/min is maintained at the beginning of the culture and then gradually increased as the culture grows. When the cells fill more than 25% of the extracapillary space, a perfusion volume of 10 ml/min must be maintained to supply adequate amounts of oxygen and nutrients. Under these conditions the recirculating medium becomes depleted very rapidly and frequent changes of the reservoir are required.

[4] C. F. W. Wolf and B. E. Munkelt, *Trans. Am. Soc. Artif. Intern. Organs* **21,** 16 (1975).

Any culture medium used for a specific cell type, grown in standard *in vitro* cultures, can be used for CCU perfusion. In this system the capacity to supply oxygen to the culture is severely limited by the low solubility of oxygen in the medium. To improve delivery, hyperbaric conditions may be employed. Fluorocarbons (3M Corp., St. Paul, Minnesota) or erythrocyte suspensions may also be used as oxygen carriers, although they are not necessary under standard conditions.

Cell growth can be detected by gross observation of the CCU. During the second week after seeding, a white haze appears and gradually obscures the capillaries. The amount of tissue present within a CCU can be estimated by measuring the total DNA content, but this requires destruction of the culture. A less precise estimation can be made by measuring the glucose and oxygen consumption of the growing cell population in the CCU. In repetitive cultures of the same cell population, the correlation between both metabolic parameters and DNA content of the CCU is reasonably reliable.

Histologic sections through the capillary bundle can be obtained by flushing the perfusion circuit for 1–2 min with warm buffered saline and then circulating a fixative solution, such as 10% formaldehyde, through the capillaries for several hours. The extracapillary medium is then replaced with a 2% agarose solution at 37° by slowly injecting the solution through one shell port while withdrawing it from the other until agarose has penetrated the whole bundle. After solidification the shell is removed and the capillary bundle can be processed for histologic examination (Fig. 1C).

## Utilization of the Culture Units

The major theoretical reason for developing the CCU was to reproduce tissue-like structures *in vitro*. In the CCU a high cell density ($100 \times 10^6$ cells/cm$^3$) is obtained, much higher than under standard *in vitro* culture conditions; the pericellular microenvironment is preserved; the exchange of metabolites occurs through a capillary-like network; and the same culture can be studied for several weeks. The importance of these parameters in regard to the biological properties of cell populations has already been evaluated under a variety of conditions. Prolactin and growth hormone production were studied in rat and human cells of pituitary origin.[5] Secretory stimulation of human $\beta$ cells was studied,[6,7] and

[5] R. A. Knazek and J. S. Skyler, *Proc. Int. Symp. Growth Horm. Rela. Pept., 3rd, 1975* p. 386 (1976).

[6] W. L. Chick, A. A. Like, and V. Lauris, *Science* **187,** 847 (1975).

[7] R. A. Knazek, *in* "Pancreatic Beta Cell Culture" (E. von Waiselewski and W. L. Chick, eds.), p. 29. Excerpta Med. Found., Amsterdam, 1977.

CCU units were used in a successful attempt to reproduce an "artificial pancreas."[8-10] Bilirubin conjugation was observed with hepatic cells cultivated in CCU units, and *in vivo* transplants were also attempted.[4] Chorionic gonadotropins were produced in large amounts from human choriocarcinomas,[11-13] and carcinoembryonic antigen was also harvested from cells (LS-174) isolated from a human colon carcinoma.[13] Cells maintained in culture under standard procedures have shown two fairly general characteristics, i.e., loss of functional differentiation and shedding of cell components. Both events represent a severe limitation in correlating *in vitro* and *in vivo* conditions; the CCU represents a step toward bridging this gap.

[8] W. J. Tze, F. C. Wong, L. M. Chen, and S. O'Young, *Nature* (*London*) **264,** 466 (1976).
[9] A. M. Sun and H. G. Macmorine, *Diabetes* **25,** Suppl. 1, 339 (1976).
[10] W. L Chick, A. A. Like, V. Lauris, P. M. Galletti, P. D. Richardson, G. Panol, T. W. Mix, and C. K. Colton, *Trans. Am. Soc. Artif. Intern. Organs* **21,** 8 (1975).
[11] R. A. Knazek, P. M. Gullino, P. O. Kohler, and R. L. Dedrick, *Science* **178,** 65 (1972).
[12] R. A. Knazek, P. O. Kohler, and P. M. Gullino, *Exp. Cell Res.* **84,** 251 (1974).
[13] L. P. Rutzky, J. T. Tomita, M. A. Calenoff, and B. D. Kahan, *In Vitro* **13,** 191 (1977).

## [15] Microcarrier Culture: A Homogeneous Environment for Studies of Cellular Biochemistry

*By* WILLIAM G. THILLY and DAVID W. LEVINE

Microcarrier culture is the growth or maintenance of anchorage-dependent cells on small beads suspended in a stirred tank. The primary conception of this technology was made by A. L. van Wezel of the Netherlands.[1] He and his colleagues were responsible for several important demonstrations of feasibility.[2-4] Other laboratories have confirmed their observations.[5,6] Van Wezel and others noted a marked toxicity of the original microcarriers which was ascribed variously to binding of essential nutrients or unidentified toxic contaminants associated with manufacture. However, recent studies have revealed a simple parameter, surface

[1] A. L. van Wezel, *Nature* (*London*) **216,** 64 (1967).
[2] A. L. van Wezel, *Prog. Immunobiol. Stand.* **5,** 187 (1972).
[3] A. L. van Wezel, *in* "Tissue Culture: Methods and Applications" (P. F. Kruse, Jr. and M. K. Patterson, eds.), p. 372. Academic Press, New York, 1973.
[4] D. van Hemert, D. G. Kilburn, and A. L. van Wezel, *Biotechnol. Bioeng.* **11,** 875 (1969).
[5] C. Horng and W. McLimans, *Biotechnol. Bioeng.* **17,** 713 (1975).
[6] R. E. Spier and J. P. Whiteside, *Biotechnol. Bioeng.* **18,** 659 (1976).



ISBN 0-12-181958-2

charge density, which, when optimized, permits the facile attachment and growth of animal cells on microcarriers without any of the toxic effects previously noted.[7-9]

The advantages of microcarrier culture are based on provision of a large surface area in relation to the volume of the growth vessel used. This may be understood by comparing roller bottles to microcarriers in terms of the surface area afforded. A standard 500 $cm^2$ roller bottle requires 100 ml of nutrient medium. One milliliter provides 5.0 $cm^2$ of growth surface. In our microcarrier cultures, 1 ml of medium provides a minimum of 30 $cm^2$ of growth surface, and the opportunity exists for increasing this number by 2 or 3 times. Thus, for most cell types, a 5-liter spinner flask would replace a facility using 300 roller bottles. While the simple increase in surface area is important, other advantages are offered as a result of the suspended, stirred configuration of microcarrier culture. Possibly the most important advantage is the provision of a homogeneous environment for the cell population, in which the time-dependent changes in microenvironment which occur in culture plates and minimally agitated roller bottles are reduced by simple mixing of the cellular microenvironment and the bulk liquid of the culture. We suspect that students of cell physiology are performing many studies without knowledge of microenvironmental changes or their effects on cellular behavior. Microcarrier culture permits the investigator to increase the rate of exchange between the cell surface and the bulk liquid. Thus, the properties of the cellular macro- and microenvironments can approach identity. The bulk properties (pH, redox potential, concentrations of excreta and nutrients, etc.) become controllable by the experimenter with the facility afforded in bacterial, yeast, or suspension-adapted animal cell culture. When needs arise for amounts of cellular material in the range of 10–100 g, scale-up can be accomplished easily using very simple equipment.

## Source of Microcarriers

Microcarriers are obtained from Flow Laboratories, Inc., Rockville, Maryland. A usable small-scale procedure for their synthesis is based on that of Hartman[10] in which diethylaminoethylchloride · HCl is reacted

[7] D. W. Levine, D. I. C. Wang, and W. G. Thilly, *in* "Cell Culture and Its Applications" (R. T. Acton and J. D. Lynn, eds.), pp. 191–216. Academic Press, New York, 1977.

[8] D. W. Levine, J. S. Wong, D. I. C. Wang, and W. G. Thilly, *Somatic Cell Genet.* **3,** No. 2, 149–155 (1977).

[9] D. W. Levine, D. I. C. Wang, and W. G. Thilly, *Biotechnol. Bioeng.* (submitted for publication).

[10] M. Hartmann, U.S. Patent 1,777,970 (1930).

with the glucose of cotton fiber under alkaline conditions as previously described.[8]

Uncharged, unhydrated beads of cross-linked dextran are obtained commercially from Sigma Chemical Co., St. Louis, Missouri, as the gel chromatography support Sephadex G50M. The commercial product is sieved, and the dry bead fraction having diameters between 50 and 75 $\mu$m is used for subsequent steps. It should be noted that the commercial products variously described as DEAE-Sephadex and marketed both by Pharmacia Fine Chemicals, Piscataway, New Jersey, or Sigma Chemical Company, have been found unsuitable for use as microcarriers in our hands. The limitations of these products have been extensively discussed elsewhere.[1–8]

Diethylaminoethylchloride · HCl (DEAE · HCl) is obtainable from Sigma Chemical Company. Before use, it is twice recrystallized from reagent-grade methylene chloride. Sodium hydroxide used in the reaction must also be reagent-grade material.

One gram of dry sieved beads is hydrated with 10 ml of distilled water. Ten milliliters of distilled water which contain 10 mmol DEAE · HCl and 15 mmol NaOH are added to the hydrated beads. The suspension is agitated briskly at 60° for 60 min. Microcarriers are separated from the reaction mixture by filtration and are washed with 1 liter of distilled water.

The reaction is essentially complete in 60 min and binds 2.0 mmol of DEAE to 1 g of dry cross-linked dextran beads. This corresponds to a microcarrier with an exchange capacity of 2.0 meq/g cross-linked dextran. To produce a carrier with a lower degree of DEAE binding, similar reaction conditions are employed but reaction periods of less than 1 hr are used. For example, to produce carriers having an exchange capacity of 1.0 meq/g cross-linked dextran, the above reaction mixture is agitated on a shaking water bath for 3 min at 60°.

The synthetic product has been found to result in excellent culture productivity for a number of different cell types: primary or secondary cultures, fibroblasts, or epithelial cells. The charge density of 2.0 meq/g dry dextran bead is optimal for growth of secondary chick fibroblasts.[8] It is probable that other cell types will require slightly different charge densities to achieve optimal results. Our experience to date indicates that a range of 0.5–2.5 meq/g dry dextran bead may be profitably explored when a new cell type is approached.

After synthesis, microcarriers should be titrated to determine their exact charge capacity as part of a quality-control procedure. The carriers formed by the reaction of 1 g of dry cross-linked dextran with DEAE · HCl are thoroughly washed with 0.1 *N* HCl (4 × 50 ml) to allow saturation of all the exchange sites with chloride ions. The carriers are rinsed with $10^{-4}$ $N$ HCl (4 × 50 ml) to remove the unbound chloride. The beads are rinsed with aqueous sodium sulfate (10% w/w), and the effluent is collected. This procedure replaces bound chloride ions with sulfates, releasing the chlorides into the effluent solution. The chloride is titrated with 1 $M$ silver nitrate in the presence of dilute potassium chromate as an indicator. The carriers are washed with 30-ml portions of 10% sodium sulfate solution until the amount of released chloride in the effluent is less than 0.1 m$M$. Typically, three to four washes are required. After titration, the carriers are again washed thoroughly with water (500 ml) and then with phosphate-buffered saline (PBS). The carriers are suspended in 100 ml of PBS and autoclaved.

### Initiation of the Culture

The microcarrier culture bottle is simply a glass or plastic vessel containing a suspended stirrer in which the provision is made to avoid grinding surface contacts with the liquid culture. Simple bottles which are particularly convenient for microcarrier work are available from Wilbur Scientific, Boston, Massachusetts, in the volume range of 50–1000 ml. As mentioned, microcarriers are sterilized by autoclaving a suspension of physiological saline buffered at pH 7.

The culture is begun by the combination of microcarriers, e.g., 5 mg dry weight beads per milliliter culture medium, with the desired volume of culture medium and the introduction of the cells to be used (freshly trypsinized or otherwise collected) as a suspension into the culture bottle. The bottle is gassed to the desired $CO_2$ concentration if necessary and placed on a magnetic stirrer motor (50–90 rpm) and allowed to incubate at 37°. This process is indeed as simple as it must seem to the reader acquainted with cell culture practice.

### Observation of the Culture

The use of the microcarriers at 5 mg dry weight dextran per milliliter will provide about 40,000 beads/ml. Seeding at 300,000 cells/ml assures distribution of cells to all carriers. Within a few minutes, most cells will be attached to microcarriers as illustrated for secondary chick fibroblasts in Fig. 1A. Samples of stirred cultures are simply withdrawn by pipetting using nonwettable pipettes to avoid adherence of beads to hydrophilic surfaces. Plastic pipette tips (Rainin Instrument Company, Brighton, Massachusetts) are the simplest solution to bead adherence. Samples are observed directly under a microscope. In a few hours, cells will have attained their normal culture morphology as illustrated in Fig. 1B. Growth

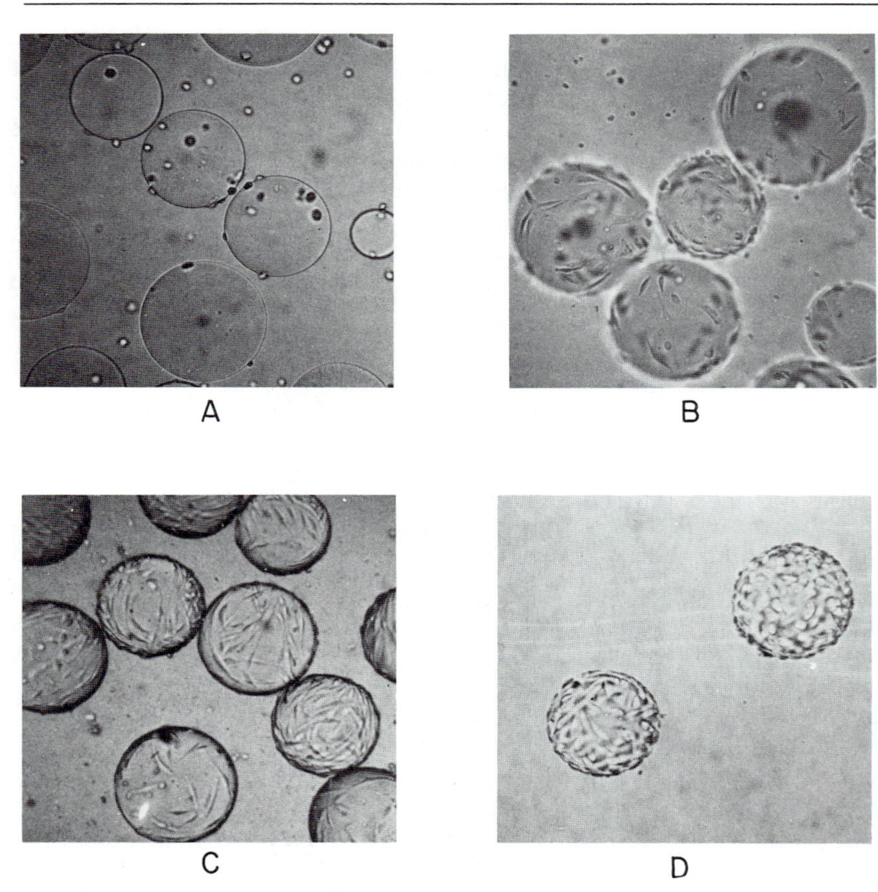


FIG. 1. Attachment and growth of secondary chick embryo fibroblasts on microcarriers. (A) Thirty minutes after seeding. Five grams per liter microcarriers, 2.0 meq/g dextran. Initial concentration of 300,000 cells/ml and approximately 40,000 beads/ml. Only 10% of seeded cells can be found in the supernatant liquid 30 min after seeding under stirred conditions. (B) Twelve hours after seeding. All cells have assumed typical fibroblastic morphology. (C) Forty-eight hours after seeding. Cell number has increased almost 4 times relative to seeding density. (D) One hundred twenty hours after seeding. Beads are covered with a confluent cell layer.

is easily observed microscopically as seen in Fig. 1C. Confluence is attained as on other culture surfaces (Fig. 1D).

Cell counting is easily performed using a hemocytometer and the technique for enumeration of nuclei reported by Sanford *et al.*[11] A 1-ml

[11] K. K. Sanford, W. R. Earle, V. S. Evans, J. K. Waltz, and J. E. Shannon, *J. Natl. Cancer Inst.* **11**, 773 (1951).

sample of the culture is centrifuged and resuspended in 1 ml of 0.1 *M* citric acid containing 0.1% crystal violet, in which it is allowed to incubate at 37° for 1 hr, pipetted up and down gently, and loaded onto a hemocytometer. The hypotonic solution leads to cellular but not nuclear lysis, and the crystal violet stains the nuclei a dark purple to permit easy enumeration. The beads are too large to enter the hemocytometer counting chamber and do not interfere with the count.

The counting procedure of removing cells from carriers by trypsinization, sieving to remove beads, and counting cells electronically should also work easily, but we have not yet diverted sufficient attention to develop a specific protocol.

Figure 2 shows the growth curve for secondary chick fibroblasts shown in Fig. 1A–D. Saturation is attained by the fifth day at a cell concentration of 4 × $10^6$ cells/ml. Within 3 days of reaching confluence, cell death and detachment are evident. A 50% replacement of medium is made 48 hr after culture initiation.

Similar results, short or nonexistent lag times, rigorous growth to confluence, and high final cell densities have been observed for a variety of cell types, both fibroblastic or epithelial. Figure 3 shows the appearance of (A) Chinese hamster ovary (CHO) cells in exponential growth prior to attaining confluence and (B) HeLa cells at approximately the same growth

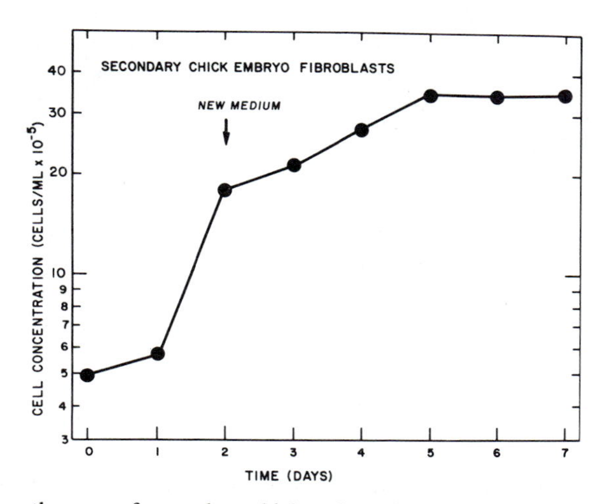


FIG. 2. Growth curve of secondary chick embryo fibroblasts. Initial seeding at 300,000 cells/ml with 5 g/liter microcarriers at 2 meq/g dextran. After an initial lag phase (typically shorter than observed in this experiment) exponential growth to about 2 × $10^6$ cells/ml is followed by linear growth to saturation between 3–4 × $10^6$ cells/ml. Use of 10 g/liter microcarriers with timed partial medium changes will bring chick embryo fibroblast concentrations to 1.0–1.2 × $10^7$ cells/ml.

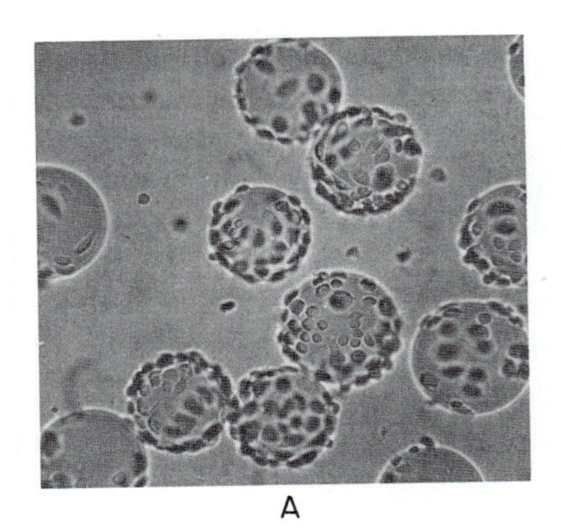

A

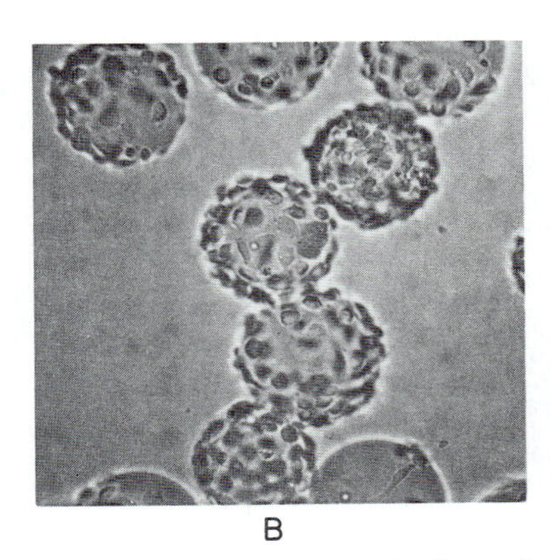

B

FIG. 3. Appearance of epithelial cell lines on microcarriers. Cultures at 5 g/liter microcarriers, 2.0 meq/g dextran. Both cultures are in exponential growth phase. (A) Chinese hamster ovary cells; (B) HeLa cells.

state. CHO, HeLa, murine cl 1, and human embryonic lung fibroblasts (HEL299) have demonstrated vigorous growth without any refeeding after seeding. However, some cell types such as African green monkey kidney (AGMK) or human foreskin fibroblasts (FS-4) have not demonstrated equally vigorous growth (Fig. 4).

To date, no dependence on medium type or formulation has been observed. At this time, it seems that media suitable for growth of cells in roller bottles or petri dishes are equally suitable when used with the microcarriers described in this article.

### Harvesting

The mode of harvesting will vary depending on individual research needs. The simple washing of the microcarriers with buffered saline, trypsin exposure, and separation of cells from beads by differential sedimentation rate or by sieving is easy and effective. Increased shear can be achieved in a number of ways such as increased impeller velocity, pumping through a small orifice,[6] or other strategems.

Some workers have asked about removing cells without trypsin for biochemical studies in which loss of external cellular components is undesirable. We have been unable to devise a means of preventing such losses in harvest, but note that considerable cellular material remains on microcarriers and standard glass or plastic dishes when either trypsinization or strictly mechanical means are used to remove cells from the attachment substrate. An alternate approach to this problem might be the em-

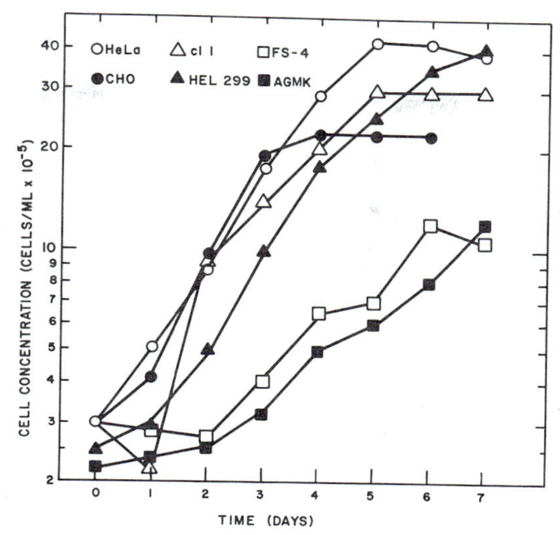


FIG. 4. Growth curves for various cell types. All were seeded at 2–3 × $10^5$ cells/ml in 100-ml cultures, 5 gm/liter microcarriers at 2.0 meq/g dextran. HeLa—human epithelial-like, aneuploid; CHO—hamster epithelial-like, pseudodiploid; cl 1—murine epithelial-like, aneuploid; FS-4—human fibroblast, diploid; AGMK—African green monkey, epithelial-like, diploid.

ployment of a synthetic growth surface of which a known component would be severed by enzymic treatment, leaving the cell surface or its associated exudate unaltered.

## Viral Studies

Viruses of interest to investigators may remain physically within a cell, be budded continuously from the cell surface, or be expressed into the liquid support medium by cell lysis. In the case of a cell-contained virus, e.g., Marek's disease, the cells simply can be stripped from the beads as in cell harvest and the cell suspension used in biochemical studies or as the basis of a vaccine.

Some budding viruses, e.g., RNA tumor viruses, present a special challenge to virologists in that loss of infectious units with time is often a problem necessitating frequent or continuous harvesting. Microcarrier culture offers a ready solution to these problems. The motor drive of a microcarrier culture, and a pump connecting the culture to a refrigerated collection bottle, can be connected to a timer. The culture is grown to an appropriate cell density, infection is accomplished, and when viral expression commences, the stirring motor is turned off, the microcarriers allowed to settle for a few minutes, the supernatant pumped into the collection bottle, fresh medium pumped into the culture, and stirring begun again. This cycle is repeated at intervals appropriate to the cells and viruses involved.

Viruses expressed into the culture medium lytically are handled similarly. Stirring ceases, and the supernatant liquid above the settled beads is pumped into a receiving container.

One consideration in the use of microcarrier culture is that the viral growth cycles may not be the same in these microcarrier suspension cultures as they are on roller bottles or plates, possibly because of differences in microenvironmental conditions. Thus, preliminary experiments optimizing for time of infection, latent period, and harvest are necessary. Viruses on microcarriers that have been studied in this manner include Sindbis, polio, Moloney murine leukemia, and vesicular stomatitis.[12] Rabies virus has been produced in excellent yield,[13] and a number of commercially important veterinary vaccine viruses are apparently produced in good yield.

## Interferon and Other Cellular Products

Theoretically, interferon or other secretory products could be produced in a manner directly analogous to the budding viruses on a semicontinuous basis. Initial interferon studies[14] have found that using the same conditions of cell type, growth media, and superinduction, the interferon production on a per-cell basis was the same for cells in microcarrier culture or in roller bottles.

There seem to be no obvious reasons to expect that other cellular products such as peptide hormones or secretory enzymes could not be made or studied to advantage through the use of microcarrier systems. It should be noted, however, that the microcarriers described in this article are porous, positively charged hydrogels. Some products having a low molecular weight or a negative charge may be expected to associate with the microcarriers and possibly require special elution procedures.

## Mitotic Cell Selection

One limitation on study of the mammalian cell division cycle has been the difficulty of obtaining a sufficient number of synchronously dividing

[12] D. J. Giard, W. G. Thilly, D. I. C. Wang, and D. W. Levine, *Appl. Environ. Microbiol.* **34,** 668–672 (1977).

[13] B. Mered, personal communication.

[14] D. J. Giard, D. H. Loeb, W. G. Thilly, D. I. C. Wang, and D. W. Levine, *Biotechnol. Bioeng.* (submitted for publication).

cells. One way of obtaining cells of sufficient synchrony is to selectively detach the relatively nonadherent mitotic cells from their attachment substrate. This method, devised by Terasima and Tolmach,[15] yields suspensions of cells, some 80–95% of which are mitotic. At this writing, we have extended this concept to microcarriers. Microcarrier cultures of Chinese hamster ovary cells in exponential growth ($\sim 10^6$ cells/ml, 5 mg/ml microcarriers) are simply agitated more vigorously by increasing the spinner speed for 10 min after 2 hr of growth in the presence of 0.08 $\mu$g/ml colcemid. About 85% of the cells freed into suspension are mitotic as measured microscopically, and the increase in cell number observed soon after plating agrees with this estimate. No doubt other cell lines will lend themselves to this approach. Since 1–2% of the total population has been recovered in our experiments with CHO cells, 100 ml of culture yield between 1 and $2 \times 10^6$ mitotic cells. Of course, scale-up is possible.

## Biochemical Engineering

Microcarriers appear to offer a ready means for large-scale use of anchorage-dependent animal cells. The special concerns of such usage are beyond the scope of this article, which is intended to offer the basic details of this approach for the consideration of cell biologists and biochemists who may be hampered by the difficulty and expense of present means of cell culture. However, engineering questions are discussed in a separate publication.[9]

## Comments

Our collective knowledge of the *nutritional* substrates for growth of bacteria, yeasts, and higher plant and animal cells has advanced considerably beyond our knowledge of how living cells use—and why they sometimes require—*attachment* substrates for cellular growth toward division or for growth in a particular physiological mode. Our present understanding may be likened to that of the early nutritional biochemists who knew many of the nutrients that were necessary for the growth of bacteria or young animals, but did not yet comprehend the specific enzymic systems which used the nutrients as substrates or required the various vitamins as catalysts. In this presentation of methodology, which is *sufficient* for the growth of anchorage-dependent animal cells, we recognize that the underlying molecular mechanisms for the specific requirements are essentially unknown. Perhaps the definition of these requirements will lead to hypotheses useful in the design of experiments that can

[15] T. Terasima and J. Tolmach, *Exp. Cell Res.* **30,** 344 (1963).

result in better understanding of a cell's relation to its attachment substrate.

Of course, investigators have been growing cells on solid substrates for some 75 years. Blood clots, collagen, glass, treated plastics, various cross-linked proteins, and other materials, including stainless steel, have been applied with varying degrees of success. The petri dish, the T-flask, and the roller bottle used by cell researchers today are primarily made of polystyrene treated with ozone via plasma discharge exposure and carry a net negative charge of approximately 0.3–0.5 meq/$cm^2$,[16] creating a wettable surface, but it is not clear that hydrophilicity alone is both necessary and sufficient for cell attachment and growth. Wettable surfaces can, of course, be created with positively charged substituents, as discussed in this article. Our ignorance of specific mechanisms prevents us from defining which specific microenvironmental parameters are necessary for cell attachment and growth.

### Acknowledgments

The research reported in this chapter was supported by National Science Foundation Grant #77-15463PCM and National Cancer Institute Grant #NIH-2-RO1-CA-15010-04, and it was performed in the M.I.T. Cell Culture Center. We gratefully acknowledge the technical assistance of John Y. Ng and Charles L. Crespi in the preparation of growth curves and photography, which are original with this article.

[16] N. C. Maroudas, *J. Theor. Biol.* **49,** 417 (1975).

# [16] Mass Culture of Mammalian Cells

*By* WILLIAM F. MCLIMANS

## Introduction

The purpose of this article is to provide a guide to the mass culture of mammalian cells for those investigators who require milligram or gram quantities of cells for biochemical studies. Thus, some points may appear controversial or even offensive to the "cell culturist" anxious to protect his domain. The "state of the art" and need are such, however, that an effort must be attempted, even at that risk. The rewards are great for those who would exercise control over their own source of cells.

The approach is complicated by the multiple configurations of equipment and methods that are available. Basically, however, there are only two effective mass-scale systems; one employs the suspension culture of established cell lines, and the other and newer technique, is that of direct



ISBN 0-12-181958-2

growth in suspension of primary cells that are anchored to small "beads" or microcarriers. Since both systems have been extensively employed in our laboratories, we will present our operative methods. These are methods that may not be optimal, superior, or even equal to others, yet they are methods that we know "first hand" to be reliable. It should be emphasized that, as a biologic system, there is no universal culture media, no universal procedure, and certainly no universal cell system.

No matter which system is used, detailed attention must be directed toward the "tooling up" procedures—the preparative operations, i.e., glassware, media preparation, reagent selection, water quality, etc. Accordingly, these prerequisite, operative procedures will be dealt with in the first section, prior to discussing the actual operations of primary vs. cell line mass culture configurations. Little attention will be directed toward equipment configurations and/or applications, since these have been covered in other reports, such as that of Nyiri.[1]

## Preparative Operations

### *Water*

Extensive experience has demonstrated conclusively that the single most important constituent of the culture media is proper-quality water. It is recommended that wide-mouth bottles of sterile, pyrogen-free water, as purchased for intravenous use in man (*a*),[2] be employed in the direct formulation of all culture media and/or reagents, as directly added to the culture environment. This entails an expense of about $0.66/liter of media,[3] a minor cost, compared to the uniformity of culture results. In large systems—100 liters or more—this cost might be prohibitive. In such cases, triple glass-distilled water may be used, but this requires care in preparation as well as periodic checks to certify the absence of pyrogens prior to use for media formulation. Pyrogen checks are particularly important following installation of new stills, after cleaning or repair of stills, and in the instance of stills which have not been used for a long period. On a day-to-day basis, stills should be drained each evening and "steamed" out prior to use each morning.

Triple glass-distilled water should be employed as the final rinse for all glassware, as well as for the preparation of various noncritical reagents and solutions. In no instance should the distilled water be collected in

[1] L. K. Nyiri, *Biotechnol. Bioeng. Symp.* **3,** 31–40 (1972).

[2] The italicized lower case letters in parentheses refer to vendors whose names are summarized in Table I.

[3] Prices are estimates as of January 1978.

TABLE I
SOURCES: MATERIALS AND EQUIPMENT

| Letter ref. | Source |
|---|---|
| *a* | Travenol Laboratories, Inc., 1 Baxter Parkway, Deerfield, Illinois 60015 |
| *b* | Difco Laboratories, Detroit, Michigan |
| *c* | Flow Laboratories, Inc., 1710 Chapman Avenue, Rockville, Maryland 20852 |
| *d* | Grand Island Biological Company (GIBÇO), 3175 Staley Road, Grand Island, New York 14072 |
| *e* | Microbiological Associates, 5221 River Road, Bethesda, Maryland 20016 |
| *f* | Colorado Serum Company, 4950 York Street, Denver, Colorado 80216 |
| *g* | Fiske Associates, Inc., Bethel, Connecticut |
| *h* | Cryogenic Supply Co., Inc., Quinby Park, Hamburg, New York 14075 |
| *i* | The Matheson Co., Inc., East Rutherford, New Jersey |
| *j* | Armour Pharmaceuticals, Inc., P.O. Box 511, Kankakee, Illinois 60901 |
| *k* | Bellco Glass, Inc., 340 Edrudo Rd., Vineland, New Jersey 08360 |
| *l* | Ace Glassware, Inc., Vineland, New Jersey / 54 Moddy St., (P.O. Box 425), Ludlow, Massachusetts 01056 |
| *m* | Virtis Co., Route 208, Gardiner, New York 12525 |
| *n* | New Brunswick Scientific Co., Inc., 1130 Somerset St., New Brunswick, New Jersey 08903 |
| *o* | Fermentation Design, Inc., Div. of New Brunswick Scientific Co., Bethlehem, Pennsylvania 18017 |

anything other than Pyrex glass or plastic containers specifically reserved for this purpose. Distilled water which has been stored without sterilization for more than 6 hr should not be used.

### *Chemicals*

Biologic grade or USP chemical reagents should be used in the formulation of all tissue culture preparations. Certification of the source and purity of the reagents should be furnished by the vendor. A certificate of analysis of amino acids and vitamin mixtures should be obtained prior to purchase of bulk lots.

The use of commercially prepared liquid culture media is not recommended for the following critical reasons. One has no control over the quality of the water component. The formulation structure is too rigid to permit critical adjustment of pH, osmolarity, and glucose concentration at various levels. Some media components, such as penicillin and glutamine, are labile and thus undergo continuous changes in concentration as a result of degradation.[4] We have found the preweighed, dry powder, media mixes without salts, glucose, antibiotics, and/or bicarbonate, in 1-, 5-, 10-, or 100-liter batches to be ideal for our investigations (*b, c, d*).

*Serum*[5]

An important variable of cell culture media is the serum component.[6,7] Commercially available serum is usually procured by random bleeding of slaughter-house animals (*b, c, d, e*). Relatively little control is exercised, over the selection of donors, as based on the animal history, i.e., exposure to infectious agents, toxic grasses (bishydroxycoumarin), pesticides, etc. The problem can be resolved to a degree by obtaining test aliquots from a large serum lot. Following the demonstration of efficacy of the serum aliquot in supporting cell growth and generation time, the entire lot may be reserved for future use. This has proven to be a satisfactory procedure in many laboratories. More costly, but superior, is the commercial production of sera from animal herds maintained by the producer with absolute control over the history and exposure of animals to such things as antibiotic-containing feed and diethystilbestrol (*f*).

[4] G. L. Tritsch and G. E. Moore, *Exp. Cell Res.* **28,** 360–364 (1962).

[5] Careful certification of the serum requirements to the vendor may be made in terms of the following typical specifications: (1) All sera should be prepared from blood collected under aseptic conditions. (2) Periodic inspection of bleeding procedures should be made by the vendor. (3) Insofar as possible, efforts should be made to preclude the bleeding of animals which are: suspect of having a disease or derived from a herd known to contain any sick animals; known to have been treated with drugs or vaccines for 30 days prior to bleeding; known to have had access to or to have been exposed to antibiotic-containing feed, pesticides, or insect sprays for 30 days prior to bleeding. The vendor should furnish a statement as to the feasibility and the contemplated degree, if any, of his compliance with this specification. (4) Preservatives, decolorizing agents, or foreign materials should not be added to sera. (5) Sera should be immediately frozen, stored, and delivered, in the frozen state, without prior heat inactivation. (6) Immunoelectrophoresis may be performed to verify source and normal protein distribution. (7) Physical and chemical specifications of serum may include a protein concentration of at least 3.0%; hemoglobin content not above 25 mg/100 ml (Mollison); total solids not below 5.5%; and nonprotein nitrogen not greater than 0.7 mg/ml. The specific gravity should not be less than 1.02, with a pH range within the limits of 7.0–7.5. The serum should not be turbid, should be low in lipids, and should be totally free of visible surface "fatty ring." (8) The final serum product should be critically tested to insure sterility. The sera should be demonstrated to be PPLO and virus-free, noncytotoxic, and able to support the growth of primary and cell line tissue cultures.

[6] K. Honn, A. Singley, and W. Chavin, *Proc. Soc. Exp. Biol. Med.* **149,** 344–347 (1975).

[7] S. Fedoroff, V. J. Evans, H. E. Hopps, K. K. Sanford, and C. W. Boone, *In Vitro* **1,** No. 3, 161–167 (1972).

TABLE II
STOCK SALT SOLUTIONS

| Stock number | Stock reagent | Final (×1) (mg/liter) | Stock (×10) (g/liter) | Stock (ml/liter) |
|---|---|---|---|---|
| 1 | NaCl | 6460 | 64.6 | 90[a] |
| 2 | $CaCl_2$ (anhyd.) | 100 | 10.0 | 10[b] |
| 3 | KCl | 400 | 40.0 | 10 |
| 4 | $MgSO_4 \cdot 7\ H_2O$ | 97.7 | 9.77 | 10 |
| 5 | $NaH_2PO_4H_2O$ | 580 | 58.0 | 10 |
| 6 | Phenol red | 2 | 20.0 mg | 2 |
| | | *Glucose* | | |
| 7 | Glucose | 1–4 g/liter<br>Stock of 100 mg/ml or 100 g/liter in water. | | |
| | | *Antibiotic* | | |
| 8 | Gentamycin | 50 mg/ml stock: use 0.5 ml/liter = 0.025 mg/ml = 25 μg/ml. | | |
| 9 | Mycostatin | 500,000 μg/vial + 10 ml $H_2O$ = 50,000 μg/ml stock: use 2.0 ml stock/liter for final conc. of 100 mg/ml. | | |
| | | *Buffer adjust stock* | | |
| 10 | $NaHCO_3$ | 5 g/100 ml or 5% | | |
| | | *Osmolarity adjust stock* | | |
| 11 | NaCl (mos *M*) | 1 mg/ml | | |

[a] Stock number 1: NaCl is added at lower level than that normally used so as to permit osmolarity adjustment later.

[b] In some suspension-type cultures the deletion of $Mg^{2+}$ and $Ca^{2+}$ ions is recommended in order to reduce the tendency of some cell types to clump.

## *Media Preparation*

We prepare culture media as follows. The stock salt solutions, as detailed in Table II, are prepared in a volume of 1 liter with distilled water as the diluent. The stock solutions are either freshly prepared and used immediately, or stored as sterile solutions. Solutions are sterilized by filtration in order to avoid possible contamination of the stock solutions during autoclaving with steam containing volatile boiler preservatives.[8] Sterilization by autoclaving in units fitted for steam, as generated from distilled water, is acceptable. Care must be exercised during the filtration to wash the filter by discarding the first portion of fluid passing through the filter; this precaution will remove filter-bonding material, detergent, or water that may have been retained in the filter.[9]

[8] G. E. Gifford, *J. Bacteriol.* **80,** No. 2, 278–279 (1960).

[9] R. D. Cahn, *Science* **155,** 195–197 (1967).

We have found McCoy's 5A to be satisfactory basic media. Extensive modifications in the salt and sugar levels are frequently and readily made. Thus, the sugar concentration may be reduced to 100 mg/100 ml and the bicarbonate and/or osmolarity levels may be varied.

*Media pH*

A problem in the preparation of tissue culture media is the uncontrolled variability of the final pH under the actual experimental conditions. This, particularly in the instance of bicarbonate buffer systems, results from: variation in the bicarbonate levels, as found in various serum components; the intrinsic buffer capacity of the media mixture; variation in the actual concentration of the $CO_2$ overlay (see section on culture gas); and the adjustment of pH at room temperature during media preparation whereas the cells and medium are actually maintained at 36.5°. Thus, if medium is prepared at 25° with the pH adjusted to 7.30, the actual pH on incubation at 36.5° will be 7.40. For $CO_2$/bicarbonate systems in this range a differential of 0.087 pH units per degree C increase is observed.

The critical importance of pH shifts to cellular function has been described in detail.[10] For several years, we have used a very simple "tonometer" for pH adjustment that consistently has yielded culture media poised at any preset or desired pH level (Fig. 1). The tonometer consists of a 100-ml water-jacketed flask, via which the contained medium is held at the desired culture temperature—36.5°. A Teflon-coated spin bar is used to facilitate equilibration. The upper configuration of the tonometer possesses two ports for the direct flow and circulation of the selected $CO_2$–air mixture over the stirred flask contents. Normally, the same cylinder of gas is used for the tonometer as for the actual cultures. A third indwelling port is provided as a thermometer well. Finally, a silicone stopper is added to both seal the chamber and allow insertion of a combination pH electrode the tip of which is centered in the middle of the medium. A vaccine stoppered port penetrates the stopper and permits direct addition via a tuberculin syringe (2–5 ml) of 5% $NaHCO_3$ without loss of equilibration gas.

The bicarbonate is added step by step, being sure that the equilibration level has been reached at each step before the next addition. Extrapolation from the 50–100 ml tonometer volume to the actual batch volume yields the correct amount of 5% $NaHCO_3$ required for any desired pH, at any incubation temperature, with any gas mixture, despite variations in the buffer capacity of the medium. Its prime disadvantage is that it re-

[10] S. Gailiani, W. F. McLimans, A. Nussbaum, F. Robinson, and O. Roholt, *In Vitro* **12**, 363–372 (1976).

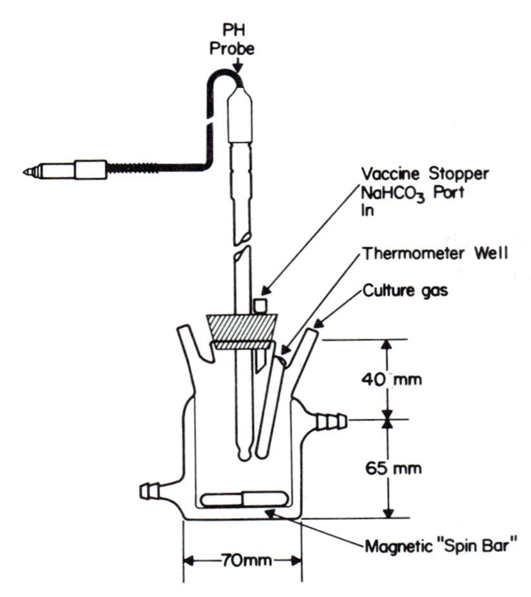


FIG. 1. Schematic of the tonometer.

quires about 2 hr to complete the titration and "careless overshoot" can create problems.

*Osmolarity Adjustment*

After adjustment of the pH, medium osmolarity is measured with an appropriate osmometer. Our instrument (*g*) has a precision of $\pm 1.0$ mos$M$/kg water. The balance of NaCl required to achieve the desired osmolarity is calculated in the following manner.

1. Desired or normal osmolarity of healthy persons = $291.8 \pm 2.6$ mos$M$/kg water.[11]

2. Measure the osmolarity of experimental media and calculate the amount of a special stock NaCl (1 mg/ml) which must be added to achieve the desired osmolarity. Thus: 1 mg NaCl/ml = 1 ml stock (mos$M$) = 32 mos$M$ increase. Therefore:

$$\frac{D - O}{32} = X$$

[11] G. H. Hobika, J. L. Evers, and R. Oliveros, *J. Med.* **7**, No. 1, 13–31 (1976).

where $D(\text{mos}M)$ = Desired mos$M$
$O(\text{mos}M)$ = Observed mos$M$
$X$ = ml of stock NaCl (mos$M$) to be added per milliliter of medium.

*Contamination Control*

A vast effort has been expended in pursuit of knowledge relating to the cultured mammalian cell. Much of it may be of little real value, since the studies were conducted on cells of questionable origin, cells contaminated with microbial agents, or indeed comparative studies of different cell lines, later proven to be all of one common origin. Therefore, great care and time must be expended in the constant check and monitoring of the culture system for contaminants, be they bacterial, viral, fungal, or other mammalian cells being carried in the laboratory.

Critical and mandatory control procedures encompass (1) adherence to strict aseptic techniques, as practiced in any bacteriology laboratory, as well as (2) certifying the source of the "seed" culture in regards to a detailed history of the cell line, histologic characteristics, growth curve, banded marker chromosomes, and glucose 6-phosphate dehydrogenase profile. The extent of prior attempts to isolate and/or demonstrate microbial agents should be certified as well. Certainly, aliquots of the culture should be submitted periodically to experts in the above noted fields for a characterization check.

Periodically, antibiotic-free media should be prepared and incubated in culture flasks to check the efficiency of all operational procedures from media preparation to "cell-less" cultures. This insures "in-house" control of techniques. Ideally, all cultures should be antibiotic-free, but this cannot be achieved in many of the facilities currently available to the biochemist.

The problem is of sufficient importance that methods for its evaluation are covered in detail in this volume [2] and [12].

*Culture Gas*

While the major goal remains the simplification of the tissue culture "tool," meaningful studies demand that the mammalian cells be cultured under physiologic conditions that are nontoxic and reproducible. In these terms, one of the most important parameters of that environment is the culture gas. Thus, for example, culture buffers operative without bicarbonate utterly fail to provide sufficient $CO_2$ and, therefore, must depend on

TABLE III
GAS CONTAMINANTS IN COMMERCIAL BOTTLED GAS

| | Cylinders, research grade (ppm) | | Estimated level in gas overlay of cell culture (ppm) |
|---|---|---|---|
| | $O_2$ | $CO_2$ | |
| Carbon monoxide | | <50 | 5% $CO_2$ < 2.5 ppm |
| Nitrous oxide | <1 | | 20% $O_2$ < 0.2 ppm |
| Hydrocarbons (as methane) | 16 | | 20% $O_2$ < 3.2 ppm |

the cells to provide this gas for the environment.[12] Indeed, the buffer capacity of the $CO_2$/bicarbonate system is perhaps the least important of its many metabolic control configurations.[12] Accordingly, in our laboratories we routinely use the $CO_2$/bicarbonate buffer system.

Specialty gas mixtures, medical or biologic grade, are available for use as culture gas. However, wide variations may exist in the actual $pCO_2$ levels of commercial gas mixtures. Such variation can be extensive if certified gas mixtures are not obtained. In our laboratories, infrared analyses of the actual $CO_2$ content of 27 tanks of uncertified 5% $CO_2$ in air showed a mean variation of 0.9% from the tank value noted by the supplier. The actual range of variation was from 0.1–3.3% with an overall span of 3.2%. Very low values, 2–3%, may be encountered if the tanks have not been properly mixed by rolling.

Gaseous contaminants may include particulate matter, microbial agents, and oil. These are easily removed by in-line filtration. Other contaminants such as hydrocarbons, nitrous oxide, and carbon monoxide are of concern as toxic or potentially mutagenic agents (Table III).

It is recommended that specialty gas mixtures be procured with clear specifications as to the permissible level of contamination and variation of component concentration, i.e., less then 3 ppm moisture, less than 2 ppm CO, etc. Certification by lot to lot analyses is much less expensive than tank to tank and just as effective.

## Suspension Culture—Cell Lines

A number of investigators have demonstrated that established mammalian cell lines proliferate in agitated fluid suspensions. Initial tech-

[12] W. F. McLimans, *in* "Growth, Nutrition and Metabolism of Cells in Culture" (G. H. Rothblat and V. J. Cristofalo, eds.). Vol. 1, Chapter 5, pp. 137–162. Academic Press, New York, 1972.

niques were introduced through the imaginative efforts of Earle *et al.*,[13] who reported successful growth of the L strain (No. 929) fibroblast utilizing a conventional rotary shaker equipped with special flasks and media. Confirming studies with similar techniques were conducted by Kuchler and Merchant.[14] Prior to Earle, Owens *et al.*[15] had demonstrated that the deBruyn mouse lymphosarcoma MB (T-86157) could be cultured readily in "tumbling tubes." Despite the fact that only small volumes were used in this latter investigation, it did suggest that the ability of cells to proliferate in submerged culture may not be restricted to one specific type of cell or method, an observation since amply confirmed. Thus, a monkey kidney cell strain could be propogated in roller tubes rotated around their horizontal axis at 40–50 rpm.[16] Concurrently, good growth of ascites tumor cells was demonstrated in hexagon-shaped roller tubes.[17] Cherry and Hull[18] found that a fluid suspension contained in a round-bottom flask and agitated by a suspended magnetic stirrer permitted growth of the LLMC strain. Brown[19] obtained good growth of mammalian cells employing a "wrist shaker," and McLimans *et al.* reported satisfactory growth of a number of cell lines in spinner culture (25–100 ml),[20–22] in the New Brunswick fermentor (1.5–5 liters),[23] and in a stainless-steel, water-jacketed, impeller-agitated fermentor (20 liters).[24] Additionally, each system was used effectively for the production of viral particles and complement-fixing antigen,[25] as well as for growth of anterior pituitary cells of human origin.[26]

[13] W. R. Earle, J. C. Bryant, and E. L. Schilling, *Ann. N.Y. Acad. Sci.* **58,** 1000–1011 (1954).

[14] R. J. Kuchler and D. J. Merchant, *Proc. Soc. Exp. Biol. Med.* **92,** 803–810 (1956).

[15] O. von H. Owens, M. K. Gey, and G. O. Gey, *Proc. Am. Assoc. Cancer Res.* **1,** 41 (1953).

[16] A. F. Graham and L. Siminovitch, *Proc. Soc. Exp. Biol. Med.* **89,** 326–327 (1955).

[17] A. K. Powell, *Cancer Res. Campaign, Ann. Rep.* **32,** 125 (1954).

[18] W. Cherry and R. N. Hull, *Tissue Cult. Assoc. Meet., 1956* p. 9 (1956).

[19] A. Brown and F. M. Hardy, "Growth of Venezoelan Equine Encephalomyletis Virus in Monolayer and Fluid Suspension Cultures of 'L' Cells," Maryland Soc. Am. Bacteriol., Ft. Detrick, Frederick, 1957.

[20] W. F. McLimans, E. V. Davis, F. L. Glover, and G. W. Rake, *J. Immunol.* **79,** No. 5, 425–433 (1957).

[21] F. Giardinello, W. F. McLimans, and G. Rake, *Appl. Microbiol.* **6,** No. 1, 30–33 (1958).

[22] E. V. Davis, F. Glover, and W. F. McLimans, *Proc. Soc. Exp. Biol. Med.* **97,** 454–456 (1958).

[23] W. F. McLimans, F. E. Giardinello, E. V. Davis, C. J. Kucera, and G. W. Rake, *J. Bacteriol.* **74,** No. 6, 768–774 (1957).

[24] D. W. Ziegler, E. V. Davis, W. J. Thomas, and W. F. McLimans, *Appl. Microbiol.* **6,** No. 5, 305–310 (1958).

[25] M. T. Suggs, H. L. Casey, D. D. Sligh, A. R. Fodor, and W. F. McLimans, *J. Bacteriol.* **82,** No. 5, 789–791 (1961).

[26] W. F. McLimans and S. Gailani, in preparation.

A summary of the literature,[27] as reported from nine laboratories, revealed that 33 cell lines, derived from 14 tissues of human origin, have been propagated in suspension-type systems. Additionally, at least 11 cell lines originally established from monkeys, rabbits, mice, and rats have been propagated in the submerged culture system. Failure to obtain satisfactory growth has been reported in the instance of monkey kidney lines, LLC-Mk and LLc-MK, and the diploid line WI-38.

Accordingly, most established cell lines can be grown as discrete units in suspension-type systems employing a variety of conditions, equipment, and tissue culture media. Subsequent to the original reports of scale-up to 5–20 liter cultures about 15 years ago,[23,24] mass cultivation of such cells was reported by several laboratories in 5–200 liter systems.[28]

*Selection of Mass Culture Systems*

Mass suspension cell culture systems are based on the "state of the art," as borrowed from the bacterial physiologist. The simplest configurations of vessels range from the 30–1000 ml spinner flasks to a 5–15 liter glass fermentor, or to a 20–200 liter stainless-steel fermentor. The material presented here is for the most part limited to glass systems of less than 20 liters volume, with a configuration such that they can be operational in the ordinary laboratory.

Selection of the appropriate size of a fermentor system entails careful evaluation of the critical needs of the investigator, cost, available reagent supplies, washroom facility, sterile rooms, hoods, etc. As a rule of thumb, one calculates the grams of cells, wet weight, required per week and assumes a biweekly harvest of $\frac{1}{2}$ of the fermentor volume or a weekly harvest equal to the volume of the unit. The maximum cell population attainable, while a function of the cell and its environment, is determined in the "batch" system by the rate of lactate production and the decrease in pH with concomitant definition of the time interval between medium changes. Usually this will be found to be at the peak of $1–2 \times 10^6$ cells/ml. Since $1 \times 10^6$ cells equals approximately 1 mg (wet weight), then the production of 1 g of cells per week would require a harvest twice each week of 500 mg or $500 \times 10^6$ cells. Since the harvest should not exceed $\frac{1}{2}$ of the fermentor volume, a fermentor volume of 1000 ml will yield a harvest population of $1 \times 10^6$ cells/ml, a total of 500 mg of cells. Thus, the

[27] Data based on a literature survey made by the author in 1964. The list of successful cultures has, of course, been considerably expanded since that time.

[28] G. E. Moore, J. W. Kullen, H. A. Franklin, and N. Kinsley, *in* "Germ-Free Biology: Experimental and Clinical Aspects" (E. A. Mirand and N. Back, eds.), Vol. 3, pp. 343–356. Plenum, New York, 1969.

cultures may be operated continually with the desired 1 or 2 g yield of cells per week.

The cost of the above operation, exclusive of personnel and/or capital equipment, will be in the range of $13.74/liter of media or $6.13/g (wet weight) of cells. The simplest and safest procedure in this example would be to use two or more 500-ml spinner cultures; thus, accidental microbial contamination of one vessel would not long curtail the ongoing investigations.

If much larger quantities of cells are required—10–100 g/week—then conventional types of fermentors must be used for "batch" operation. This may involve fermentor volumes of 10–100 liters. Moore *et al.*[28] successfully operated, on a batch basis, a commercial, stainless-steel soup kettle modified for the routine propagation of lymphoid-type cell lines. One should consider multiple carboy-type spinner systems prior to investment of large sums in more sophisticated fermentors.

Since yield of cells is the prime objective, the most efficient large-scale fermentors are those that operate on a continuous or chemostat-type principle. Such systems, while requiring extensive instrumentation, allow cell populations of 6–10 $\times$ $10^6$ cells/ml (Table IV). This not only produces savings in cost of medium, but also reduces enormously the size of the fermentor necessary for the same weekly yield as a "batch" process. Another major advantage is the more uniform yield of cells than with the "batch" system.

Basic considerations for operation of a suspension cell culture include the following: (1) the minimal population of cells per milliliter required to initiate a culture. This varies with cell lines and may be as low as 25 $\times$ $10^3$/ml; routinely we use 2–6 $\times$ $10^5$ cells/ml. (2) Calcium and magnesium levels may be reduced, if excessive clumping occurs. (3) Although spinner agitation speeds are a function of vessel size and impellor design, speed should generally be kept low, between 50–150 rpm, in order to avoid excessive shear forces. (4) The inactivated serum, i.e., heated to 56° for 30 min, is routinely used at a concentration of 10% although lower levels may be acceptable. (5) The incubation temperature is normally 36.5°. Care must be taken to insure lack of excessive localized heat transfer from the matnetic drive assembly.

The simplest procedure for monitoring the growth response of the cells is the direct counting technique (see this volume [10]). Good cultures should display readily cellular viabilities of at least 80% or better. Anything less than this suggests an unsatisfactory environment. At times, viability and state of culture may be markedly improved by trypsinization of the entire cellular contents and transfer to a new spinner.

Morphologic assessment of suspension cultures can be made best by

TABLE IV

A COMPARISON OF COSTS (LESS LABOR) OF BATCH VS. CONTINUOUS PROCESS OF PRODUCTION OF MAMMALIAN CELLS AS REFLECTED BY CELL POPULATION, FERMENTOR VOLUME, AND SERUM CONCENTRATION IN CULTURE MEDIUM[a]

| | Batch[b] | | Continuous[d] |
|---|---|---|---|
| Fermentor volume (liters) | 1 | 10 | 10 |
| Maximum cell population ($\times 10^6$/ml) | 1.4 | 1.4 | 10 |
| Day of first harvest | 2–3 | 2–3 | 5–6 |
| Number of harvest days | 11 | 11 | 8 |
| Total cells harvested ($\times 10^9$) | 3.85 | 38.5 | 800 |
| Grams cell yield/week (g) | 2.24 | 22.5 | 467 |
| Total media used (liters) | 6.5 | 65.0 | 120 |
| Total serum (10%) used (liters) | 0.65 | 6.5 | 12 |
| Cost of serum-free media (at $1.46/liter) | $8.54 | $85.41 | $157.68 |
| Cost of serum | 5.20 | 52.00 | 96.00 |
| Total media cost[c] | 13.74 | 137.41 | 253.68 |
| Cost/gram of cells ($) | 6.13 | 6.11 | 0.54 |

[a] All calculations are based on 24-hour generation time for 12 days following establishment of the culture on day 0.

[b] Batch system calculations based on harvest of $\frac{1}{2}$ culture every day.

[c] Media cost based on 66¢/liter for Baxter Water, 78.5¢/liter for dry mix, 10¢/liter for salts and antibiotics, and $8.00/liter for serum at concentration of 10%, i.e., price levels obtaining January 1978.

[d] Continuous systems are calculated to reflect conditions at the same cell population level as batch and under conditions wherein cell population was maintained at about ×7 this level. Such levels are realistic with some cell lines. We have operated units at 8–10 × $10^6$ cells/ml, while Earle has reported short-term propagation at levels as high as 20 × $10^6$ cells/ml.

establishing conventional monolayer subcultures and by periodically determining the efficiency of plating with attention to histologic types of colonies.

In "batch-type" systems, the culture pH should be checked daily and efforts made to maintain it above 6.9. This is achieved best by the daily addition of fresh media in an amount generally equal to 10–20% of the total volume. Except for very high cell populations, this procedure will help to maintain growth and circumvent possible exhaustion of a substrate such as glucose.

### Suspension Culture—Primary Cells

A critique of suspension culture systems must include observations relating to the morphologic and biochemical similarity of established cell lines: abnormal chromosome patterns, malignancy (real or imagined), as well as the undefined culture parameters, i.e., cell dissociation, culture surface, gas, and nutritional environment. It is therefore important to attempt to improve the definition and control of the culture environment so that primary cells, representative of those found in the tissue of origin, may be maintained *in vitro* with meaningful and reproducible growth and function.

One may define an ideal culture system for the mass growth of mammalian cells as "a closed-system suspension culture configuration, capable of supporting the direct growth of normal, primary cells with long-term maintenance of specialized cellular function; as a system achieved with cells that are minimally exposed to foreign substances or environments in their transition from the *in vivo* to the *in vitro* state; as a system achieved with cells that are grown to high population densities, under conditions such that the state of the culture may be critically assessed and controlled via daily monitoring; and finally as a system with an operational efficiency such that the cost and availability of media components are held at realistic levels."

One approach could well be to achieve growth of primary cells directly in a homogeneous suspension system in which the definition and control of the culture environment is easily established—a steady-state culture system. During the last decade, a great deal of experimental work concerned with the environmental control of mammalian cells in suspension-type cultures has been reported. However, in all instances the investigations were seriously limited by being restricted in application to established cell lines.

As a consequence, the reports from van Wezel and his associates[29–31] that mammalian cells could be grown on the surface of a microcarrier, such as DEAE-Sephadex beads, while being stirred in a suspension system were of great interest. Accordingly, a serious evaluation of the potential of microcarrier suspension for direct culture of primary cells from differentiated tissue was made in our laboratories. It is here used as a prototype to define and illustrate the appropriate techniques, although these methods are undergoing rapid change as new "beads" become available.[32] The technique is also discussed in this volume [15].

Calf anterior pituitary cells (CAP) were employed as the differentiated cell type. They were successfully cultivated in the microcarrier suspen-

[29] A. L. van Wezel, *Nature (London)* **216,** 64–65 (1967).

[30] A. L. van Wezel, *in* "Tissue Culture: Methods and Application" (P. F. Kruase, Jr. and M. K. Patterson, Jr., eds.), Chapter 2, pp. 372–377. Academic Press, New York, 1973.

[31] P. van Hemert, D. G. Kilburn, and A. L. van Wezel, *Biotechnol. Bioeng.* **11,** 875–885 (1969).

[32] C-b. Horng and W. F. McLimans, *Biotechnol. Bioeng.* **17,** 713–732 (1975).

sion system. Details regarding the type of cell growth, kinetics of cell proliferation, inhibitory effect of high bead concentrations, and growth-enhancing activity by a cell factor have been presented.[32] The reported experimental work appears to have established the necessary background for a direct approach to the long-term, steady-state culture of primary differentiated cells at high population densities in suspension culture.

The calf anterior pituitary gland, immediately after removal, was isolated, decapsulated, and soaked in a trypsin medium—McCoy's 5A containing 6.250 IU/ml of Tryptar (*j*). After transport to the laboratory, the tissue was minced and trypsinized in a stirred 50-ml spinner flask containing 15 ml of "tryptar" medium. After 6 hr of trypsinization, the cells were readily dissociated with a Pasteur pipette. Typically, from one pituitary gland, 2–7 × $10^7$ cells were obtained with a viability above 95%. In other instances, e.g., mammary and prostate, cultures were initiated better directly from a fine tissue mince mixed with the prepared "beads"; trypsin appeared to be deleterious to cells derived from these organs.

McCoy's 5A medium supplemented with 10% heat-inactivated fetal calf serum was used as growth medium for the pituitary cell cultures. The media was adjusted to a pH of 7.4 with a gas overlay of 2% $CO_2$ in air for normal tissue and 5% $CO_2$ for tumor tissue. The osmolarity was controlled at 291–315 ± 10 mos*M*.

DEAE-Sephadex was, in the main, used as a microcarrier, although other beads are available.[33] Prior to use, the beads are sequentially washed with 0.5 *N* NaOH, distilled water, 0.5 *N* HCl, distilled water, and a phosphate buffer solution, and sterilized by autoclaving at 121° for 20 min. They are stored at 4°. Just prior to use, the phosphate buffer is replaced with culture medium.

Suspension cultures are carried out, for example, in a spinner flask of 50–100 ml capacity (*k,l*). The cultures are incubated at 36.5° with a magnetic spin bar adjusted to a speed of about 100 rpm. The spinner flasks are continuously gassed via an overlay of 2% $CO_2$ in air. The usual culture volume is 30 ml. The pH of the culture fluid is monitored daily. If the pH decreases below 7.0, the culture is usually fed with 10–20 ml of fresh growth medium. Every 3–4 days, or daily, depending on the activity of the culture, 1 ml of the cell/bead suspension is removed and mixed with 3 ml of phosphate buffer solution contained in a petri dish. The percentage of beads with cells growing on their surface is checked by microscopic observation. Subsequently, the beads with attached cells are washed with a phosphate buffer solution and exposed to 4 ml of 0.1 *M* citric acid while being incubated at 37° for 1 hr. Following the incubation, the released

[33] D. W. Levine, J. S. Wong, D. I. C. Wang, and W. G. Thilly, *Somatic Cell Genet.* **3**, No. 2, 149–155 (1977).

nuclei are washed from the bead/cell surface with the citric acid solution, stained with gentian violet, and counted in a hemocytometer or by an electronic counter; method for counting nuclei follows the procedure originally developed by Sanford *et al.*[34]

Interestingly, "naked beads" added to a monolayer culture will pick up and anchor cells. Conversely, "beads" with anchored cells when transferred from suspension to a flat culture surface will initiate monolayer-type clones and monolayers.

For monitoring the morphology and cell types, petri dish cultures are established from a dilution of the microcarrier suspension cultures. Feeding is done by replacing the acidic spent medium with 3 ml of fresh growth medium at 2–4 day intervals. Quantitation of colonies, as well as histologic evaluation, is made after 14 days of incubation in a $CO_2$ chamber or bell jar incubator. This constitutes a critical control procedure!

The DEAE-Sephadex bead is transparent in growth medium. Thus, the morphology of the attached cells can be fixed by the usual methods and monitored directly via the light or electron microscope.

As noted, petri dish cultures are employed for preliminary assessment of cell growth under various conditions. Immediately after inoculation, cells are observed to attach to the bead surface and form a round mass for about 2 days. Thereafter, the cells appear to spread out and begin to proliferate. By the 4th day, a partial monolayer of cell growth on the bead surface can be observed. After 2 weeks of incubation, nearly all of the beads are completely covered by cell layers. It appears that cells may move from one bead to an adjacent bead, leading eventually to the formation of bead clumps with cell layers serving as a bridge; cells are predominantly of epithelial morphology when attached to the bead surface. Electron microscopy of cells from these cultures shows the presence of active mitochondria, endoplasmic reticulum, and tight cell junctions. Some cytoplasmic projections have been noted between bead and cell, which probably indicates an active anchorage site, rather than passive physical-chemical attachment of the cells on the bead surface.

Suspension cultures in spinner flasks have been routinely successful. Within certain limits of bead concentration, cells proliferated in the suspension system with growth on the order of magnitude of a 10-fold cell population increase in 8 days. Contrary to the expectation that higher bead concentration should provide greater surface area for cell growth and a higher incidence of cell–bead collision and attachment, it appears that with a constant cell inoculum, higher bead concentrations may inhibit cell growth. This inhibitory effect, as would be expected, can be reduced

[34] K. K. Sanford, W. R. Earle, V. J. Evans, H. K. Waltz, and J. E. Shannon, *J. Natl. Cancer Inst.* **11**, 773–795 (1950–1951).

by using proportionately higher quantities of inoculum. There might be a critical cell-to-bead ratio below which the cells will not show sustained net growth. By using a rough estimation of $1.6 \times 10^5$ beads/ml in the bead bed, this ratio was calculated to be about 1 : 5 for the calf anterior pituitary cells.

In our hands, direct primary cell culture with the microcarrier suspension system and batch feeding can routinely yield a cell population of about $5 \times 10^6$ cells/ml. However, with such a dense population, nutrient feeding becomes the limiting factor. Even feeding 3 times a day, one cannot keep up with the extreme shifts in pH. To circumvent this problem, we have devised a continuous-feed culture system with automatic pH monitoring and feedback loop, as based on use of a modified "Spin-Filter" (*m*). Other systems may be equally applicable (*n,o*).

Employing the "Spin-Filter," a 500-ml microcarrier (DEAE) culture of primary anterior pituitary cells was continuously maintained, at a high cell population (4–7 $\times 10^6$ cells/ml), for 57 days with excellent control of the culture parameters. Repetitive subcultures to petri dishes were routinely successful.

Cell debris does not cause significant problems in healthy cultures. However, freely suspended cells do appear. Whether or not these represent newly divided cells as released from bead surfaces remains to be established. At times, freely suspended cells may reach a population as high as $3 \times 10^5$/ml and have a viability above 85%. If such a cell population is truly being continuously released from the beads, it might well be possible to consider these methods for synchronized cell culture studies.

Other systems and configurations for cell culture have been demonstrated, i.e., capillary tubes,[35] plastic or cellular sheets or tubes,[36] multiplate fermentors,[37] and even columns of Sephadex.[38] They are not considered here, since they all appear to suffer from difficulty in quantitating growth rates and responses. There is also a marked tendency for formation of clumps or cellular masses. Such clumps may, in turn, serve as foci of necrosis and transformation. Vital to any system is the facility of meaningful monitoring techniques with which to assess the true kinetic or histogenic profile of the culture. This, in turn, must depend on obtaining truly representative aliquots.

It is the author's opinion that the microcarrier technique will, in time, prove to be the best tool for this task.

[35] R. A. Knazek, P. M. Gullino, P. O. Kohler, and R. I. Dedrick, *Science* **178**, 65–67 (1972). See also this volume [15].

[36] M. D. Jensen, D. F. H. Wallach, and P. Sherwood, *J. Theor. Biol.* **56**, 443–458 (1976).

[37] R. Weiss and J. B. Schleicher, *Biotechnol. Bioeng.* **10**, 601–616 (1968).

[38] B. Bergrahm, *Biotechnol. Bioeng.* **10**, 247–251 (1968).

## Acknowledgment

I acknowledge the invaluable and tireless assistance of Chi-byi Horng, Barbara Kwasniewski, Angeline Wasielewski, Frances O. Robinson, Robert Zeigel, Elizabeth Repasky, W. "Sam" McLimans, Peter Hasenpusch, and William Beers in the investigative phases that form the bases of this article. I thank Ann M. Gannon for the editing and other assistance. The United States Public Health Service supported with these grants-in-aid various facets of this work: 5 R01-ES00030-05, 89013-01, NIH-NCI-G-72-3866, 1 R01 GM22532-01, and 5 R26 CA1865203.

# [17] Propagation and Scaling-Up of Suspension Cultures

*By* RONALD T. ACTON, PAUL A. BARSTAD, and ROBERT K. ZWERNER

## I. Introduction

One major consideration in cell culture is whether a given cell line grows in suspension or must attach to a matrix. Initial attempts at growing cells *in vitro* dealt mainly with anchorage-dependent types. Following the early demonstrations[1,2] that cells would multiply in suspension cultures, many investigators recognized the potential of this approach and sought to increase the efficiency of the process (see Perlman *et al.*[3-6] for review). From the beginning animal cell suspension culture was appealing due to the anticipation of utilizing fermentation techniques similar to those used in cultivating microorganisms. Although the approach is similar in principle the difference in growth parameters between prokaryotic and eukaryotic cells poses unique problems.

An approach will be described here for the growth of cells in suspension culture from 1 ml up to 200 liters.[6] Although the method was initially derived for the propagation of lymphocytes,[7] it has been found amenable to mastocytoma, hepatoma, L cells, and various nervous tissue derived cell lines. Since cell culture requires a high level of attention to detail,

[1] O. V. H. Owen, M. K. Gey, and G. O. Gey, *Ann. N.Y. Acad. Sci.* **58,** 1039 (1954).

[2] W. R. Earle, R. L. Schilling, J. C. Bryant, and V. J. Evans, *J. Natl. Cancer Inst.* **14,** 1159 (1954).

[3] D. Perlman, *Proc. Biochem.* **2,** 42 (1967).

[4] G. E. Moore, *Methods Cancer Res.* **5,** 423 (1970).

[5] R. C. Telling and P. J. Radlett, *Adv. Appl. Microbiol.* **13,** 91 (1970).

[6] R. T. Acton, P. A. Barstad, R. M. Cox, R. K. Zwerner, K. S. Wise, and J. D. Lynn, *in* "Cell Culture and Its Application" (R. Acton and J. Lynn, eds.), p. 129. Academic Press, New York, 1977.

[7] R. K. Zwerner, C. Runyan, R. M. Cox, J. D. Lynn, and R. T. Acton, *Biotechnol. Bioeng.* **17,** 629 (1975).





ISBN 0-12-181958-2

even tne more mundane aspects of the approach are described. Experience has taught that inability to grow a particular cell line frequently lies in the failure of an individual to adhere precisely to the prescribed steps rather than in the method itself.

## II. Cell Culture

### *1. Recovery of Cells*

The initial procedure for culturing continuous lines depends on the state in which cells are received. Cells may be transported either as viable cultures or in the frozen state. If cells are received as a growing culture one simply determines density and viability and places them in fresh medium. If received in the frozen state cells may be either stored in liquid nitrogen or placed directly into culture. The ampule containing the frozen cells is thawed by immediately immersing in a 45° water bath. The ampule should be agitated to insure that the thawing process is completed in less than 1 min. The ampule is immersed in 70% ethanol to reduce the possibility of contamination and transferred to a vertical flow laminar hood. The neck of the ampule is scored with a small sterile file and the cells transferred to a sterile 15-ml polystyrene screw-cap disposable centrifuge tube (Corning #25310) using a sterile Pasteur pipette. Ten volumes of culture medium at 4° are added and the tube is capped and inverted several times. The cells are pelleted by centrifugation at 400 g for 5 min. It is important to conduct these procedures immediately after thawing the cells in order to remove the cryoprotective agent as quickly as possible. Regardless of the designated concentration of cells contained in the ampule the cell density and viability should be determined at this time. The cells are then apportioned to sterile 25 $cm^2$ or 75 $cm^2$ polystyrene tissue culture flasks (Corning #25100 or 25110, respectively), and the cell concentration is adjusted to approximately $1 \times 10^6$ cells/ml with medium at 37°. An aliquot is evaluated for microbial contamination. The flasks with slightly loose caps are then placed in a humidified $CO_2$ incubator at 37°. Each day thereafter the cells are enumerated and evaluated for viability, and the cell concentration is adjusted by adding new medium. When the culture volume allows, an aliquot of cells should be frozen and stored in liquid nitrogen to serve as a source of future stocks and reference point. In the event the cells grow slowly special attention must be given to include centrifugation every other day and resuspension in fresh medium, assessing the use of other media or nutrient supplementation and variation in cell density.[7]

### 2. *Optimizing Cell Growth*

When cells appear to be growing well attempts should be made to optimize the rate and density of growth. Maintaining pH within a narrow range and sustaining the cells in the exponential stage of growth greatly enhances the yield of cells or cell products per unit volume of medium. In addition other factors influence the efficacy of cell production *in vitro*.

#### A. Media

Several excellent media are available for the growth of established cell lines. RPMI-1640 supplemented with horse serum (HS) or fetal calf serum (FCS) (Flow Laboratories, Inc.) has been demonstrated to sustain a wide variety of cells in culture and is routinely used in our laboratory. While certain cell lines may have unique nutritional requirements for growth, most established cell lines grow well in several media and serum supplements.[8] For example, the BW5147 cell line was obtained from the Cell Distribution Center, Salk Institute, in Dulbecco's Eagle's Medium supplemented with 10% FCS (Flow Laboratories, Inc.). It readily adapted to RPMI-1640 supplemented with either 10% FCS or 10% HS. The cell line grew nearly as well in Minimal Essential Medium (MEM) supplemented with insulin or zinc and 2% HS as in RPMI-1640 with 10% HS. Subsequently, it was demonstrated that BW5147 grew as well in RPMI-1640 + 2% HS.[9] By trial and error and evaluation of growth curves one should be able to derive an economical culture medium which generates the maximum number of cells per unit volume in the shortest possible time.

#### B. Vessel Configuration

One aspect of cell culture frequently overlooked in suspension culture is the size and configuration of the spinner flask. When cells are brought from the frozen state to active culture they are first grown in a tissue culture flask in a humidified $CO_2$ incubator. This provides acceptable pH, $CO_2$, and oxygen conditions for growth. Frequently cells exhibit reduced levels of growth when transferred to a spinner flask due to the dimensions of the vessel.[7] We routinely overlay the spinner flask immediately after inoculation of the seed culture with 5% $CO_2$ + 95% air after which the cap is tightly replaced; thus, the mixture of gases in the head space of the

[8] J. L. Williams, T. H. Stanton, and R. M. Wolcott, *Tissue Antigens* **6**, 35 (1975). See also this volume [5].

[9] P. A. Barstad, R. K. Zwerner, and R. T. Acton, *Protides Biol. Fluids, Proc. Colloq.* **25**, 637 (1978).

vessel must be sufficient to sustain growth until the cells are evaluated again. As one progresses to larger volumes the amount of oxygen available to the cells becomes a limiting factor. For volumes between 50 ml and 6 liters the Bellco-type spinner flasks (Bellco Glass, Inc.) offer the largest volume above the culture and provide the best system for suspension culture. For volumes above 6 liters the configuration of vessels from most manufacturers are such that a gas mixture must be added continuously to obtain optimal growth. For some cell lines a fermentation-type vessel in which pH and dissolved oxygen are continuously monitored and controlled is more effective. These vessels are available in sizes from 1 liter to 14 liters with various optional instrumentation (New Brunswick Scientific Co., Inc., Edison, New Jersey).

### C. Defining Growth Parameters

In most suspension culture operations cells are propagated under semicontinuous culture conditions. This mode of operation represents essentially a "feast–famine" situation in which cells are inoculated into fresh medium at a relatively low density and overlayed with a gas mixture of $CO_2$ and air. Growth continues until a build-up of metabolic products or a depletion of nutrients slows or stops cell growth. In order to efficiently produce cells *in vitro,* the factors limiting cell growth must be understood and ultimately controlled. An examination of the events which occur in a theoretical, semicontinuous suspension culture of mammalian cells indicates which parameters can be controlled to optimize growth. A growth curve of mammalian cells in suspension culture is a semilogarithmic sigmoidal plot of cell density (cells per milliliter of culture fluid) as a function of time. Figure 1 is a theoretical mammalian cell growth curve. Segment 1 of the curve is called the "lag phase" of growth. Cells in this phase undergo little or no division; i.e., the slope of the growth curve for the lag phase is approximately 0. In segment 2 the slope of the curve changes from 0 to a positive value, as the cells enter the terminal lag phase. In this phase cell division has begun and its rate is constantly increasing. When the slope of the curve reaches a maximum value (segment 3), the cells are in the exponential phase of growth and are undergoing division at the maximum rate. The cells continue growing at this rate until they enter the "early stationary phase," segment 4. In this phase the cells' division rate decreases. This decrease may be due to nutrient depletion or accumulation of deleterious waste products. The cells will eventually enter the "stationary phase" of growth, segment 5, where the slope of the curve once again approaches 0. If the cells remain in the stationary phase too long, division ceases and a net decrease in the cell number results.

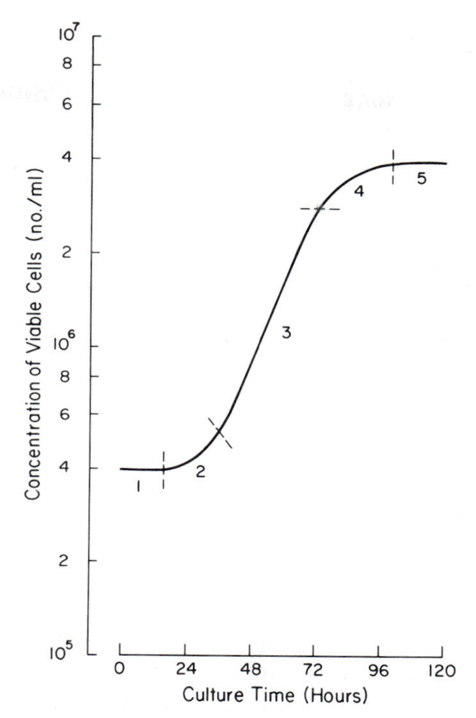


FIG. 1. Theoretical growth curve of mammalian cells in suspension culture.

Figure 2 illustrates that mammalian cells grown in suspension culture do indeed perform as just described. This figure defines the growth properties of the murine lymphoblastoid cell line S49.1 in a 4-liter spinner flask. In this experiment the culture was started at a concentration of approximately $4.8 \times 10^5$ cells/ml. For 24 hr after inoculation the cells divided very slowly, as expected of cells in the lag phase of growth. Soon after this interval, the cells entered the exponential phase of growth where an average doubling time of 17 hr was observed. At the maximum density of $2.3 \times 10^6$ cells/ml, cell division ceased and a reduction in cell number was observed over the next 14 hr. Thus, the cells had entered the stationary phase of the growth curve (segment 5, Fig. 1). Cells taken from the stationary phase and reinoculated into fresh medium generally fail to divide until 24–48 hr have elapsed.[7] However, if cells are placed in fresh medium from the exponential or early stationary phase of growth, they continue to divide at a rapid rate and do not enter a lag phase.

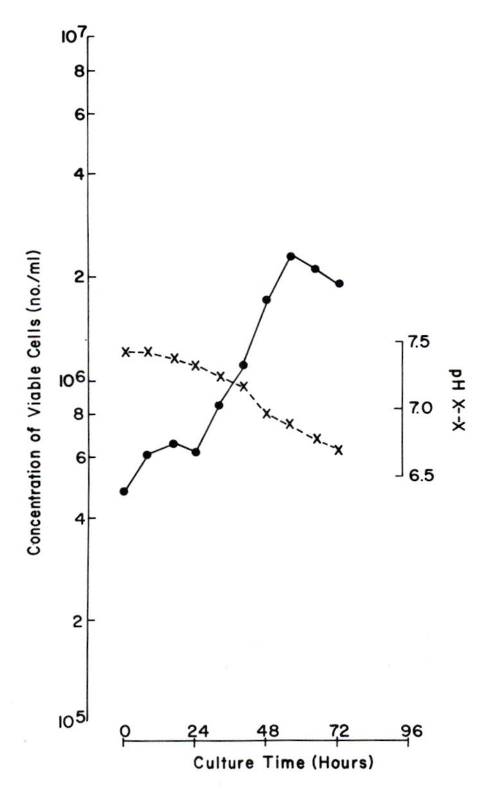


FIG. 2. Growth of S49.1 in a 4-liter spinner flask (Wheaton Scientific). The medium was RPMI-1640 supplemented with 10% HS. The culture was overlayed with a mixture of 5% $CO_2$ + 95% air and capped immediately.

A set of conditions has been established which proves useful for the growth of cell lines in suspension culture.[6,10,11] When culturing mammalian cells according to this scheme, cells are maintained in concentrations between late exponential and early stationary phases of growth by the addition of fresh media.

In practice one may establish a growth curve and continuous culture conditions by the following method:

1. Begin cultures at several initial densities starting at some concentration, e.g., $10^5$ cells/ml, using rapidly dividing cells. The optimal inoculation point is the lowest initial concentration that results in a very short or no lag phase.

[10] P. A. Barstad, S. L. Henley, R. M. Cox, J. D. Lynn, and R. T. Acton, *Proc. Soc. Exp. Biol. Med.* **155,** 296 (1977).

[11] R. T. Acton and J. D. Lynn, *Adv. Biochem. Eng.* **7,** 85 (1977).

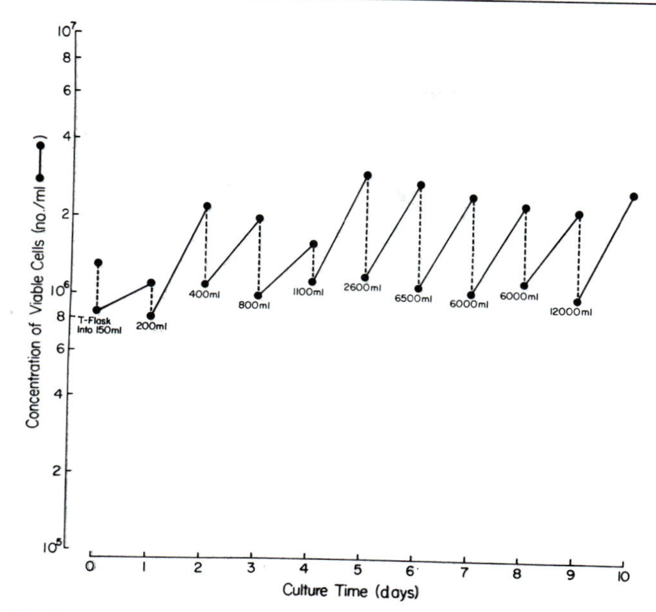


FIG. 3. Scaling-up the growth of S49.1 from a tissue culture flask to growth in a 12-liter culture vessel. Growth from the 150-ml stage to the 2600-ml stage was conducted in Bellco spinner flasks. Growth from 6.5-liter to 12-liter was conducted in a 14-liter New Brunswick fermentor (New Brunswick Scientific Co.).

2. Begin cultures at the optimal inoculation point.
3. Monitor the slope of the exponential phase.
4. When the slope decreases to 50% of the maximum value, add fresh media to dilute cells to the optimal inoculation density once more.

By adhering to these guidelines it is possible to culture cells in suspension in a reproducible fashion as shown in Fig. 3.[12] The S49.1 cell line was rapidly scaled up from a tissue culture flask to a 12-liter suspension culture. After an initial adjustment to the 250-ml spinner flask, the cells were maintained in the exponential phase of growth with doubling times averaging 17 hr. For most of the 10-day culture period, maximum cell densities ranged between $2–3 \times 10^6$ cells/ml.

### 3. *Evaluation of Culture Status*

Throughout the culture period several parameters can be utilized to evaluate progress. The number of cells per unit volume and viability should be determined at least every 24 hr. At times it may be informative

[12] J. D. Lynn and R. T. Acton, *Biotechnol. Bioeng.* **17,** 659 (1975).

to determine cell size distribution as well as to evaluate some phenotypic properties if this latter capability is available. These parameters can be measured in the following manner.

### A. Determination of Cell Density and Viability

Upon removal from the magnetic stirrer and before the cells settle, 0.5 ml of the culture is aseptically removed by use of a sterile 1-ml pipette and placed in a 12 × 75 mm RTU culture tube (BD #7813 Becton Dickenson and Co. Rutherford, New Jersey). A 100-μl aliquot is mixed with 100 μl of a solution of 0.4% trypan blue in normal saline (#525 Grand Island Biological Co., Grand Island, New York). This high concentration of trypan blue is necessary to counteract the affinity of the dye for proteins in solution.[13] An alternative dye is erythrosine B which has a greater affinity for nonviable cells than the soluble proteins in the culture fluid.[13] Cells are then counted in a standard 0.1-mm deep hemocytometer bright-line counting chamber (American Optical Corp. Buffalo, New York). By incorporating one of the two vital stains into the diluting fluid as described the viability of cells can be obtained simultaneously during cell enumeration. In the hemocytometer viable cells will appear as bright circles whereas dead cells appear as dark circles, often with irregular edges.

### B. Phenotypic Analysis

Several cell surface components have been shown to vary in expression during various stages of the cell cycle.[14-18] Although it is impractical to maintain cells in suspension culture in synchronous growth, this condition is approachable by close attention to the culture procedures previously described. If the main interest, for example, is cell surface components, it is important to evaluate the particular property of interest during growth in order to determine the optimal time to harvest cells for maximal yield. For example, Thy-1 alloantigen expression on murine lymphoblastoid cells is lowest during the lag and stationary phases of growth,[7,14] and maximal expression occurs when cells are maintained in the exponential phase of growth. One method to quantitate the amount of cell surface component on a cell is the absorption of antiserum as detected by the

[13] H. J. Phillips, *in* "Tissue Culture Methods and Application" (P. Krase, Jr. and M. Patterson, Jr., eds.), p. 406. Academic Press, New York, 1973.

[14] R. K. Zwerner and R. T. Acton, *J. Exp. Med.* **142,** 378 (1975).

[15] M. Cikes, *J. Natl. Cancer Inst.* **45,** 979 (1970).

[16] M. Cikes and G. Klein, *J. Natl. Cancer Inst.* **49,** 1599 (1972).

[17] R. A. Lerner, M. B. S. Oldstone, and N. R. Cooper, *Proc. Natl. Acad. Sci. U.S.A.* **68,** 2584 (1971).

[18] M. A. Pellegrino, S. Ferrone, P. G. Natali, A. Pellegrino, and R. A. Reisfeld, *J. Immunol.* **108,** 573 (1972).

cytotoxicity assay.[19-21] By comparing the ability of cells to absorb a standard antiserum during various stages of the growth curve, one can determine the exact stage for harvest that yields the highest concentration of a particular cell surface component.[7,22]

As previously indicated, there are a number of factors that affect the rate of cell growth. In the initial phase of optimizing cell growth one should determine which culture medium produces the maximum number of cells per unit volume in the shortest possible time. However, it is also important to determine if the culture medium chosen also yields the highest quantity of the cell component under study. The type of serum supplement used in culturing cells can influence the expression of murine lymphoblastoid cell surface components. For example, the L251A cell line had a 15-fold greater expression of the murine thymus leukemia antigen (TL) when grown in medium supplemented with 10% fetal calf serum instead of 10% horse serum.[14] Although it is currently difficult to ascertain at this point what factors in serum affect the expression of cell surface components, it is important to evaluate the culture medium selected with regard to the particular phenotypic property of interest.

### 4. *Quality Control*

#### A. Microbial Contamination

Contamination by microorganisms is the most frequent problem in cell culture. It is important throughout the maintenance of cells in culture to routinely monitor the sterility of media as well as the culture.[23] We have found it convenient to evaluate cultures and media by inoculation into thioglycolate broth and streaking onto blood agar plates (Grand Island Biological Co.) that are held at 37°. In addition, the cultures are streaked onto a tube of Sabouraud's dextrose agar (Grand Island Biological Co.) and held at room temperature as well as 37°. Most contaminants are detected by these procedures within 24–48 hr.

Mycoplasmal contamination is more difficult to access but can significantly alter the phenotype or growth properties of cells.[24] Some of these

[19] P. Gorer and P. O'Gorman, *Transplant. Bull.* **3**, 142 (1956).

[20] E. A. Boyse, M. Mijazawa, T. Aoki, and L. J. Old, *Proc. R. Soc. London, Ser. B* **170**, 175 (1968).

[21] H. Wigzell, *Transplantation* **3**, 423 (1967).

[22] M. A. Pellegrino, S. Ferrone, and A. Pellegrino, *Proc. Soc. Exp. Biol. Med.* **139**, 434 (1972).

[23] "Difco Manual," 9th ed. Difco Laboratories, Detroit, Michigan, 1977.

[24] M. F. Barile, *in* "Cell Culture and Its Application" (R. Acton and J. Lynn, eds.), p. 291. Academic Press, New York, 1977.

organisms can be detected by direct culture procedures in which aliquots of the cell culture are inoculated into mycoplasma broth and agar (Flow Laboratories, Inc., Mycoplasma Isolation Kit #3006800). Cells also can become contaminated with nonculturable strains of mycoplasma which dictate the use of indirect procedures to detect their presence. These include binding of fluorescent antibodies specific for various strains of mycoplasma, uracil/uridine ratios, and DNA staining procedures, as well as a variety of other less commonly used approaches.[24]

### B. Karyology of Cells

When a cell line is to be utilized for long periods of time, a periodic analysis of cell surface markers and karyotype should be performed. When several cell lines are being carried in the laboratory at one time, it is important to guard against cross-contamination of the cell lines. In addition to cell surface phenotype analysis, another criteria for monitoring cell populations is karyotypic analysis.[25] The fact that the chromosome constitutions of cells differ between species as well as certain cell lines allows one a convenient means of monitoring cross-contamination. There are available a number of procedures for preparation and analysis of cell karyotypes.[25]

### C. Freezing and Storage

After the optimal growth parameters for a given cell line have been established, aliquots of the culture should be frozen and cryopreserved. Aliquots of cells are taken while in exponential growth, centrifuged at 400 *g* for 10 min, and resuspended at 4° in either fetal calf serum or horse serum containing 5% dimethylsulfoxide (Fisher Scientific Co.) to a concentration of 1–5 × $10^7$ cells/ml in sterile, 2-ml, prescored cryule ampules (Wheaton 200 Brand #651486, Wheaton Scientific). The tip of the ampule is sealed in a hot flame by use of the pull-seal method. The ampules are frozen at a controlled rate of 1°/min in a BF-4-1 freezing chamber regulated by a Linde BF-6 controller (Union Carbide Corp., Linde Division). If a controlled-rate freezing device is not available the ampules can be placed in a small Styrofoam box lined with cotton. The lid is taped into place, and the carton is placed in a −70° freezer overnight. After freezing, the ampules are stored in liquid nitrogen. The next day, an ampule or two of the frozen cells are placed in culture to evaluate the freezing process by determining viability, growth properties, and possible microbial contamination. Periodically throughout the maintenance of a given stock of cells,

[25] T. C. Hsu, *in* "Tissue Culture Methods and Application" (P. Krase, Jr. and M. Patterson, Jr., eds.), p. 764. Academic Press, New York, 1973.

aliquots should be frozen for future reference points. It is also important to maintain numerous vials of each cell line under cyropreservation, preferably in multiple locations, to protect against loss of the line should contamination or other accidents occur depleting the cultured stock.

### 5. *Summary*

The general approach discussed here is one that should serve as a guide for the culture of almost any type of cell in suspension culture. Following the initiation of cells in culture, it is advised that optimal conditions for growth be determined. In order to maintain the integrity of the cells and to protect the time invested in their growth it is also mandatory that one effect adequate quality-control measures to assure that cells are maintained in a healthy state, free from contamination, and that aliquots are stored for future reference and use. By adherence to these relatively simple rules, one will find that cells in culture can be a most valuable tool in molecular biology.

### Acknowledgments

This work was supported by U.S. Public Health Service Grants CA 15338, CA 18609, and CA 13148 from the NCI, IM33C from the American Cancer Society, GB-43575X from the Human Cell Biology section of the NSF and the Diabetes Trust Fund. The work of R. T. A. was done during the tenure of an Established Investigatorship of the American Heart Association. P. A. B. was supported in part by National Service Award F32-AI 05423 from the NIAID. R. K. Z. was supported in part by National Research Service Award T32-GM 07561 from the NIGMS.

## [18] Harvesting the Products of Cell Growth

*By* ROBERT K. ZWERNER, KIM S. WISE, and RONALD T. ACTION

Rapid and efficient harvesting of cultured cells and their products is a major problem associated with biological investigations of cells. Depending on the particular needs of an investigator this process may be far more costly and time consuming than growing cells. The process described here is designed to provide subcellular organelles and products such as virus and mycoplasma in the spent medium by harvesting rapidly and in high yield.[1,2]

[1] M. J. Crumpton and D. Snary, *Contemp. Top. Mol. Immunol.* **3**, 27 (1974).
[2] D. F. H. Wallach and P. Sun Lin, *Biochim. Biophys. Acta* **30**, 211 (1973).



ISBN 0-12-181958-2

## I. Acquisition of Subcellular Organelles

Figure 1 summarizes the product acquisition scheme. Cells are separated from the medium by centrifugation at 650 *g* for 20 min in a Beckman J6 centrifuge utilizing 1-liter screw-cap polyproplylene centrifuge bottles (Beckman Instruments, Inc., No. 878564) in a JS-4.2 SW rotor. The spent medium is decanted and stored at 4° or processed further as described below. The noted speed of centrifugation is effective in forming a solid but not tightly packed cell pellet thereby permitting decantation of the culture medium with minimal cell loss. Centrifugation at greater gravitational forces packs cells too firmly, causing problems in their resuspension, and

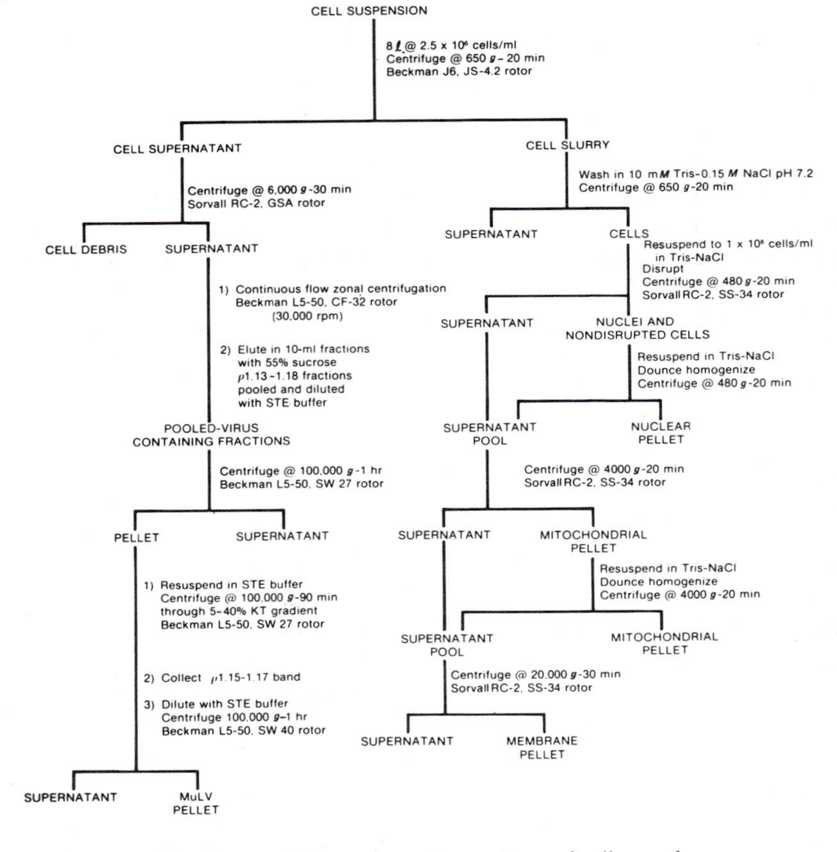


FIG. 1. Acquisition scheme for products of cell growth.

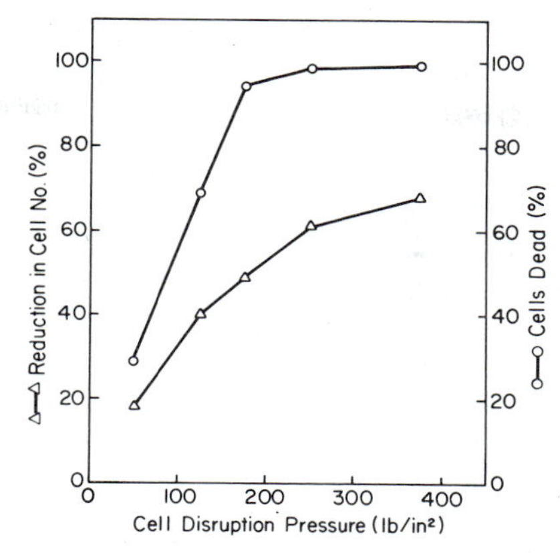


FIG. 2. The continuous and controlled-rate disruption of EL4 murine lymphoblastoid cell line. A suspension of $1 \times 10^8$ cells/ml was pumped at a rate of approximately 2 liters/hr over a range of back pressures on the disrupting valve.

may damage the cells due to shearing effects. The cells are resuspended in one-half the original volume of Tris-NaCl buffer (10 m*M* Tris chloride and 0.15 *M* NaCl at pH 7.4) and centrifuged as before to remove residual culture medium. Following this procedure 95% or more of the cells in the starting suspension culture can be harvested with a viability greater than 95% as determined by trypan blue exclusion. The cells are resuspended in Tris-NaCl buffer to a concentration of $1 \times 10^8$ cells/ml.

To obtain subcellular organelles, cells must be disrupted in a manner preserving the integrity of these structures and allowing effective separation into pure fractions. We have found the Stansted Cell Disrupter equipped with a model AO 612 air-driven hydraulic pump and a model 716 disrupting valve (Energy Service Co., Washington, D.C.) an excellent instrument for this purpose. This instrument provides a means for the continuous and controllable disruption of a wide variety of cell types.[1,3,4] Disruption is caused by the shearing effect of the fluid as it passes under pressure through an orifice whose dimensions are controlled by air pressure on a piston. For each cell line a calibration curve must be determined to ascertain a disrupting pressure that results in the highest yield of the desired product. Figure 2 represents a typical controlled disruption profile for the murine lymphoblastoid cell line EL4. Disruption is measured by the severity of total cellular disintegration and cell viability as

[3] B. M. Wright, A. J. Eduardo, and V. E. Jones, *J. Immunol. Methods* **4,** 281, (1974).

[4] R. T. Acton, P. A. Barstad, R. M. Cox, R. K. Zwerner, K. S. Wise, and J. D. Lynn, *in* "Cell Culture and Its Application" (R. T. Acton and J. D. Lynn, eds.), p. 129. Academic Press, New York, 1977.

determined by trypan blue exclusion. The cell suspension was pumped through the orifice at pressures varying between 50 and 375 psi. At higher pressures there was increased total disruption as evidenced by a reduction in the number of cells discernable under phase microscopy. With increased pressure, a large proportion of the remaining cells took up trypan blue, indicating membrane damage. The optimal yield of plasma membranes from several murine lymphoblastoid cell lines was obtained under conditions that result in 50% reduction in cell number with the remaining cells being less than 10% viable. For the EL4 line the cell disruption pressure of choice as observed from the plot in Fig. 2 was found to be 200 psi.

Following disruption of the cells, separation of the subcellular components is accomplished by differential centrifugation in 50-ml capped centrifuge tubes (Fisher Scientific Co., Cat. #5-529C) using a Sorvall RC-2 centrifuge with a SS-34 head. Nondisrupted cells and nuclei are first removed by centrifugation at 480 *g* for 20 min. The resulting pellet is washed twice by resuspending in Tris-NaCl buffer with a Dounce homogenizer with a loose-fitting Teflon pestle. This step disperses clumped material and releases entrapped subcellular organelles from nondisrupted cells. The initial and subsequent wash supernatant fluids are combined and centrifuged at 4000 *g* for 20 min. The resulting pellet is washed twice in Tris-NaCl buffer, with homogenization, and a mitochondrial pellet generated. The supernatant fluids from this step are combined and centrifuged at 20,000 *g* for 30 min resulting in a microsomal pellet containing plasma membrane. Should one desire, the latter supernatant can be subjected to centrifugation at 100,000 *g* for 90 min in an L5-50 Beckman centrifuge with a Beckman Type 45 fixed-angle titanium rotor, resulting in a ribosomal pellet. The final supernatant liquid, containing a variety of soluble enzymes, is also available for further purification procedures.

The interests of our laboratory are directed mainly toward lymphocyte plasma membrane cell surface components which are utilized as an indication of the efficiency for the procedures described. As can be seen in Table I, this fractionation procedure results in a 45% yield of plasma membrane as judged by analyzing the Thy-1 differentiation alloantigen, a T-lymphocyte cell surface marker;[5] purification is almost 10-fold. The uniformity of this membrane preparation may be assessed by isopycnic sedimentation techniques. This may be accomplished by resuspending the microsomal pellet in STE buffer (0.15 *M* NaCl, 1 m*M* ethylene diaminetetraacetic acid, 0.01 *M* Tris chloride; pH 7.4) and subjecting the sample to isopycnic centrifugation on linear gradients of 5–40% (w/w) potassium tartrate in STE buffer, using a Beckman SW27 swinging-bucket rotor in a

[5] R. K. Zwerner, P. A. Barstad, and R. T. Acton, *J. Exp. Med.* **146,** 986 (1977).

TABLE 1
DISTRIBUTION OF THY-1.1 ACTIVITY IN FRACTIONS OF DISRUPTED BW5147 CELLS[a]

| Fraction | Total protein (mg) | Total activity (Thy-1 units) | Protein (%) | Purification | Yield (%) |
|---|---|---|---|---|---|
| Disrupted cells | 2296 | 45.6 | — | — | — |
| 400 *g* Supernatant | 1019 | 38.5 | 44 | 1.8 | 79 |
| 400 *g* Pellet | 1085 | 16.3 | 47 | 0.7 | 34 |
| 4000 *g* Supernatant | 800 | 29.0 | 35 | 1.7 | 60 |
| 4000 *g* Pellet | 95 | 7.4 | 4 | 3.7 | 15 |
| 20,000 *g* Supernatant | 796 | 6.7 | 35 | 0.4 | 14 |
| 20,000 *g* Pellet | 105 | 21.9 | 5 | 9.9 | 45 |

[a] Cell disruption pressure of 275 psi was utilized.

Beckman L5-50 centrifuge at 4° for 90 min at 100,000 g. The gradients are eluted in 1-ml fractions. Densities are determined by refractometry (Abbe Refractometer, Fisher Scientific Co., Cat. No. 13-964), the fractions dialyzed against phosphate-buffered saline (PBS), and the volumes adjusted to 2 ml with PBS. The protein content and relative activity of a membrane marker should be determined in each fraction.

Density gradients have not proven advantageous in enhancing plasma membrane purification from murine lymphoblastoid cell lines. As shown in Fig. 3, Thy-1.1 activity is associated with the predominant protein band in a potassium tartrate gradient at $p = 1.14–1.16$ g/cm$^3$. Although complete recovery of the total Thy-1.1 activity applied to the gradient could be achieved, no significant additional purification over the microsomal pellet was obtained. By this criterion the microsomal pellet represented a rather homogeneous preparation.

## II. Acquisition of Components from the Culture Medium

In addition to providing quantities of subcellular components, suspension cultures of mammalian cells offer a rich source of products normally released by cells into growth medium that can be purified by appropriate procedures. One major advantage of isolating cellular products from the spent medium is that acquisition prior to cell damage simplifies purification and minimizes enzymic degradation of cellular products associated with cell-disruptive procedures. Suspension culture medium of certain murine lymphoblastoid cell lines can be a ready source of endogenous murine leukemia virus (MuLV). In addition, mycoplasmas, either intentionally or unintentionally present in lymphoblastoid cell lines, also reside

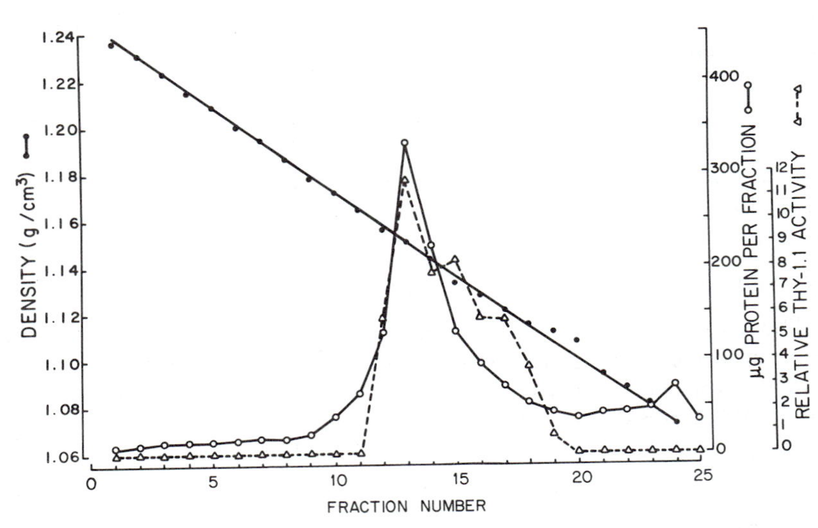


FIG. 3. Isopycnic centrifugation of purified BW5147 membranes on a potassium tartrate gradient, showing coincidence of Thy-1.1 antigenic activity with protein in membrane-containing fractions.

in the cellular supernatant of infected cultures. The isolation and partial purification of these two components from BW5147 cells is outlined to illustrate fundamentals of processing supernatant products.

### 1. *General Considerations of Virus Acquisition*

Certain conditions must be met for successful recovery of endogenous MuLV from murine lymphoblastoid cultures. First, the cells must produce virus in sufficient quantities. Second, culture medium requirements should be determined that insure good cell viability, high virus production, and lack of interference during purification. The type and concentration of serum is important in meeting these criteria. As an illustration, BW5147 cells grow well in medium supplemented with 2% fetal calf serum (FCS), but virus production is much reduced compared to levels obtained with medium supplemented with 10% FCS. In addition 2% FCS increases the fragility of cells, which leads to excessive fragmentation during harvesting that physically and enzymically interferes with virus purification. Horse serum is inappropriate because of its high lipid content that decreases virus yield during purification and results in highly impure virus preparation. For this cell line, medium containing 5% FCS is an appropriate compromise resulting in high virus production and efficient harvest of cells. A third general condition for purification of supernatant products is that cell contamination be closely monitored. Mycoplasmas that are not cultivatable by ordinary techniques[6] will grow in suspension cultures of lymphoblastoid cells. Although we have found no indication that mycoplasma contamination of BW5147 cell cultures decreases MuLV production, the purification of virus is hampered by the presence of mycoplasma in supernatant liquid from the culture.

### 2. *Isolation Procedures*

The low-speed cell supernatant fluid obtained from the initial centrifugation step (see Fig. 1) is further centrifuged in 250 ml polypropylene bottles (Fisher Scientific Co., Cat. No. 5579) at 6000 g for 30 min at 4° in a Sorvall RC-2 centrifuge with a GSA rotor to remove residual cells and cellular debris.

A step gradient is formed in a Beckman CF-32 continuous-flow zonal rotor by sequential addition of STE buffer (200 ml), 15% (w/w) sucrose (120 ml), 43% sucrose (110 ml), and 50% sucrose until STE buffer is eluted from the rotor. Taking care not to disturb the sediment, the clarified supernatant liquid is removed and continuously pumped at 2.5 liters/hr into the rotor spinning at 30,000 rpm. After the sample has been applied, STE buffer is pumped through the rotor for about an hour to allow partial equilibration of the sedimenting virus. The gradient is then eluted from the rotor by pumping 55% sucrose into the bottom of the gradient, and 10-ml fractions are collected. Virus is found distributed in the gradient at densities between 1.13–1.18 g/cm$^3$. The fractions are pooled, diluted, and centrifuged in cellulose nitrate tubes (Beckman No. 302237) at 100,000 g in a Beckman SW27 bucket rotor at 4° for 60 min.

If cultures are contaminated with mycoplasma, this pellet may contain large quantities of these organisms. Virus can be further purified by isopycnic centrifugation on linear gradients of potassium tartrate. Prior to this step, thorough dispersion of the 100,000 g pellet in STE buffer is necessary and may be carried out either by vigorous homogenization in a Dounce homogenizer fitted with a Teflon pestle or by gentle sonication. The latter technique should be carefully controlled and possibly avoided if structural integrity of the virus is essential. We have observed that envelope components of MuLV may be disrupted by excessive sonication. This suspension is applied to a linear gradient of 5–40% (w/w) potassium tartrate in STE buffer, and isopycnic centrifugation is conducted in cel-

[6] M. F. Barile, *in* "Cell Culture and Its Application" (R. T. Acton and J. D. Lynn, eds.), p. 291. Academic Press, New York, 1977.

lulose nitrate tubes (Beckman No. 302237) at 100,000 g in a Beckman SW27 bucket rotor at 4° for 90 min. Potassium tartrate offers an advantage over sucrose gradients in the shorter time needed to reach equilibrium, due to the lower viscosity of this solution. A visible virus band can be seen at a density between 1.15–1.17 g/cm$^3$. Mycoplasma from infected cultures band at $p$ = 1.20–1.24 g/cm$^3$.[7] The virus bands removed from gradients by aspiration with a Pasteur pipette or by gradient elution are diluted 1:3 with STE in Beckman cellulose nitrate tubes (No. 331101), and centrifuged at 100,000 g for 1 hr in a Beckman SW40 rotor at 4° to remove tartrate and concentrate the virus. Identification of virus is routinely performed by electron microscopy using negative staining with phosphotungstic acid. For additional virus identification, or more sensitive determination of virus distribution on tartrate gradients when the viral band is not visible, immunological methods[8] or determination by UV absorption may be employed.

### 3. *Other Methods*

A second method for treating large volumes, i.e., 5 liters or more, requires concentration of the cell supernatant fluid (Fig. 1) prior to clarification. Filtration with a Diaflo DC10 continuous recirculating hollow-fiber ultrafiltration system (Amicon Corp.) with a 100,000 nominal molecular weight cut-off provides a rapid procedure for concentrating the macrosolute up to 50-fold. The concentrate can be clarified and, depending on the final volume obtained, the virus may be purified by zonal centrifugation or by tartrate gradient isopycnic centrifugation as previously described.

For acquisition of smaller molecular weight supernatant components, hollow-fiber ultrafiltration can provide a simple method for concentrating molecules of different sizes by sequential filtration through fibers of varying retention limits.

If small amounts of virus are needed or if the cell line is an excellent producer of virus, smaller volumes of culture medium may suffice. In that case, the cell supernatant is clarified at 6000 g as noted. However, virus from the clarified supernatant liquid can be sedimented by centrifugation in 71-ml polycarbonate bottles (Beckman No. 339168) at 100,000 g for 45 min at 4° in a Beckman type 45 Ti fixed-angle rotor. The virus pellet is then dispersed and applied to a tartrate gradient for final purification.

[7] K. S. Wise, G. H. Cassell, and R. T. Acton, *Proc. Natl. Acad. Sci. U.S.A.* **75**, 4479 (1978).

[8] K. S. Wise and R. T. Acton, *Protides Fluids, Proc. Colloq.* **25**, 707 (1977).

## III. Comments

The procedures described were designed to allow the effective purification of cells, subcellular organelles, and components from suspension culture medium. By varying this general approach one should be able to harvest products from the growth of various cell types in suspension culture in volumes of 1–12 liters. When several products are utilized from a given suspension culture it becomes a very economical approach and enhances the possibility of gaining further information on the molecular biology of the cell.

### Acknowledgments

This work was supported by U.S. Public Health Service Grants CA 15338, CA 18609, and CA 13148 from the NCI, IM33C from the American Cancer Society, GB-43575X from the Human Cell Biology section of the NSF, and the Diabetes Trust Fund. R. K. Z. and K. S. W. were supported in part by an institutional National Research Service Award No. T32 GM07561 from the NIGMS. This work by R. T. A. was done during tenure of an Established Investigatorship of the American Heart Association.

Section II

## Specialized Techniques

A. Metabolism
*Articles 19 through 25*

B. Genetics, Hybridization, and Transformation
*Articles 26 through 31*

C. Virus Preparation
*Articles 32 through 36*

# [19] Cell Cycle Analysis by Flow Cytometry[1]

*By* J. W. GRAY and P. COFFINO

Growth of cell populations can be analyzed by considering individual cells to progress sequentially through a series of compartments or phases. The processes of mitotic division and DNA synthesis can be recognized by light microscopy and by incorporation of [$^3$H]thymidine followed by autoradiography, respectively. Application of these methods led to a description of the cell cycle composed of four phases: mitosis ($M$), gap 1 ($G_1$) preceding DNA synthesis, synthesis of DNA ($S$), and, finally, gap 2 ($G_2$) following DNA synthesis and preceding the next mitosis.

Experimental methods are often required to answer the following questions about a population of tissue culture cells: What is the fraction of cells in each phase? At what rate are cells progressing from one phase to the next? What is the locus of action in the cycle of an agent that perturbs cell cycle progression?

Techniques that depend on light microscopy and autoradiography have been used extensively to answer these questions. The method of autoradiography (see this volume [22]) and the analysis of labeling index and mitotic index experiments[2] have been described in this series. Flow cytometric (FCM) methods provide an alternative method of analysis. Cells are stained with fluorescent dyes that bind to DNA. By exciting the dye and measuring the emitted fluorescence of single cells as they flow in single file at high rates, the DNA content of each cell can be measured and its location in the cell cycle thereby inferred. The development of specific and convenient staining techniques and the commercial availability of sophisticated instruments have made this method of analysis increasingly accessible to biologists.

First we shall review flow cytometric principles, and then we shall describe selected methods for the preparation of cultured cells for flow cytometry. Three examples of the use of flow cytometry in cell cycle studies will be given.

[1] This report was prepared as an account of work sponsored by the United States Government. Neither the United States nor the United States Energy Research & Development Administration, nor any of their employees, nor any of their contractors, subcontractors, or their employees, makes any warranty, express or implied, or assumes any legal liability or responsibility for the accuracy, completeness or usefulness of any information, apparatus, product or process disclosed, or represents that its use would not infringe privately owned rights.

[2] J. B. Kurz and D. L. Friedman, this series, Vol. 40, p. 44.

METHODS IN ENZYMOLOGY, VOL. LVIII ISBN 0-12-181958-2

TABLE I
COMMERCIALLY AVAILABLE FLOW CYTOMETERS AND FLOW SORTERS

| Company[a] | Instrument type | Address |
|---|---|---|
| Becton-Dickinson FACS | Flow sorter[4]<br>Flow cytometer | 506 Clyde Ave., Mountain View, California 94040 |
| Ortho Instruments | Flow cytometer[5,7]<br>Flow sorter | 410 University Ave., Westwood, Massachusetts 02090 |
| Coulter Electronics, Inc. | Flow sorter[6] | 590 West 20th St., Hialeah, Florida 33010 |

[a] Reference to a company or product name does not imply approval or recommendation of the product by the University of California or the U.S. Department of Energy to the exclusion of others that may be suitable.

## Flow Cytometry Principles

A variety of flow cytometric instruments are available commercially (Table I). The capabilities of these instruments are varied; procedures that can be accomplished readily with one instrument might be impossible with another. Therefore, in selecting an instrument for a particular application, one should understand the principles involved in flow cytometry and know which capabilities are common to all flow cytometers or sorters

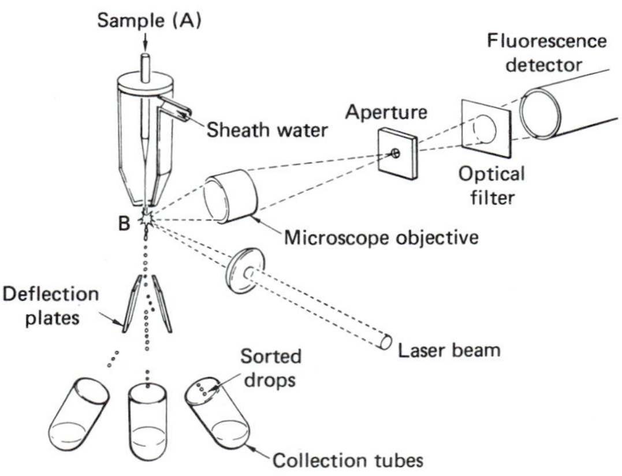


FIG. 1. Schematic representation of a flow sorter. Fluorescently stained cells in aqueous suspension are introduced into the sorter chamber (A). Cells traverse the chamber and emerge one at a time in a liquid jet, where they are illuminated by a laser beam (B). The resulting fluorescence is collected by a microscope objective and projected through an optical filter (to remove scattered laser light) onto a photomultiplier. The photomultiplier and associated amplifier produce a voltage pulse whose height is proportional to the fluorescence intensity. The height of the voltage pulse is the criterion for cell sorting. Drops containing cells to be sorted are charged as they separate from the liquid jet. Empty drops or drops containing unwanted cells are not charged. The charged droplets are separated from the others as they fall through an electric field generated by charge deflection plates.

and which are machine-specific. Generally speaking, in a flow system such as the sorter illustrated in Fig. 1, fluorescently stained cells are transported in a liquid medium, one by one, through an intense light that excites the dye. The resulting fluorescence that is measured and recorded is proportional to the amount of the cellular component (in this discussion, DNA) to which the dye is bound. The rate of cell analysis is usually about $10^3$ cells/sec. In addition, cells can be sorted (separated) on the basis of their fluorescence.

A brief discussion of flow cytometric concepts, including cell transport, fluorescence excitation and detection, cell sorting, and data aquisition, emphasizing features that are important in DNA-content measurements, follows. Flow cytometric principles have been reviewed in detail by Van Dilla and Mendelsohn[3] and by Herzenberg *et al.*[4] (also see Kamentsky *et al.* and Göhde[7]).

### *Cell Transport*

All flow cytometers operate on the principle that stained cells in aqueous suspension will flow with the suspending medium. By channeling this flow properly, the cells can be forced through a region where their optical properties can be measured. Usually the medium containing the cells is surrounded by a sheath of similar fluid.

Figure 2 shows a liquid jet emerging from a nozzle into the air; this occurs in most flow sorters. The black stream was produced by injecting ink into the cell sorter instead of a cell suspension. In an instrument such as the one pictured here, cells are measured immediately after the jet is formed. In other instruments, cells are injected into glass capillary tubes or water-filled chambers where they are measured. An intense light source illuminates the cells in the sample stream. If the diameter of the sample stream is not sufficiently small, some cells may be illuminated less intensely than others and thus the fluorescence frequency distributions will be distorted. The sample stream diameter can be decreased by increasing the pressure on the sheath relative to that on the sample. When the cell concentration is low, the number of cells analyzed per second should not be increased by increasing the sample stream diameter, because of the danger of improper illumination. Instead, the cell sample should be concentrated.

[3] M. A. Van Dilla and M. L. Mendelsohn, *in* "Flow Cytometry and Sorting." Wiley, New York (in press).

[4] L. A. Herzenberg, R. G. Sweet, and L. A. Herzenberg, *Sci. Am.* **234,** 108–117 (1976).

[5] L. A. Kamentsky, *Adv. Biophys. Med. Phys.* **14,** 83 (1973).

[6] A. Brunsting, *in* "Flow Cytometry and Sorting," pp. 79–88. Wiley, New York (in press).

[7] This flow cytometer operates according to somewhat different principles than those described in this review. For more information, see W. Göhde, *in* "Fluorescence Techniques in Cell Biology," pp. 79–88. Springer-Verlag, Berlin and New York, 1973.

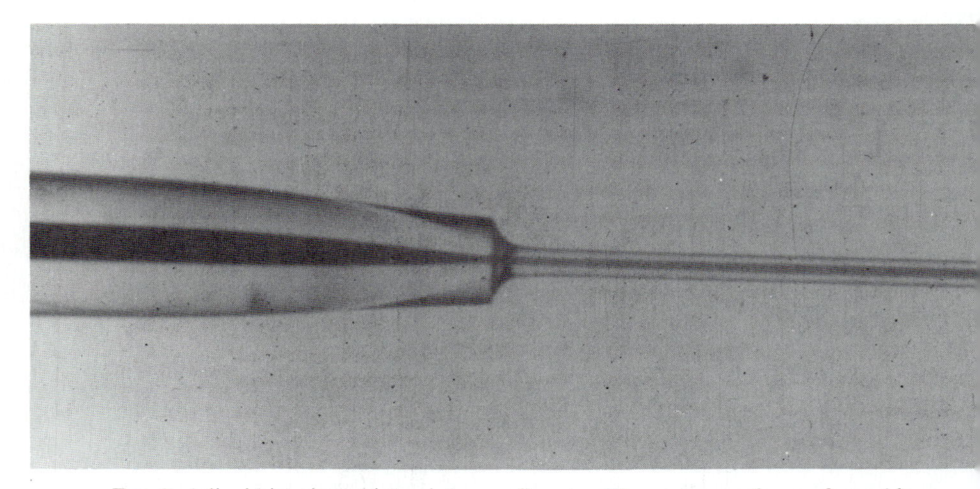

FIG. 2. A liquid jet ejected into air by a cell sorter. The sorter nozzle was formed by a drawn glass capillary with a 50-$\mu$m inside diameter. The black stream was formed by injecting ink in place of the cell suspension; the confinement of the ink to the center of the jet is clearly visible. In practice, the sample stream should be even smaller than the ink stream shown in the figure.

*Fluorescence Excitation and Collection*

This is a complex subject to discuss in specific terms because of the different requirements of the many DNA stains currently in use. We will describe only the requirements for the dyes propidium iodide (PI) and chromomycin A3 (CA3). Detailed staining protocols for these reagents are given below.

The stain content in a cell is measured as the cell flows at constant velocity through an intense light. The dye is stimulated by the light, and the resulting fluorescence is detected and is assumed to be proportional to the amount of dye in the cell; this is not always true for asymmetric cells.[8] The wavelength of the exciting light should be matched as closely as possible to the excitation maximum of the dye (see Fig. 3 for corrected excitation and emission spectra of the dyes CA3 and PI bound to DNA).[9] The excitation maximum of CA3 is about 430 nm, while that of PI is about 535 nm.

The light source of most flow cytometers is an argon-ion laser; an exception is the Ortho flow cytometer[7] which uses an arc lamp. The laser

[8] B. L. Gledhill, S. Lake, L. L. Steinmetz *et al.*, *J. Cell. Phys.* **87,** 367–375 (1976).

[9] These data were kindly supplied by Dr. R. Langlois, Lawrence Livermore Laboratory, Livermore, California.

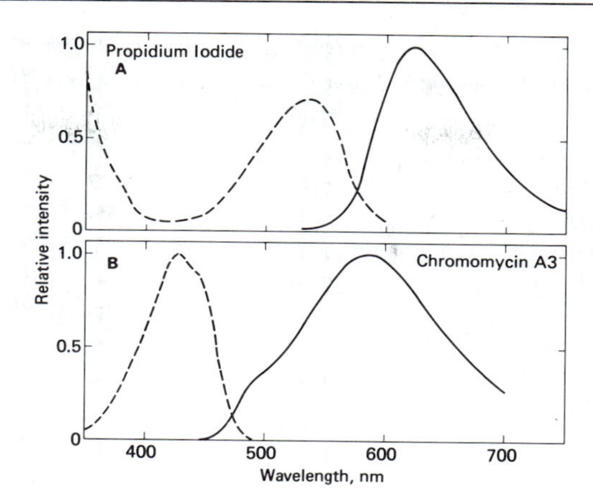


FIG. 3. Excitation and emission spectra[9] for the dyes propidium iodide (A) and chromomycin A3 (B). Dashed lines represent excitation spectra and solid lines emission spectra.

should be adjusted to emit light at 457 nm to excite CA3 and at 514 nm to excite PI. A system with a smaller laser might operate only at 488 nm. PI can be excited at this wavelength, but CA3 cannot.

The light collected from each cell as it traverses the exciting light beam is due partly to cellular fluorescence and partly to scattered laser light. To measure only fluorescence, the scattered light is removed by an optical filter that transmits light at wavelengths longer than that illuminating the cell. It is important to select a filter that transmits at longer wavelengths than the laser light because the scattered light may be several orders of magnitude more intense than the fluorescence. At the same time, a filter should be chosen that transmits as much fluorescence as possible. After the fluorescence from a cell passes through the optical filter, it strikes a photomultiplier tube and produces an electrical pulse, the amplitude of which is proportional to the fluorescence intensity. Each pulse is amplified and its amplitude digitized and added to the memory of a multichannel analyzer. The analysis of a large number of DNA-stained cells results in a frequency distribution of cellular DNA content. The DNA distributions of S49 cells stained by the CA3 and PI procedures are shown in Fig. 4. The flow cytometer was adjusted by changing the gain of the pulse amplifier so that the mode (highest point) of each DNA distribution was 50 on the $X$ axis. This adjustment can be made in a variety of ways on different instruments: by changing the pulse amplifier gain, the photomultiplier high voltage, or the laser intensity. If a choice is available, an adjustment of the amplifier gain is preferable; every effort should be made

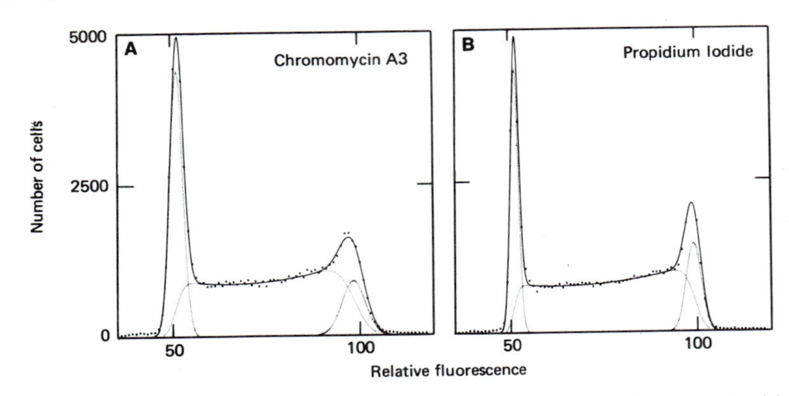


FIG. 4. Fluorescence distributions of S49 mouse lymphoma cells stained with (A) chromomycin A3 and (B) propidium iodide. The data points are solid circles. The solid lines are the result of least squares best fits to the data; each solid line is the sum of the dotted component curves.

to avoid changing the laser power. The $G_1$ peak mode should be adjusted to the same location at the start of each experiment to standardize data.

Most flow cytometers can measure and record more than one parameter for each cell. For example, both the dye fluorescence and the scattered laser light might be recorded to form a two-parameter distribution, the dye fluorescence proportional to DNA content and the scattered light proportional to cell size. The exact relationship between scattered light and cell size is difficult to quantitate; it depends on the light collection, angle, and aperture, index of refraction of the suspending fluid, etc. Usually, however, the scattered light increases monotonically with cell volume.

### *Cell Sorting*

The selection of cells on the basis of their DNA content can be coupled to tracer technology. For example, the cell cycle location of a radioactively labeled cohort of cells can be determined by sorting cells according to their varying DNA contents (and hence position in the cell cycle) and assaying the fractions for radioactivity by autoradiography or liquid scintillation counting.

Commercially available cell sorters use the electrostatic deflection principle illustrated in Fig. 1 to fractionate cells. The cells are ejected into air and their stain content measured immediately. The cells continue down the liquid jet until the jet breaks into droplets. When a cell to be sorted reaches this point, a brief voltage pulse is applied to the jet, and the droplet containing the cell is charged when it separates from the liquid column. The distance from the measurement point to the droplet breakoff must be constant during operation of the sorter. This distance determines the time delay between cell measurement and the application of the charging pulse and is stabilized by vibrating the flow chamber at high frequency, e.g., 30 kHz. The distance can be disturbed by a change in the pressure producing the liquid jet, by the occurrence of a bubble within the flow nozzle, or by a piece of debris lodged in the nozzle. Should one of these events occur, the wrong droplets will be charged. Since such malfunctions are not uncommon, the droplet breakoff point should be checked frequently.

## Methods

### *Cell Dispersal*

All flow cytometric techniques require analysis of individual cells. Therefore, prior to analysis, the population must be dispersed into a suspension of single cells.

Cells grown in suspension culture usually require no dispersal, but those grown in monolayer cultures must be dissociated. To accomplish this, the medium is decanted and the monolayer gently rinsed with 5 ml cold PBS (solution 1).[10] The PBS is decanted, 2 ml of trypsin (solution 2) are added, and the cells incubated for about 5 min at 37°. The incubation should be restricted to the minimum time sufficient to free the cells from the flask. The cells will leave the monolayer in sheets. This process can be hastened by striking the culture flask sharply with the heel of the hand. Dispersal of clumped cells will be aided by gently passing the suspension in and out of a pipette. After dispersal, 8 ml of culture medium containing serum are added to inhibit trypsin and the mixture is transferred to a plastic centrifuge tube. The cells from the monolayer culture now should be a suspension of single cells. Subsequent steps are identical for cells grown in monolayer or in suspension. The cells are centrifuged (250 g for 5 min in this and subsequent steps), the supernatant liquid aspirated, and the cells resuspended vigorously in 10 ml cold PBS with a vortex mixer. Centrifugation and aspiration are repeated and the cell pellet is vigorously resuspended in the residual fluid and fixed by adding 5 ml of cold 70% ethanol dropwise during continuous agitation with a Vortex mixer. After fixation and cell suspension can be stored almost indefinitely at 4°.

[10] All working solutions are described in the Appendix.

*Staining*

In our experience, the simplest and most reliable DNA stain is the dye chromomycin A3 (Calbiochem). The staining procedure is adapted from that of Crissman and Tobey who used the dye mithramycin.[11] Mithramycin and chromomycin A3 have similar properties, and chromomycin A3 was chosen because of its availability and high purity.

Cells should be fixed in 70% ethanol for at least 30 min before staining. A suspension containing approximately $5 \times 10^6$ fixed cells is centrifuged and the supernatant liquid removed by aspiration. The cells are resuspended by a Vortex mixer in the CA3 staining solution (solution 3). After 30 min of incubation in the dark at room temperature, the cells are ready for flow cytometric analysis. This procedure is elegant in its simplicity; unfortunately CA3 fluorescence cannot be excited by some flow cytometers (see discussion of flow cytometry above).

When CA3 cannot be used, propidium iodide (PI) can be substituted with a slight modification of the staining procedure.[12] Cells must be treated initially with RNase because the dye binds to all nucleic acids. To accomplish this, approximately $5 \times 10^6$ cells from the fixative are removed by centrifugation and resuspended with a Vortex mixer in RNase–buffer solution (solution 4) and incubated for 15 min at 37°. The cells are centrifuged and the supernatant solution aspirated. Then the cells are resuspended by vigorous agitation by a vortex mixer in the PI staining solution (solution 5). The stained cells should remain in the dye solution for 30 min at room temperature before analysis.

*Sorting for Radioactivity Assay*

Cells for autoradiographic analysis can be collected directly on microscope slides. At least 1000 cells should be sorted onto each slide. When the cells are sorted, the slides are air dried. Prior to autoradiography, the cells are fixed to the slides in a series of three successive 5-min ethanol baths (95%, 95%, 100%) and then washed for 1 min in distilled water to remove salt crystals (the sheath fluid in sorters is commonly isotonic saline).

Alternatively, cells may be analyzed by liquid scintillation counting for incorporated radioactivity. They are sorted directly onto glass-fiber filters to minimize the cell loss that occurs if the sample is collected in one vessel and transferred to filters. When the sorter is adjusted properly, the collection efficiency is greater than 95%, so that an accurate measurement of the radioactivity per sorted cell can be made.

[11] H. A. Crissman and R. A. Tobey, *Science* **184,** 1297–1298 (1974).

[12] W. Dittrich and W. Göhde, *Z. Naturforsch., Teil B* **24,** 360–361 (1969).

Filters containing the sorted cells are washed in a Millipore filtration apparatus with sequential 10-ml portions of 10% cold trichloroacetic acid (TCA), 5% cold TCA, and 70% ethanol and then loaded into scintillation vials. To each vial is added 125 μl water and 1 ml tissue solubilizer (NCS, Amersham); after 1 hr, 10 ml of scintillation fluid (Liquifluor, New England Nuclear) are added. The vials are then ready for counting.

## Applications

Estimates of cell cycle parameters such as the fractions of cells in the $G_1$, $S$, and $G_2 + M$ phases, average durations and variabilities of the three phases, and rates of DNA synthesis can be obtained from flow cytometric data. However, proper estimation methods might require relatively sophisticated computer techniques. We make no attempt to discuss data analysis techniques in detail; instead, we refer the interested reader to a review article by Gray *et al.*[13] for further information and references.

### *Analysis of Phase Fractions*

The most common application of flow cytometric techniques related to the cell cycle is the determination of the fraction of cells in the $G_1$, $S$, and $G_2 + M$ phases. This information is obtained from DNA distributions such as those shown in Fig. 4, measured for asynchronous, exponentially growing S49 mouse lymphoma cells from the same culture; one aliquot was stained with CA3 and the other with PI. In each distribution, the peak at ×1 DNA content (relative fluorescence, 50) is produced by diploid, $G_1$ phase cells. The peak at ×2 DNA content (relative fluorescence, 100) is produced by $G_2 + M$ phase cells, and the intermediate continuum is produced by $S$ phase cells in which varying amounts of DNA have replicated. The areas under each of these regions of DNA distribution are proportional to the fractions of cells in the corresponding cell cycle phase.

The $G_1$, $G_2 + M$, and $S$ phase areas are determined by simultaneously fitting the DNA distribution with the sum of two normal distributions and a broadened second-order polynomial.[14] The normal distributions approximate the $G_1$ and $G_2 + M$ phases, and the polynomial approximates the $S$ phase. The estimated fractions in each phase are shown in Table II.

Phase fraction analysis can be used in the assessment of cellular growth conditions. For example, FCM measurements and cell counts were made in parallel to assess the effect of dilution of suspension cultures

[13] J. W. Gray, P. N. Dean, and M. L. Mendelsohn, *in* "Flow Cytometry and Sorting." Wiley, New York (in press).

[14] P. N. Dean and J. Jett, *J. Cell Biol.* **60**, 523–527 (1974).

TABLE II
ESTIMATED FRACTIONS OF CELL CYCLE PHASES IN A CULTURE OF ASYNCHRONOUS S49 MOUSE LYMPHOMA CELLS

| | Propidium iodide | Chromomycin |
|---|---|---|
| $G_1$ | 0.20 | 0.27 |
| $S$ | 0.68 | 0.63 |
| $G_2M$ | 0.12 | 0.10 |

of Chinese hamster ovary (CHO) cells on the growth rate and the distribution of cells in the cycle. Figure 5A presents the growth curves for two exponentially growing cultures. One culture was maintained as a control; the second was established by a 4-fold dilution of one-fourth of the control culture at time zero. The control culture was started 48 hr, i.e., almost four population-doubling times, prior to dilution to ensure the decay of any synchrony that might have been induced initially. The growth curves of the two cultures, based on frequent sampling, indicated that both grew exponentially for at least 24 hr after the dilution step. At later times the cultures entered stationary phase with saturation densities greater than $1 \times 10^6$ cells/ml.

Because of the apparent continued exponential growth after dilution, the noticeable perturbation in the DNA distribution of the diluted culture (Fig. 5B) was unexpected. The control culture contained a relatively constant proportion of cells in the $G_1$ phase, both before and after the time

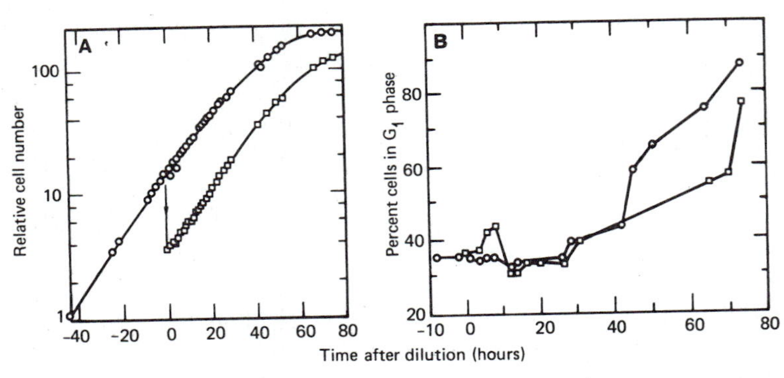


FIG. 5. The response of Chinese hamster ovary cells to dilution of the culture. At −44 hr a spinner flask was established with 1100 ml culture medium and allowed to reach asynchronous, exponential growth (○—○). At time zero, 275 ml of this control culture was poured into a second spinner flask containing 825 ml of fresh 37° medium (□—□), giving a 1:4 dilution. Cell counts (A) and estimates of the fraction in the $G_1$ phase (B) were made periodically for both diluted and undiluted cultures.

zero point and up to about 26 hr. This uniformity indicated asynchrony and was in agreement with the growth curve. In contrast, in the diluted culture the fraction of cells in the $G_1$ phase increased and reached a peak at about 8 hr after dilution. The $G_1$ fraction then quickly declined, returning to the control value at about 17 hr, more than one generation time after the dilution. By 30 hr the fraction of cells in the $G_1$ phase in the control culture began to increase noticeably while the growth curve showed no obvious change in slope. When growth in both cultures slowed and entered stationary phase, the fraction of cells in $G_1$ phase increased dramatically, as expected, because many conditions of growth restriction normally produce $G_1$ (or $G_0$) phase arrest.

*Perturbed Population Analysis*

We can take advantage of the time-dependent changes in the DNA distributions resulting from perturbations to study the effect of the perturbing influence on cell cycle progressions. The effect on other cellular properties such as size can be assayed as well by measuring another parameter for each cell simultaneously. Figure 6 illustrates the results of measuring DNA content (CA3 fluorescence) and size (90° light scatter) of S49 mouse lymphoma cells at several time points after addition of dibutyryl cyclic AMP (dbcAMP) to an asynchronous, exponentially growing culture. The qualitative effect of the treatment was a $G_1$ phase block. Most of the cells have the DNA content of the $G_1$ phase after incubation for 25 hr; their size is about the same as that of early $G_1$ cells in the unperturbed population. Before entering the $G_1$ block, cell growth appears unaffected; the cells cycle and increase in size at the normal rate. These observations can be quantitated by computer modeling. On the right-hand side of Figure 6 is a series of distributions of DNA content vs. size predicted for the time-dependent response of S49 cells to dbcAMP. The model was generated using the parameters listed in Table III (the reader is referred to Gray[15] for a detailed description of a similar one-parameter computer model). These parameters were adjusted so that the computer simulations matched the data; the parameters obtained from this process are assumed to be valid describers of the population.

The second parameter in such dual-parameter analyses is limited only by one's imagination. For example, the recent development of a variety of fluorescent enzyme substrates by Dolbeare and Smith[16] points to the practicality of simultaneous assessment of DNA content and enzyme content.

[15] J. W. Gray, *Cell Tissue Kinet.* **9,** 499–516 (1976).

[16] F. A. Dolbeare and R. E. Smith, *in* "Flow Cytometry and Sorting." Wiley, New York (in press).

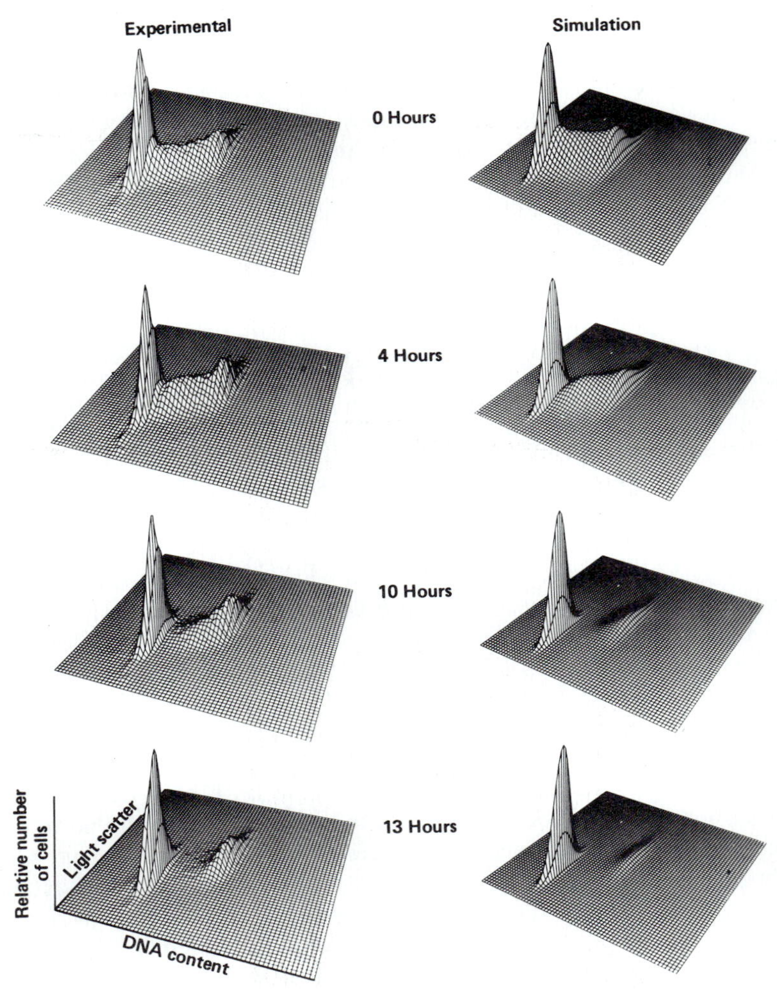


FIG. 6. Light-scatter and DNA-content distributions of S49 mouse lymphoma cells responding to the administration of dbcAMP at zero time. Experimental distributions are shown on the left and distributions simulated according to the parameters in Table III are shown on the right.

## *Combined DNA Distribution–Radioactive Tracer Analysis*

Another sophisticated and powerful application of flow cytometry to cell cycle studies involves the measurement of radioactive tracer uptake as a function of cell cycle position. This has been accomplished in the past by measuring tracer uptake in synchronized cell populations (see this

TABLE III

KINETIC PARAMETERS ASSOCIATED WITH THE RESPONSE OF S49 MOUSE LYMPHOMA CELLS TO dbcAMP[a]

| | Average phase duration (hours) | Coefficient of variation |
|---|---|---|
| $G_1$ | 2.1 | 0.25 |
| $S$ | 12.1 | 0.25 |
| $G_2 + M$ | 2.5 | 0.25 |

[a] The intensity of 90° light scatter was assumed to increase linearly with cell volume; the light scatter from mitotic cells is twice that from $G_1$ cells. The effect of the dbc AMP in the model was assumed to be an immediate and complete block in early $G_1$ phase. The rate of DNA synthesis as a function of DNA content was described by a normal distribution with a coefficient of variation of 0.4 and the mean centered in mid-$S$ phase. (This choice in the model reflects the data in Fig. 7.)

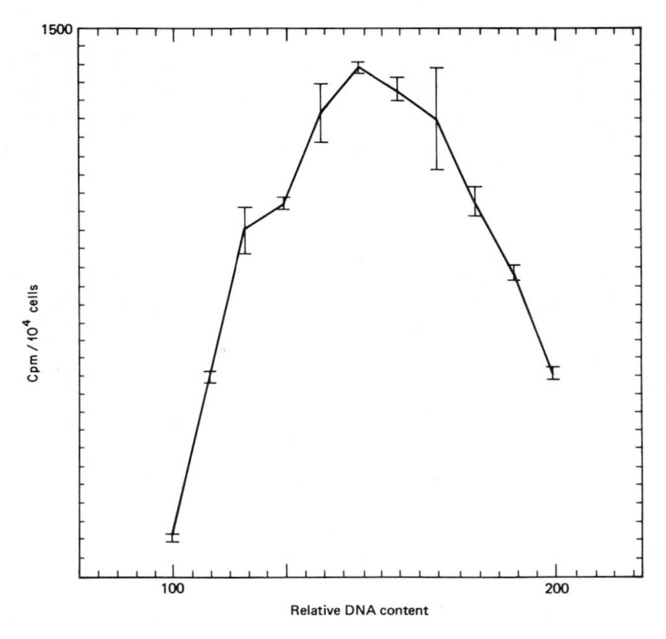


FIG. 7. Incorporation of [$^3$H]TdR into cellular DNA as a function of DNA content. The [$^3$H]TdR was administered to asynchronous, exponientially growing CHO cells as a 30-min pulse. The [$^2$H]TdR uptake was assayed by liquid scintillation counting. The bars span replicate measurements on cells from the same culture. The solid lines connect the means of the measurements at each DNA-content region.

volume [20]). Such studies were limited by the synchrony techniques and could not be applied *in vivo*. The use of flow sorting to select prelabeled cells according to size and/or DNA content eliminates the need for synchronization.

An example of such tracer techniques is illustrated by the technique used in the estimation of the rate of DNA synthesis as a function of DNA content for CHO cells.[17] Prior to sorting, cells were pulse labeled for 30 min with 0.5 $\mu$Ci/ml tritiated thymidine ([$^3$H]TdR) (specific activity ~20 Ci/m*M*). The cells were then fixed, stained with CA3, and processed through a cell sorter where $10^4$ cells were sorted from each of ten DNA-content regions into which the *S* phase was divided. The sorted cells from each region were collected on glass filters and the radioactivity measured by liquid scintillation counting. Figure 7 shows the results of these measurements. The rate of [$^3$H]TdR uptake is maximal in mid-*S* phase, decreasing to minimal values during early and late *S* phase.

Application of the methods described here and related techniques provides a rapid and convenient means for answering questions related to cell cycle kinetics that would be otherwise difficult or impossible to study. The rapidity of the fixation and staining methods makes it possible to monitor results while experiments are in progress. The recent development of fluorescent DNA-staining techniques that do not affect cell viability[18] contributes further to the utility of these methods. Somewhat balancing these advantages, however, are the instrumental complexity and the considerable hardware cost of this technology, which may limit its accessibility for the occasional user.

## Appendix: Working Solutions for Cell Dispersal and Staining

*Solution 1—PBS*

| *Component* | *Amount* |
|---|---|
| NaCl | 800 mg |
| KCl | 200 mg |
| $Na_2PO_4$ | 1150 mg |
| $KH_2PO_4$ | 200 mg |
| $H_2O$ | 1000 ml |

[17] J. W. Gray, Y. S. George, K. Brown, P. Dean, in preparation.
[18] D. J. Arndt-Jovin and T. M. Jovin, *J. Histochem. Cytochem.* **25**, 585–589 (1977).

*Solution 2—Trypsin*

| *Component* | *Amount* |
|---|---|
| NaEDTA | 4 ml (stock solution: 1 g NaEDTA/100 ml $H_2O$) |
| Trypsin (Difco Bacto-trypsin #0153-60) | 1 vial reconstituted in 10 ml PBS |
| PBS | 186 ml |

The final solution should be sterilized by filtration through 0.2 μm filters.

*Solution 3—CA3 Stain*

| *Component* | *Amount* |
|---|---|
| Chromomycin A3 | 10 mg |
| $MgCl_2 \cdot 6 H_2O$ | 1.5 g |
| $H_2O$ (4°) | 500 ml |

The chromomycin A3 will not dissolve in warm water.

*Solution 4—RNase*

| *Component* | *Amount* |
|---|---|
| RNase (Sigma; 71 Kunitz units/mg) | 10 mg |
| $Na_2HPO_4 \cdot 7 H_20$ | 55.6 mg |
| $Na_2HPO_4$ | 168.9 mg |
| $H_2O$ | 10 ml |

*Solution 5—PI*

| *Component* | *Amount* |
|---|---|
| Propidium iodide | 10 mg |
| $H_2O$ | 1000 ml |

### Acknowledgments

The authors gratefully acknowledge helpful technical comments from Y. S. George. This work was performed under the auspices of the Department of Energy, Contract #W-7405-ENG-48 with support from USPHS grant 5R0114533, NIH grant GM16496, and NSF grant PCM 75-06764. P. C. is the recipient of a Research Career Development Award from the NIH, Institute of General Medical Sciences.

# [20] Cell Synchronization

*By* TSUKASA ASHIHARA and RENATO BASERGA

The growth of cells in cultures can be divided into three stages: (1) a lag period, immediately after the inoculum; (2) a growth phase, during which the cell number increases rapidly and usually in an exponential fashion; and (3) a plateau or stationary phase, during which the cell number remains constant.

In exponentially growing populations the cells are distributed asynchronously throughout the cell cycle in its four phases—$G_1$, $S$, $G_2$, and $M$ (mitosis). Under conditions restrictive for growth, some cell lines of rodent (3T3) or human (WI-38) origin will accumulate in a $G_0$–$G_1$ stage of the cell cycle, i.e., the cells have a $2n$ content of DNA ($G_1$ content), do not traverse the cell cycle, and, when stimulated to resume growth, will enter first $S$ and subsequently mitosis.[1] However, most cell lines, especially virally transformed cell lines, are asynchronously distributed throughout the cell cycle, even when growth is markedly slowed down by nutritional deficiencies.[2] Since metabolic processes are different in different phases of the cell cycle, synchronized populations of cells are not only useful but almost necessary if one wishes to study cell-cycle related events. As an example we can cite the case of histone mRNA, which is absent from the cytoplasm of $G_1$ cells, is abundant in the cytoplasm of $S$ phase cells, and rapidly disappears again from the cytoplasm when the cell proceeds from $S$ to $M$.[3] Clearly, when one wishes to study the regulation of histone mRNA synthesis and processing one must have synchronized populations of cells.

Theoretically, at least, cells can be synchronized in any one phase of the cell cycle, including $G_0$. However, the most common sites of arrest are $G_1$, $G_0$, and $M$, while $G_2$ arrest is difficult to achieve and $S$ phase arrest is

[1] F. Wiebel and R. Baserga, *J. Cell. Physiol.* **74,** 191 (1969).

[2] J. C. Bartholomew, H. Yokota, and P. Ross, *J. Cell. Physiol.* **88,** 277 (1976).

[3] M. Melli, G. Spinelli, and E. Arnold, *Cell* **12,** 167 (1977).




ISBN 0-12-181958-2

often lethal. Ideally synchronized populations of cells in culture should meet these criteria: (1) they should be perfectly synchronized at a specific point of the cell cycle; (2) the procedure used for synchronizing cells should have little or no effect on the metabolic processes of the cell; and (3), especially for the biochemist, the method should allow the harvesting of synchronized cells in large quantities. One can quickly say that no synchronizing procedure fully achieves these three goals, but there are a few methods to be related below, that come close to the proposed goals.

We shall discuss the degree of synchrony that can be obtained with various synchronizing procedures, the effect that these procedures have on the metabolic processes of cells in culture, and the amount of cells that can be obtained by various synchronizing procedures. As a general rule, we act on the proposition that $10^8$ cells are sufficient for most biochemical determinations. In fact, a great number of biochemical determinations can be carried out accurately with $10^7$ cells, but $10^6$ cells constitute the lower limit for biochemical determinations even if methods are scaled down considerably. This number of cells, i.e., a million, roughly corresponds to 7–15 $\mu$g of DNA, depending on the ploidy of the cells under consideration. However, simpler determinations, as for instance incorporation of [$^3$H]uridine into total RNA or of [$^3$H]thymidine into DNA, can be carried out with as little as $10^4$ or $10^5$ cells.

## Evaluation of the Degree of Synchrony

There are several methods for determining the degree of synchrony of populations of cells in culture.

### *1. Autoradiography with [$^3$H]thymidine*

Unfortunately, light microscopy combined with [$^3$H]thymidine autoradiography is still the best method for an analysis of the cell cycle and an evaluation of the degree of synchrony. We say unfortunately because autoradiography is a tedious and time-consuming procedure and also because most biochemists have an ancestral fear of the microscope. Yet autoradiography gives more information than any other method that is currently used for an analysis of the cell cycle. By looking at a coverslip of cells labeled with [$^3$H]thymidine one can obtain information not only about the number of cells in DNA synthesis, but also about the number of cells in mitosis. At the same time, one can ascertain whether the population is homogenous or heterogeneous, whether there are dying or dead cells, and whether the cultures are contaminated or not with mycoplasma.

The technique for autoradiography is not complex. It has been de-

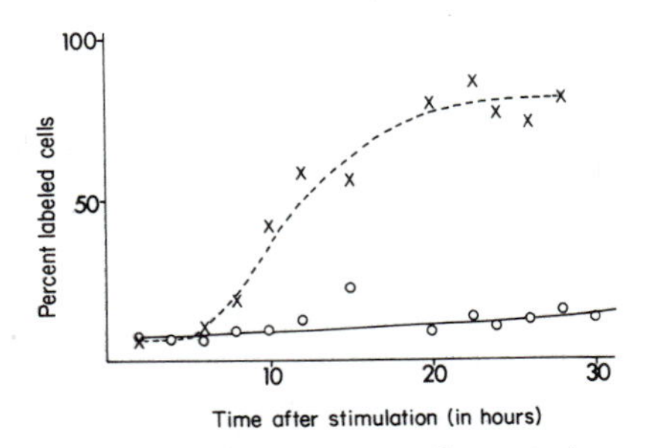


FIG. 1. Cumulative percentage of cells in DNA synthesis, after quiescent K12 cells were stimulated to proliferate by serum. K12 cells, a ts mutant that arrests in $G_1$ at the nonpermissive temperature, were continuously exposed to [$^3$H]thymidine. The ordinate gives the cumulative percentage of cells labeled by [$^3$H]thymidine, and the abscissa gives the time after stimulation, at the permissive (X---X) or at the nonpermissive (○—○) temperatures.

scribed in detail in a monograph by Baserga and Malamud[4] and is presented in this volume [22]. Briefly, to determine the degree of synchrony the cells can be continuously exposed to [$^3$H]thymidine or, alternatively, pulse labeled with [$^3$H]thymidine for a brief period of time, 30 min or so. In the former case the concentration of [$^3$H]thymidine (6.7 Ci/m*M*) in the medium should not exceed 0.1 $\mu$Ci/ml. We use, satisfactorily, a concentration of 0.05 $\mu$Ci/ml. In the latter case a concentration of 0.5 $\mu$Ci/ml is permissible. The cells are fixed at the desired intervals after the synchronization procedure, usually in Carnoy, and then autoradiographed with Eastman Kodak NTB nuclear emulsion (at the concentrations mentioned above, an exposure time of 7–10 days is usually sufficient), developed, fixed stained with hematoxylin and eosin, and scored at the light microscope.

A typical result with this procedure is illustrated in Fig. 1, which shows the cumulative labeling index of K12 cells synchronized by serum deprivation. Serum-deprived cells (48 hr in 0.5% serum in coverslip cultures) were stimulated by addition of fresh medium plus 10% serum and immediately exposed to [$^3$H]thymidine, 0.1 $\mu$Ci/ml. Duplicate coverslips were then fixed at the intervals indicated on the abscissa, autoradiographed, and analyzed. As one can see from Fig. 1, the entry of serum-stimulated K12 cells into *S* is not sharply synchronized. At the time of serum stimulation most of the cells are in $G_0$–$G_1$, and about 20 hr after stimulation, 80% or more are in *S* phase. However, they are not synchronized at the same point. In other words, if one wishes to obtain cells synchronized in the same *phase* of the cell cycle ($G_1$ or *S*), then the method

[4] R. Baserga and D. Malamud, "Autoradiography. Techniques and Application." Harper (Hoeber), New York, 1969.

described in Fig. 1 has achieved a reasonable degree of synchronization. But if one wishes to obtain cells synchronized at a *precise point* of the cell cycle, then this procedure is not sufficient. This is a problem that affects most of the synchronizing procedures used in most laboratories, with but a few exceptions that will be described below. The reader is cautioned to be exceedingly careful in interpreting the synchronization curves that are actually published. In many instances the authors, consciously or subconsciously, increase the appearance of synchronization by simply shortening the abscissa and lengthening the ordinate. This, of course, results in very steep curves of entry into $S$. The reader always should check carefully the numbers on the abscissa, which usually refer to the time, and he will find that some of the steepest curves published still have slopes of about 8–10 hr between the first and last cells entering $S$.

For this reason, we insist that autoradiography data be plotted on arithmetic scales as illustrated in Fig. 1, although it has become fashionable lately to plot the exit of cells from $G_1$ on a semilogarithmic scale.[5,6] As every biochemist knows,[7] logarithmic or semilogarithmic plots tend to hide the scattering of data and complex kinetics and, especially in our case, can give a false impression of the degree of synchronization.

### 2. *Incorporation of* [$^3$H]*thymidine into DNA*

A quick method for evaluating the degree of synchrony, much quicker than autoradiography, consists in determining by liquid scintillation counting the amount of [$^3$H]thymidine (or [$^{14}$C]thymidine) incorporated into acid-insoluble material.[8] Although quicker, this method is not as informative as autoradiography. The procedure is the same except that instead of fixing the cells and processing them for autoradiography the cells are harvested after extensive washing with a balanced salt solution. The DNA from the harvested cells is extracted by the method of Scott *et al.*,[9] and the radioactivity is assayed by liquid scintillation counting using a Triton X-100 Toluene liquifluor.[10] Alternatively, the cells are simply extracted with perchloric acid to remove the acid-soluble fraction, and then they are dissolved in a liquid scintillation vial and counted. We find this procedure useful as a screen for determining whether a method is worthwhile pursuing or is simply hopeless. However, after screening a

[5] R. F. Brooks, *Cell* **12,** 311 (1977).

[6] L. J. De Asua, M. K. O'Farrell, D. Clingan, and P. S. Rudland, *Proc. Natl. Acad. Sci. U.S.A.* **74,** 3845 (1977).

[7] N. R. Cozzarelli, *Annu. Rev. Biochem.* **46,** 641 (1977).

[8] E. Stubblefield, R. Klevecz, and L. Deaven, *J. Cell. Physiol.* **69,** 345 (1967).

[9] J. F. Scott, A. P. Fraccastoro, and E. B. Taft, *J. Histochem. Cytochem.* **4,** 1 (1956).

[10] M. S. Patterson and R. C. Greene, *Anal. Chem.* **37,** 854 (1965).

given procedure by this method we always prefer to determine the extent of synchrony in a more accurate way by autoradiography. The advantage is that, while the autoradiographs are being exposed and prepared, one can proceed with other experiments.

*3. Flow Microfluorimetry*

This is a much faster method in which the distribution of cells over the cell cycle can be obtained with as little as $5 \times 10^4$ cells. The method is essentially based on the determination of the amount of DNA per cell[11] by flow microfluorimetry.[12] It divides a population of cells into cells with a $G_1$ content of DNA, cells with a $G_2$ or $M$ content of DNA, and cells in between, with variable amounts of DNA depending on the position they have in the $S$ phase. By computer processing of data, the percentage of cells in each of these three phases can be calculated with a high degree of accuracy. The disadvantage is that it gives somewhat less information than autoradiography and requires expensive instrumentation. However, the instrument also can be used for more sophisticated endeavors, e.g., the simultaneous determination of the amount of RNA and DNA per cell, as described by Darzynkiewicz *et al.*[13] (see also [19]).

## Methods for Synchronizing Cells in Culture

*1. Physical Methods*

A. SYNCHRONIZATION BY MITOTIC DETACHMENT

The procedure described is a slight modification of that of Terasima and Tolmach.[14]

*Procedure.* As an illustration, we present the procedure used for mitotic synchronization of K12 cells, a ts mutant of the Chinese hamster line WG, originally isolated by Roscoe *et al.*[15] and characterized by Smith and Wigglesworth.[16]

K12 cells are grown in Dulbecco's Modified Eagle's Medium containing 10% (v/v) calf serum, at the permissive temperature of 34°. Exponen-

[11] T. T. Trujillo and M. A. Van Dilla, *Acta Cytol.* **16,** 26 (1972).

[12] D. M. Holm and L. S. Cram, *Exp. Cell Res.* **80,** 105 (1973).

[13] Z. Darzynkiewicz, F. Traganos, T. Sharpless, and M. R. Melamed, *Proc. Natl. Acad. Sci. U.S.A.* **73,** 2881 (1976).

[14] T. Terasima and L. J. Tolmach, *Exp. Cell Res.* **30,** 344 (1963).

[15] D. H. Roscoe, H. Robinson, and A. W. Carbonell, *J. Cell. Physiol.* **82,** 333 (1973).

[16] B. J. Smith and N. M. Wigglesworth, *J. Cell. Physiol.* **84,** 127 (1974).

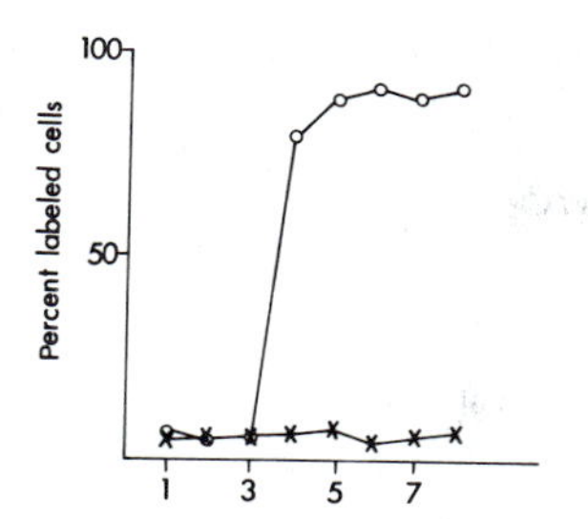


FIG. 2. Cumulative percentage of cells in DNA synthesis in K12 cells collected by mitotic detachment and replated at either the permissive (O—O) or the nonpermissive (X—X) temperatures. Cells were exposed to [$^3$H]thymidine from the time of replating.

tially growing populations are detached from the plate with 0.25% trypsin in Hanks's solution for 2 min. They are plated in 100-mm dishes (surface of 78.5 cm$^2$), at a concentration of $1.5 \times 10^6$ cells per dish. By microscopic observation of the dishes with an inverted microscope, the time at which mitotic figures are most frequent can be determined. For K12 cells, this time interval is about 18 hr after plating. Mitotic cells are collected by shaking the plates gently. The yield is about 5–8% of the total number of cells, and the percentage of mitotic cells in the collected fraction is 90% or more. The mitotic cells are then plated in 35-mm dishes, at a density of $2 \times 10^4$ cells/cm$^2$. The degree of synchronization can be established by continuous labeling with [$^3$H]thymidine at a concentration of 0.1 $\mu$Ci/ml, added at the time of plating, and by counting the number of labeled cells at various intervals after plating (see above). The results of such an experiment are shown in Fig. 2.

*Degree of Synchrony*. The degree of synchrony (Fig. 2) is satisfactory and such cell populations can be used for a study of $G_1$ events and events in the $S$ phase. The labeling index at the beginning is less than 5%. At the permissive temperature, 80% of the cells at 34° go from $G_1$ into $S$ in a period of only 2 hr. It is possible to follow such cell populations and to determine how this degree of synchrony is maintained by observing the wave of mitoses that follows the period of DNA synthesis. With separate cultures it also can be shown that the number of cells doubles in a rather brief interval, corresponding to the wave of mitoses. A note of caution is in order: When mitotic cells are plated, the number of cells plated is calculated from the number of mitotic cells that one actually seeds in different dishes. Within 15 min after plating most of the mitotic cells have completed mitosis, so that the number is already doubled. This should be kept in mind if one wishes to determine the increase in cell number at a later period.

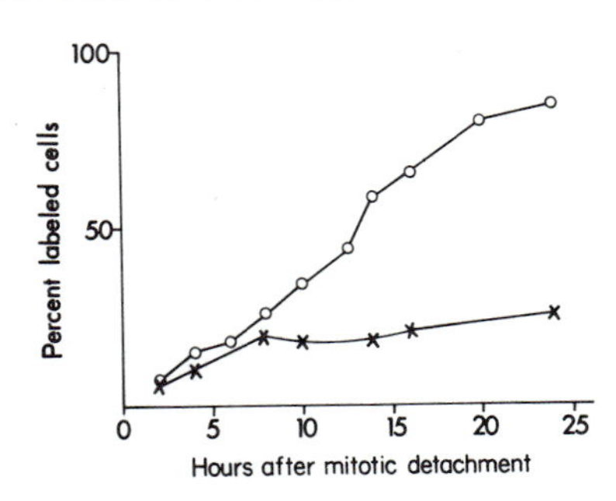


FIG. 3. Cumulative percentage of cells in DNA synthesis in ts AF8 cells collected by mitotic detachment and replated at either the permissive (O—O) or the nonpermissive (X—X) temperatures. The cells are synchronized in a *phase* of the cell cycle (at the nonpermissive temperature they are this side of the ts block) but not at a precise point.

In general, the degree of synchrony decreases as cells leave the *S* phase. This has been discussed in detail in a review by Nias and Fox,[17] to which the reader is referred.

The degree of synchrony varies from one cell line to another, as shown in Fig. 3, which refers to a mutant of BHK cells called AF8 cells, originally described by Burstin *et al.*[18] These cells were treated in essentially the same manner as K12 cells, except that the peak of mitoses occurred at 30 hr after plating instead of 18 hr. The collected mitotic cells were plated and their entry into *S* determined as before. Notice the gradual slope of the curve of entry into *S*. K12 cells have a shorter cell cycle time than AF8 cells, and this may explain the differences in the degree of synchronization. However, the reasons why some cell lines synchronize quite well after mitosis (see, e.g., HeLa cells in the original paper by Terasima and Tolmach[14] and in the review by Nias and Fox[17]), while others synchronize poorly even by mitotic selection, are not really understood.

*Variations*. The first important variation to keep in mind is that the time of harvesting varies among cell lines. With the procedure described above, since we use a very high density, one can collect a reasonable number of cells, i.e., as much as $10^5$ cells/100-mm dish. However, the peak of mitoses after plating varies among cell lines, and there is no simple rule that allows the investigator to select the ideal timing without prior experimentation. If one wishes a rule, one could say that the peak of mitoses generally occurs in most cell lines between 18 and 30 hr after plating, but it is imperative that the investigator should carefully monitor

[17] A. H. W. Nias and M. Fox, *Cell Tissue Kinet.* **4,** 375 (1971).
[18] S. J. Burstin, H. K. Meiss, and C. Basilico, *J. Cell. Physiol.* **84,** 397 (1974).

the plated dishes for the appearance of mitoses every time that he deals with a different cell line.

The yield of mitotic cells can be increased in a variety of ways. Increase in the yield is often desirable because, as mentioned above, a dish will yield only about $10^5$ cells. This is sufficient for experiments in cell biology, but biochemical studies demand larger numbers of cells; one can easily calculate that 100 100-mm dishes would be necessary to obtain $10^7$ mitotic cells. To increase the yield of mitotic cells several techniques are available. The most popular one is the use of Colcemid. This technique was first devised by Stubblefield and Klevecz[19] who found that Chinese hamster cells could be treated for 2 hr with Colcemid, 0.06 $\mu$g/ml; the effect was to arrest cells in metaphase in a reversible manner. These investigators also used a brief period of trypsinization, 45 sec at 4°, resulting in an increased yield of about 8–12% of the total population. In general, the use of Colcemid for brief periods (2–3 hr) has no remarkable effect on the biochemical events in mitotic cells or in synchronized $G_1$ and $S$ phase cells. However, we occasionally have observed that cells treated with Colcemid had different kinetics of entry into $S$ phase from cells that had not been so treated. Again, a simple rule cannot be given, and the investigator will have to determine for each cell line whether the increased yield of mitotic cells that can be obtained with Colcemid is not counterbalanced by possible toxic effects to the mitotic cells. Nias and Fox,[17] for instance, found that Colcemid treatment beyond 2 hr leads to an increased number of aberrant mitoses; we have similar experience with several cell lines.

Another way of increasing the yield of mitotic cells is to use repeated harvesting, a procedure first introduced by Petersen *et al.*[20] With this procedure the monolayers were shaken at 10-min intervals, and each harvest was rapidly cooled to 4°. Again, previous experiments had shown that cells stored at 0° for a period of 4 hr are still capable of completing mitosis in a manner identical to that of control cultures when plated in warm medium. Repeated shaking at 10- or 30-min intervals can therefore yield a larger number of mitotic cells. These may be stored at 4° without loss of viability for a period of up to 8 hr, and the collected mitotic cells can then be plated. Undoubtedly a combination of Colcemid and repeated harvesting leads to an increased yield of mitotic cells. Repeated shaking, however, may decrease the purity of the preparation, i.e., the percentage of mitotic cells in the mitotically detached population. Provided one moni-

[19] E. Stubblefield and R. Klevecz, *Exp. Cell Res.* **40,** 660 (1965).

[20] D. F. Petersen, R. A. Tobey, and E. C. Anderson, *Fed. Proc., Fed. Am. Soc. Exp. Biol.* **28,** 1771 (1969).

tors the degree of synchrony, these techniques can be used effectively to increase the yield of mitotic cells.

In our experience, as well as in the experience of other investigators, Colcemid is less toxic than colchicine. Other mitotic inhibitors are not suitable. For instance, vinblastine is quite effective in arresting cells in mitotis but the arrest, unfortunately, is irreversible.[21]

Another method of collecting cells in mitosis[22] consists of gentle shaking in medium free of calcium. It has been known that calcium-free medium decreases the cell substrate binding and therefore favors the detachment of cells. The problem with the low calcium method is that nonmitotic cells also can be easily detached.

*Advantages and Disadvantages.* The advantage of mitotic detachment is that this is probably the technique that gives the highest degree of synchrony without the use of toxic drugs. The synchrony is not only phase-specific, but is almost point-specific since the duration of metaphase–anaphase is usually of the order of 30 min. If the degree of synchrony is not satisfactory, as in the case of the AF8 cells shown in Fig. 3, one should remember that this is inherent to certain cells in culture. Terasima and Tolmach[14] had originally observed that the asynchrony that develops during the course of a single cell cycle arises mainly during the $G_1$ period. Thus with HeLa cells the fraction of cells in DNA synthesis began to increase 7 hr after mitosis and continued to increase until 14 hr after replating. The curve of entry into $S$ was paralleled by an increase in cell number commencing some 10 hr later. Synchronization in $S$ can be improved by adding hydroxyurea during the $G_1$ period (see below). The technique is simple and if care is taken to use warmed media, the effect on cells is minimal.

Among the disadvantages of the mitotic synchronization method, the only two that we can think of are the low yield of mitotic cells and the fact that quite obviously this technique is applicable only to cells that can be grown in monolayers. The low yield can be remedied by using the proper number of culture dishes. For cells that only grow in suspension other techniques are necessary for their synchronization.

### B. Other Physical Methods

Other physical methods for the synchronization of cells in culture have not been very successful. In general they are based on gradient techniques that hopefully separate, on the basis of volume selection, cells in different phases of the cell cycle. Terasima and Tolmach[14] have indeed shown that, during the mitotic cycle, cells double in volume so that it should be possi-

[21] N. Bruchovsky, A. A. Owen, A. J. Becker, and J. E. Till, *Cancer Res.* **25**, 1232 (1965).

[22] E. Robbins and P. C. Marcus, *Science* **144**, 1152 (1964).

ble to select cells from different points of the cycle on the basis of volume, and thereby produce a synchronized population. In practice, pure populations of $G_1$, $S$, or $G_2$ cells have not been obtained, although some of the physical methods used have produced enriched populations. Linear sucrose gradients, Ficoll gradients, and gradients in fetal calf serum in phosphate-buffered saline have been used. The several methods have been discussed at length in the review by Nias and Fox[17] who concluded that with all the physical methods "a high degree (more than 90%) of synchronization has been demonstrated only in cells harvested from monolayer cultures by the mitotic selection technique." Two recently introduced methods, one by Everson *et al.*[23] and the centrifugal elutriation method of Mitchell and Tupper,[24] are described in more detail.

### C. Ficoll Gradient According to Everson *et al.*[23]

This has been used to separate human lymphoblastoid cells growing in culture. Since these cells are of interest because they are the only human diploid cells that grow indefinitely, the technique for separation on a Ficoll gradient is given below, as described by Everson *et al.*[23]

A 5–20% (wt/w) Ficoll linear continuous gradient is generated in a density gradient generator, resulting in a total volume of 80 $cm^3$ contained in cylindrical polycarbonate tubes 3 cm in diameter and 10.5 cm in length. On top of this gradient 10 $cm^3$ of 5% (w/w) Ficoll are layered to modify the initial slope of the gradient. A suspension of cells in Hanks's balanced salt solution is then carefully layered on top of this 5% Ficoll buffer zone. The cell load can reach approximately $5 \times 10^7$ cells, suspended in a final concentration of $1 \times 10^7/cm^3$.

Centrifugation is carried out at 4° in a PR.2 centrifuge using a swing-out rotor at 80 g for 20–25 min. Fractions are collected by placing a stainless-steel tube through the center of the gradient to the bottom and sampling via a polyethylene tubing with a polystatic pump.

The Ficoll is prepared to a starting concentration of 40% (w/w) with water. It is then subsequently adjusted to a final concentration of 20%, with an equal volume of double-strength Hanks's balanced salt solution. This 20% Ficoll solution is then diluted 1 : 4 with isosmolar Hanks's balanced salt solution to obtain a 5% (w/w) Ficoll solution. The two concentrations are used to generate the necessary gradient.

*Degree of Synchrony.* The degree of synchrony is modest, as can be seen from the original figures in the paper by Everson *et al.*[23] However, reasonably good populations of $G_1$ and $S$ can be obtained.

[23] L. K. Everson, D. N. Buell, and G. N. Rogentine, Jr., *J. Exp. Med.* **137,** 343 (1973).
[24] B. F. Mitchell and J. T. Tupper, *Exp. Cell Res.* **106,** 351 (1977).

*Advantages and Disadvantages.* The disadvantage, as mentioned above, is the modest degree of synchrony; the major advantage is, as with the isoleucine method (see below), the possibility of obtaining a large bulk of cells for biochemical studies.

D. CENTRIFUGAL ELUTRIATION

This method uses the Beckman JE6 elutriator rotor and the elutriation system described in the literature of the manufacturer, Spinco publication DS-125B. With centrifugal elutriation, the sedimentation rate of cells reflects their size and consequently their position in the cell cycle since, generally speaking, cells in $G_1$ are smaller than cells in other phases of the cycle. With this method, Mitchell and Tupper[24] have isolated different fractions of 3T3 and SV40-3T3 cells. The results are not spectacular because of considerable overlapping, but some synchronization is obtained. In general, the centrifugal elutriation method enriches certain fractions in $G_1$ cells, but there is considerable overlapping and the other phases are totally indistinguishable.

## 2. *Chemical Methods*

A. SYNCHRONIZATION BY ISOLEUCINE DEPRIVATION

*Procedure.* The procedure described below is the original one by Ley and Tobey.[25] Chinese hamster cells (CHO) are cultured in suspension as a monodispersed population in spinner flasks. The cells are usually propagated in F10 medium, supplemented with 10% calf serum and antibiotics. According to Ley and Tobey[25] the original F10 formula has been modified in their laboratory by omitting stock solutions D and E (which contain iron, copper, zinc, and calcium) and also sodium pyruvate. The growing cells are synchronized by transfer to F10 medium prepared without isoleucine and glutamine; 10% calf and 5% fetal calf sera are included after they are dialyzed against 10 volumes of Earl's balanced salt solution for 6 days at 3° with changes of salt solution every other day. In this isoleucine-deficient F10 medium, CHO cells rapidly reach a stationary phase; the stationary phase consists of cells arrested in the $G_0$–$G_1$ phase of the cell cycle. Arrested cells remain viable in the $G_0$–$G_1$ phase for 60 or more. Cell cycle traverse is promptly resumed when isoleucine and glutamine are added to the deficient F10 medium. The concentrations of glutamine and isoleucine that are added to the arrested cultures are 40 $\mu M$ and 4 $\mu M$, respectively. Upon addition of these amino acids, the cells divide in a fairly synchronous fashion as early as 30 hr later.

[25] K. D. Ley and R. A. Tobey, *J. Cell Biol.* **47,** 453 (1970).

*Degree of Synchrony.* The degree of synchrony is fair. The cells resume cell cycle traverse and enter DNA synthesis with a slope that varies from 1–12 hr after resuspension in medium containing isoleucine and glutamine, i.e., the first cells start entering DNA synthesis 1 or 2 hr after resuspension and the last cells enter DNA synthesis 12 hr later. This degree of synchrony is acceptable if one wishes to study a large number of cells in $G_1$ or in the $S$ phase with a small percentage of out-of-phase cells.

*Advantages and Disadvantages.* The obvious advantage of this procedure is that it allows the synchronization of cells in suspension and therefore makes available large quantities of cells at a reasonable cost. Since cells in the stationary phase can reach a concentration, at least for CHO, of $1.5 \times 10^5$ cells/ml it can easily be seen that 500 ml of such a suspension will yield as many as $10^8$ cells. Among the disadvantages is that the technique is not applicable to all cell lines, not even in suspension culture. A number of investigators have found that the level of synchrony for a variety of cell lines is not satisfactory when the isoleucine deprivation method is used.

### B. Synchronization by Serum Deprivation and Hydroxyurea

*Procedure.* The detailed technique for synchronization by serum restriction and hydroxyurea is as follows. Cells are plated at a density of $5 \times 10^5$ cells per 100-mm dish for 3 days. The medium is removed, the plates are washed thoroughly, and the cells are exposed to their usual medium containing 1% (or 0.5%) serum for 48 hr. After 48 hr, the serum-deficient medium is replaced by a medium containing 10% donor calf serum. Six hours later, concentrated hydroxyurea stock solution is added to each dish to reach a final concentration of 1.5 m$M$ hydroxyurea. After the cells are blocked with hydroxyurea for 14 hr the medium is removed, the plates are washed, and new medium without hydroxyurea is added. During the entire process, prewarmed washing solutions and medium are always used. These precautions are taken to avoid the effect of temperature-change shock on the synchronization of cells.

*Degree of Synchrony.* With BHK cells and derivatives of BHK cells, the above-mentioned procedure gives a good synchrony and little toxicity (Fig. 4).[26] Upon removal of hydroxyurea, the cells quickly enter $S$ phase in a synchronized fashion. The procedure is therefore satisfactory if cells synchronized in $S$ are desired. The amount of DNA per dish increases by about 80%, which is about as much as one can get with hydroxyurea using concentrations that will synchronize the cells at the $G_1/S$ boundary. Lower concentrations will let cells go through the $S$ phase, although slowly; higher concentrations are more toxic.

[26] H. Chang and R. Baserga, *J. Cell. Physiol.* **92,** 333 (1977).

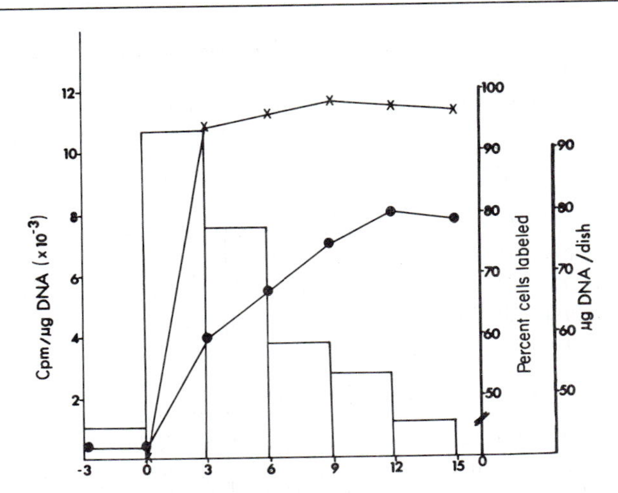


FIG. 4. Synchronization of AL106 by a combination of serum restriction and hydroxyurea (HU) treatment. AL106 cells were plated on 100-mm petri dishes for 3 days. Cells were then cultured in medium containing 1% serum for 2 days. Fresh medium containing 10% serum was then given and HU (1.5 m*M*) was added 6 hr later. HU was removed at zero time on the abscissa. Cells were pulsed with [$^3$H]dT for 3 hr at the times indicated for determination of specific activity. Cells were continuously labeled for autoradiography. Bar, specific activity of DNA; ●—●, µg DNA/dish; X—X, % labeled cells. (Reprinted, with permission, from Chang and Baserga.[26])

*Variations*. The time periods given do not apply to all cells, as one can easily understand. Some cells may require more than 48 hr of serum restriction to achieve quiescence. Other cells may have a longer prereplicative phase and, therefore, hydroxyurea must be added at a later time. Generally speaking, one should try out the optimal conditions for each cell line. Our suggestion is that one should not use more than 14 hr of exposure to hydroxyurea; otherwise toxicity can be very high. Similarly, it should be emphasized that the final hydroxyurea concentration of 1.5 m*M* can apply to certain cell lines and not to others. For instance, AF8 do better at 1.25 m*M* whereas a hybrid between AF8 and LNSV, called AL106, does better at 1.5 m*M*. Again, the investigator is advised to test the optimal concentration of hydroxyurea which will induce a $G_1/S$ boundary block with a minimal amount of toxicity.

Mironescu and Ellem[27] have suggested that pretreatment of serum-restricted cells with either hydroxyurea or cytosine arabinoside, before they are stimulated to proliferate, increases the degree of synchrony of entry into *S* phase. The effect has no relationship to the role of hydroxyurea or cytosine arabinoside as specific inhibitors of DNA synthesis,

[27] S. Mironescu and K. A. O. Ellem, *J. Cell. Physiol.* **90**, 281 (1976).

since the serum-restricted cultures were mitotically quiescent; and a similar enhanced response can be induced in density-inhibited cultures by only 2 hr of prestimulation exposure to cycloheximide, an inhibitor of protein synthesis. Since we do not have any direct experience of the ability of this treatment to increase the synchrony of density-inhibited cells, the reader is referred to the original paper[27] for details.

*Advantages and Disadvantages.* This is an excellent method for obtaining cells synchronized at the $G_1/S$ boundary and the subsequent $S$ phase. With an appropriate choice of manipulations (length of serum deprivation, percentage of serum during the restriction period, amount of and length of exposure to hydroxyurea), toxicity can be low.

### C. Synchronization by Double-Thymidine Block

The double-thymidine block has been found to be unreliable as the only synchronizing agent. The concentration of thymidine, which in randomly growing cultures inhibits the rate of cell division by more than 90%, allows a considerable degree of DNA synthesis.[28] Therefore, the concentrations of thymidine commonly employed to produce cell synchrony do not arrest the cells at the $G_1/S$ boundary, but allow slow progress into $S$. However, the double-thymidine block can be used in conjunction with mitotic selection to obtain populations of synchronized cells. For instance, in the case of HeLa cells the following procedure is applicable.

HeLa cell cultures in suspension in the exponential growth phase are treated for 16 hr with thymidine at a concentration of 2 m$M$. They are then grown for 8 hr in regular medium and then again for 16 hr in medium containing 2 m$M$ thymidine. The cells are then plated in plastic dishes and after 4 hr Colcemid is added. Beginning 2 hr after the addition of Colcemid, cells in mitosis are collected by shaking at half-hour intervals and pooled as described above. This method allows the use of thymidine for a partial synchronization and of mitotic selection for sharp synchronization. At the same time, the damage done to cells by thymidine is repaired during the plating period between the removal of the second thymidine block and the addition of Colcemid. Again, note that it is not necessary to use Colcemid, although several experiments have shown that Colcemid at proper concentrations does not interfere with macromolecular synthesis and other processes of HeLa cells in culture.

### D. Synchronization by Calcium Deprivation

A decrease in the concentration of calcium may have a synchronizing effect, especially on normal untransformed cells such as WI-38. Boynton

[28] G. P. Studzinski and W. C. Lambert, *J. Cell. Physiol.* **73,** 109 (1969).

*et al.*[29] have shown that, when the concentration of calcium in the extracellular medium is reduced to 10 $\mu M$, WI-38 go into a state of quiescence and do not enter DNA synthesis. It may take more than 48 hr to bring most of the WI-38 cells out of the cycle. Increasing the calcium concentration to 1.35 m$M$ at this point immediately produces a burst of DNA synthesis which is highly synchronous. The rapidity with which this burst occurs suggests that a decreased amount of calcium may synchronize the cells very closely to the onset of DNA synthesis. It should be noted that as usual transformed cells cannot be synchronized by this method but continue to grow even in very low concentrations of calcium.[29]

### E. Other Chemical Methods

A number of other chemical methods for synchronizing cells in culture have been proposed at one time or another and have already been discussed in the review by Nias and Fox[17] and elsewhere. Most of them have only historical interest, while others may be useful in very specific situations but are not generally applicable. For instance, partial synchronization can be achieved by suicide with high concentrations of high specific activity [$^3$H]thymidine. Cells in $S$, which have incorporated huge amounts of [$^3$H]thymidine, are selectively killed. Clearly, such methods have only a marginal interest for the biochemist.

## Recommendations

These can be summarized as follows:

1. Synchronization by mitotic detachment is the best single method. It is excellent for studying cells synchronized in mitosis, $G_1$, and $S$. Its limitations are the number of cells that can be obtained and the fact that it is not applicable to cells in suspension.

2. To select cells in $S$ or $G_2$, a combination of serum deprivation followed by serum stimulation in the presence of hydroxyurea is recommended. It gives larger yields than mitotic detachment, but again it is limited to monolayer cultures.

3. To obtain large quantities of cells in $G_1$ or $S$, the isoleucine-deprivation method is very good for some cell lines in suspension cultures. For other cell lines, one may try a double-thymidine block followed by mitotic detachment (if the cells can be plated).

4. Other methods, at present, give only partially satisfactory results. However, some of these methods are amenable to improvements that should considerably enhance their value.

[29] A. L. Boynton, J. F. Whitfield, R. J. Isaacs, and R. Tremlay, *J. Cell. Physiol.* **92,** 241 (1977).

# [21] New Techniques for Culturing Differentiated Cells: Reconstituted Basement Membrane Rafts

*By* LOLA M. REID and MARCOS ROJKIND

This article describes techniques, currently under investigation, that should eventually provide routine methods for culturing differentiated cells, whether normal or malignant. Indeed at present the techniques are sufficiently developed that they can greatly lengthen the time that differentiated cells retain their tissue-specific functions *in vitro*.

The need for model systems of differentiated mammalian cell types to be used in biological and clinical research has not been sufficiently satisfied because of the inherent difficulties in establishing differentiated cells, especially normal cells, *in vitro*. Epithelial cells have proven particularly difficult to establish in cell culture, a fact that has severely hampered certain fields of research including some important cancer studies. Since 85–90% of all malignancies are carcinomas, malignancies of the epithelium, the need for normal and malignant culture models of epithelial cells is acute.

Over many years standard methods have evolved for keeping explanted cells alive and functioning. Most methods used currently, as for example those described in this volume, are for organ culture and for cell culture. Analysis of these techniques and the response of cells to them reveals methodologic limitations and certain requirements for cellular differentiation and growth. To facilitate the discussion we shall define several of the terms used in culture studies:

*Organ culture:* maintenance of organs dissociated from their central vascular supply but with the organ, as an entity, left intact.

*Tissue culture:* culture of tissues or fragmented organs. The "sociocellular" relationships of the tissue architecture are preserved.

*Cell culture:* culture of an individual cell type divorced from other cell types.

*Subcell culture:* maintenance of one or more subcellular components of homogenized cells.

A marked difference in response of explanted cells occurs when they are cultured by cell rather than organ or tissue culture techniques. Methods which retain tissue architecture permit retention of tissue-specific functions including hormone and pharmacological responses. However, the tissue normally degenerates within a few weeks due primarily to difficulties in vascularizing all the cells within the tissue. Cell culture procedures overcome this limitation since they use explanted tissues which are




ISBN 0-12-181958-2

disaggregated into single cells. The cells are adapted to grow as a monolayer on treated plastic, or as a cell suspension, and can be maintained in culture indefinitely. Nutrients are supplied by a liquid medium of defined basal composition supplemented with one of various sera. Ideally, one isolates a clonal cell line, i.e., a single cell whose progeny are maintained in continuous cell culture. There is genetic uniformity, easier maintenance of the cells, and reduction of variables associated with a multicell culture system. Nevertheless, cell culture procedures usually result in distortion of cellular phenotype and karyotype; normal cells rarely adapt as permanent cell lines without developing abnormal karyotypes[1-3] or losing tissue-specific functions;[4-8] malignant cells adapt more easily than do benign tumor cells or normal cells;[4,8] and fibroblasts or stromal components are selected preferentially over epithelial cells.[7] The difficulties of establishing differentiated cells in cell culture have been attributed to many causes. These include an inadequately defined basal medium[9-12] (see also this volume [5]); inadequately defined hormone requirements;[13] the static conditions of cell culture in which nutritional and oxygen gradients develop and limit the growth and functioning of cells with strict nutritional and oxygen requirements;[6,14,15] and loss of or damage of cell–cell junctions, perhaps essential in growth and/or differentiation or both by the cell culture procedures of mechanical and enzymic

[1] J. L. Biedler, *in* "Human Tumor Cells in Vitro" (J. Fogh, ed.), p. 359. Plenum, New York, 1975.

[2] H. L. Goldblatt, L. Friedman, and R. L. Cencher, *Biochem. Med.* **7,** 241 (1973).

[3] R. Parshad and K. K. Sanford, *J. Natl. Cancer Inst.* **47,** 1033 (1971).

[4] R. M. Cailleau, *in* "Human Tumor Cells in Vitro" (J. Fogh, ed.), p. 70. Plenum, New York, 1975.

[5] D. J. Giard, S. A. Aaronson, G. J. Todaro, P. Arnstein, J. H. Kersey, H. Dosik, and W. P. Parks, *J. Natl. Cancer Inst.* **51,** 1417 (1973).

[6] R. Knazek, P. Gullino, P. Kohler, and R. Dedrick, *Science* **178,** 65 (1972).

[7] G. H. Sato, L. Zaroff, and S. Mills, *Proc. Natl. Acad. Sci. U.S.A.* **46,** 963 (1960).

[8] Y. Yasumura, A. H. Tashjian, Jr., and G. Sato, *Science* **154,** 1186 (1966).

[9] R. G. Ham, *Proc. Natl. Acad. Sci. U.S.A.* **53,** 288 (1965).

[10] A. Leibovitz, *in* "Human Tumor Cells In Vitro" (J. Fogh, ed.), p. 23. Plenum, New York, 1975.

[11] W. McKeehan, W. G. Hamilton, and R. G. Ham, *Proc. Natl. Acad. Sci. U.S.A.* **73,** 2023 (1976).

[12] C. Waymouth, *in* "Growth, Nutrition and Metabolism of Cells in Culture" (G. H. Rothblat and V. J. Cristofalo, eds.), Vol. 1, p. 11. Academic Press, New York, 1972.

[13] G. Sato and L. Reid, *in* "Biochemical and Mode of Action of Hormones" Series II. (H. V. Rickenberg, ed.), Ser. II, Chapter 7. Univ. Park Press, Baltimore, Maryland, 1978.

[14] J. Leighton and R. Tchao, *in* "Human Tumor Cells In Vitro" (J. Fogh, ed.), p. 23. Plenum, New York, 1975.

[15] W. F. McLimans, D. T. Mount, S. Bogtich, E. J. Crouse, G. Harris, and G. E. Moore, *Ann. N.Y. Acad. Sci.* **139,** 190 (1966).

dissociation into single cell suspensions.[16–19] Undoubtedly all of these have contributed to the impasse in maintaining differentiated cells in cell culture. Yet despite some progress on these various fronts, the goal of routinely culturing normal differentiated cells remains elusive. It is likely that still other considerations must be made to achieve this goal.

We believe that substantial progress in culturing differentiated cells, including normal cells, will be made possible by study of cell–cell interactions within the tissue matrix. The relevance of tissue architecture to proliferation and functioning of cells has not yet been fully ascertained, but the studies discussed below suggest that some of the interactions within the tissue matrix are critical for normal cellular physiology. An almost ubiquitous and, therefore, probably primary relationship within tissues is that between the epithelium and mesenchyme. *In vivo,* normal epithelial cells capable of proliferation or long-term survival are attached to a basement membrane which, in turn, is associated with mesenchymally derived cells, most commonly fibroblasts. The basement membrane, a layer of secretion located between and produced by the epithelium and the mesenchymal cells, is assumed to contain the extracellular subset of factors affecting the epithelial–mesenchymal interactions.

The epithelial–mesenchymal relationship has been studied extensively in developmental systems since the 1950s.[20–26] Many laboratories have shown that differentiation of and the direction of differentiation of the epithelium are dependent on factors derived from the mesenchyme. More recently, Green and his colleagues have found that differentiation of skin epithelium depends on factors derived from the mesenchyme.[27] Martin found that an embryonal carcinoma cell line proliferates as a stem cell as long as it is cultured on a feeder layer and differentiates when it is sepa-

[16] I. Fentiman, J. Taylor Papadimitriou, and M. Stoker, *Nature (London)* **264,** 760 (1976).
[17] W. R. Loewenstein, *Proc. Can. Cancer Res. Conf.* **8,** 162 (1969).
[18] N. S. McNutt, R. A. Hershberg, and R. S. Weinstein, *J. Cell Biol.* **51,** 805 (1971).
[19] J. D. Pitts and R. R. Burk, *Nature (London)* **264,** 762 (1976).
[20] G. R. Cunha, *Int. Rev. Cytol.* **47,** 137 (1976).
[21] P. Davies, A. C. Allison, and C. J. Cardella, *Philos. Trans. R. Soc. London* **271,** 363 (1975).
[22] R. Fleischmajer and R. Billingham, eds., "Epithelial-Mesenchymal Interactions." Williams & Wilkins, Baltimore, Maryland, 1968.
[23] C. Grobstein, *Science* **11,** 52 (1953).
[24] R. L. Pictet, and W. L. Rutter, *in "Cell Interactions in Differentiation"* (M. Karkinen-Jaafkelainen, L. Fafen, and L. Weiss, eds.), p. 339. Academic Press, New York, 1977.
[25] J. W. Saunders, Jr., M. T. Gasseling, and M. D. Gfeller, *J. Exp. Zool.* **317,** 39 (1958).
[26] F. L. Vaughan and I. A. Bernstein, *Mol. Cell. Biochem.* **12,** 171–179 (1976).
[27] H. Green, J. G. Rheinwald, and T. Sun, *in* "Shape and Surface Architecture," p.493. Allen R. Liss, Inc., New York, 1977.

rated from the feeder cells.[27a] The type of stromal cell associated with the epithelium is also important as shown by Sakakura and her associates who demonstrated that the differentiation of mouse mammary gland depends on association with mammary stroma but not other types of stroma.[28]

Conclusive evidence for epithelial–mesenchymal interactions in adult tissues is not yet available. However, such a relationship is suggested by the histopathological observation that the anatomical relationship between the epithelium and the mesenchyme exists in normal and benign tumor tissue but is missing in malignant tissues. Malignancy is, in fact, defined in solid tissues when the basement membrane is disrupted[29,30] and when either epithelial cells alone (carcinoma) or mesenchymal cells alone (sarcoma) proliferate with concomitant loss of the epithelial–mesenchymal association. This observation may explain, in part, why malignant cells establish in cell culture more easily than do normal cells; normal cells may be interdependent with other cells in the tissue matrix, whereas malignant cells may be qualitatively or quantitatively independent of these interactions.

Some of the secretions of the epithelium and the mesenchyme found in the basement membranes are known to be important for the maintenance of cultured epithelial cells. Best studied of the basement membrane components are the collagenous proteins, adhesion proteins, and glycosaminoglycans.

Collagen has been used frequently for establishing primary cultures since most normal epithelial cells will attach more efficiently to collagen than to other cell culture substrates.[4] This is true despite the fact that basement membrane collagen (Type IV collagen) is genetically distinct from the collagen typically used in cell culture (Type I). More detail on collagen types is given in subsequent sections. Floating collagen substrates have been used by Pitot's group for normal liver cells[31] and by Pitelka's group[32] for normal mammary epithelium to extend the survival of tissue-specific functions of those epithelial cells for up to a month. The advantages of floating collagen substrates over collagen plates (in which collagen is present at the bottom of the plate) are thought to be increased oxygen tension at the air–media interface and the ability of epithelial cells to contract the raft and, thereby, to undergo a change in cell shape apparently essential for differentiated cellular functions.[31–33]

[27a] G. Martin, personal communication.

[28] T. Sakakura, Y. Nishizuka, and C. J. Dawe, *Science* **194,** 1439 (1976).

[29] L. Liotta, J. Leinerman, P. Catanzaro, and D. Rymbrandt, *J. Natl. Cancer Inst.* **58,** 1427 (1977).

[30] S. Read, *Guy's Hosp. Rep.* **123,** 53 (1974).

[31] G. Michalopoulos and H. C. Pitot, *Exp. Cell Res.* **94,** 70 (1975).

[32] J. T. Emerman and D. R. Pitelka, *In Vitro* **13,** 346 (1977).

[33] T. Allen and C. S. Potten, *Nature (London)* **264,** 545 (1976).

Adhesion proteins such as LETS (Large External Transformation Substance) are present; also called fibronectin and CSP (Cell Surface Protein), this is a high-molecular-weight protein attached to cell surfaces and known to bind cells to collagen or to the plastic dishes.[8,34–41] Digestion of plasma membrane components by nonspecific proteases reduces the amount of adhesion proteins such as LETS[42,43] that have been implicated in adhesion phenomena.[35] Lack of these proteins on the cell surface and in the basement membranes is correlated with changes in cell morphology,[44] alterations in the organization of the cytoskeleton,[44,45] cell motility,[46] and alterations in the growth pattern in culture.[47] Hepatocytes isolated after perfusion of the liver with bacterial collagenase show profound alterations in morphology[45] that appear to return to normal after several hours in culture. Bacterial collagenase has been used because of its high specificity in digesting the collagen in the stroma of tissues.[48] However, most commercial preparations of collagenase are contaminated with nonspecific proteases that could be attacking the adhesion proteins on cells. Purified bacterial collagenase is ineffective in releasing hepatocytes from rat liver (unpublished observations), suggesting that the release of hepatocytes from the collagen stroma requires enzymic digestion by enzymes other than, or in addition to, collagenase. LETS protein is decreased in many malignantly transformed cells[35] and disappears from the cell surface of fibroblasts dissociated by trypsin;[42,43] it reappears approximately 10 hr after the cells are plated in a serum-supplemented medium.[43]

Glycosaminoglycans, also called mucopolysaccharides, are polymers

[34] E. Engvall and E. Russlahti, *Int. J. Cancer* **20,** 1 (1977).
[35] R. O. Haynes, *Biochem. Biophys. Acta* **458,** 73 (1976).
[36] M. Höök, K. Rubin, A. Oldberg, B. Obrink, and A. Veri, *Biochem. Biophys. Res. Commun.* **70,** 726 (1977).
[37] R. J. Klebe, *Nature (London)* **250,** 248 (1974).
[38] K. M. Yamada, S. S. Yamada, and I. Pastan, *Proc. Natl. Acad. Sci. U.S.A.* **72,** 3158 (1975).
[39] K. M. Yamada, S. S. Yamada, and I. Pastan, *Proc. Natl. Acad. Sci. U.S.A.* **73,** 1217 (1976).
[40] K. M. Yamada, D. Schlesinger, D. Kennedy, and I. Pastan, *Biochemistry* **16,** 5552 (1977).
[41] M. Zimmerman, T. Devlin, and M. Pruss, *Nature (London)* **185,** 315 (1960).
[42] P. Bornstein and J. F. Ash, *Proc. Natl. Acad. Sci. U.S.A.* **74,** 2480 (1977).
[43] J. Wartiovaara, E. Linder, E. Rouslahti, and A. J. Vaheri, *J. Exp. Med.* **140,** 1522 (1974).
[44] D. Goldman, C. Chang, and J. F. Williams, *Cold Spring Harbor Symp. Quant. Biol.* **39,** 601 (1974).
[45] J. C. Wanson, P. Drochmans, R. Mosselmans, and M. F. Ronveaur, *J. Cell Biol.* **74,** 858 (1977).
[46] M. H. Gall and C. W. Boone, *Exp. Cell Res.* **70,** 33 (1972).
[47] E. Martz, H. M. Phillips, and M. S. Steinberg, *J. Cell Sci.* **16,** 401 (1974).
[48] S. Seifter, and E. Harper, *in* "The Enzymes" (P. D. Boyer, ed.), 3rd ed., Vol. 3, p. 649. Academic Press, New York, 1971.

of sulfated or acetylated disaccharides that are often associated with proteins forming complexes referred to as proteoglycans. The glycosaminoglycans and proteoglycans are present in all connective tissues where they form a gelatinous charged matrix. The best known and most studied of the glycosaminoglycans and proteolgycans include hyaluronate (skin, aorta), dermatan sulfate (skin, tendon, aorta), chondroitin sulfate (cartilage, bone, cornea, notochord, skin), heparan sulfate (aorta, lung, liver), and heparin (lung, liver, skin). Their chemical structures, biosynthesis, specific tissue localization, variations among species, and evolution are reviewed elsewhere.[49]

In developmental systems it has been shown that glycosaminoglycans and proteoglycans together with collagenous proteins can have a significant influence on cellular differentiation,[50,51] proliferation,[52] aggregation,[53] and migration.[54] Hyaluronate stimulates aggregation of chick embryo cells.[54] In the development of the cornea and the notochord, hyaluronate is thought to play a role in collagen fibril formation that in turn induces cellular migration. Further development of these tissues involves replacement of hyaluronate with sulfated glycosaminoglycans that are the typical polyanionic glycans of the adult tissues. Development of the mouse salivary gland requires the presence of chondroitin sulfate at the cell surface. Specific enzymes that eliminate the chondroitin 6-sulfate from the developing tissues arrest their development; morphogenesis can be reactivated by renewed biosynthesis of the proteoglycan.[51,55–57]

In adult tissues the functions of the proteolgycans and glycosaminoglycans in cellular differentiation and proliferation are uncertain. Kraemer's studies[58–60] indicate that mammalian cells in cell culture are coated with a variety of glycoproteins and polyanionic glycans among which the most common is heparan sulfate. Kraemer hypothesizes that

[49] M. B. Mathews, "Connective Tissue: Marcromolecular Structure and Evolution." Springer-Verlag, Berlin and New York, 1975.

[50] E. A. Balazs and B. Jacobson, *in* "The Amino Sugars" (E. A. Balazs and R. W. Jeanloz, eds.), Vol. 2A, pp. 281–309. Academic Press, New York, 1966.

[51] M. R. Bernfield, S. D. Banerjee, and R. H. Cohen, *J. Cell Biol.* **52,** 674 (1972).

[52] M. Lippman, *in* "Epithelial-Mesenchymal Interactions" (R. Fleischmajer and R. E. Billingham, eds.), p. 208. Williams & Wilkins, Baltimore, Maryland, 1968.

[53] B. Pessac and V. Defendi, *Science* **175,** 898 (1972).

[54] B. P. Toole and R. L. Trelstad, *Dev. Biol.* **26,** 28 (1971).

[55] M. R. Bernfield and S. D. Banerjee, *J. Cell Biol.* **52,** 664 (1972).

[56] R. H. Cohn, S. D. Banerjee, and M. R. Bernfield, *J. Cell Biol.* **73,** 464 (1977).

[57] R. H. Cohn, J.-J. Cassiman, and M. H. Bernfield, *J. Cell Biol.* **71,** 280 (1976).

[58] P. M. Kraemer, *in* "Biomembranes" (L. A. Manson, ed.), Vol. 1, p. 67. Plenum, New York, 1971.

[59] P. M. Kraemer, *Biochemistry* **10,** 1437 (1971).

[60] P. M. Kraemer, *Biochemistry* **10,** 1445 (1971).

heparan sulfate is crucial to some cellular process(es).[60] Chiarugi and Vannucchi[61] also focused on heparan sulfate as perhaps a significant factor in the differentiation of cells. However, no conclusive evidence exists confirming these suggestions, and the possible roles of the glycosaminoglycans await further experimentation.

On the basis of these previous studies it seems reasonable that basement membrane components might facilitate cultures of differentiated cell types. We propose that new culture procedures should be developed for differentiated cells that involve utilization of these basement membrane components and co-culturing of multiple, interacting cell types representing a subset of the relationships existing in the tissue matrix. Generic characterization of the procedures might be called "sociocell" culture. One such technique which we are developing simulates epithelial–mesenchymal relationships and utilizes reconstituted basement membrane rafts on which are floated epithelial cells over primary cultures of mesenchymal cells normally in association with the epithelial cells. Others have already shown that use of collagen rafts for primary cultures of normal tissues can significantly enhance the retention of tissue-specific functions of normal cells in cell culture.[31,32] We have modified the raft's composition to more nearly resemble the *in vivo* substrate for epithelial cells, the basement membrane. Then we float the rafts over primary cultures of mesenchymal cells in order to provide other epithelial–mesenchymal factors that may be labile, may be taken up by the cells, and/or are the result of ongoing interactions between the two cell types. Our efforts to define the raft's composition are focusing on three of the major constituents of the basement membrane: collagen types, adhesion proteins, and glycosaminoglycans. Since normal epithelial cells may be in contact with other collagen types, we are assaying the efficacy of collagen rafts made with collagen types other than Type I. Adhesion proteins, and glycosaminoglycans. Since normal epithelial cells may be in contact with other collagen types, we are assaying the efficacy of cell types that have lost their adhesion proteins due to neoplastic transformation. A possible complication with some cell types is the cellular secretion of proteases; addition of adhesion proteins to such cells may or may not facilitate attachment, depending on the amount of proteolytic activity present. Addition of glycosaminoglycans, e.g., dermatan sulfate, keratan sulfate, chondroitin sulfate, and heparan sulfate, is being pursued to define empirically their participation in the proliferation and/or functioning of epithelial cells.

Since normal differentiated epithelial cells, e.g., endocrine cells, rarely undergo mitosis *in vivo* in an adult organism, it is expected that success at

[61] V. P. Chiarugi and S. Vannucchi, *J. Theor. Biol.* **61**, 459 (1976).

defining the reconstituted basement membrane rafts will result in the cells' survival and functioning in "sociocell" culture but not in growth. The reverse might be expected for stem cells such as the skin epithelium or the stem cells from the bone marrow. For stem cells, an effective basement membrane raft and feeder layer may result in proliferation of the cells without differentiation. Differentiation might occur when the stem cells are detached from the raft.

The techniques presented here are: (1) preparation of the collagen raft, a modified procedure of the original one developed by Michalopoulos and Pitot;[31] (2) purification procedures of the several collagen types and other basement membrane components; and (3) the procedures for preparing epithelial cell suspensions and feeder layers of mesenchymal cells. Although the efficacy of some of these procedures has been tested with only a few cell types, we believe they will form the basis for future procedures for culturing many differentiated normal and malignant cell types.

## Culture Techniques Simulating the Epithelial– Mesenchymal Relationship

The techniques involve purification of collagenous proteins, adhesion proteins, and glycosaminoglycans; mixing and gelling of these components into a substrate; preparation of pure epithelial cell suspensions; addition and attachment of these cells to the substrates; detachment of substrates to form rafts; preparation of feeder layers of stromal cells; and transfer of rafts plus epithelial cells to dishes where they float over the primary cultures of mesenchymal cells. For short-term studies (3–4 weeks), one can dispense with the feeder layers and use only the rafts of reconstituted basement membrane as substrates for maintaining differentiated epithelial cells.

### A. *Purification of Collagenous Proteins and Other Basement Membrane Components*

#### 1. Preparation of Collagen Extracts

The collagen class of proteins is a heterogeneous group of proteins that nevertheless is characterized generally by a unique amino acid composition (about 30% glycine, 20% proline, 20% hydroxyproline, and a variable content of hydroxylysine). To date, three distinct collagens have been isolated from interstitial tissues and characterized on the basis of the uniqueness of their individual polypeptide chains. These are designated as Type I, Type II, and Type III collagens. Type I collagen has a chain composition of $[\alpha 1(\mathrm{I})]_2\alpha_2$; it is the only collagenous component of bone, tendon, and tooth. This type of collagen is also present with other collagens in skin, liver, heart, and kidney, but is absent from normal hyaline and elastic cartilages. Type II collagen has a chain composition of $[\alpha 1(II)]_3$; it is the only collagenous component of hyaline and elastic cartilages. Type III collagen has a chain composition of $[\alpha 1(III)]_3$; it is present in most tissues containing Type I collagen except for bone, tendon, and tooth.[62–64] Collagen also has been isolated from structures identified morphologically as basement membranes. Despite suggestions that these collagens are homogeneous with a chain composition of $[\alpha 1(IV)]_3$, the evidence is more convincing that basement membranes constitute another class of proteins containing collagenous sequences, ranging in molecular weights from 25,000–100,000.[65–67] Collagens isolated from basement membranes differ in amino acid composition from interstitial collagens in that they contain 3-hydroxyproline, relatively more hydroxylysine, and relatively less alanine. Recently tissues containing Type I and Type III collagens have been found also to contain collagenous components similar in amino acid composition to those found in basement membrane collagens.[68,69] Since these collagens are extracted from tissues and not from just the basement membranes alone, they are referred to as basement membrane-like collagens. The basement membrane-like collagens as well as Type III collagen are solubilized mainly by limited proteolysis with pepsin.

Collagen used as a cell culture substrate[21,70,71] or as a raft[31,32] is obtained from skin or tendon by extraction with neutral salt solutions or dilute acid. These extracts contain primarily Type I collagen that has been shown to substantially lengthen the survival in culture of liver cells[31] and mammary epithelium.[32] Since other types of collagen are part of the *in vivo* substrate for epithelial cells (e.g., in normal liver, Type III collagen[72,73] and basement membrane-like collagens[72] rather than Type I collagen are in close association with the hepatocytes), these collagens also

[62] P. M. Gallop and M. A. Paz, *Physiol. Rev.* **55,** 418 (1975).
[63] K. I. Kivirikko and L. Risteli, *Med. Biol.* **54,** 159 (1976).
[64] E. J. Miller, *Mol. Cell. Biochem.* **13,** 165 (1976).
[65] B. G. Hudson and R. G. Spiro, *J. Biol. Chem.* **247,** 4229 (1972).
[66] B. G. Hudson and R. G. Spiro, *J. Biol. Chem.* **247,** 4239 (1972).
[67] T. Sato and R. G. Spiro, *J. Biol. Chem.* **251,** 4062 (1976). See also, this volume [6].
[68] E. Chung, K. Rhodes and E. J. Miller, *Biochem. Biophys. Res. Commun.* **71,** 1167 (1976).
[69] R. L. Trelstad and K. R. Lawley, *Biochem. Biophys. Res. Commun.* **76,** 376 (1977).
[70] M. B. Bornstein, *Lab. Invest.* **7,** 134 (1958).
[71] W. D. Hillis and F. B. Bang, *Exp. Cell Res.* **17,** 557 (1959).
[72] L. Biempica, R. Morecki, C. H. Wu, M. A. Giambrone, and M. Rojkind, *Gastroenterology* **73,** 1213 (1977).
[73] S. Gay, P. P. Fietzek, K. Remberger, M. Eder, and K. Kuhn, *Klin. Wochenshr.* **53,** 205 (1975).

should be used as substrates in attempts to improve the survival and/or functioning of specific epithelial cell types.

*a. Tail Tendon Collagen (Type I Collagen).* Collagen is extracted from tendon fibers dissected from the tail of the rat.[74] Approximately 1 g of tendon fibers is suspended in 100 ml of dilute acetic acid (0.01–0.25 *M*).[32] After incubation with constant stirring at 4° for 48 hr, solubilized collagen is collected in the supernatant obtained by centrifugation at 27,000 g for 30 min. The collagen thus obtained can be used without further purification for preparation of substrates or can be stored at −20° until needed. The residual tissue can be reextracted with acetic acid to improve the yields of collagen. The collagen so extracted may contain small amounts of noncollagenous material that may or may not gel together with collagen. If purified collagen is to be used, the acid-solubilized collagen can be precipitated a few times by dialysis against 0.02 *M* $NA_2HPO_4$ and redissolution in acetic acid.[74]

The concentration of solubilized collagen is estimated from the amount of hydroxyproline after acid hydrolysis;[75] it is assumed that each polypeptide chain of approximately 100,000 daltons contains 100 residues of hydroxyproline.[62] Although collagen can be estimated by the Lowry technique,[76] a collagen standard is needed; collagen produces low color yields due to its low tyrosine content.

*b. Skin Collagen (Type I Collagen).* Although skin from any animal may be used, rat, mouse, and guinea pig skins are most commonly used. The animals are killed, shaved, washed with soap and water, and thoroughly rinsed. The animals are skinned, and the skins are weighed. From 10 g of skin one obtains approximately 60–70 mg of collagen (about 1–2 mg of collagen are needed per 35 mm cell culture plate). The skins are ground and washed for 24 hr in 0.45 *M* NaCl at a ratio of 1 g of tissue to 100 ml of salt solution. Saline solutions extract small amounts of collagen, but the yields are much smaller than those obtained with dilute acid extraction. Thus, the saline wash can be used to eliminate saline-soluble contaminants, an advantage since there are more contaminants in skin than in tail. tendon, i.e., skin is tissue, whereas the tail tendon is primarily collagenous protein. The washed mince is extracted with dilute acetic acid (0.1–0.25 *M*) and purified as described for tail tendon collagen. All procedures should be carried out at 4°.

*c. Other Collagen Types.* Type III collagen and the group of basement membrane-like collagens can be solubilized by limited proteolysis with

[74] D. A. Hall, "The Methodology of Connective Tissue Research." Johnson-Brouvers, Ltd., Oxford, 1976.

[75] M. Rojkind and E. Gonzalez, *Anal. Biochem.* **57**, 1 (1974).

[76] O. H. Lowry, N. J., Rosenbrough, A. L. Farr, and R. J. Randall, *J. Biol. Chem.* **193**, 265 (1951).

pepsin at 4°.[68,77] The concentration of pepsin to be used is 0.1–1 mg/ml, and the ratio of enzyme to substrate is 1 : 100. Collagen rendered soluble after enzymic digestion for 6 hr at 4° is collected in the supernatant after centrifugation at 27,000 g for 30 min and neutralized immediately to pH 7.0 in order to inactivate pepsin. The neutralized solution is then dialyzed for 24 hr against 0.05 *M* Tris-chloride (pH 7.4) containing 0.45 *M* NaCl. The collagen that precipitates is removed after centrifugation. The clear supernatant is dialyzed for 24 hr against Tris-chloride (pH 7.4) containing 1.7 *M* NaCl. Type III collagen will precipitate and can be collected by centrifugation as described. The clear supernatant liquid obtained after removal of Type III collagen is dialyzed for 24 hr against the Tris buffer. Type I collagen will precipitate and can be collected by centrifugation. The remaining clear supernatant fluid containing a mixture of the basement membrane-like collagens is then dialyzed against 0.02 *M* $NaH_2PO_4$ in order to precipitate these collagens. The collagens so obtained are soluble in dilute acid or in neutral salt solutions.

The pepsin-solubilized collagens lack short segments of the nontriple helical portions of the individual collagen chains. Since such segments may participate in gel formation, the extent of digestion should be carefully controlled. In fact these collagens may not form firm gels by themselves so that mixtures of Type I collagen with varying proportions of other collagens should be used to create rafts with which to test the efficacy of other collagen types on cellular proliferation and functioning.

2. Adhesion Proteins (LETS)

LETS protein has an apparent molecular weight of 220,000 and is present in relatively large amounts on the cell surface of normal fibroblasts.[35] The protein is organized in a reticular fashion[42,43,78] and mediates the attachment of fibroblasts to collagen in culture.[37,79] It binds strongly to collagen and has been purified recently by affinity chromatography on collagen bound to Sepharose 4B.[34]

To prepare LETS protein, monolayers of chick embryo fibroblasts are plated in disposable 690 cm² roller bottles and fed daily. After reaching confluency, monolayers are washed at 37° with 50 ml of Hanks's balanced salt solution 4 times, and then they are rinsed 60 min with 25 ml of serum-free medium (Dulbecco's Modified Eagle's Medium or Diploid Growth Medium) containing 2 m*M* phenylmethylsulfonyl fluoride (PMSF) as a protease inhibitor. After another rinse with salt solutions, the monolayers are extracted for 2 hr with 25 ml of serum-free medium con-

[77] E. Chung and E. J. Miller, *Science* **183,** 1200 (1974).
[78] A. Vaheri, E. Ruoslahti, B. Westermark, and J. Pontén, *J. Exp. Med.* **143,** 64 (1976).
[79] E. Pearlstein, *Nature (London)* **262,** 497 (1976).

taining PMSF plus 1.0 *M* urea (ultrapure grade). After centrifugation of the mixture at 25,000 g for 15 min, LETS protein is precipitated by adding solid ammonium sulfate to 70% saturation and adjusting the pH to 7.4 with $NH_4OH$. After 30 min at 4°, the solution is centrifuged at 25,000 g for 15 min. The pellet is solubilized with 1/20 volumes of 10 m*M* cyclohexylaminopropane sulfonate (CAPS), pH 11.0, containing 0.15 *M* NaCl and 1 m*M* $CaCl_2$ at a concentration of about 1 mg/ml. The solution is adjusted to pH 11.0 with 5 *N* NAOH, dialyzed overnight against two changes of 400 volumes of CAPS buffer, and stored at −70°. The above conditions provide a concentrated, nonaggregated preparation of LETS protein. If LETS protein is to be added to collagen solutions or to substrates, the pH of the solution should be adjusted to pH 7.4 with HCl. This method was described by Yamada *et al.*[38–40] and yields approximately 50% of the LETS protein present in cultures of chick fibroblasts.

3. GLYCOSAMINOGLYCANS/PROTEOGLYCANS

Purification procedures for the various glycosaminoglycans and proteoglycans are presented by Hall[74] and in a review by Rodén *et al.*;[80] accordingly they will not be presented here.

### B. *Preparation of Collagen or Reconstituted Basement Membrane (RBM) Substrates*

About 20 years ago, undenatured collagen in neutral salt solutions at 37° was shown to "precipitate as a mass of typical cross-striated fibrils in an opalescent gel."[81] This is not to be confused with a gelatin (denatured) gel. All the methods used for the preparation of collagen gels require the removal of any acetic acid and/or equilibration of the collagen solution with a buffer at a neutral pH and containing NaCl.[82] The ideal concentration for making rigid gels is approximately 1 mg/ml. If an extract is too dilute, it should be concentrated by dialysis overnight at 4° against 0.02 *M* $Na_2HPO_4$, and the resulting collagen precipitate should be centrifuged at 27,000 g for 30 min. The pellet is suspended in a volume of dilute acetic acid solution appropriate to yield a 1 mg collagen per milliliter of solution. The acid extract should be dialyzed extensively against a neutral buffer using any buffer appropriate for subsequent experiments. Dialysis and all handling of the collagen solutions should be at 4° to prevent collagen from coming out of solution before addition to plates. For culture experiments,

[80] L. Rodén, J. R., Baker, J. A. Cifonelli, and M. B. Mathews, *in* "Methods in Enzymology" (V. Ginsburg, ed.), Vol. 28, Part B, p. 73. Academic Press, New York, 1972.
[81] J. Gross and J. Kirk, *J. Biol. Chem.* **233,** 355 (1958).
[82] J. Gross and C. M. Lapiere, *Proc. Natl. Acad. Sci. U.S.A.* **48,** 1014.

acetic acid extracts can be dialyzed against phosphate-buffered saline for 5–6 hr with several changes of saline in order to eliminate most of the acetic acid. Completion of neutralization can then be done with serum-free medium containing a pH indicator to permit following neutralization visually. The advantages of this form of neutralization are that the subsequent gels are impregnated with the medium to be used for the cells and that one avoids overshooting neutralization to an alkaline pH where collagen molecules may be cleaved. The neutralized extract is viscous and should be plated quickly onto petri or tissue culture dishes (1–2 ml per 35-mm dish and 3–4 ml per 60-mm dish). Gelled substrates should form within an hour at 37°. Collagen substrates should be sterilized by irradiating the plates with UV overnight or by irradiating with 50,000 rad of cobalt 60 or cesium 135. The large amount of irradiation is necessary to eliminate radiation-resistant bacterial spores. To minimize contamination prior to sterilization one can add antibiotics (penicillin at 0.12 mg/ml and streptomycin at 0.27 mg/ml) to the collagen extract and to the serum-free medium during the final dialysis.

To make reconstituted basement membranes (RBMs) one may add LETS, and/or any glycosaminoglycan or proteglycan, to the collagen extract during the final dialysis and copolymerize them onto the culture plates. One can also try coating the surface of the collagen gel with any of these components. The exact concentrations necessary for a given cell type are undefined at present, but approximate concentrations derived from previous culture experiments suggest 50 $\mu$g/ml for LETS[39] and 50 $\mu$g/ml for glycosaminoglycans.

## C. *Preparation of Pure Epithelial Cell Suspensions*

### 1. Cell Lines or Cell Strains

Since cell lines and cell strains are cloned populations of cells, they are a source of pure epithelial cells that is the easiest to use. Although cell lines and cell strains consist of cells that are adapted to cell culture conditions and therefore do not require for cell culture a simulated epithelial–mesenchymal relationship, there may be experiments in which feeder layers are necessary for the cells. For these circumstances one can readily use collagen rafts to facilitate separation of the epithelial cells from the feeder layer. We do not know at present whether established, functional epithelial cell lines will regain any tissue-specific functions when co-cultured with other cell types.

To prepare cell lines or cell strains for "sociocell" culture, the medium is removed from culture plates of the cells which are then washed with phosphate-buffered saline and trypsinized with a 0.1% trypsin solution at 37° for several minutes. The cells should be observed closely during trypsinization; as soon as the cells begin to detach from the plates, medium containing serum is squirted onto the plates. All serum (chicken serum is an exception) contains a trypsin inhibitor. Thus, squirting the cells with the serum-supplemented medium will complete the detachment from the plate and inactivate trypsin. The cell suspension is centrifuged in a desk-top centrifuge at 900 rpm at 4°, the supernatant liquid is removed, and the pellet of cells resuspended in cold (4°) medium containing serum. The cell suspension should be kept on ice until ready for addition to the collagen or RBM substrates.

2. TRANSPLANTABLE TUMORS

Transplantable carcinomas can be cleaned of nonepithelial cells by transplanting the tumors into immunologically suppressed, xenogenic hosts for one to two passages (see this volume [31]). Carcinomas transplanted into a xenogeneic host undergo replacement of their stroma and vascular cells by host stroma and vascular cells. Treatment of primary cultures or cell suspensions of transplanted tumors with anti-serum against host cells will provide suspensions or plates of cells containing only carcinoma cells. For example, nonmouse carcinomas can be transplanted into athymic nude mice, whose T-cell deficiency permits heterograft transplantation.[83,84] Tumors successfully transplanted will contain tumorigenic epithelial cells associated with mouse stroma and vascular components.[85] The tumors may be excised, minced finely to release cells, and the cell suspension treated with antimouse antiserum. Procedures for making antimouse antiserum (or antiserum against any host used for the tumor transplantations) are given below.

*a. Preparation of Rabbit Antimouse Antiserum* (*or Antiserum for Other Hosts*). Mouse cells (host cells), ideally a cell line derived from mouse cells, are grown to confluence in culture, subcultured, and $2 \times 10^8$ cells injected subcutaneously into each of several rabbits. Ten days later, the rabbits receive booster shots containing $2 \times 10^8$ of the same cells. Seven days later the rabbits are bled and the serum assayed for its titer against mouse (host) cells. The rabbits are given booster shots every 2 weeks and test bled for antibody titer until the titers are high (the minimum dilution to achieve 99% kill should be 1 : 100). The antiserum is tested against the host cells used as the antigen and against the tumor cells used as a control. Rabbit blood is collected in a sterile Falcon conical tube, allowed to clot at room temperature for 2 hr, and then refrigerated overnight. The clot is

[83] C. O. Povlsen and J. Rygaard, *Acta Pathol. Microbiol. Scand., Sect. A* **79,** 159 (1971).

[84] J. Rygaard, "Thymus and Self: Immunobiology of the Mouse Mutant Nude." Wiley, New York, 1973.

[85] L. Reid and S. Shin, *in* "The Nude Mouse in Experimental and Clinical Research," (J. Fogh and B. Giovanella, eds.), pp. 313–351. Academic Press, New York, 1978.

centrifuged at 1000 rpm for 10 min, and the serum is decanted into a fresh sterile Falcon conical tube. Varying dilutions from 1 : 4 to 1 : 500 are made of antiserum in serum-free media.

*b. Test for Antibody Titer.* The antiserum at the varying dilutions is added as a 1% solution in serum-free medium along with 1% unimmunized rabbit serum as a source of complement; it is tested on both the host cells and the tumor cells. A dilution is selected that gives a high titer of activity against host cells and a minimal one against tumor cells. To test the titer, a cell suspension or plate of cells attached to a substrate is rinsed with phosphate-buffered saline, treated with the appropriate dilution of antiserum + complement, and the cells incubated at 37° for 1 hr after which they are rinsed with phosphate-buffered saline and incubated in fresh medium containing serum. Within an hour, the host cells will lyse and only tumor cells remain.

3. Primary Tumors and Normal Tissue

Enriching primary tumors or normal tissues for epithelial cells is more difficult than using either clonal cell lines or strains or transplantable tumors. If one is not concerned with having pure epithelial cells, primary cultures of normal tissues or primary tumors can be plated onto the substrates. The tumor or normal tissue is dissected from the host under sterile conditions, minced finely, and suspended in medium supplemented with serum. The suspension is added to the substrates. Attachment of the cells occurs within hours or at most overnight. Following attachment of the cells, the substrates are detached and allowed to float. Such primary cultures on floating rafts remain functional for 3–4 weeks.

For pure epithelial cell suspensions one can resort to a variety of techniques that, although not entirely satisfactory, do give enriched populations of epithelial cells. One of the more straightforward procedures is to layer dissociated cells (from mechanical dissociation, mincing or sieving, or from enzymic dissociation) onto a density gradient containing sterile Ficoll or sucrose. The cells are centrifuged at 4° at 1000 rpm or sedimented by gravity permitting the cells to separate according to size and weight. Bands are collected that are enriched in epithelial cells and centrifuged at 4° at 1000 rpm. The supernatant liquid is removed, the pellet of cells suspended in medium supplemented with serum, and the cells added to the substrates.

### *D. Attachment of the Cells to the Substrates and Release of the Substrates to Form Rafts*

Epithelial cell suspensions are added to sterilized collagen or RBM substrates and incubated overnight at 37° in an incubator flushed with 5%

$CO_2$ in air. Once the cells attach to substrates, the substrates can be released to form rafts by rimming the gelatinous layer with a sterile spatula. The substrates will float and can be transferred from one plate to another by gentle pipetting with a large-mouth pipette. In culture, the rafts with attached epithelial cells will contract to as little as one-half their original size.

### *E. Preparation of Feeder Layers: Primary Cultures of Stromal Cells*

Ideally one prepares primary cultures of stromal cells that are normally in association *in vivo* with the epithelial cells to be cultured. Thus, for pancreatic epithelial cells, one would prepare pancreatic fibroblast cultures; for breast epithelial cells, breast fibroblast cultures; and for skin epithelium, skin fibroblast cultures. Tissue is dissected under sterile conditions from the animal, minced finely, and the mince suspended in medium containing serum. The suspension is added to regular cell culture plates (plastic). The cultures gradually select for fibroblast cells.[7] When plates are confluent, they can be irradiated with 5000 rad of cobalt 60 or cesium 135 to eliminate all mitotic activity. Irradiation of the feeder layers may prove to be unnecessary since the epithelial cells are separated from the stromal cells by being on the rafts, and the irradiation may even impair some epithelial–mesenchymal interactions. This possibility is currently under investigation. The floating substrates of collagen or RBM with attached epithelial cells are transferred to and floated over these primary cultures of fibroblasts.

# [22] Autoradiography

*By* GRETCHEN H. STEIN and ROSALIND YANISHEVSKY

Autoradiography is a technique for detecting radioactivity through the formation of silver grains in a photographic emulsion. An emulsion consists of crystals of silver halide, usually silver bromide, suspended in gelatin. To increase their sensitivity, nuclear emulsions designed for autoradiography have a higher ratio of silver halide to gelatin than do emulsions designed for light photography. They have a very high efficiency for low-energy $\beta$ particles such as those emitted by tritium (18 $keV_{max}$), carbon-14 (155 $keV_{max}$), phosphorous-33 (250 $keV_{max}$), sulfur-35 (167 $keV_{max}$), and iodine-125 (35 $keV_{max}$). The efficiency is somewhat less for high-energy $\beta$ particles, such as those emitted by phosphorus-32 (1710 $keV_{max}$).

Autoradiography is based on the same principle as photography except that the energy for conversion of silver bromide to metallic silver is derived from ionizing radiation rather than from photons of light. As $\beta$ particles pass through an emulsion layer, they set electrons free in the silver bromide crystals. These crystals contain sensitivity specks, which are irregularities in the crystal lattices. The free electrons migrate to sensitivity specks and attract silver ions, which combine with the electrons to form atoms of silver. A latent image of the autoradiogram is formed when enough silver atoms accumulate at a sensitivity speck to form a nucleus of metallic silver that catalyzes the conversion of the entire crystal into a silver grain during development. Developers are reducing agents that furnish electrons to reduce silver ions to silver atoms. Because crystals that have latent images are reduced to metallic silver more quickly than other crystals, development can be stopped when only the latent images have developed into silver grains. Unreduced crystals are preferentially dissolved in the fixer, leaving behind a pattern of silver grains that denotes the presence of radioactive material.[1]

## Materials

*Nuclear Emulsions.* Eastman Kodak (Rochester, New York) manufactures four liquid emulsions. NTB, NTB-2, and NTB-3 produce large grains suitable for light microscope autoradiography. NTB grains are the largest (approximately 2700 Å diameter) and NTB-3 grains are the smal-

[1] W. D. Gude, "Autoradiographic Techniques: Localization of Radioisotopes in Biological Material." Prentice-Hall, Englewood Cliffs, New Jersey, 1968.




ISBN 0-12-181958-2

lest (approximately 2300 Å diameter) of the three.[2] NTE, which is designed for electron microscope autoradiography, produces very small grains (approximately 500 Å diameter).[3] NTB-3 is approximately twice as sensitive as NTB-2, which is twice as sensitive as NTB or NTE.[4,5] Because increased sensitivity means that less energy is needed to make an activated crystal develop into a silver grain, the background increases with the sensitivity.

Ilford Ltd. (Ilford, Essex, Great Britain) also manufactures nuclear emulsions that cover a range of sensitivities and grain sizes. Ilford L4 is frequently used for electron microscope autoradiography because it has a small grain size (1200 Å) and is much more sensitive than NTE. Recently, Kodak has produced an improved emulsion for electron microscope autoradiography, which is also more sensitive than NTE. It is called Kodak Special Product Type 129-01.

*Dipping Vessels.* We use Lab-Tek No. 4310 Cyto-Mailers as dipping vessels (see Fig. 1). These vessels are thrown away when the emulsion is discarded.

*Slide Boxes.* Small black Bakelite slide boxes (Scientific Products), which hold 25 slides, are used for storing the slides during the exposure period. Some investigators use this type of slide box as a light-tight box, provided that it is sealed with tape. However, we prefer to store the taped slide boxes inside a second light-tight box.

*Desiccant.* We include small packets of Drierite in the slide boxes during the exposure period because nuclear emulsion is most sensitive when it is dry and because moisture can cause latent image fading. The packets of Drierite, including some blue indicator Drierite, are wrapped in several layers of cheesecloth and secured with tape.

## Autoradiography Darkroom

An autoradiography darkroom should exclude all light. This demand exceeds the normal requirements of a photographic darkroom. (A model darkroom for autoradiography is described by Kopriwa.[6]) The darkroom should be inspected for light leaks when the eyes have adapted to darkness (a minimum of 15 min). Sources of light entry may be window and door frames, door steps, ventilation shafts, or where pipes pass through floors. Light may also come from electrical contacts that spark on switch-

[2] L. G. Caro, *Methods Cell Physiol.* **1,** 327–363 (1964).

[3] M. M. Salpeter, *Methods Cell Physiol.* **2,** 229–253 (1966).

[4] A. Ron and D. M. Prescott, *Methods Cell Biol.* **4,** 231–240 (1970).

[5] A. W. Rogers, "Techniques of Autoradiography." Am. Elsevier, New York, 1973.

[6] B. M. Kopriwa, *J. Histochem. Cytochem.* **11,** 553 (1963).

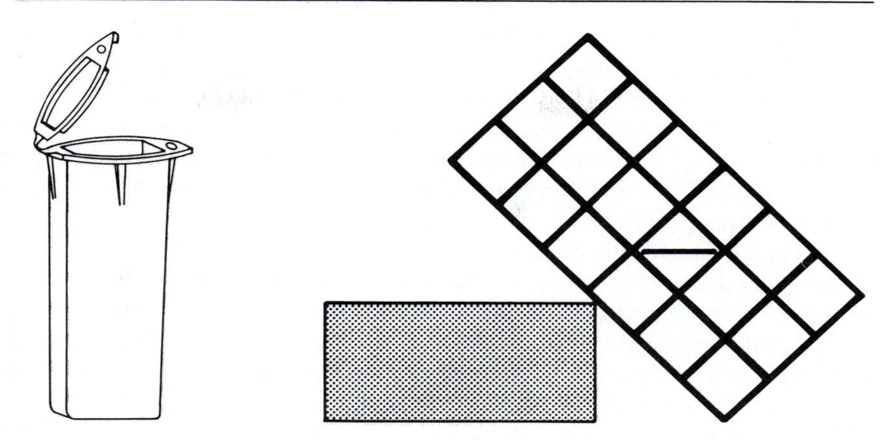

FIG. 1. A Lab-Tek Cyto-Mailer, which is used as a dipping vessel, is shown on the left. The position of a slide rack used for drying emulsion-coated slides is shown on the right.

ing, equipment containing vacuum tubes that emit light from the rear, clocks and watches that have luminous dials, and indicator lights such as those on waterbaths. These objects should be removed or taped over to shut out the light. Fluorescent lights continue to glow for an appreciable time after being turned off and, if possible, should not be used in the darkroom. The darkroom should have an entrance corridor with double doors so that one may enter or leave without letting light into the darkroom.

A safelight may be used during autoradiography according to the emulsion manufacturer's recommendations. For example, a dark red Wratten No. 2 filter can be used for Kodak NTB, NTB-2, and NTB-3 emulsions. The power of the light bulb should be no greater than 15 W. The safelight housing must be light tight so that light escapes only through the filter. Safelights and filters may be purchased from Kodak. The safelight should be mounted about 1 m above the working space. Turn off the safelight when it is not needed, especially when the slides are drying, because dry emulsion is the most sensitive.

## Basic Procedure for Autoradiography of Tissue Culture Cells

### *1. Cell Culture*

Cells are grown on 22 × 22 mm glass coverslips, thickness number $1\frac{1}{2}$, in 35-mm plastic tissue culture petri dishes or on 22 × 50 mm coverslips in four-well multiplates (Lux No. 5278, available from Flow Laboratories).

The coverslips are cleaned before use by: (1) 5 min in 95% ethanol to remove grease; (2) 5 min in 1 *N* HCl; (3) 5 min in deionized water, followed by four short rinses in deionized water; (4) two rinses in 95% ethanol; and (5) two rinses in 100% ethanol. They are sterilized in a 250° oven for 7 hr.

Cells seeded in small petri dishes and in four-well multiplates tend to settle more heavily in the middle of these vessels. This creates different growth conditions at the center and periphery. In addition, the cells may be too crowded in the center for single cell analysis by autoradiography. We have found that filling the vessels very full when the cells are seeded reduces this problem. After the cells have attached, the volume of medium may be reduced. We use 5 ml medium to seed cells in 35-mm petri dishes and 12 ml medium per well in four-well multiplates.

Tissue sections fixed to slides, or cell suspensions smeared on slides, may be processed for autoradiography in the same way as cells grown on coverslips.[1] The use of a cytocentrifuge to flatten cells in suspension onto slides is discussed in the section on self-absorption.

### 2. *Fixation*

We fix the cells by adding to the growth medium an equal volume of 3 : 1 methanol : acetic acid fixative that has been freshly prepared. After 10 min, the half-strength fixative is removed by aspiration and replaced with an equal volume of undiluted methanol : acetic acid fixative. After 10 min, the coverslips are removed and allowed to air dry. Methanol : acetic acid fixative can cause some bursting of cell plasma membranes; however, it flattens the cells well. By adding the fixative to the medium first, we have eliminated this problem for our experiments with human cells.

### 3. *Removal of Soluble Pools of Radioactively Labeled Precursors*

For many experiments using radioactive nucleotides, fixation with 3 : 1 methanol : acetic acid adequately extracts the soluble pools of these precursors. For other experiments, such as when the cells are relatively crowded (Fig. 2) or when they are labeled with radioactive amino acids, the cells are: (1) extracted with 5% trichloracetic acid (TCA) at 4° for 5 min; (2) washed 3 times with 70% ethanol, for 5 min each; (3) rinsed 2 times with 95% ethanol; (4) rinsed 2 times with 100% ethanol; and (5) air dried.[7] TCA extraction is preferable to perchloracetic acid (PCA) extraction because PCA is difficult to wash out of cells and will inactivate the emulsion if not completely removed.

[7] D. M. Prescott, *Methods Cell Physiol.* **1**, 365–370 (1964).

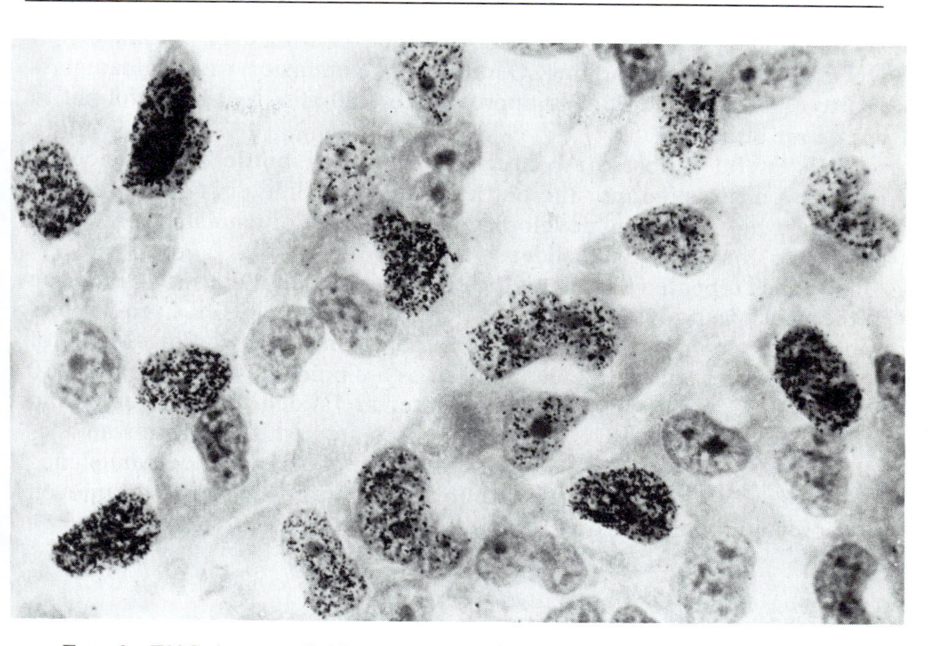

FIG. 2. T98G human glioblastoma multiforma cells were labeled with 4 $\mu$Ci/ml [$^3$H]thymidine (specific activity 50 Ci/m*M*) for 10 min. The cells were fixed in methanol : acetic acid (3 : 1) and extracted with 5% TCA. The slides were processed according to our basic procedure, using an exposure time of 3 days. The cells were stained with Giemsa stain. Silver grains are concentrated over the nuclei that synthesized DNA during the labeling period.

*4. Coverslip Mounting*

The coverslips are mounted cell side up on clean microscope slides using a drop or two of Permount (Fisher) or Euparol (Carolina Biological). Generally we let the mounting medium dry for several hours at 60°, or overnight at room temperature, before the slides are dipped.

*5. Preparation of Emulsion*

We routinely use Kodak NTB-2 for analysis of [$^3$H]thymidine incorporation into tissue culture cells because it has low background, produces grains large enough to count at ×400 magnification, and needs only 3–4 days exposure to develop 25–50 grains/labeled nucleus under our experimental conditions (0.01 $\mu$Ci/ml [$^3$H]thymidine for 24 hr).

The emulsion is a gel at room temperature and is melted in a waterbath at 39–43°. Although NTB-2 can be handled under a Kodak Wratten No. 2

safelight filter, we suggest keeping the light off whenever possible, e.g., when the emulsion is melting. Gently stir the emulsion with a clean glass rod to check its fluidity after approximately 30 min, being careful not to create air bubbles.

It is useful to test the background of a new bottle of emulsion by dipping a clean slide into the bottle, letting the slide dry for 2–3 hr in a light-tight box, and then developing it. One bottle of emulsion (118 ml) is enough to coat 1000–3000 slides.[7]

Because repeated heating of the emulsion to melt it causes an increase in the background, we dispense new emulsion into Lab-Tek Cyto-Mailers and store them at 4°. For subsequent experiments, a single Cyto-Mailer of emulsion is melted for 15 min and used as a dipping vessel. In general, 10 ml of NTB-2 and 10 ml of distilled water are added to each vessel and mixed gently. For relatively flat samples, diluted emulsion forms an adequate layer of emulsion. The major advantage of using diluted emulsion is that subsequent staining is easier. For preparations that have an uneven surface, undiluted emulsion covers better.

### 6. *Dipping and Drying*

A Cyto-Mailer of melted emulsion is held upright on the workbench in a small beaker of 39–43° water or in the waterbath. We dip one slide at a time, remove it with a slow even motion, and touch the bottom end of the slide to a paper towel to drain the excess emulsion. The slide is allowed to dry horizontally with the cell side up by supporting it in a test-tube rack that has been turned on its side and tilted at a 45° angle (Fig. 1). In this way, the wet slide is supported by its edges. The safelight is turned off while the slides are drying or the test-tube rack with slides is placed in a large light-tight box containing Drierite until the slides are dry, i.e., about 1 hr. They are placed in slide boxes containing packets of Drierite and sealed with black electrical tape to exclude moisture and light. The sealed slide boxes are placed inside a second light-tight box and exposed at 4°.

There are many variations of this dipping procedure. Two slides can be placed back to back to be dipped simultaneously and then pulled apart to dry separately.[7] Slides can be dried vertically in a test-tube rack. However, vertical drying can cause a small gradient of increasing emulsion thickness from top to bottom, and it may cause artifacts as discussed below under "Autoradiographic Background." The emulsion can be wiped off the back of the slide immediately after dipping rather than after development. A drop of emulsion can be placed on the cells and spread over the surface using the side of a 9″ Pasteur pipette tip as a roller.[8] This

[8] L. E. Allred, personal communication.

is an easy way to coat small samples, e.g., coverslip fragments. It can be used without waiting for the coverslip mounting medium to dry. For large coverslips, however, it is difficult to spread the emulsion evenly and without causing mechanically induced background.

*7. Exposure*

We expose autoradiograms at 4° to minimize latent image fading.[5] However, for short-term exposures room temperature is also satisfactory.

The duration of the exposure depends on the experimental procedure used and the grain density desired. The way to determine an exposure time is to include several test slides in the experiment. These can be developed at various times and used to estimate the correct exposure time for the experiment.

*8. Development*

We use Kodak D19 developer and Kodak Fixer 197-1746, a general-purpose hardening fixer. These solutions are stored in brown bottles and are discarded after 2 months. If the D19 turns yellow before then, it is discarded. To develop slides, unused developer and fixer are poured into staining dishes and are used at 18–20° maximum. The slides are allowed to warm to room temperature and are placed in staining racks. Slides coated with either NTB, NTB-2, or NTB-3 are developed for 2 min in D19, rinsed 10–30 sec in 1% acetic acid stop bath, and fixed for 5 min in Fixer. After fixing, the slides are rinsed gently in running tap water for 20 min, given a final rinse in distilled water, and air dried. Ideally, the running tap water should be at the same temperature as the other solutions used to avoid stressing the gelatin by temperature changes. A thermostatic regulator, which may be purchased from a photographic supply store, can be used to maintain the appropriate temperature. However, a regulator is a convenience rather than a necessity.

*9. Staining*

We use Giemsa Blood Staining Stock Solution (J. T. Baker Chemical Co.) diluted 1:25 in 0.01 *M* sodium phosphate at pH 7.1. Developed autoradiographs are stained for 50 min and destained for 5 min in the same buffer. Thereafter, the film of emulsion on the back of the slide is wiped off with a damp Kimwipe. This procedure gives good definition of the nucleus and cytoplasm without staining the nucleus so darkly that grains over it are difficult to count. Much shorter staining and destaining periods may also be used successfully. The choice of a staining procedure depends on the nature of the experiment. Other stains are discussed below.

## Other Techniques

### *Prestaining*

Prestaining a specimen before applying nuclear emulsion generally yields a clearer image than poststaining the developed autoradiogram. This is so because, in prestaining, the gelatin of the emulsion layer is not stained and the specimen stains more vividly and precisely. However, only a few stains may be used for prestaining because most stains cause increased background (positive chemography) even if they are removed before the emulsion is applied. Prestains are useful for observing morphological details that may be obscured by overlying silver grains after autoradiography. For example, chromosomes can be more easily identified before autoradiography. Stains used to detect enzymic reactions are often used as prestains because the autoradiographic process alters or destroys the enzymic reaction of interest.[5]

Basic fuchsin, used in the Feulgen procedure, and aceto-orcein are the most commonly used prestains. Even these acceptable prestains can cause a high background if they are of low purity or are old. Loss of radioactive material during staining is also a potential problem in using prestains. The hydrolysis step of the Feulgen procedure may remove some radioactivity from the specimen; however, for most qualitative studies, the loss of radioactivity is insufficient to alter the interpretation of the results.[9,10] Aceto-orcein, used as a prestain, fades during the photographic processing. Furthermore, aceto-orcein cannot be used as a poststain because it removes developed silver grains. The Feulgen procedure also cannot be used for poststaining because grains are removed during the hydrolysis step.

### *Poststaining*

An effective poststain must overcome the presence of a layer of emulsion over the specimen and the chemical changes brought about by the photographic processing. If the gelatin takes up too much stain, there will be little contrast between the specimen and the stained gelatin. If the gelatin affects the staining reagents, or provides an environment of inap-

[9] R. Baserga and K. Nemeroff, *Stain Technol.* **37,** 21 (1962).
[10] W. Lang and W. Maurer, *Exp. Cell Res.* **39,** 1 (1965).

propriate pH, or if the reactive groups on the specimen are altered, the stains may be rendered totally ineffective.

Many of the methylene blue–eosin type stains (Giemsa, Wright) are effective poststains. The procedure for Giemsa staining used in our laboratory is presented above in "Basic Procedure." Hematoxylin and eosin may be preferable for greater cytological detail although this procedure is more time consuming. There are several lists of stains that have been used with autoradiography,[11–14] and recipes for them can be found in basic histology textbooks.[15–17]

*Removal of Emulsion*

Occasionally, it may be useful to remove the emulsion from a completed autoradiograph, e.g., if the background is too high or if visualization of the underlying specimen is to be facilitated. Techniques for removal of emulsion have been described by Bianchi *et al.*[18] and Rogers.[5]

*Stripping Film*

Kodak AR-10 stripping film is a prepared layer of gelled emulsion that is 5 μm thick, has approximately the same sensitivity as NTB, and produces grains suitable for light microscope autoradiography. Its principal advantage is that it gives a uniform layer of emulsion. Its disadvantages are that it is more difficult to apply than liquid emulsion, the stripping process creates some static discharges that can cause background, and it is more difficult to stain the specimen through 5 μm of emulsion. A detailed description of the technique for using AR-10 has been presented.[5]

*Double Isotope Autoradiography*

Two isotopes that have different energies can be analyzed in a single specimen. They are distinguished by the distance that their β-particles

[11] C. P. Leblond, B. M. Kopriwa, and B. Messier, *in* "Histochemistry and Cytochemistry" (R. Wegmann, ed.). Pergamon, Oxford, 1963.

[12] J. M. Thurston and D. L. Joftes, *Stain Technol.* **38,** 231 (1963).

[13] R. Baserga, *Methods Cancer Res.* **1,** 45–116 (1967).

[14] L. K. Schneider and S. L. Pierce, *Stain Technol.* **48,** 69 (1973).

[15] S. N. Thompson, "Selected Histochemical and Histopathological Methods." Thomas, Springfield, Illinois, 1966.

[16] A. G. Pearse, "Histochemistry: Theoretical and Applied," 3rd ed. Longmans, Green, New York, 1969.

[17] G. L. Humason, "Animal Tissue Techniques," 3rd ed. Freeman, San Francisco, California, 1972.

[18] N. Bianchi, A. Lima-de-Faria, and H. Joworska, *Hereditas* **51,** 207 (1964).

penetrate the emulsion. The most frequently used combination of isotopes is tritium and carbon-14. The $\beta$ particles from tritium range in energy from zero to 18 keV and those from carbon-14 range from zero to 155 keV. Consequently, the discrimination between the two isotopes cannot be complete because approximately 15% of the $\beta$ particles from carbon-14 have the same energy as tritium's $\beta$ particles.

In the double-exposure method,[19] slides are dipped in emulsion, exposed, and developed after a period of time appropriate for the amount of tritium in the specimen. The ratio of tritium to carbon-14 should be very high, so that only a small fraction of the grains produced during this short exposure are from carbon-14. The slides are then dipped in celloidin embedding solution (no. M-4700, Randolph Products Co., Carlstadt, New Jersey) that has been diluted 1 : 2 in alcohol–ether. Finally, the slides are dipped a second time for a period appropriate for the amount of carbon-14 in the specimen. The $\beta$ particles from tritium are not sufficiently energetic to penetrate to the second layer of emulsion. Therefore, grains in the second layer are produced only by carbon-14. The focal plane of the grains in the two layers will be different and can be distinguished. Two emulsions that produce different-size grains may be used for each layer to allow easier discrimination between them. Alternatively, the silver grains in one layer can be colored by dye-coupling.[20]

A new procedure uses only a single layer of emulsion, which is expanded in glycerol after the silver grains have been developed.[21] Silver grains at different levels above the cells can be counted separately as in the double-exposure method.

### *High-Speed Scintillation Autoradiography*

High-speed scintillation autoradiography is a technique for amplifying the number of silver grains produced per radioactive disintegration. Several procedures have been published for the application of this method to cells fixed on slides.[22–26] The basic procedure involves immersion of the emulsion-coated specimen in a solution of scintillator, e.g., PPO (2,5-diphenyloxazole) plus dimethyl POPOP [14-bis-2-(4-methyl-5-phenyl-

[19] R. Baserga, *J. Cell Biol.* **12,** 633 (1962).
[20] E. O. Field, K. B. Dawson, and J. E. Gibbs, *Stain Technol.* **40,** 295 (1965).
[21] S. W. Perdue, R. F. Kimball, and A. W. Hsie, *Exp. Cell Res.* **107,** 47 (1977).
[22] R. J. Przyblyski, *J. Cell Biol.* **43,** 108a (1969).
[23] G. S. Panayi and W. A. Neill, *J. Immunol. Methods* **2,** 115 (1972).
[24] B. G. M. Durie and S. E. Salmon, *Science* **190,** 1093 (1975); B. G. M. Durie, *ibid.* **195,** 208 (1977). Includes a correction to preceding paper.
[25] A. S. Mukherjee and R. N. Chatterjee, *Histochemistry* **52,** 73 (1977).
[26] W. Sawicki, K. Ostrowski, and E. Platkowska, *Histochemistry* **52,** 341 (1977).

oxazolyl)benzene] dissolved in dioxane. The emulsion becomes impregnated with scintillator molecules. Photons are released as $\beta$ particles pass through the scintillator. These photons activate more silver crystals in the emulsion than would have been activated by the $\beta$ particles alone.

Theoretically, the efficiency of the high-speed method of tritium-labeled molecules could be 30-fold that of conventional autoradiography.[26] However, different investigators have had variable success with this technique; in some cases only a small increase in efficiency was produced and was accompanied by an increase in the background.

### *Autoradiography of Diffusible Substances*

There are several techniques for preparing autoradiographs of cells containing radioactivity in diffusible substances. Briefly, cells can be freeze-dried and placed under a film of dried emulsion.[27] Cells can be frozen and kept at low temperature until completion of the autoradiogram, thus preventing solute redistribution during the process.[28] Cells can be applied directly to dry emulsion-coated slides and air-dried although controls to monitor positive or negative chemography are necessary because the wet specimen can interact with the emulsion.[29]

## Potential Problems

### *Self-Absorption*

Increasing the thickness of a radioactive specimen beyond a certain point does not increase the number of $\beta$ particles entering the emulsion. This phenomenon is due to self-absorption of the $\beta$ particles by the specimen. The distance that $\beta$ particles can penetrate depends on their initial energies and on the density of the specimen. Because most of tritium's $\beta$ particles penetrate less than 3 $\mu$m through cellular material,[30] self-absorption is an important factor in autoradiography of tritium-labeled whole cells. Nevertheless, tritium is useful for autoradiography because it has a short pathlength and, therefore, can be localized better than more energetic isotopes.

Tissue culture cells are sufficiently large that they must be in a flat configuration for tritium autoradiography in order to minimize self-

[27] O. L. Miller, Jr., G. E. Stone, and D. M. Prescott, *Methods Cell Physiol.* **1,** 371–379 (1964).

[28] S. B. Horowitz, *Methods Cell Biol.* **8,** 249–275 (1974).

[29] W. E. Stumpf, *Methods Cell Biol.* **13,** 171–193 (1976).

[30] W. Maurer and E. Primbsch, *Exp. Cell Res.* **33,** 8 (1964).

absorption. Many types of cells flatten out on the solid substrate on which they are grown. Other types of cells are round, e.g., cells that are grown in suspension. In addition, flat cells become round when they are in mitosis and when they are removed from the substrate, e.g., with trypsin. Spreading may take from 2–12 hr, depending on the cell type, the substrate, and the culture conditions. Generally, cells spread more rapidly on tissue culture plasticware than on glassware. Coating glass with a layer of evaporated carbon enhances spreading.[31] When round cells are to be used for autoradiography, they can be flattened onto microscope slides with a cytocentrifuge (Shandon Southern Instruments, Sewickley, Pennsylvania). The chambers of a cytocentrifuge are designed so that the centrifugal force drives the cells against a slide at the back of the chamber. The cells are flattened on the slide and are then fixed as usual.

The choice of fixative affects the degree of flattening of the cells. Even cells that are well spread in culture are several micrometers thick. For example, Chinese hamster ovary cells in monolayer culture are about 7 μm thick and can benefit from additional flattening. We have found that 3 : 1 methanol : acetic acid flattens cells much more effectively than neutral formalin fixative. The ratio of methanol : acetic acid may have to be modulated to obtain an intact, well-flattened specimen; increasing the concentration of acetic acid will flatten cells but may disrupt cellular integrity.

Uniform self-absorption in flattened cells simply reduces the efficiency at which $\beta$ particles are detected by the emulsion, whereas nonuniform self-absorption can create artifacts. For example, cells increase in size during the cell cycle, and large $G_2$ cells may not be flattened to the same thickness as small $G_1$ cells. Consequently, the efficiency could be less in the $G_2$ cells. Nonuniform self-absorption is also inherent within individual cells because the density of various organelles differs: the nucleus is less dense than the cytoplasm, which is less dense than the nucleolus.[30]

*Autoradiographic Background*

The most common causes of background are listed and discussed briefly. More detailed discussions are given by Boyd,[32] Baserga,[13] and Rogers.[5]

*1. Photographic Processing.* Excessive background will occur if the temperature of the developing solutions is high, or if the duration of development is long. This problem can usually be recognized because

[31] G. A. Meek, "Practical Electron Microscopy for Biologists." Wiley, New York, 1970.

[32] G. A. Boyd, "Autoradiography in Biology and Medicine." Academic Press, New York, 1955.

background grains will be scattered randomly throughout the emulsion and will be smaller than the grains that arose from the radioactive material in the specimen.

*2. Chemography.* Chemical reactants, particularly reducing agents, present in the biological specimen or in solutions that come in contact with the undeveloped autoradiograph, can produce a latent image in the emulsion. This is termed positive chemography. The opposite, negative chemography, occurs when chemical groups cause rapid latent image fading, i.e., loss of the nucleus of metallic silver by recombination of the silver atoms with bromine atoms to form silver bromide. Chemography is uncommon in cell culture experiments unless caused by prestaining with an inappropriate stain.

Chemography due to the specimen itself may be looked for by exposing unlabeled specimens to normal emulsion and to emulsion that has been fogged by a brief exposure to light. Positive chemography will produce grain densities higher than background in the normal emulsion. These grains are often large and irregular and sometimes clumped. Negative chemography will produce faded areas having grain densities lower than the uniformly black background in the fogged emulsion.

*3. Light Exposure.* A high background may occur if the emulsion has inadvertently been exposed to light, e.g., if the storage facilities for slides are inadequate or if the darkroom has light leaks. A proper safelight is especially important because the emulsion will be exposed under it for a number of minutes.

*4. Static Discharges.* Electrostatic discharges may be emitted when tape is stripped from its roll and used to seal a slide box or when it is stripped from the slide box itself. Similarly, there may be electrostatic discharges when stripping film is stripped from its support. Sometimes the discharges may be accompanied by flashes of light. The discharges and the light can cause background, which often appears as streaks of silver grains running parallel to one another across the developed emulsion. This problem can be controlled by cutting strips of tape before the emulsion is removed from its light-tight box and by untaping slide boxes slowly. Increasing humidity and decreasing temperature in the darkroom also help reduce these effects.

*5. Pressure Effects.* Emulsion is sensitive to pressure, e.g., from scratches or fingerprints. Contraction of the emulsion as it dries also can produce a high background if it occurs too quickly or goes too far. When the emulsion layer is very thin, it may dry too fast. Drying slides coated with dilute emulsion in the horizontal position rather than in the vertical position is useful in avoiding the creation of a very thin layer of emulsion at the top of the slide.

6. *Environmental Radiation.* Radiation from the environment, e.g., cosmic rays, is usually a relatively minor source of background and is acceptable for most work. However, if desired, the background may be reduced by shielding the emulsion and coated slides.

7. *Other.* Occasionally a latent image speck will form spontaneously. This will occur more often with the more sensitive emulsions. Hence, the choice of emulsion should be based on the least sensitive one that is adequate for the experiment. Although the rate of formation of spontaneous background increases with age of the emulsion, we find a reasonable background for 2–3 months after the expiration date suggested by the manufacturer. However, one should always dip a test slide in the emulsion and check it before use in experiments in which a low background is important.

Finally, everything from the coverslips used for cell culture to the darkroom itself should be meticulously clean. Dust, spilled chemicals including developer and fixer, and materials that come in contact with the emulsion all may cause background. Glass, plastic, or high-grade stainless-steel utensils are generally safe to use in contact with the emulsion but other metals, particularly copper, are to be avoided.

### Acknowledgments

We wish to thank Dr. David M. Prescott, Dr. John Heumann, and Dr. Dorothy Pumo for their help in preparing this manuscript. It was written while we were supported by American Cancer Society grants NP-156B (G.H.S.) and PF 1292 (R.Y.).

## [23] Induction and Production of Interferon

*By* ROBERT M. FRIEDMAN

Many cells in culture can be stimulated to produce at least some interferon. This is a convenient property because it permits the study of a useful cell product with a specific biological activity, inhibition of virus growth. Stimulation of interferon by cells in culture appears to be due to induction of a protein that is not ordinarily made. Recent studies have suggested that interferon production is controlled by a repressor[1] and is determined by specific loci on chromosomes that have in some cases been identified.[2]

[1] R. L. Cavalieri, E. A. Havell, J. Vilĉek, and S. Pestka, *Proc. Natl. Acad. Sci. U.S.A.* **74,** 4415 (1977).

[2] P. P. Creagan, Y. H. Tan, S. Chen, and F. H. Ruddle, *Fed. Proc., Fed. Am. Soc. Exp. Biol.* **34,** 2222 (1975).





ISBN 0-12-181958-2

In general, it is best to use confluent monolayers of cells to produce interferon. Cells that have been aged for a week after confluence in monolayer cultures are often better producers of interferon than are fresh, recently trypsinized cells.[3] In some diploid cultures the production of interferon varies with the passage level of the cells; with excessive passages the ability of a diploid cell line to produce interferon declines significantly.[4] In this case it is, of course, important to know the history of the cells being used. In many systems, however, the passage history of the cells does not affect the yield of interferon. This is certainly true of heteroploid lines such as mouse L cells, which are usually excellent interferon producers at any passage level.

Interferon production in tissue cultures may be stimulated by viral and nonviral inducers. Some viruses are usually excellent interferon inducers, the best in general being among the group A arboviruses (also called togaviruses) and the paramyxoviruses.[4] With these viral inducers interferon production is usually maximal by 18–24 hr after induction. The group A arboviruses are good inducers whether active or heat inactivated, and a multiplicity of infection of 20–40 virus particles per cell usually suffices to give optimal interferon production; too high a multiplicity of virus may be inhibitory. The best generally available inducer in this group is Chikugunya virus, which may be obtained from the American Type Culture Collection (ATCC, 12301 Parklawn Drive, Rockville, Maryland). Chikungunya virus is usually propagated by intracerebral inoculation of suckling mice with about $10^4$ plaque-forming units (pfu) of virus. After 40 hr the mice are killed and their brain tissue homogenized to yield a 10% suspension in tissue culture medium. The virus can be assayed by plaque formation in Vero cells. Mouse brain suspensions produced in this manner usually have titers well in excess of $10^9$ pfu/ml.

To produce interferon with Chikungunya virus, human RSTC-2 cells are infected at a virus : cell multiplicity of 30 for 1 hr at 37°. Cultures are washed 5 times and incubated for 22 hr in fresh medium with 2% serum. In order to inactivate residual virus that would interfere with the assay for interferon, the culture fluids are harvested and treated with drops of *N* HCl until the pH is 2. The medium is immediately brought back to neutrality with *N* NaOH. In cultures induced by this means, yields of interferon are of the order of 1500 mouse international reference units/ml if the RSTC-2 cells are treated with actinomycin D (0.1 μg/ml) for 1 hr before infection with virus (see below on superinduction). Much lower interferon yields are obtained from RSTC-2 cells if actinomycin D pretreatment is omitted, but in many cell systems high interferon titers can be stimulated

[3] E. A. Havell and J. Vilĉek, *Antimicrob. Agents & Chemother.* **2**, 476 (1972).
[4] M. W. Myers and R. M. Friedman, *J. Natl. Cancer. Inst.* **47**, 757 (1971).

by infection with Chikungunya virus in the absence of antimetabolite pretreatment.[4]

Among the paramyxoviruses, Newcastle disease virus (NDV) is most frequently employed as an interferon inducer. NDV is an excellent inducer in most systems other than chick cells. This is probably due to replication of NDV with resultant severe cytopathic effects in chick cells. In other systems NDV does not multiply significantly but does induce high titers of interferon. NDV also may be obtained from the American Type Culture Collection and, for this purpose, it is probably best to use the B1 vaccine strain which has little toxicity even for chickens. The virus can be propagated in 9–10 day old embryonated chick eggs. The infected embryos are harvested after 5 days when the yields are about 1000 hemagglutinin units (HAU)/ml. One HAU is equivalent to about $10^6$ virus particles. NDV can be assayed easily by its capacity to hemagglutinate fowl red blood cells.

Interferon induction by NDV in human or mouse cells can be effected with either active or inactivated virus. NDV is rapidly inactivated by ultraviolet (UV) light (5000 erg/sec, 10–20 sec at 18 cm). In human cells, 20 HAU/$10^6$ cells have been used to induce interferon production; in other systems, UV-inactivated virus is often employed. A multiplicity of 300 plaque-forming units of UV-inactivated virus per cell (virus titer as measured *before* inactivation) is required to give optimal yields in mouse L cells. Cells are infected and handled in the manner described for production with Chikungunya virus. Since NDV is far more pH-stable than Chikungunya virus inactivation must be carried out with great care. To be on the safe side, it is best to leave NDV preparations at pH 2 for 4 days. With mouse L cells, titers of interferon as high as 10,000 international mouse reference units/ml have been obtained by this procedure.[5]

For most purposes the only nonviral inducer of interferon that should be considered is polyriboinosinic : polyribocytidylic acid (poly IC). Poly IC is an inducer in many tissue culture systems, and excellent preparations of poly IC are available from several commercial sources. It is important that the instructions of the manufacturer be followed carefully in attempting to dilute poly IC preparations. Otherwise, the viscosity of the polymer will result in an unevenly mixed solution. On the other hand, excessive shaking may shear the molecule and this can alter its properties. A range of 0.1–100 μg/ml should be tried to determine the optimal concentration of poly IC necessary. With poly IC induction, interferon yields are usually maximal 8 hr after induction. In order to obviate removal of poly IC from the interferon preparation obtained, it is best to treat cells with the in-

[5] A. J. Hay, J. J. Skehel, and D. C. Burke, *J. Gen. Virol.* **3**, 175 (1968).

TABLE I
SUPERINDUCTION OF INTERFERON IN HUMAN FIBROBLAST CELL CULTURES

| Cell line (reference) | Treatment | Time (hours) | Interferon yield |
|---|---|---|---|
| RSTC-2 (4) | poly IC (50 μg/ml) | | |
| | + DEAE-dextran (25 μg/ml) | 0–1 | 500[a] |
| RSTC-2 | poly IC + DEAE-dextran | 0–1 | 10,000[a] |
| | + puromycin (20 μg/ml) | 0–4 | |
| | + actinomycin D (1 μg/ml) | 1–4 | |
| FS-4 (3) | poly IC (100 μg/ml) | 0–2 | 100[b] |
| FS-4 | poly IC | 0–2 | |
| | + cycloheximide (50 μg/ml) | 2–6 | 10,000[b] |
| | + actinomycin D (1 μg/ml) | 5.5–6 | |

[a] Yield in reference units per milliliter at 8 hr after inducer addition.
[b] Yield at 24 hr after inducer addition.

ducer for only an hour and then to wash the cells 3–5 times with saline or medium. In addition, DEAE-dextran (10–100 μg/ml) may be used to increase poly IC adsorption. The yield of interferon in human FS-4 cells induced with poly IC alone (20 μg/ml) is usually less than 1000 units/ml.[3,4]

Several steps may be taken with poly IC induction to increase interferon yields, but the single most effective way to do this is a superinduction procedure that employs actinomycin D and cycloheximide or puromycin.[3,4] The timing of the addition of antimetabolites is critical. Monolayers of cells confluent for at least 6 days are usually used, especially if interferon is being induced in diploid cells. After the cultures are washed with buffered saline, Eagle's Medium (MEM, serum-free) containing poly IC (5–20 μg/ml) and cycloheximide is added. After 5 hr, actinomycin D (final concentration 1–5 μg/ml) is added to the medium for 1 hr and, 6 hr after induction, the medium is removed and discarded. The cells are washed 4 times with saline and MEM with 0.2% human plasma protein (U.S.P.) or 2% fetal calf serum is added. The medium containing interferon is collected 30 hr after the onset of the production run. The yields of interferon in FS-4 or RSTC-2 human diploid cells with this procedure are at least 10,000–20,000 reference units/ml so that a very significant potentiation of interferon production may be obtained. However, it is important to note that superinduction with antimetabolites does not work in all systems. Mouse L cells, for instance, are not superinductable for interferon production. Table I reviews the results obtained with superinduction of interferon in two systems.

Priming is another practical procedure that has yielded high titers of interferon with either viral or nonviral inducers.[3,6] In this context priming means exposure to interferon several hours before an inducer is added. Primed cells are often much better interferon producers than nonprimed cells. In the case of human diploid FS-3 cells, preincubation with 100 units/ml of interferon for 16 hr followed by induction with 100 $\mu$g/ml of poly IC results in a yield of 512 units/ml of interferon; unprimed FS-3 yielded only 8 units/ml in the same study. However, superinduced FS-3 cells produced over 10,000 units/ml. Unfortunately a combination of superinduction and priming did not give rise to significantly higher titers of interferon than did superinduction alone. Some cell lines cannot be primed, and in a few instances interferon treatment actually inhibits interferon production.

Once interferon is obtained it is necessary to assay it in order to determine how many biological units are present. International standards for mouse, rabbit, and human interferons can be obtained in small quantities from the Resources Research Branch, NIAID, Westwood Building, NIH, Bethesda, Maryland. These are necessary to standardize individual laboratory interferon preparations that in turn should be employed in each individual assay as standards.

Two relatively fast biological assays have been described for interferon that employ rapidly growing viruses (Sindbis virus or EMC virus) with a wide spectrum of infectivity; a simple hemagglutinin assay is also available.[7,8] These are the generally employed procedures that are easiest to arrange and can be applied to assay a variety of interferons.

[6] R. M. Friedman, *J. Immunol.* **96,** 872 (1966).

[7] H. K. Oie, C. Buckler, C. Uhlendorf, D. A. Hill, and S. Baron, *Proc. Soc. Exp. Biol. Med.* **140,** (1972).

[8] P. Jameson, M. A., Dixon, and S. E. Grossberg, *Proc. Soc. Exp. Biol. Med.* **155,** 173 (1977).

## [24] Evaluation of Chemical Carcinogenicity by *in Vitro* Neoplastic Transformation

*By* NOËL BOUCK and GIAMPIERO DI MAYORCA

The malignant transformation of cultured mammalian cells in response to chemical carcinogens, first observed[1] and documented[2] many years ago, has now been demonstrated in a number of different cells with a wide variety of *in vitro* assays for transformation. Of the many combinations of

[1] W. R. Earle and A. Nettleship, *J. Natl. Cancer Inst.* **4,** 213 (1943).

[2] Y. Berwald and L. Sachs, *Nature (London)* **200,** 1182 (1963).





ISBN 0-12-181958-2

cell type and assay procedure that have been recorded, one of the more rapid and reliable is described in this article;[3] it uses a fibroblast line of baby hamster kidney cells, BHK 21/cl 13,[4] and assays for its transformation after a brief treatment by suspected carcinogen by testing the ability of the treated cells to form colonies in soft agar.[5]

This established BHK cell line offers advantages over cells newly placed in culture in that cells grow rapidly (12-hr doubling time at 37°), transform with high frequency in response to a variety of carcinogens,[3,6–8,9,10,11] clone readily to give homogeneous populations, and are amenable to culture for periods of time sufficient to carry untreated controls to the end of an experiment. The technique is sometimes criticized because the cells may have already undergone one or more steps in the process leading to malignant transformation, being "immortal" whereas normal diploid cells are not. However, in defense of the method it should be noted that recently cloned BHK cells are not tumorigenic *in vivo* at moderate doses,[12] whereas their virally and chemically[8] transformed derivatives are. Thus the transformation induced *in vitro* clearly represents a very crucial step in any series of changes that may be required for tumorigenesis.

The soft agar assay for *in vitro* transformation, although slightly more cumbersome than some others, has the advantage of selecting in a single step for cells capable of anchorage-independent growth. This particular characteristic of the transformed phenotype of a fibroblast cell is the only selective one that is consistently correlated with *in vivo* tumorigenicity.[13]

## Cells and Culture Conditions

A culture of BHK 21/cl 13 should be obtained, preferably from someone currently working with it, at as low a passage level as possible. It

[3] N. Bouck and G. di Mayorca, *Nature (London)* **264,** 722 (1976).

[4] M. Stoker and I. Macpherson, *Nature (London)* **203,** 1355 (1964).

[5] I. Macpherson and L. Montagnier, *Virology* **23,** 291 (1964).

[6] Y. Ishii, J. A. Elliott, N. K. Mishra, and M. W. Lieberman, *Cancer Res.* **37,** 2023 (1977).

[7] R. F. Newbold, C. B. Wigley, M. H. Thompson, and P. Brookes, *Mutat. Res.* **43,** 101 (1977).

[8] G. di Mayorca, M. Greenblatt, T. Trauthen, A. Soller, and R. Giordano, *Proc. Natl. Acad. Sci. U.S.A.* **70,** 46 (1973).

[9] I. F. H. Purchase, E. Longstaff, J. Ashby, J. A. Styles, D. Anderson, P. A. Lefevre, and F. R. Westwood, *Br. J. Cancer* **37,** 873 (1978).

[10] J. A. Styles, *Br. J. Cancer* **36,** 558 (1977).

[11] J. Ashby, J. A. Styles and D. Anderson, *Br. J. Cancer* **36,** 564 (1977).

[12] O. Jarrett and I. Macpherson, *Int. J. Cancer* **3,** 654 (1968).

[13] S. Shin, V. H. Freedman, R. Risser, and R. Pollack, *Proc. Natl. Acad. Sci. U.S.A.* **72,** 4435 (1975).

grows well when cultured in Dulbecco's Modified Eagle's medium[14] (DME; see GIBCO catalogue) containing 10% autoclaved Difco Bacto tryptose phosphate broth (TPB) and 10% calf serum in a humidified 10% $CO_2$ incubator at 37°. (Hereafter this combination is referred to as complete medium.) When the cell density is very low, the serum concentration is increased to 20%.

## Selection of a Low Background Subclone of BHK

In order to obtain maximum sensitivity from a transformation assay with this cell line, it is necessary to clone it so as to remove the background of spontaneous transformants and to select a clone in which they will not arise again with too high a frequency.

1. Trypsinize a culture of BHK cells, dilute, and plate them at a very low density; prepare, for example, 20–30 100-mm dishes of 200 cells per dish, using complete medium with 20% serum. Incubate the dishes at 37° for 8–11 days until very heavy single colonies can be seen.

2. Carefully select about 10 of the colonies which, when compared to their neighbors, appear to be unusually light and flat and have especially feathery edges. Pick the colonies, trypsinizing them within a cloning cylinder secured with autoclaved silicone grease, into small, 30-mm dishes containing complete medium with 20% serum. Allow them to grow and passage them whenever they become 60–70% confluent.

3. As soon as there are sufficient cells, about passage 3, freeze several vials of each clone and also test an aliquot of each by assaying the cloning efficiency in liquid and agar. Plate 200 cells/60-mm dish containing complete medium with 20% serum and plate $10^6$ cells/60-mm dish in soft agar as described below under "Soft Agar Assay." Choose for subsequent experiments a clone that gives a reasonable plating efficiency in liquid (about 20%) and the minimal plating efficiency in agar (less than one colony per $10^6$ cells).

When growing the cloned cells for use in carcinogenesis experiments, culture cells from the frozen vials, using them prior to passage 7. Maintain the cells at densities of less than 70% confluency so as to avoid providing any selective advantage to spontaneously arising transformants that can grow more effectively than normal cells in crowded cultures.

## Carcinogen Treatment

1. Prepare a fresh carcinogen stock solution, 1–10 mg/ml, in water or in dimethylsulfoxide, depending on solubility. Sterilize by filtration. Di-

[14] J. D. Smith, G. Freeman, M. Vogt, and R. Dulbecco, *Virology* **12**, 185 (1960).

lute the stock solution in TD buffer[14] (0.8% NaCl, 0.038% KCl, 0.01% $Na_2HPO_4$, 0.3% Trizma base, and HCl to pH 7.0 to give a range of concentrations that are double the final desired doses. The final concentration of dimethylsulfoxide in contact with the cells should be less than 2%.

2. Harvest by trypsinization cells that have been grown from the selected clone, wash them by centrifugation in TD buffer containing 2% calf serum, and resuspend them at $2 \times 10^6$ cells/ml in TD buffer without serum.

3. Mix equal volumes of cell suspension and diluted carcinogen, usually 1–3 ml each, in sterile tubes containing small magnetic stirring bars and stir very gently for 1 hr at 37°. Include a zero-dose control tube containing solvent and cells but no carcinogen.

4. At the end of the hour, dilute the cell–carcinogen mixture with 5–10 volumes of complete medium containing 20% serum, recover the cells by centrifugation, and wash and resuspend them in fresh medium with 20% serum.

5. To determine the amount of cell killing due to the carcinogen treatment, dilute an aliquot of the treated cells and plate in quandruplicate in complete medium with 20% calf serum at densities of 200, 2000, or 20,000 cells/60-mm dish. The number of cells added per dish depends on the expected killing. It is safest to plate at several cell densities from a given dose. Count colonies arising on these plates after incubation for 7 days at 37°.

6. Plate another aliquot of the treated and washed cells at a concentration of about $2 \times 10^5$ cells/100-mm dish in complete medium with 20% serum and grow at 37° for 4 days. If, during this time, any dishes become more than 70% confluent, the cells should be subcultured. At the end of this expression period, cultured cells from each dose should be assayed in soft agar as described below.

### The Soft Agar Assay for Transformation

1. For each culture to be assayed, prepare 10 60-mm dishes, each containing 5 ml of complete agar medium. Complete agar medium consists of 40% double-strength DME, 10% Difco tryptose phosphate broth, 10% calf serum, and 40% of an agar solution. The agar solution is prepared by mixing 1.275 g of Difco Bacto agar per 100 ml of glass-distilled water. The solution is sterilized by autoclaving and cooled to 45° in a water bath. The other ingredients are warmed to 45° and mixed with the agar. The complete agar medium can then be distributed with a 5-ml automatic Cornwall syringe from the vessel in the water bath to culture dishes.

These agar base layers should be allowed to solidify at room temperature and, if possible, used within the hour. The ingredients for a second flask of complete agar medium of sufficient volume to provide more than 11 ml per culture to be assayed should be ready for mixing just prior to the actual plating of the cells. For this second flask, filter-sterilize the calf serum to remove any clumps that could provide a growing surface for the normal cells. Keep all agar medium at 45° until used.

2. Harvest by trypsinization the cells that have been cultured from each carcinogen dose, and either strongly pipette them several times through a drawn out Pasteur pipette or force them through a 20-gauge syringe needle in order to remove clumps of normal cells that can form spurious colonies in agar. Suspend the cells at a concentration of $10^6$ cells/ml in complete medium made with filter-sterilized calf serum.

3. Assay an aliquot of the suspended cells for liquid cloning ability by plating 200 cells/per dish in quadruplicate in complete medium with 20% serum. Count visible colonies which arise after 7 days of incubation at 37°.

4. Assay a second aliquot of the suspended cells for ability to clone in soft agar by mixing 5.5 ml of the suspended cells with 11 ml of complete agar medium and pipetting the resulting slurry on top of the prepared agar plates, using 1.5 ml slurry per dish. This results in 10 dishes of $5 \times 10^5$ cells per dish with the cells suspended in a final agar concentration of 0.34%. The cell concentration, number of dishes, etc. may be varied, but the final agar concentration must be exact. Incubate the plates at 37° for $2\frac{1}{2}$–3 weeks in a well-humidified incubator until visible colonies arise. The transformed cells will form colonies that grow progressively to greater than 0.2 mm diameter and can be counted by eye or with a bacterial colony counter without staining.

5. In order to be assured that the clones seen in agar represent stable transformants, a small sample of them may be picked from the agar plates with a sterile Pasteur pipette, suspended in DME, pipetted strongly to free each colony from the surrounding agar, and transferred to dishes with one-tenth volume of TPB and one-fifth volume of calf serum. The resulting cultures should be cloned again to eliminate any normal cells inadvertently picked up with the transformed colony. After 4–5 passages, the clones may be retested by assaying them in soft agar, using $10^3$ cells per dish. Stable transformed lines should clone in agar with an efficiency of 1–40%.

If desired, it is possible to test the *in vivo* tumorigenicity of these clones in Syrian hamsters.[12,15]

[15] M. Stoker, *Virology* **18**, 649 (1962).

### Presentation and Interpretation of Data

*1. Survival Curve.* From the liquid platings done on the day of carcinogen treatment, calculate the surviving fraction and graph its log versus carcinogen dose. A shoulder in this curve may indicate repair of carcinogen damage; a sharp break in slope suggests that cell killing is going on by more than one process, sometimes due to toxic carcinogen breakdown products.

*2. Transformant Induction.* Calculate from agar plate counts the number of transformants per cell plated, divide by the liquid plating efficiency (determined at the time the cells were put into agar), and correct by subtracting the average number of spontaneous transformants estimated from the zero-dose platings, to give an estimate of the transformation frequency per survivor. A graph of the log of this value versus dose should rise rapidly and will usually plateau when the rates of induction of new transformants and cell killing come into balance.

It is also expected that a graph of the log of the surviving fraction against the transformants per survivor would, at least at low doses, be linear.[16] Estimate the number of transformants per cell initially treated. When the log of this value is graphed against dose, the resulting curve should rise significantly from its starting point, indicating an absolute increase in the number of transformants with increasing dose. The curve should then fall in parallel with the survival curve, indicating that the transformants are as sensitive to killing by the carcinogen as the remainder of the population. These calculations assume that during the growth period between carcinogen treatment and agar assay the transformants do not enjoy any growth advantage over the normal cells in the population. This assumption is reasonable if the cultures are not allowed to become crowded since chemically transformed BHK derivatives do not multiply faster than their normal parents.[8] The potencies of different carcinogens can be compared by comparing the transformants per survivor at 37% survival.

### Variations

If the action of a particular carcinogen requires that the cells be actively growing during exposure, they can be treated on plates instead of in suspension. The problem with treating growing cells is that it provides an opportunity for the carcinogen to act as a selective agent and enrich the proportion of preexisting transformants.

[16] R. J. Munson and D. T. Goodhead, *Mutat. Res.* **42,** 145 (1977).

BHK can activate effectively a variety of carcinogens, and its hydrolytic enzymes are present at a good constitutive level. However, if there is any doubt about the capacity of the cells to activate a given carcinogen, a liver extract[17] may be added before and/or during the treatment with the carcinogen.[10]

A period of growth is usually required between carcinogen treatment and challenge with soft agar to attain the maximum expression of transformation by newly induced transformants. Recently, however, a BHK subclone has been reported[6] that apparently is able to carry out the divisions necessary for maximum expression while suspended in soft agar, thereby eliminating the growth period in liquid culture. Substituting Noble agar[6,10] or agarose[18] for Difco Bacto agar may also allow one to skip the growth period in liquid. This is useful for large screening programs, although the number of background colonies is high and maximum transformation frequencies may not be achieved.

[17] B. N. Ames, W. E. Durston, E. Yamasaki, and F. D. Lee, *Proc. Natl. Acad. Sci. U.S.A.* **70,** 2281 (1973).

[18] I. Macpherson, *in* "Tissue Culture Methods and Applications" (P. F. Kruse and M. K. Patterson, eds.) p. 277. Academic Press, New York (1973).

# [25] Independent Control of the Local Environment of Somas and Neurites

*By* ROBERT B. CAMPENOT

This article describes a newly devised, three-chamber culture system that has been used for studying the development of sympathetic neurons dissociated from superior cervical ganglia of newborn rats. The neuronal somas are placed in a central chamber and send their neurites across tightly sealed barriers into separate side chambers. The medium in the three chambers can be independently altered at any time during the growth of the neurons, so that the action of agents such as nerve growth factor (NGF) can be tested separately on the somas and proximal portions of the neurites (central chamber) or on the distal portions of the neurites (side chambers). Moreover, the neurons readily can be stimulated chronically by passing current across the barriers.

## Description of Three-Chamber Culture Dishes

The three chambers are formed by a Teflon divider (Fig. 1) joined with silicone grease to the collagen-coated coverslip floor of a modified 35-mm



ISBN 0-12-181958-2

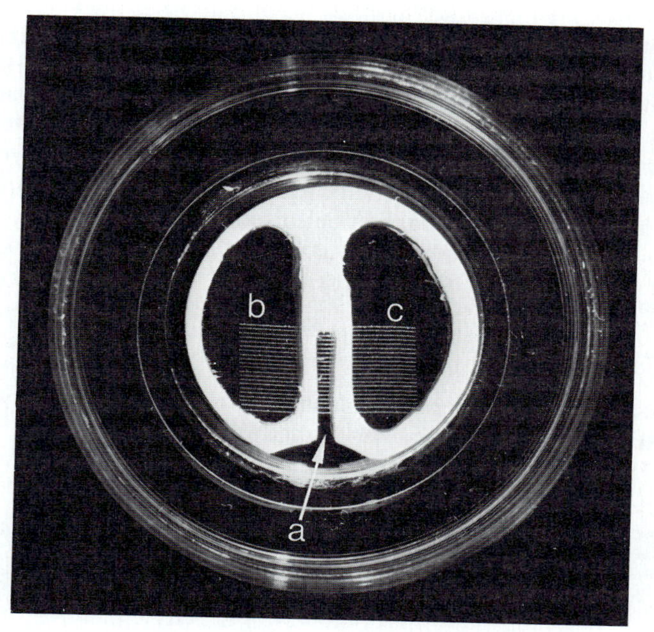


FIG. 1. A petri dish (35 mm, Falcon, viewed from above) divided into three chambers by a Teflon divider (7.5 mm high) sealed to the floor of the dish (a collagen-coated coverslip) with silicone grease. The floor of the narrow (1 × 5 mm) central chamber (a), in which the neurons are plated, is transected by 20 parallel scratches that serve to guide the growing neurites into the left (b) and right (c) side chambers.

petri dish. The grease is applied to the entire face of the divider that is to mate with the coverslip, and the two are pressed together. Before assembling the system, 20 parallel scratches are made in the collagen-coated coverslip (Fig. 1); since the growing neurites tend not to cross these scratches and are thus confined to the collagen-coated channels between them, the scratches guide the growing neurites into the side chambers (see below). Just before assemblying the system, a drop of L-15 $CO_2$ medium containing Methocel (composition given in this volume [53]) is placed on the scratched region of the coverslip; this prevents the silicone grease on the divider from adhering to the coverslip in this region. After applying the drop of medium, the Teflon divider and coverslip are gently pressed together so that the scratched, medium-coated region spans the narrow central chamber and extends into the side chambers to the left and right (Fig. 1). Although the silicone grease does not adhere to the coverslip in this region, it prevents significant bulk flow of medium between the central and side chambers; fluid-level differences of about 5 mm between the central and side chambers are supported for 4 days without appreciable equalization.

The somas of principal neurons dissociated from superior cervical ganglia of newborn rats (procedures are given in this volume [53]) are then plated into the narrow central chamber. The neurites grow parallel to the scratches and traverse the barriers, presumably in a thin film of medium between the grease and the coverslip. The neurites emerge into the side chambers after about 3 or 4 days in culture (see Fig. 2).

## Construction and Use of the System

### *Modified Petri Dishes*

Plastic petri dishes (35 mm, Falcon) are provided with polystyrene coverslip floors in a manner similar to that previously described (see this volume [53]). Briefly, a hole 22 mm in diameter is cut in the bottom of the

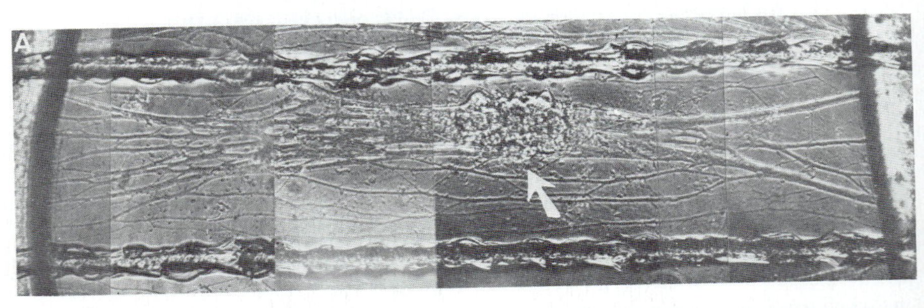


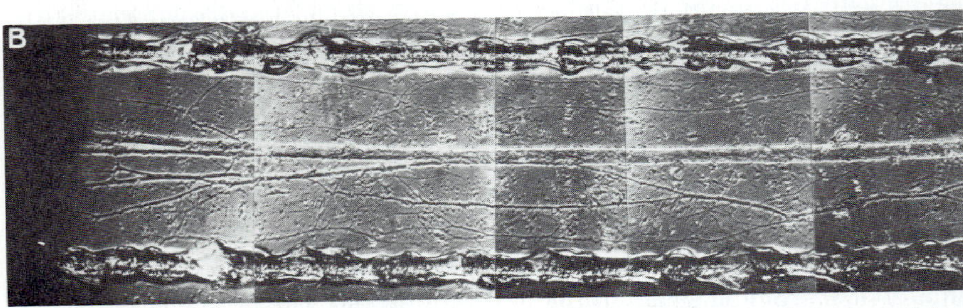


FIG. 2. Photomicrographs of 25-day-old neurons grown in a three-chamber culture dish. The upper montage (A) spans the central chamber and shows a cluster of somas (arrow) and bundles of neurites on one of the collagen channels in the central chamber. The scratches that border the 200-$\mu$m-wide channel are at the upper and lower edges, and the silicone grease seals are at the left and right edges of the montage. The lower montage (B) shows the neurite bundle on the same collagen channel emerging into the right chamber. The grease is at the left edge of this montage.

petri dish, and the coverslip is bonded over the hole on the outside suface of the dish with Sylgard 184 encapsulating resin (Dow Corning). A day or two before plating the neurons, the coverslip is coated with a dried collagen film (see this volume [53]). Twenty parallel scratches about 200 $\mu$m apart are made in the collagen-coated coverslip with a rake made by cementing together 20 insect pins; a rubber mat under the coverslip helps to make the depth of the scratches uniform. The dishes are then sterilized with ultraviolet light.

*Teflon Dividers*

Teflon dividers are machined from $\frac{3}{4}$-in. rod stock. They are 7.5 mm in height; the central chamber is about 1 mm wide and 5 mm long. The side chambers hold about 0.5 ml of medium; the central chamber communicates with the rest of the dish that holds 1–2 ml. The slotted shape of the central chamber protects the neuronal somas from disturbance when the medium sloshes during handling of the dish.

The first time the dividers are autoclaved, they are distorted and their faces must be sanded flat with fine sandpaper; subsequent autoclaving does not distort them further. Before each use, the dividers are wiped clean of silicone grease and sequentially washed briefly in Nochromix (Godax Laboratories Incorporated), in 95% ethanol for 24 hr, and in flowing, deionized water for several hours; they are then autoclaved in glass-distilled water, rinsed several times in glass-distilled water, and finally placed in a glass petri dish and sterilized by autoclaving. They must be completely dry before use.

*Silicone Grease Preparation*

Silicone grease (Dow Corning high-vacuum grease) is loaded into a 1-ml glass syringe provided with a 22-gauge hypodermic needle cut flat at the end. The grease and syringe are sterilized by autoclaving. If the grease is applied to the divider before sterilization and then autoclaved, a skin forms on the grease and the results are poor.

*Assemblying the System*

The component parts of the system are assembled under sterile conditions in a hood, with the aid of a dissecting microscope. The Teflon divider is held by a hemostat with right-angle jaws under the microscope. Silicone grease is applied to the entire face of the divider. A petri dish, with a drop of medium on the scratched region of the coverslip, is inverted and placed gently on top of the greased divider so that the scratched, medium-coated region crosses the central chamber (see Fig. 1). Final seating of the divider is accomplished by gently pressing the coverslip with a pair of fine forceps until the silicone grease is seen to adhere to the coverslip everywhere but in the medium-coated region. In the medium-coated region, the grease is pressed firmly against the coverslip but does not adhere. Then the hemostat is turned over, and the assembled system is released from its jaws. The side chambers, but not the central chamber (see below), are filled with medium, and the lid is placed on the petri dish.

*Plating the Neuron*

Dissociated sympathetic neurons from newborn rats are suspended in L-15 $CO_2$ medium and loaded into a sterile syringe. The Methocel in the medium prevents neurons from settling during the plating procedure. Approximately 50 $\mu$l of cell suspension are injected into the central chamber. To help confine the suspension to the central chamber, a small dab of silicone grease is usually placed on the coverslip at the entrance during the assembly procedure. The cultures are incubated overnight to allow the neurons to settle on the coverslip. Medium, 1–2 ml, is added to the main body of the petri dish, ensuring that confluence is established with the medium in the central chamber. Within 2 days after plating, neurites on the channels between the scratches in the central chamber have reached the grease seals to the left and right. After 3–4 days the neurites emerge from the seals into the side chambers (see Fig. 2).

*Effectiveness of the Seal*

Silicone grease seals prepared in the above manner will support a 5-mm difference in fluid levels between the central and side chambers for at least 4 days without appreciable equalization. However, they have low electrical resistance, and thus are permeable to ions. Evidently, a thin film of medium remains between the grease and the coverslips. Doubtless, molecules diffuse through this film from one chamber to another, and unequal fluid levels between the chambers must produce at least a slight bulk flow of medium. Tracers have not yet been used to assess the effectiveness of the seals, but experiments on the local effects of NGF on the neurons suggest strongly that biologically significant amounts of this protein do not cross the barriers.[1] (a) It was found that neurites will not enter a chamber to which no NGF has been added even though the chamber containing the somas is supplied with a high concentration of NGF. (b)

[1] R. B. Campenot, *Proc. Natl. Acad. Sci. U.S.A.* **74,** 4516–4519 (1977).

NGF withdrawal from a chamber into which the neurites have grown stops the growth of the neurites; it also appears to result in slow degeneration of the neurites, even though the somas and proximal portions of the neurites continue to have access to NGF. (c) When NGF is withdrawn from the chamber containing the somas, the neurons appear to survive only if their neurites have crossed into chambers where NGF is continuously available. In these experiments, the fluid level in the NGF-free chambers was always kept at a higher level than that in adjacent chambers that contained NGF, thereby eliminating transfer of NGF by bulk flow.

### *Long-Term Electrical Stimulation of the Neurons*

Although the barriers do not have high electrical resistance, they restrict current flow enough so that current passed between the chambers can stimulate the crossing axons to produce action potentials. This was shown by recording the membrane potential of the somas in the central chamber with intracellular microelectrodes while passing current between the chambers with platinum wires immersed in the medium. In virtually every neuron, action potentials were produced by pulses of 1 msec duration and amplitudes of 1–5 V; in most cases, the threshold was slightly less than 1 V. The stimulation was apparently effective on a long-term basis without any damage to the neurons. One culture was stimulated for 3 days at 1/s (pulse amplitude of 4 V). After the 3 days, the threshold voltage for action potentials was about 1 V. In another culture, stimulation at 4 V and 1/s was begun the first day after the neurons were plated, i.e., before the neurites had crossed the barriers; the neurites crossed the barriers while the stimulation was in progress, indicating that the stimulation was not harmful. Using this method of stimulation, it has been shown that electrical activity during development can determine whether sympathetic neurons develop cholinergic or adrenergic transmitter function.[2]

## Acknowledgments

This work was performed in the laboratory of Edwin J. Furshpan and David D. Potter. Support was provided by National Institutes of Health Research Grants NS-02253, NS-03273, NS-11576, and Training Grants NS-07009 and NS-07112.

[2] P. A. Walicke, R. B. Campenot, and P. H. Patterson, *Proc. Natl. Acad. Sci. U.S.A.* **74,** 5767–5771 (1977).

# [26] Mutant Isolation

*By* LARRY H. THOMPSON

During the last decade somatic cell genetics has developed into a recognized discipline—largely because of the development of techniques for somatic cell hybridization and methods for the isolation of a variety of mutant phenotypes. Already the spectrum of mutant cell lines available is sufficient to have a substantial effect on approaches to studying metabolic regulation and growth control in animal cells. In this article the term "mutant" will be used to refer to heritable phenotypic changes that appear to arise from changes in DNA structure, as evidenced by stability, response to mutagens, the presence of altered gene product, and other criteria. Despite some earlier concerns, an epigenetic basis for variant phenotypes has received little documentation and does not appear to be a major complicating factor in the isolation and use of mutations. Most mutant lines that have been isolated behave in a reproducible way and are suitably stable in phenotype, provided a minimal number of technical precautions are taken.

The diploid nature of somatic cells undoubtedly remains a major barrier to isolating mutants. Since the great majority of mutations are recessive when tested in somatic cell hybrids, such mutations will only be expressed in autosomal genes when both alleles are altered. There is some evidence suggesting that certain established cell lines, particularly the well-studied Chinese hamster ovary (CHO) lines, may contain significant hetero- or hemizygosity.[1,2] The precise extent of this functional haploidy, however, is still unclear since other evidence indicates that the cells behave essentially as diploid.[3] In general, when planning a mutant isolation, one should assume that the gene coding for the protein (or RNA) molecule of interest is probably present in two copies. This constraint does not imply that mutants for such loci cannot be isolated, but it does suggest that a *much greater* effort will be required than in the case of an X-linked function such as the commonly studied hypoxanthine phosphoribosyltransferase (HPRT).

Because many mutants are found at very low frequency, even after potent mutagenesis, the probability of a successful isolation may be directly proportional to the magnitude of the experiment, i.e., to the number

[1] L. Siminovitch, *Cell* **7,** 1 (1976).

[2] L. Siminovitch and L. H. Thompson, *J. Cell. Physiol.* **95,** 361 (1978).

[3] M. J. Siciliano, J. Siciliano, and R. M. Humphrey, *Proc. Natl. Acad. Sci. U.S.A.* **75,** 1919 (1978).


ISBN 0-12-181958-2

of cells screened. Herein lies a major difficulty. At best, one can handle about $10^8$ cells per liter of medium under selective conditions; in monolayer culture this number of cells would typically be distributed among 50 or more 100 mm-petri dishes. These plating requirements set a practical lower limit on detectable mutant frequencies at $\simeq 10^{-9}$, i.e., one mutant from 500–1000 dishes (5–10 liters of medium). Thus, the isolation of mutations, especially of the temperature-sensitive conditional class, at loci that require simultaneous alteration in both alleles will often be very difficult. However, for certain "recessive" mutations it is possible to obtain the homozygous (−/−) state by a two-step process because the heterozygous (+/−) state is selectable on the basis of a slight difference in phenotype (such as degree of drug resistance, discussed subsequently for the *aprt* locus) from the wild type. Only a few mutations having dominant (or incompletely dominant expression) have been identified; these include resistance to α-amanitin, ouabain, Methotrexate, ricin, and colchicine.

## Choice of Cell Line and Culture Requirements

### *Chinese Hamster Ovary Cells as a Model System*

The biochemist who wants to select mutants and is not committed to a particular cell system should choose a line that combines the most favorable genetic properties with ease of handling. Chinese hamster cells are commonly used because of their favorable karyotypic properties. The Chinese hamster ovary (CHO) line has been one of the most popular, partly because of its early introduction as a system for isolating auxotrophic mutations.[4] To date, more than 50 different recessive mutations have been isolated in CHO cells, including purine and pyrimidine auxotrophy,[5,6] lectin resistance,[7] and temperature-sensitive conditional mutations for aminoacyl-tRNA synthetases.[8] This variety of mutations has established the genetic utility of CHO strains and provides substantial background information for attempts to isolate new mutant classes.

Although extensive chromosomal rearrangement is evident in the near-diploid CHO cell line, the karyotype appears to be reasonably stable from cell to cell within a clone, and upon repeated subcloning.[9] This stability should minimize the opportunity for phenotypic variation poten-

[4] F.-T. Kao and T. T. Puck, *Proc. Natl. Acad. Sci. U.S.A.* **60**, 1275 (1968).
[5] D. Patterson, *Somat. Cell Genet.* **2**, 189 (1976).
[6] D. Patterson and D. V. Carnright, *Somat. Cell Genet.* **3**, 483 (1977).
[7] P. Stanley and L. Siminovitch, *Somat. Cell Genet.* **3**, 391 (1977).
[8] G. M. Adair, L. H. Thompson, and P. A. Lindl, *Somat. Cell Genet.* **4**, 27 (1978).
[9] R. G. Worton, C. C. Ho, and C. Duff, *Somat. Cell Genet.* **3**, 27 (1977).

tially associated with chromosomal change. Mutant lines that have been carefully examined were found to resemble closely the wild type.[9]

Another attractive feature of CHO cells is that they are not unreasonably fastidious in terms of culture requirements, and they can be grown under a variety of conditions. A high plating efficiency, an important property for quantitative mutant isolation, is routinely achievable. Unlike most other hamster lines, CHO cells can be grown in liquid suspension culture, in soft agar, on top of soft agar,[10] as well as in monolayer. In $\alpha$-MEM medium lacking nucleosides (K. C. Biological, Inc., Cat. No. DM-325) but containing 10% fetal bovine serum CHO cells have a doubling time of about 12 hr both in monolayer and suspension culture.

Not surprisingly, a number of different wild-type CHO strains exist, having different karyotypes, growth rates, morphology, ease of growth in suspension, and other properties. This variety should be kept in mind when choosing a strain for mutant isolation. There is as yet no evidence that any particular strain is generally more mutable than another, except in some specific instances where certain clones behave as though they are hemizygous at a specific autosomal locus that was not intentionally subjected to selection pressure.[2]

### *Suspension vs. Monolayer Culture*

Since suspension culture methodology is less common than monolayer culture, it is worth emphasizing here some of the major advantages of liquid suspension as it relates to mutant isolation: (1) In suspension, cells such as CHO are grown either in roller tubes that revolve on a drum, or in spinner flasks in which the cells are kept suspended by stirring bars. Since the cells exist as single cells, culture growth rate can be monitored easily by sampling aliquots. This means that cultures for experimentation can be readily standardized with reference to cell concentration. (2) The production of bulk quantities of cells for mutant characterization is more readily achievable because large spinner cultures are commercially available. (3) Precise temperature control is much easier in water baths than in forced-air incubators; the selection and characterization of temperature-sensitive mutants depend on good control, especially for restrictive temperatures.

### *Potential Problems with Media and Sera*

Since the difficulty of quality control of serum-supplemented culture medium is probably the greatest limitation in doing quantitative cell cul-

[10] M. W. Konrad, B. Storie, D. A. Glaser, and L. H. Thompson, *Cell* **10**, 305 (1977).

ture, a few comments are in order. Not only should the serum supplement be pretested, but it should be tested after dialysis if the use of dialyzed serum is anticipated. Dialysis often impairs the growth-supporting activity of serum. It is not uncommon for mutant lines (auxotrophic, temperature-sensitive, etc.) to grow more poorly than wild type in certain serum lots. Thus, it is important to select lots of sera that provide vigorous growth of wild-type cells.

*Strategy and Timing for Mutant Isolation*

In terms of typical biochemical experiments, cell culture and mutant isolation procedures may seem slow and tedious. Because of the length of time required to obtain results, the importance of good incubators, rigorous sterile technique, and high-quality culture medium should not be underestimated. The essential steps for isolation of mutants are given in Table I along with an indication of the approximate minimum time required for each step. It is advisable to freeze cultures at intermediate stages, such as before recloning, to guard against loss from contamination or other problems. Although cells can be stored temporarily at −80°, liquid nitrogen temperature is required for maintaining viability for a period of years.

TABLE I
GENERALIZED OUTLINE FOR MUTANT ISOLATION

| Steps in the isolation of a mutant | Approximate number of cell generations |
|---|---|
| 1. Treatment of culture with mutagen (mutagenesis) | 1 |
| 2. Growth of the cells for fixation and expression of mutations | 6 |
| 3. Treatment of culture with selective agent and outgrowth of presumptive mutant colonies | 12 |
| 4. Growth of isolated colonies to mass cultures ($2 \times 10^7$ cells) | 12 |
| 5. Subculture and growth of clones for biochemical assay and other tests for drug sensitivity, mycoplasma, etc. | 2 |
| 6. Recloning of good mutants to ensure genetic homogeneity | 24 |
| 7. Freezing of multiple aliquots of freshly cloned mutants, preferably in liquid nitrogen | — |
| Total time | 57 generations (43 days[a]) |

[a] Assuming an average generation time of 18 hr.

## Induction of Mutants

### *Should a Mutagen Be Used?*

A mutagen is not recommended unless it is necessary in order to obtain the desired phenotype from a reasonable number of cells. Potential complications with the mutagen may occur from DNA damage at many other sites than the target gene; such secondary mutations may affect the growth rate or other properties of the cells, thereby interfering with the characterization of the mutant phenotype. When a mutagen is not used, it is important to remember that the frequency of a particular mutation will be higher, on the average, in cultures that have not been recently cloned or diluted heavily.[11] However, in most instances a mutagen will be necessary to elevate the mutant frequency to a detectable level.

### *Choice of Mutagen and Conditions of Exposure*

Commonly used mutagens for mammalian cells include MNNG (*N*-methyl-*N*′-nitro-*N*-nitrosoguanidine), MMS (methyl methanesulfonate), EMS (ethyl methanesulfonate), ENU (ethyl nitrosourea), and ultraviolet (UV) light. EMS has been the most popular because it is easy to use and has proved reliable for enhancing the mutation frequency at a large number of loci, producing deficiency mutations in which enzyme activity is lost (such as 6-thioguanine resistance or numerous auxotrophies) as well as phenotypes that are presumed to involve primarily missense mutations (temperature sensitivity, or reduced affinity of a protein for an inhibitor such as $\alpha$-amanitin or Methotrexate).

All chemical mutagens should be handled carefully since most are known to be carcinogenic. The following standard procedure for EMS mutagenesis of CHO cells is recommended as an example:

1. An exponentially growing suspension culture is collected by centrifugation and resuspended in fresh medium at a concentration of $2\text{–}3 \times 10^5$ cells/ml.

2. EMS (Sigma Chemical Co.) (density = 1.2 g/ml), diluted to a concentration of 1% in water or serum-free medium, is added to the culture to give a final concentration in the range of 200–300 $\mu$g/ml. The exact concentration depends on the particular clone of CHO and the composition of the medium, such as the presence of certain ribonucleosides in the medium, which may enhance the sensitivity to mutagen. For an optimal frequency of mutant induction without excessive damage to the cells, a surviving fraction after exposure of 0.1 or higher should be used.

[11] L. H. Thompson and R. M. Baker, *Methods Cell Biol.* **7**, 209 (1973).

3. After incubation for 16 hr the mutagen is removed. This is accomplished by centrifugation and resuspension in buffered saline to rinse the cells, and then suspension in fresh medium at a reduced cell concentration to allow for growth of the culture.

*Requirement for Mutation Expression*

A period of recovery from the mutagen is necessary to allow fixation of the mutations and outgrowth of the surviving cells. Full expression of a mutant phenotype generally requires dilution of preexisting gene product by successive cell divisions. The number of divisions necessary for maximal expression of new mutations varies considerably with the phenotype being selected. However, in most cases a 50- to 100-fold increase in cell number (5–7 doublings) should be adequate. An excessively long interval will result in the selective loss of mutants if their growth rate is less than that of wild type.

During this mutation expression interval it is very important to maintain a culture of adequate size by low ratio dilution during subculture so that mutants are not lost. Thus, when the survival is low the size of the culture should be expanded at the first subculture so that mutants have an opportunity to multiply. For the actual selection, some fraction of the total number of cells can then be screened. If the selection efficiency for mutants is expected to be low, e.g., 10% as in [$^3$H]amino acid suicide discussed in a subsequent section, it is desirable to screen a proportionately larger number of progeny cells than the number of survivors after mutagenesis treatment since a large fraction of the mutants will be killed by the selection.

*Independent Cultures*

It is often important to know that individual mutants represent independent mutational events and not siblings. In order to achieve this condition, the experiment should be designed so that separate cultures are mutagenized or that a large culture is subdivided immediately following mutagen treatment before there has been opportunity for mutations to become replicated as siblings within the population.

## Direct Selection of Drug-Resistance Mutants

As a general class of mutants, drug-resistance mutations are among the easiest to isolate because the selective situation favors the direct outgrowth of the mutant clones. This approach can be used in any situation in which a toxic compound can be directed at a specific target molecule, including cases in which the mechanism of toxicity is unknown. Examples of these mutants that have been isolated in CHO cells include resistance to lectins, Methotrexate, colchicine, α-amanitin, histidinol, emetine, 8-azaadenine (or 2,6-diaminopurine), ouabain, hydroxyurea, cycloheximide, 5-bromodeoxyuridine, 6-thioguanine (or 8-azaguanine), azetidinecarboxylic acid (proline analog), diphtheria toxin, and excess adenosine.

In all of these cases mutants were isolated as clones growing in monolayer under plating conditions. The isolation of resistant cells by continuous exposure and subculturing in the presence of the drug is not recommended because of the difficulty of interpreting the number of genetic changes associated with a phenotype. Since many laboratories will be interested in isolating new classes of mutants, the present discussion will focus on the principles and problems involved in drug selections rather than the use of particular drugs.

*Choice of Drug Concentration*

Because a discussion of this point has been given previously,[11] only the most important aspects will be reiterated here. As indicated above, drug selections are accomplished in monolayer in order to ascertain the individual mutant clones by visual examination and microscopy. A drug concentration is chosen at which wild-type cells are satisfactorily killed while the mutant cells of interest survive with high efficiency. The use of too high a concentration may well result in a failure to obtain any mutant colonies. This can be illustrated nicely with the results that are obtained in selections carried out for 8-azaadenine resistance, which is normally associated with loss of activity of the autosomal enzyme, adenine phosphoribosyl transferase. A mutant cell in which one of the two alleles is altered to produce defective enzyme is usually expected to have ≥50% of wild-type transferase activity; such a heterozygous mutant remains drug-sensitive. However, the mutant will be slightly more resistant than wild-type cells—probably, at most, 2-fold in terms of drug concentration. On this basis the heterozygous mutant can be discriminated from the wild type if the drug dose is chosen carefully. At a surviving fraction of $\simeq 10^{-5}$ in the drug ($\simeq$8 μg/ml in dialyzed fetal calf serum), a sizeable percentage of the large colonies will actually possess increased resistance, and will thus be presumptive heterozygotes with respect to the enzyme.[11a] Under these conditions, the choice of drug dose is influenced by the number of alleles one expects to be associated with the gene locus under considera-

[11a] G. E. Jones and P. A. Sargent *Cell* **2**, 43 (1974).

tion. In this example of selecting a presumptive heterozygote, it is essential that the drug concentration be carefully adjusted for the desired survival value. It is also necessary to demonstrate explicitly an increased drug resistance and reduced enzyme activity for the presumptive mutant colonies after they have been isolated since some colonies may not be mutant for statistical reasons,[11] and resistance may involve other loci.

Often, more than one gene locus is involved in the mutation to resistant phenotypes, as in the case of Methotrexate,[12] diptheria toxin,[13] or colchicine;[14] with each agent, isolation of mutations in the intracellular target protein is complicated by the presence of permeability variants. Since different genetic classes of resistance may occur at different drug concentrations, the drug dose must be chosen accordingly.

*Details of Plate Inoculation and Incubation*

Standard procedure involves inoculating 100-mm plastic petri dishes with cells in drug medium and incubating until macroscopic clones are isolatable. The maximum number of cells per dish that can be plated must be arrived at empirically since it depends on such factors as the cell type, rate of cell lysis, and extent of metabolic interaction between mutant cells in contact with wild type. The phenomenon of contact feeding (metabolic cooperation) can greatly suppress the recovery of mutant colonies when the cells are too dense; this has been a common complication with purine analogue resistance selections.[15] In general, the cell inoculum for a 100-mm dish with 20 ml of culture medium will be in the range of $1 \times 10^5$ to $1 \times 10^6$ cells. After the dishes are filled with drug medium, the cells are introduced in a small volume, e.g., 0.1–1 ml of a concentrated suspension. It is important to distribute the cells uniformly on the dish. This is accomplished by gently sliding the dishes back and forth in one direction, then applying the same motion in the orthogonal direction. This manipulation should be done very soon after pipetting the cells onto the dish because they can attach within minutes, even at room temperature. Caution must be exercised since medium spilled to the outside of the dish is an invitation for contamination.

Especially with CHO cells, it is important not to move the dishes during the colony formation interval because of the severe problem of satellite colony formation. If a dish containing colonies is moved and then returned to the incubator, one typically gets many migrating cells that

[12] W. F. Flintoff, S. V. Davidson, and L. Siminovitch, *Somat. Cell Genet.* **2,** 245 (1976).
[13] T. J. Moehring and J. M. Moehring, *Cell* **11,** 447 (1977).
[14] J. E. Aubin, A. Chase, F. Sarangi, and V. Ling, *J. Cell Biol.* **72,** No. 2, Part 2, 298a (1977).
[15] C. M. Corsaro and B. R. Migeon, *Exp. Cell Res.* **95,** 39 (1975).

were loosely attached or floating over the colony but then reattached elsewhere on the dish. The dishes should be moved only if it is absolutely necessary to replenish the medium because of drug instability or other problem. The presence of large numbers of dead cells on the dish is not esthetically pleasing but generally does not interfere with the selective conditions and can be ignored.

Mutant colonies that are to be isolated for characterization should be taken while fairly small, approximately 1 mm in diameter, to minimize the risk of cross-contamination associated with large colonies on the dish. At a normal growth rate, CHO cells produce this size colony in about 6–7 days, but mutants under drug selection may require incubation of 2–3 weeks. In any mutant selection, control dishes with cells in drug-free medium should be set up with 200–300 cells per dish in order to determine the plating efficiency. This control is important for calculating the mutant frequency, and it gives an indication that the incubation conditions were valid. Control dishes are usually stained sooner than selection dishes and should, therefore, be placed on separate incubator trays to avoid the problem of colony dispersal mentioned above. Procedures for isolating the colonies are discussed in the final section.

### *Multistep Selections*

When either more than one gene or more than one allele can be altered to effect resistance, it may be desirable to produce highly resistant cells by performing additional selections at increasing drug concentration on clonal isolates. For example, it might be necessary to obtain a permeability mutant initially, before selecting for an alteration in an intracellular molecule. Mutagenesis can be carried out as needed at each round of selection.

## Indirect Selection of Conditionally Lethal or Auxotrophic Mutations

Auxotrophic mutants and conditionally lethal (mainly temperature-sensitive, ts) mutants must be isolated either by nonselective procedures such as replica plating, or by schemes that preferentially kill the wild-type cells during temporary exposure of the population to selective conditions that allow for reversible arrest of the mutants. Most of the selective agents that have been used are ones that act during the *S* phase of the cycle; included are cytosine arabinoside, 5-fluorodeoxyuridine, tritiated thymidine (high specific activity), and 5-bromodeoxyuridine (BrdUrd) followed by exposure to visible light. The last method, developed by Puck and co-workers,[16] has been very successful for the isolation of many different auxotrophic mutants defective for steps in the synthesis of purines and pyrimidines, cholesterol, unsaturated fatty acids, amino acids, and carbohydrate metabolism. The BrdUrd/light selection methodology has been described in detail[17] and will not be reviewed here. However, it should be noted that a recent modification of the procedure greatly reduces the time required for light exposure and may also improve the selection efficiency.[18] The DNA-binding dye, 33258 Höechst, added to cultures 2–3 hr before light exposure, greatly sensitizes the BrdUrd-substituted DNA to photolysis. As a general class of mutants, auxotrophic mutants have been particularly useful because they are often readily identifiable.

Nearly all of the methods for selecting mutants that are temperature sensitive (ts) for growth are similar in principle to the auxotrophic selections, but the mutants obtained usually prove to be extremely difficult to characterize in terms of the molecular defect. Generally speaking, mutants that cannot be identified biochemically are of somewhat limited usefulness. For this reason, procedures for the isolation of *nonspecific* ts mutants will not be detailed here. The reader can refer to previous discussions that describe the use of tritiated thymidine[11] and 5-fluorodeoxyuridine[19] as selecting agents for such mutants.

### *Selection of Protein Synthesis Mutants with [$^3$H]Amino Acids*

The selection procedures for mutants with defects in protein synthesis are the only methods that have been developed that recover specifically ts mutants for a particular component of metabolism. Two kinds of procedures have been used: (1) killing of wild-type cells with tritium suicide from [$^3$H]amino acid incorporation,[20] and (2) killing by the incorporation of cytotoxic amino acid analogs.[21] The first approach will be described in detail here since it has been used extensively on CHO cells to generate a large number of mutants.[8] Although the procedures were designed with the intention of selecting a broad spectrum of protein synthesis mutations, essentially all of the mutants obtained to date from such selections have been identified as aminoacyl-tRNA synthetase mutants.[22] Despite the var-

[16] T. T. Puck and F.-T. Kao, *Proc. Natl. Acad. Sci. U.S.A.* **58,** 1227 (1967).
[17] F.-T. Kao and T. T. Puck, *Methods Cell Biol.* **8,** 23 (1974).
[18] G. Stetten, S. A. Latt, and R. L. Davidson, *Somat. Cell Genet.* **2,** 285 (1976).
[19] C. Basilico and H. K. Meiss, *Methods Cell Biol.* **8,** 1 (1974).
[20] L. H. Thompson, C. P. Stanners, and L. Siminovitch, *Somat. Cell Genet.* **1,** 187 (1975).
[21] J. J. Wasmuth and C. T. Caskey, *Cell* **9,** 655 (1976).
[22] L. H. Thompson, D. J. Lofgren, and G. M. Adair, *Cell* **11,** 157 (1977).

ious degrees of temperature sensitivity displayed by these mutants, nearly all have an altered dependency for the corresponding amino acid. Thus, the recovery of particular classes of synthetase mutants can be favored or discouraged by changing the levels of the appropriate amino acids during the selection. In the procedure described below, temperature sensitivity is treated as the primary phenotypic property, and emphasis is placed on testing for altered amino acid responses to identify the mutants.

1. CHO suspension cultures at 34° in exponential growth at a concentration of $3 \times 10^5$ cells/ml, mutagenized and allowed to express as described in a previous section, are used to initiate the selection. A minimum of $1 \times 10^8$ cells should be used, but several times this number may be required to obtain several different classes of mutants.

2. The cells are harvested by centrifugation, rinsed thoroughly with a large volume of phosphate-buffered saline, and resuspended at a concentration of $3 \times 10^6$ cells/ml in selection medium. This medium must be lacking in the amino acid that is to be added as the selecting agent (see below); therefore, *dialyzed* fetal bovine serum is used, at the regular concentration of 10%. The other 19 amino acids are normally reduced to 10% of their standard concentrations (×0.1) since mutant frequencies appear to be lower when normal amino acid levels are used. When a particularly common class of synthetase mutant, e.g., asparagyl-tRNA synthetase, is to be excluded, its selection efficiency is specifically reduced by raising the concentration of the cognate amino acid to ×1 in the selection medium. (See Adair *et al.*[8] for an example of this effect.)

3. The culture in selection medium is placed at the restrictive temperature of 39.5° (±0.2° control or better) for 4 hr.

4. The [$^3$H]amino acid used for selecting ([$^3$H]valine, [$^3$H]lysine, or others) at a specific activity of ~18 Ci/mmol (available from Amersham-Searle in sterile aqueous solution at 1 mCi/ml concentration) is added directly to the culture at a final concentration of 20 μCi/ml and incubation at 39.5° continued.

5. After 4 hr incubation with the [$^3$H]amino acid, the culture is centrifuged and resuspended at $1 \times 10^6$ cells/ml in regular medium and incubated at the permissive temperature of 34° for 1 hr to promote recovery of protein synthesis in mutant cells and to remove the isotope from the intracellular pool.

6. The culture is placed on a roller wheel (Model TC7, New Brunswick Scientific) at 2–4° for 4 days to allow the accumulation of radiation damage sufficient to kill the wild type to a surviving fraction of $\sim 1 \times 10^{-7}$. The nonspecific killing from cold storage for mutant or wild type is minimal, i.e., less than 20%.[8]

7. The cells are plated into 100-mm petri dishes containing 30 ml of

medium by pipetting directly 1.0 ml of cell suspension into each dish and dispersing it.

8. The dishes are incubated at 34° for 16–18 days before isolating the colonies. (See section below on clone isolation.) Other versions of the above procedure can presumably be designed to obtain a specific synthetase mutation by conducting the experiment at a single temperature and altering a particular amino acid concentration to a low level in the selective medium, and to a high level during mutagenesis expression and colony formation. Such a procedure should select for specific amino acid hyperauxotrophs—mutants that require an abnormally high concentration of an amino acid for growth. These will likely prove to be temperature sensitive also in many cases.

*Screening for Mutant Phenotypes*

The testing of the colonies is generally done in two stages. The first test identifies those colonies that are ts mutants, and it may also indicate whether there is an altered amino acid dependency. Each colony is picked and transferred into a tube containing 6 ml of regular $\alpha$-MEM medium, pipetted several times for dispersal, and inoculated as 2-ml aliquots into three replicate wells (CoStar 24-well cluster dish 2534) in separate trays. All replicates are first incubated at 34° for 4–5 days to allow for cell multiplication, but to a density substantially subconfluent. One tray remains as the control at 34° in standard medium while the other two are placed at 39.5°. One of the latter is kept in regular medium and the other changed to medium in which all 20 amino acids are reduced to $\times 0.1$; some mutants die only in medium with reduced amino acids. The responses in $\times 1$ and in $\times 0.1$ amino acids at 39.5° are closely compared for an indication as to whether a mutant is ts, hyperauxotrophic, or both. Cultures under test should be monitored daily for the first 3 days by microscopic examination for cessation of cell division and lysis. Complete cell lysis by 24 hr is characteristic of strong protein synthesis mutants. Wells at 39.5° that exhibit growth inhibition or lysis are compared to the control well at 34° before terminating the test at about day 7. It is important to monitor the rate of lysis carefully in the two media at 39.5° since an early difference may reflect an amino acid dependency that will not be seen subsequently if lysis has occurred in both media. Typically, about 20% of the colonies are mutants under these conditions.

More extensive amino acid tests on clones identified as mutants usually suggest which aminoacyl-tRNA synthetase is defective. This identification is based on either (1) a protective effect from an excess of a particular amino acid at 39.5° (or growth stimulation at 34°), or (2) growth inhibition/killing resulting from reduced amino acid at either 34° or 39.5°, depending on the mutant. Some mutants die only under the doubly restrictive conditions of 39.5° and reduced amino acid. These tests are done by raising or lowering individual amino acids by a factor of 10 compared to the standard $\alpha$-MEM concentrations. The tests are also performed in the 24-well cluster trays by inoculating with 20 different media containing elevated (or reduced) amino acids. The remaining four wells are used for control medium with normal or 10-fold uniformly reduced amino acids. To each well, $5 \times 10^3$ cells are added in a small volume of medium or saline. Thus, for a given mutant as many as four trays are set up including both $\times 10$ and $\times 0.1$ amino acids at each temperature. The wells are stained after about 7 days incubation at 39.5° and 8 days at 34°. An abnormal response to a particular amino acid is evident from the staining intensity.

The preparation of media with reduced amino acids is greatly facilitated by reconstituting an amino acid-free basal medium with concentrated, sterile stock solutions of the 20 amino acids. Eighteen of the amino acids can be prepared as $\times 400$ stock solutions; these are stored at room temperature or 2°, except for asparagine and cysteine which are frozen. Glutamine is stored as a frozen stock solution at $\times 100$ and, for convenience, tyrosine is treated as a "$\times 400$" liquid stock although it is insoluble and forms a fine suspension. The $\times 400$ stocks can be combined further, three or four at a time, to $\times 100$ stocks.

## Replica Plating Procedures

The lack of simple, rapid methods for replica plating mammalian cell colonies has been a major limitation in isolating auxotrophic, temperature-sensitive, DNA repair, and other classes of mutations that may be difficult to obtain by selective methods that enrich for nongrowing cells. For cell lines that grow in suspension, some attempts have been made to replicate colonies *on* soft agar using velveteen cloth in a manner similar to that used so successfully with microorganisms on agar. However, the only method[23] introduced to date that appears to have potential is one that involves making a replicate of colonies attached to a petri dish by using a fine-mesh nylon cloth that is laid down on the colonies for a period of 3–4 hr. Cells from each colony attach to the cloth, which is then transferred to a second dish for further incubation. It appears that few laboratories have seriously evaluated this method; probably conditions will have to be optimized for each cell line used. Our laboratory has just begun to utilize this approach; it appears that for CHO cells a polyester cloth (Pecap HD7-10 Super, Tetko Inc.) may work better than nylon

[23] T. D. Stamato and L. K. Hohman, *Cytogenet. Cell Genet.* **15,** 372 (1975).

because it is more rigid and less water absorbing. With this procedure it seems possible to screen the colonies from $10^4$–$10^5$ mutagenized cells when as many as 500–1000 colonies per dish are replicated. While it is not clear that more than one useful replicate can be made, even one would facilitate the isolation of mutants such as the UV-sensitive clone of CHO-Kl cells that has been reported.[24]

## Mutant Colony Isolation and Cloning

### *Isolating Colonies from Petris Dishes*

As indicated earlier, when colonies are being taken from petri dishes as presumptive or prospective mutants, it is preferable to do this when they are about 1 mm in diameter; larger colonies are more prone to shed loose cells. Our standard procedure involves removing the colonies with a Pipetman autopipette, using the 0.2 ml plastic disposable tips that are autoclavable. The colonies should be rinsed twice with serum-free medium to remove loose cells *as soon as* they are taken from the incubator, and then circled with a marking pen. Each colony is removed simply by scraping the circled area with the pipette tip while applying simultaneous suction by slow release of the plunger. This effectively removes the cells with little damage. A dish should be rinsed once or twice again before each additional colony is taken to minimize the chance of cross-contamination. The purity of colonies isolated in this manner is sufficiently high for effective mutant screening when done with the precautions indicated. The method is very simple and amenable to the isolation of large numbers of colonies using standard laboratory equipment. Another, perhaps more rigorous, method involves isolating each colony with a stainless-steel cylinder by using autoclaved vacuum grease, and then trypsinizing the cells.[17,25]

### *Cloning in Multiwell Plastic Trays*

To ensure genetic homogeneity, it is desirable to clone rigorously cultures that have been identified as mutant. This is done conveniently with 96-well cluster trays having good optical quality, e.g., the CoStar (#3596) or Nunc (#N-1480) trays. Wells are inoculated with 0.2 ml of a cell suspension in medium such that about one well in every 3 to 4 receives a viable cell. The low concentration minimizes the probability of getting two cells in one well. When the colonies are recognizable with an inverted

[24] T. D. Stamato and C. A. Waldren, *Somat. Cell Genet.* **3,** 431 (1977).
[25] T. T. Puck, P. I. Marcus, and S. L. Cieciura, *J. Exp. Med.* **103,** 653 (1956).

microscope, they should be examined and marked so that double colonies that might become overlapping are excluded. When the colonies have reached macroscopic size, they are removed by trypsinization and transferred to a larger culture vessel.

### Acknowledgment

This work was performed under the auspices of the U.S. Department of Energy, Contract No. W-7405-ENG-48.

### Notice

This report was prepared as an account of work sponsored by the United States Government. Neither the United States nor the United States Energy Research & Development Administration, nor any of their employees, nor any of their contractors, subcontractors, or their employees, makes any warranty, express or implied, or assumes any legal liability or responsibility for the accuracy, completeness or usefulness of any information, apparatus, product or process disclosed, or represents that its use would not infringe privately-owned rights.

# [27] Karyotyping

*By* RONALD G. WORTON and CATHERINE DUFF

## Introduction

Karyotyping should not be an endeavor restricted to the cytogeneticist. Rather, it is an extremely valuable research tool for anyone who works with cell cultures, regardless of the primary field of interest.

Most biologists know that the genome of higher plants and animals is organized into discrete chromosomes, that the chromosomes can be set out in groups to form a karyotype, and that the karyotype is characteristic of the species. Many will be less familiar with the fact that in permanently established cell lines there can be extensive karyotypic variation from line to line derived from the same species, or from cell to cell within a line. The biochemist, therefore, should be aware of this genetic heterogeneity in order to properly interpret cell culture data, and may on occasion use this heterogeneity to advantage in designing new experiments.

When tissues are explanted and put in culture, the cells that grow generally display the karyotype of the individual as, for example, in direct bone marrow cultures, short-term lymphocyte cultures, or primary fibroblast cultures. However, when cell cultures become established the rigid control over the karyotype is generally lost, and chromosomal variation




ISBN 0-12-181958-2

becomes commonplace. Because of the variation in chromosome number one usually characterizes an established line by its modal number rather than its absolute number of chromosomes. In general, when the modal number is close to the normal diploid number the variation about the mode is less than when the modal number is greatly elevated. The former are often termed pseudodiploid or quasidiploid cell lines, whereas the latter are often referred to as heteroploid cell lines.

In addition to numerical variation, many cell lines undergo extensive chromosomal rearrangement (deletion, translocation, inversion, etc.) resulting in new "marker" chromosomes. While the details of such structural variation in the karyotype will be of greatest interest to the somatic cell geneticist, the concept of karyotypic evolution ought to be of some concern to all who work with cell cultures as model systems. As seen below marker chromosomes can be very useful in research with cultured cells.

### *Uses of Karyotyping*

*1. Gene Dosage.* In many types of biochemical research, including, for example, studies on regulation of enzyme synthesis, it may be desirable to know whether the cell line is diploid, pseudodiploid, or heteroploid. Clearly, such studies will be more valid when carried out with diploid cells rather than pseudodiploid or heteroploid cells. Furthermore, information is rapidly accumulating on the human chromosomal map location for many enzymes so that in the future the biochemist may want to know the details of which chromosomes or chromosomal parts are duplicated or deleted in order to estimate the role that gene dosage may play in the regulatory phenomenon under investigation.

*2. Identification of Cell Lines.* Marker chromosomes generated spontaneously by structural rearrangement in cell cultures are usually characteristic of the line, and three or four such unique markers can provide an unequivocal identification test for a cell line. That such an identification test is important was realized a few years ago when it was found that a number of presumably different cell lines, thought to have been derived from a variety of different human tissues, were not different lines at all, but were in fact sublines of the heteroploid HeLa cell line derived many years ago from a human cervical carcinoma. Much of the proof for this startling discovery consisted of the finding of several marker chromosomes, characteristic of HeLa cells, in the other cell lines.[1] Contamination of one cell line by another is a potentially enormous problem for the investigator who studies a tissue-specific enzyme or tissue-specific cellu-

[1] W. A. Nelson-Rees and R. R. Flandermeyer, *Science* **191**, 96 (1976).

lar function. Since most established lines do carry one or more characteristic marker chromosomes, karyotyping provides a handy method for verifying that one is working with the correct cell line. Indeed, in many instances the karyotype may be the only means available to check the identity of a cell line.

3. *Monitoring of Cell Lines.* In many experiments, the cell line being studied comes in deliberate contact with other cell lines, with subcellular preparations derived from other cell lines, or with viruses grown in other cell lines. Examples of direct cell contact would include enzyme cross-feeding experiments or the passage of tumor cells through animals. Experiments involving contact with subcellular preparations such as chromosome transfer experiments carry the danger of cross-contamination by live cells from the donor line. In all these cases karyotyping provides a method for monitoring cell lines to confirm that cross-contamination has not taken place in the course of an experiment.

4. *Verification of Hybrid Formation.* Cell hybrids (described in this volume [28]) are extensively used for biochemical studies. In inter-species hybrids the chromosomes provide a definitive set of markers to confirm that the putative hybrid is in fact a hybrid of the desired type. In intra-species hybrids, the use of parental cell lines with different characteristic marker chromosomes can often provide a means of verifying that the hybrid is the desired one.

5. *Cytogenetic Research.* In the four examples cited above karyotyping is used as an aid in a research project that is not cytogenetic in nature. Other work is concerned with gene mapping or karyotype evolution in which karyotyping plays a major role. These projects often involve the routine procedures described in detail below. They may also make use of more highly specialized procedures outlined briefly below but the details for which are beyond the scope of this article.

### *Chromosome Banding and Other Specialized Procedures*

Prior to 1970, chromosomes only could be stained uniformly over their length (solid staining) (Fig. 1A). Since then, however, several procedures have been developed for differentially staining certain regions of the chromosomes to give them a banded appearance (Fig. 1B). Standard banding techniques in common use include: (1) Q-banding, or quinacrine banding, in which chromosomes are stained with quinacrine, a fluorescent compound, resulting in patterns of alternating bright and dull fluorescence along the chromosome; (2) G-banding or Giemsa banding, a technique in which chromosomes are treated with trypsin followed by staining with Giemsa (a complex mixture of dyes) producing a series of dark staining

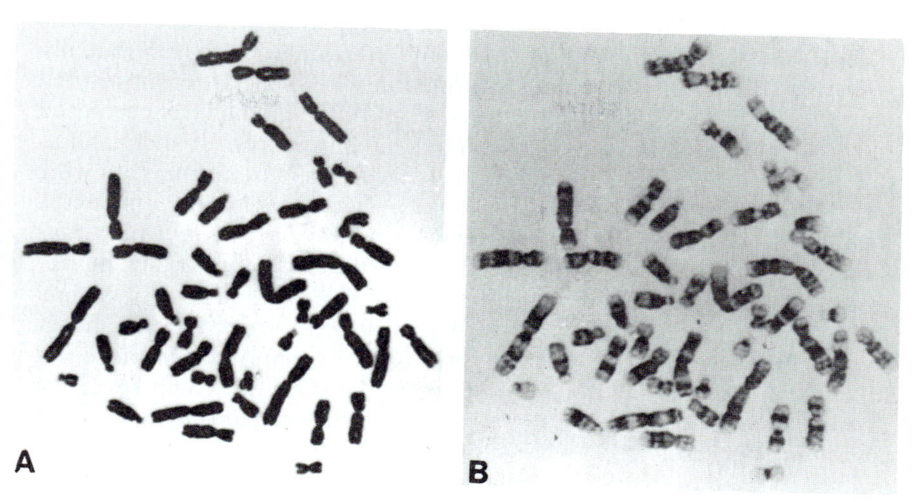


FIG. 1. A human metaphase spread stained first by solid staining (A), then destained by placing the slide in 95% ethanol for 10 min, and G-banded (B).

bands (Fig. 1B) that correspond to the bright quinacrine bands; (3) R-banding or reverse banding, a technique involving heat treatment of the slides and Giemsa staining that results in banding patterns in which the alternating light and dark regions are the reverse of those seen with G-banding.

In this article we will deal with solid staining and G-banding in considerable detail, and the reader is referred to the original literature or to review articles[2-5] for details of Q- and R-banding.

In addition to the three major banding techniques there are several other specialized procedures that will not be detailed in this article although they are mentioned briefly here to provide an idea of the available technology. These include: (1) C-banding, a procedure for specifically staining the centromere regions of chromosomes as well as regions of constitutive heterochromatin;[6] (2) a technique in which a fluorescent dye is used to stain centromeres of mouse but not human chromosomes, useful for identifying the species of origin of chromosomes in human–mouse

[2] T. C. Hsu, *Annu. Rev. Genet.* **7,** 153 (1973).
[3] P. Pearson, *J. Med. Genet.* **9,** 264 (1972).
[4] O. J. Miller, D. A. Miller, and D. Warburton, *Prog. Med. Genet.* **9,** 1 (1973).
[5] B. Dutrillaux and J. LeJeune, *Adv. Hum. Genet.* **5,** 119 (1975).
[6] F. E. Arrighi and T. C. Hsu, *Cytogenetics* **10,** 81 (1971).

hybrids;[7] (3) a technique employing incorporation of bromodeoxyuridine (BUdR) into DNA (in place of thymidine) allowing one to differentially stain and therefore to distinguish chromatin with one BUdR substituted strand from that with 0 or 2 substituted strands; the method has been used to study replication kinetics of the individual chromosome bands[8] and to differentially stain sister chromatids of the same chromosome allowing quantitation of sister chromatid exchanges.[9] Other specialized procedures include (4) nucleolus organizing region (NOR) staining in which ammoniacal silver is used to specifically stain the nucleolus organizing regions of chromosomes containing the sites of the 18 S and 28 S ribosomal RNA genes;[10] (5) cytological hybridization, a procedure in which radioactively labeled RNA or DNA is hybridized to denatured chromosomes followed by autoradiography to identify the sites of complementary chromosomal DNA;[11] and finally (6) a series of immunological procedures employing fluorescently labeled antinucleoside antibodies to identify, for example, chromosomal sites of high methylcytosine concentration.[12]

## Detailed Methods

In this section we provide details for preparing microscope slides of air-dried metaphase chromosome spreads, and the detailed techniques for solid staining and for G-banding. The solid staining requires less expertise than G-banding and is adequate for counting chromosomes or for identifying uniquely shaped marker chromosomes. The G-banding procedure is more demanding and requires higher-quality metaphase spreads, but it gives considerably more karyotypic information than does solid staining (see Fig. 1A and B)

Preparation of high-quality chromosome spreads suitable for banding contains an element of "art," and because of this the preparations usually become better with practice. As is the case for the typical enzyme purification procedure, quality varies slightly from day to day, and the optimal procedures vary somewhat from one cell type to another.

### *Slide Preparation*

There are basically five steps in the preparation of high-quality slides of air-dried metaphase cells. These are (1) optimization of cell growth to

[7] I. Hilwig and A. Gropp, *Exp. Cell Res.* **75,** 122 (1972).
[8] S. A. Latt, *Proc. Natl. Acad. Sci. U.S.A.* **70,** 3395 (1973).
[9] S. A. Latt, *Science* **185,** 74 (1974).
[10] S. E. Bloom and C. Goodpasture, *Hum. Genet.* **34,** 199 (1976).
[11] J. G. Gall and M. L. Pardue, this series, Vol. 21, p. 470.
[12] B. W. Lubit, T. D. Pham, O. J. Miller, and B. F. Erlanger, *Cell* **9,** 503 (1976).

achieve a high mitotic index, (2) treatment with a mitotic inhibitor to allow cells to accumulate at mitosis, (3) swelling of the cells in a hypotonic environment, (4) fixation of the cells, and (5) spreading and drying of the mitotic cells on a slide. Some general considerations with regard to each step are described below.

*1. Optimization of Growth.* For cultured cells, a high mitotic index is usually easy to achieve. For some cell lines, changing the medium the day before making chromosome preparations may be helpful. Rapidly growing lines are usually subcultured 1–2 days before and slower growing lines 3–4 days before "harvesting" the cultures for chromosome studies.

*2. Metaphase Arrest.* Mitotic spindle inhibitors such as colchicine or deacetylmethyl colchicine (Colcemid) are normally used for blocking cells in metaphase. Exposure time depends on the length of the cell cycle (longer treatments for slower growing cells), but 0.5–2 hr is usually sufficient. The concentration of Colcemid is not critical: we have found 0.1–1.0 $\mu$g/ml adequate for many cell lines. Higher concentrations can cause excessive contraction of the chromosomes reducing the clarity of banding patterns.

*3. Hypotonic Swelling.* Swelling of the cells is a critical step and is essential to achieve well-spread chromosomes. A variety of hypotonic solutions are in common use, but 0.075 *M* potassium chloride is the most popular and is a good choice for the initial attempt. Treatment time in "hypo" varies from one cell line to another, but 15–45 min is standard and, if periods greater than 45 min are required, the hypotonic solution might be diluted to increase its effectiveness.

*4. Fixation.* This is another critical step, and it always follows the hypotonic treatment. The near universal fixative is 3 parts methanol to 1 part glacial acetic acid. It is usually made fresh each day and kept cold until use. It is always added slowly to prevent cells from clumping. Some protocols suggest removal of the hypotonic solution from the cells and subsequent addition of fixative, while others call for slow addition of the fixative to the hypotonic solution containing the cells. Both methods work, but a given cell line may respond better to one or the other method. To insure complete fixation the fixative is changed 3–4 times before drying the cells onto a slide.

*5. Air Drying.* Spreading of the cells on a slide is also a crucial step. If not correctly performed either the mitotic cells will fail to spread and the chromosomes will remain in a heap in the cell or they will spread too much with rupture of the cell membrane and release of the chromosomes from the cell. Cells are spread by drying them onto a slide. Optimum spreading of the cells seems to be related to the drying of the fixative, which in turn depends on the temperature and relative humidity in the laboratory. Drying time can be manipulated by blowing on the slide, fanning the slide in the air, or placing the slide on a warm surface or hot plate. We prefer a combination of blowing and moderate heat. The old standard procedure of flame drying yields chromosome spreads that are difficult or impossible to band.

Below we provide details of two techniques that are used routinely in our laboratory. One is designed for handling cell suspensions and is suited to cultured lymphocytes, long-term lymphoblast cultures, or any cell line that grows in suspension. Any monolayer culture also can be trypsinized and then handled as a suspension culture. The second technique is an *in situ* method whereby monolayer cultures are processed right on their growth surface in specially designed culture chambers that consist of a microscope slide base with removable plastic chamber walls.

While the *in situ* method is not as generally applicable as the suspension method, we present it because we prefer it in certain circumstances and use it routinely for some cell types. For example, it generally requires fewer cells and is less wasteful of cells than the suspension method, an advantage for a slow-growing line where mitotic cells are rare or for human amniotic fluid cells where the results of karyotyping are required as quickly as possible. However, it has one disadvantage in that the quality of the metaphase spreads is quite sensitive to overcrowding of the cells on the slide making it a little more difficult for the beginner to obtain good spreads.

The suspension method described below applies specifically to long-term human lymphoblast cultures and the *in situ* method to human fibroblast cultures. Suggested modifications in these basic protocols for a number of other cell lines follow.

### A. The Suspension Method (As Applied to Human Lymphoblasts)

*Equipment*

Pasteur pipettes and rubber bulbs
Suction apparatus with collecting flask
Centrifuge, low-speed, refrigerated if available
Centrifuge tubes—10–15 ml siliconized glass, or plastic
Microscope slides (1 × 3 inch slides with ground-glass surface at one end), precleaned. Before use place several slides in a beaker of ice water.
Hot plate (e.g., Corning model PC 351) covered with paper towel and set to maintain 100 ml of water in a 100-ml beaker at 60–70°

*Solutions*

Colcemid (Grand Island Biological Co., #521L), 10 μg/ml, stored at 4°

Hypotonic: 75 m*M* KCl warmed to 37° (5 ml per culture)

Fixative: 3 parts methanol to 1 part glacial acetic acid, made fresh on day of harvest and kept on ice in a tightly stoppered vessel (20 ml per culture)

*Steps*

1. Optimization of Growth. Two days before harvesting the culture for chromosome spreads, subculture so that cells will be in mid to late exponential growth phase on the day of harvest.

2. Metaphase Arrest. To 3 ml of lymphoblast culture add 2 ml of fresh medium and 0.1 ml Colcemid. Incubate at 37° for 30 min.

3. Hypotonic Swelling. Centrifuge cells at approx. 1000 rpm (200 *g*) for 8 min. Remove all but 0.2 ml of supernatant with suction apparatus. Resuspend cells in this small volume, add 5 ml of warm hypotonic solution, and invert to mix. (Do not pipette cells because swollen cells are fragile.) Incubate at 37° for 15 min.

4. Fixation. Centrifuge cells at 1000 rpm for 8 min. Remove all but about 0.2 ml of supernatant, and resuspend cell pellet in this small volume by gentle agitation (no pipetting). Add first 1 ml of cold fixative slowly, drop by drop with constant agitation. Add another 4 ml of fixative and leave on ice for 15 min. Repeat fixation twice as above, but the fixative can be added more quickly on the 2nd and 3rd fixation. Allow 15 min between successive fixations. At this point cells can be stored overnight at 4° or slides can be made immediately.

5. Air Drying. Centrifuge cells at 1000 rpm for 8 min. Remove supernatant and suspend cells in about 0.5 ml of cold fresh fixative. (Make fresh fixative if cells have been stored overnight). Shake excess water and ice from a microscope slide and, using a Pasteur pipette, drop 3–4 drops of cell suspension onto the slide. Blow gently over the surface of the slide to spread the cell suspension thinly, wipe excess liquid from the underside of slide, and place on hot plate. Leave 5 min to dry thoroughly. Store slides at room temperature in slotted trays or boxes so that surfaces do not touch. (Note: It is our practice to prepare one slide and check it in the microscope under phase contrast so that adjustments can be made on subsequent slides if cells are too crowded or if spreading is poor—see Table I.)

TABLE I
COMMON PROBLEMS IN SLIDE PREPARATION AND SOME SUGGESTED SOLUTIONS

| Problem | Suggested solutions |
|---|---|
| Very few metaphase cells on slide despite presence of many nonmitotic cells | For *in situ* method mitotic cells may be washed off so reduce time in hypotonic *or* handle more gently *or* blow less vigorously on slide. For either method, the problem could be poor growth so increase time in Colcemid *or* change medium the day before harvest to stimulate growth *or* adjust time between subculture and harvest to take advantage of any partial synchronization induced by the subculture. |
| Very few metaphase cells and nuclei without cytoplasm (extensive cell lysis) | Reduce time in hypotonic *or* if your cells are especially sensitive to hypotonic swelling increase salt concentration in the hypotonic *or* handle cells more gently avoiding pipetting. |
| Many broken cells with scattered chromosomes | Blow less vigorously on slide. |
| Underspread cells with overlapped chromosomes or with chromosomes embedded in a thick layer of cytoplasm (Fig. 3A) | Blow more vigorously on slide *or* increase time in hypotonic up to 40 min *or* reduce salt concentration of hypotonic *or* if your cells are especially resistant to hypotonic swelling add trypsin to the hypotonic at the concentration used for subculture. |
| Overcondensed chromosomes or widely spaced chromatids | Reduce Colcemid concentration to 0.1 μg/ml *or* reduce time in Colcemid. |

## B. The *In Situ* Method (As Applied to Human Fibroblasts)

*Equipment*

Lab-Tek chambers (single chamber, #4801, Lab Tek Products, Miles Laboratories)
Pasteur pipettes and 22-gauge needle
Suction apparatus with collecting flask
Hot plate (e.g., Corning, model PC 351) covered with paper towel and set to maintain 100 ml of water in a 100 ml beaker at 60–70°)

*Solutions*

Colcemid (Grand Island Biological Co., #521L) 10 $\mu$g/ml, stored at 4°
Hypotonic: 0.075 *M* KCl, warmed to 37° (2 ml per culture)
Fixative: 3 parts methanol to 1 part glacial acetic acid, made fresh on day of harvest and kept on ice in a tightly stoppered vessel (10 ml per culture)

*Steps*

1. Optimization of Growth. Two days before harvesting, subculture cells from the flasks or plates in which they normally grow into two or more Lab-Tek chambers. The cells should cover no more than 5–10% of the growth surface after they are attached, and the final volume of medium is 4 ml. In order to prevent cells from settling primarily around the edges, first add medium to the chamber and then pipette the concentrated cells into the center of the chamber and rock gently to distribute outwards. Incubate for about 2 days at 37°. (Slow-growing strains may require 3 or more days.)

2. Metaphase Arrest. On the day of harvest, cultures should be nonconfluent (about 10–20% of surface covered by cells—see Fig. 2A) and actively growing (several rounded-up mitotic cells or doublets should be seen). Add 0.2 ml Colcemid and incubate for 90 min.

3. Hypotonic Swelling. With as little agitation as possible so as not to detach loosely adhering mitotic cells, carry chamber slides from incubator. Very slowly, aspirate all of the culture medium by placing a 22-gauge needle connected to a suction flask into one corner of the chamber. Add 2.0 ml of warm hypotonic solution drop by drop over the surface of the chamber. Incubate for 30 min at 37°.

4. Fixation. Remove chamber slides very carefully from incubator and, without removing any of the hypotonic solution, add 2.0 ml of cold fixative drop by drop over the surface of the chamber. Let sit on the bench for 20 min. Repeat fixation 3 times more at 15-min intervals, each time aspirating all the fixative and replacing it with 2–3 ml fresh fixative. The 15-min interval is a minimum time, and longer times can be used if more convenient. Slides can be dried immediately or stored at 4° overnight.

5. Air Drying. Aspirate all the fixative, remove chamber walls from slide, blow gently over the surface of the slide to spread the cells, wipe excess fixative from the underside of the slide, and place slide on hot plate to dry. Leave at least 10 min to dry thoroughly. Remove any rubber cement remaining after removal of chamber walls and store slides at room temperature in slotted trays or boxes so that slide surfaces do not touch.

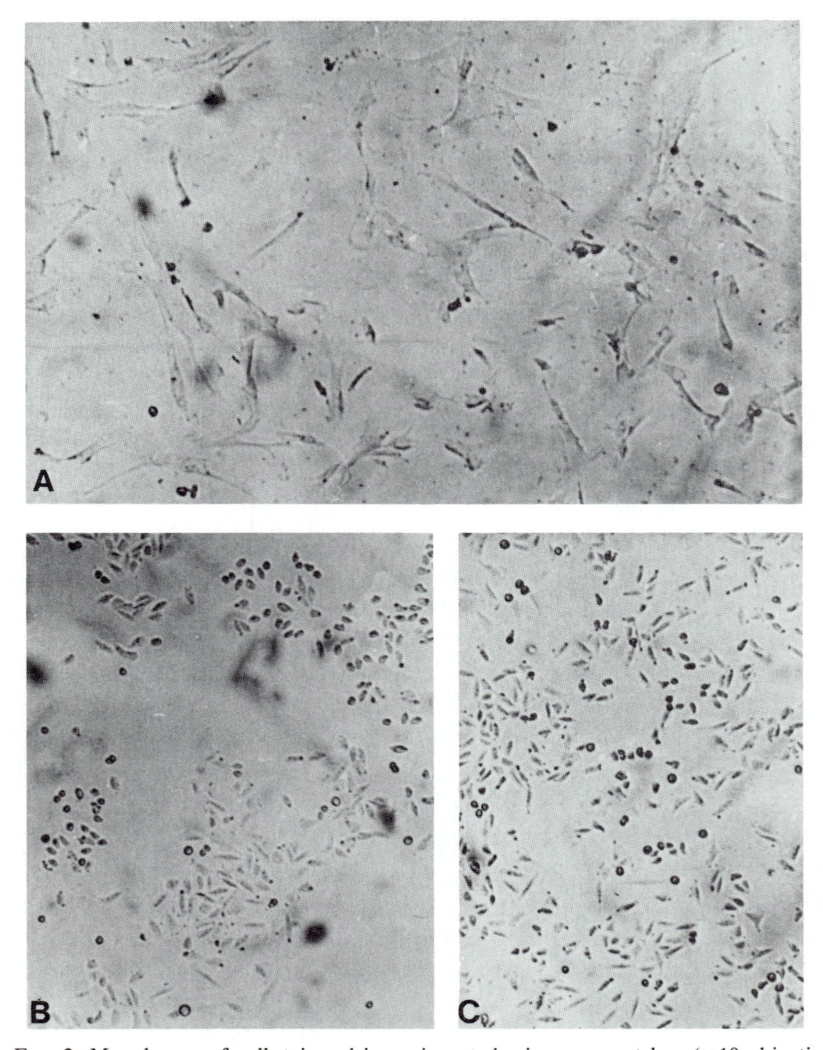


FIG. 2. Monolayers of cells viewed in an inverted microscope at low (×10 objective) power. (A) human fibroblasts in a Lab-Tek chamber at a cell density suitable to begin harvest by the *in situ* procedure. (B) CHO cells in a Lab-Tek chamber at a cell density suitable to begin harvest by the *in situ* procedure. (C) CHO cells in a flask at a cell density suitable to be trypsinized and harvested by the suspension protocol (note many rounded-up mitotic cells and doublets indicating extensive mitotic activity). The cell density is slightly less in (B) than in (C) since cells need room to spread by the *in situ* procedure.

## C. Suggested Modifications for Other Cell Lines

Since different cell lines can differ considerably in their nuclear size, amount of cytoplasm, number and size of chromosomes, and response to a hypotonic environment, any single protocol will not work well for all cell lines. We have had experience with a number of cell lines of human, hamster, and mouse origin and summarize below some general guidelines for them.

Most primary cultures of fibroblast-like cells, regardless of the species or tissue of origin, can be harvested with the *in situ* procedure described above for human skin fibroblasts.

Human amniotic fluid cells are primarily of two types, fibroblast-like and epithelial-like, and these cultures respond well to the same *in situ* procedure. In fact, for amniotic fluid cultures set up for prenatal diagnosis it is common to initiate the cultures in Lab-Tek chamber slides and to examine the chromosomes of individual colonies of cells in these primary cultures. This has two advantages. It allows diagnosis to be made in the shortest possible time (no subculturing is required), and it provides information regarding the colony of origin of any cell with an unusual karyotype.

The CHO cell line, a permanent line of Chinese hamster ovary origin, grows in suspension culture and can be harvested with the suspension method described above but with a longer period in Colcemid (60–90 min).

CHO cells also grow in monolayer but are less well attached than primary fibroblasts; at mitosis the cells round up and are easily dislodged from the surface (Fig. 2B and C). Thus, while CHO can be harvested by the *in situ* method (at the cell density shown in Fig. 2B), great care has to be taken to avoid washing away all the mitotic cells during the hypotonic and fixation steps. An alternative procedure, starting from a monolayer (at the cell density shown in Fig. 2C), is to treat for 90 min with Colcemid, then tap or shake the flasks (or plates) to detach many of the mitotic cells, removing the medium with detached cells to a centrifuge tube. Any remaining mitotic cells can be recovered by lightly trypsinizing: the trypsin solution used for subculture is added for 30–60 sec. After pooling the trypsinized cells with the mitotic cells, we centrifuge, resuspend in hypotonic solution, incubate at 37° for 30 min, and add an equal volume of fixative to the hypotonic solution before the next centrifugation. The cells are fixed twice more, and slides are prepared according to the suspension protocol.

A number of other commonly used cell lines of Chinese hamster origin (V-79, GM-7, M3-1, 18-1, DON), mouse origin (3T3, L, LMTK), or human origin (HeLa, HT 1080) have characteristics similar to CHO, or intermediate between CHO and primary fibroblasts. We have harvested all of these successfully using either the *in situ* method, or starting from a monolayer (one 75 $cm^2$ flask, 1 day before becoming confluent), trypsinizing, and following the suspension method. Colcemid treatment is usually 60–90 min and hypotonic treatment about 30 min.

Hybrid cell lines, whether of intra-species or inter-species origin, usually have characteristics intermediate between the two parental cell types and can often be harvested with the procedure used for the parental line that is most fibroblast-like.

## *Staining Techniques*

### A. Solid Staining

While many different stains have been used to give uniform staining of the chromosomes, Giemsa stain seems to be the most common and is the one we use routinely. We use commercially prepared concentrated Giemsa (Giemsa stain, Original Azure Blend Type, Harleco #620). Like many cytogenetic labs we dilute into Gurr's buffer at pH 6.8 (Gurr's buffer tablets, Searle Diagnostic, England, 1 tablet per liter of distilled water) but 10 m*M* potassium phosphate at pH 6.8 works equally well.

For solid staining, set up 3 coplin jars (glass jars with flat sides and ridges that hold 5 slides) or 50-ml beakers as follows:

1. 50 ml of Gurr's buffer (or 10 m*M* phosphate at pH 6.8) containing 1.5 ml of concentrated Giemsa;
2. 50 ml of water;
3. 50 ml of water.

Place slides into the stain for 3 min and rinse by dipping 8–10 times in each beaker of water. Shake off excess water and stand on end on a paper towel to drain and dry.

If chromosomes are not sufficiently dark the slide can be returned to the stain for a few minutes. The diluted Giemsa can be used to stain up to about 50 slides and is discarded at the end of the day.

### B. Giemsa (G) Banding

Giemsa banding or G-banding was so named because it was the first banding procedure described to utilize this stain. However, it is clearly a misnomer because the Giemsa stain is not the critical ingredient in the procedure and also because Giemsa can be used for solid staining (above) as well as in other banding procedures. In the original G-banding technique the critical step was incubation of the slide at high temperature.

This has largely been replaced by incubation of the slide in a crude trypsin solution, and that has been our standard banding method for several years.

The trypsin is a crude preparation (Bacto trypsin, #0153, Difco Laboratories, Detroit, Michigan) dissolved in 10 ml of distilled water and stored frozen in 1-ml portions. The Giemsa stain is the same as that used for solid staining (Original Azure Blend Type, Hareleco #620). The Giemsa has a shelf life of only about 3 months at room temperature; thereafter it may be used for solid staining but not for G-banding. As for solid staining, the Gurr's buffer in which the stain is diluted can be replaced with 10 m*M* phosphate at pH 6.8.

A prerequisite for good G-banding is a high-quality slide. Chromosomes should be straight and slender with sister chromatids lying close together. Cells should be well spread with few overlapped chromosomes, but the cells should not be broken. Even on a good slide, only one metaphase spread out of 10 or 20 will fit these criteria. Therefore, many mitotic cells should be provided on the slide. These properties can be checked by phase-contrast microscopy before proceeding with banding.

Our technique, modified only slightly from the original trypsin–Giemsa technique of Seabright,[13] is done at room temperature by passing the slide through a series of 7 coplin jars or 50-ml beakers as follows:

1. 50 ml of 0.15 *M* NaCl—brief rinse
2. 50 ml of 0.15 *M* NaCl containing 1 ml of trypsin stock solution—30–120 sec
3. 50 ml of 0.15 *M* NaCl—brief rinse (dip slide 8–10 times)
4. 50 ml of 0.15 *M* NaCl—brief rinse
5. 50 ml of Gurr's buffer (or 0.01 *M* phosphate buffer, pH 6.8) containing 1.5 ml of concentrated Giemsa—1–2 min
6. 60 ml of distilled water—brief rinse (dip slide 8–10 times)
7. 50 ml of distilled water—brief rinse

Slides are placed on end on a paper towel to drain and dry.

Since the time required in trypsin depends to some extent on how the slides were prepared, especially the temperature of the hot plate, a few trial slides may be necessary.

Following either solid staining or G-banding, slides are ready for microscopic study. We do not bother to mount a coverslip on the slides since the presence of a coverslip does not seem to significantly affect the quality of the microscopic images. When using oil-immersion lenses we place the immersion oil on the surface of the slide and then remove it immediately after use by dipping the slide into fresh xylene (xylol) several times. The

[13] M. Seabright, *Lancet* **2,** 971 (1971).

new imersion oils, if left on the slide, will extract the stain resulting in fading of the image.

If a coverslip is to be mounted, add a drop of Permount (Fisher Scientific Co., #SO-P-15), or a 50:50 mixture of Permount and xylene. Coverslips can be removed if necessary by soaking the slide overnight in xylene.

*Suggestions for Improving Suboptimal Slides*

For a simple chromosome count with solid staining you should encounter little difficulty. Even with suboptimal slides, some of the metaphase spreads should have chromosomes that are sufficiently distinct for counting. The only possible problems might be too few cells if the starting culture is small or a low proportion of metaphase cells if the culture was not actively growing. Of course, these problems can only be corrected by altering the culture conditions.

For G-banding or any other banding technique the requirement for high-quality slides is greater, and the novice may therefore experience difficulty in the beginning. Since all cell lines are not alike, even an experienced cytogeneticist will sometimes fail to achieve good banding with a new cell line. In these cases, examination of the rejected slides may give important clues regarding appropriate adjustments in technique. Before rejecting a set of slides, however, keep in mind that some slides can be better than others and that on any given slide only a fraction of the metaphase cells will be well spread and optimally banded. It is therefore wise to scan a few representative areas of several slides before rejecting the whole preparation.

In Tables I and II some of the common problems encountered in slide making and in banding are listed together with a few suggestions for their correction. Figure 3 illustrates a few of the problems.

*Microscopy and Photography*

Because of the great variety available, it is not possible to provide details of microscopic and photographic procedures. We therefore outline briefly some general principles and leave the details to the information booklets describing individual equipment. The following comments apply to solid-stained or G-banded preparations, but not to slides stained with fluorescent compounds.

The first requirement for a chromosome study is a good binocular laboratory microscope. Two objective lenses should be sufficient, one low power (×10 or ×16) for scanning the slide to look for metaphase spreads,

TABLE II
COMMON PROBLEMS IN BANDING AND SOME SUGGESTED SOLUTIONS

| Problem | Suggested solution |
|---|---|
| Darkly stained but unbanded chromosomes | Increase time in trypsin *or* decrease time in stain. |
| Individual chromatids partially banded, but separated (undertrypsinized—Fig. 3D) | Increase time in trypsin (chromosomes swell causing the two chromatids to come together and knobs on chromatids to coalesce into bands) *or* if chromatids are too far apart or too condensed the problem is one of slide preparation—see Table I. |
| Lightly stained swollen chromosomes with ragged edges (over trypsinized—Fig. 3C) | Reduce time in trypsin; *or* this can happen even after 10 sec in trypsin if slides are not dried with heat. |
| Lightly stained chromosomes with indistinct bands—sometimes chromosomes appear to be embedded in cytoplasm (Fig. 3A) | Decreasing trypsin time may help *or* usually problem is due to underspreading during slide preparation—see Table I. |
| Halo effect—pale chromosomes with dark outlines | Reduce time in trypsin; *or* reduce time on hot plate during slide making; *or* reduce number of cells on slides (seed Lab-Tek chamber with fewer cells in case of *in situ* technique). |
| Failure to band despite long treatment times in trypsin and well-spread slides | Try a new batch of Giemsa stain especially if old batch is over 3 months old. Some batches of Giemsa give poor banding even when new so test different lot numbers. |

and one high power (×100) (oil immersion) for counting, examining, and photographing chromosomes. It is useful if the low-power objective is a phase-contrast lens (with matching phase rings in the condenser) so that unstained slides can be scanned by this means. By simply removing the phase ring from the condenser the same objective lens can be used with bright field illumination to scan stained slides.

If photographic records of the chromosomes are required then the microscope must be equipped with a camera. An automated photomicroscope is not necessary. If you already own or have access to a good laboratory microscope, then chances are a 35-mm camera can be mounted

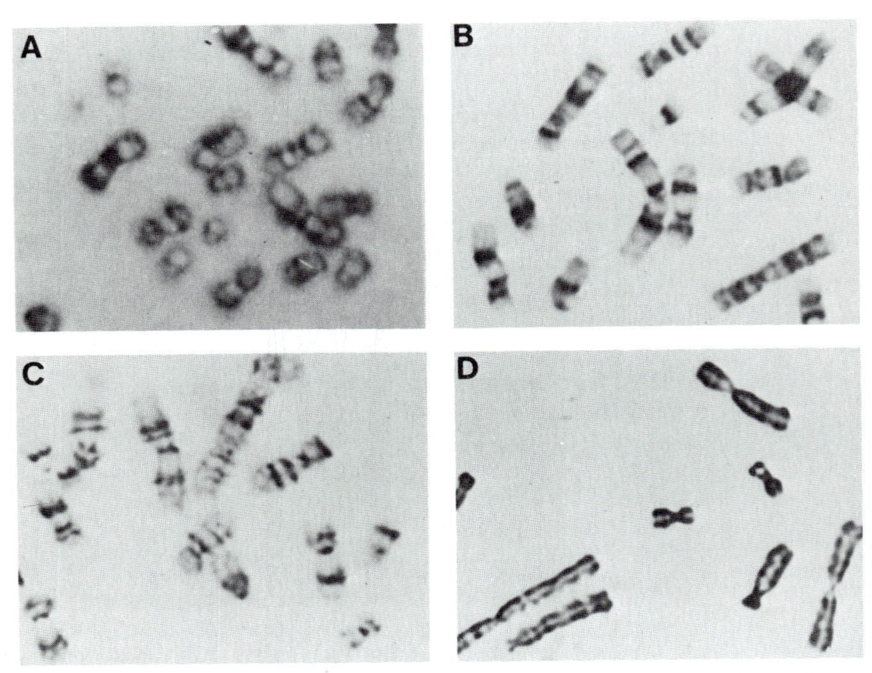


FIG. 3. (A) A portion of a poorly spread cell, trypsin banded. The thick layer of cytoplasm over the chromosomes causes background staining (blue through the microscope, grey on the print) and prevents the chromosomes from becoming clearly banded. It also refracts the light making it impossible to focus clearly on the chromosomes. (B) A portion of a well-spread, clearly banded cell for comparison with (A), (C), and (D). (C) A portion of an over-trypsinized cell. (D) A portion of an under-trypsinized cell.

on the top at a reasonable cost. Automatic film advance is an expensive luxury and not necessary in most cases. Automatic exposure control is also unnecessary since the amount of light reaching the film is relatively constant from one metaphase spread to another, and from slide to slide. Thus, once exposure time has been determined for a particular combination of microscope, camera, and film, it need not be altered. A green filter is commonly used and gives sharper images on black and white film.

The choice of film is not critical and any medium-contrast black and white film should provide adequate pictures. If you develop your own films a local photographic supplier should be able to provide advice on the choice of film and developer. One combination that we have used successfully is Tri-X (Kodak) film developed in Microdol X (Kodak) developer according to the manufacturer's instructions. Fixation in Rapid Fix (Kodak) and wash procedures are standard.

For printing of pictures (approximately 5 × 7 inches) any photographic enlarger that handles 35-mm film can be used. We print on Kodak single-weight Ektamatic SC + N + A paper (contrast grade 2 without filters) and develop automatically in an Ektamatic Processor (Kodak). In the absence of automatic processing equipment prints can be made on any medium-contrast (contrast grade 2 or 3) paper and developed in trays according to instructions with the developer.

Whether processing one's own films or sending them out for processing, one or more test films must be obtained. Choose a typical metaphase spread and photograph it several times using two or three different light levels aiming for an exposure time of 1–3 sec. Once the best exposure time for given equipment is determined, note all the settings, e.g., condenser diaphragm opening, aperture diaphragm opening, current to light source, choice of filters, and auxilliary magnification lenses, and use these standard conditions for all future photographic work.

*Arranging a Karyotype*

With only a little experience solid-stained chromosomes can easily be counted in the microscope with high-power magnification (total magnification ×1000–1400). However, for any further analysis, such as identification of specific marker chromosomes, or determination of the number of copies of individual chromosomes, it is not only necessary to use banding techniques but it is usually necessary to photograph and print the chromosome spreads and then to cut out the individual chromosomes in order to arrange them in a karyotype.

Figure 1 shows two photographs of the same human metaphase cell, first solid stained and then G-banded. It is clear that counting chromosomes is much easier with solid staining than with G-banding, and that matching up and identifying homologous chromosomes in the banded cell would be difficult without actually cutting out the individual chromosomes and arranging them in a karyotype.

Figures 4, 5, and 6 are normal diploid (male) karyotypes of human, Chinese hamster, and mouse, respectively. These three are presented because many of the animal cell cultures in common use are derived from these three species. The arrangement of the human karyotype was standardized in 1971 at the Paris Conference on Standardization in Human Cytogenetics. Not only was each chromosome assigned a specific number, but all the bands were also numbered and a shorthand system was described for noting chromosomal abnormalities.[14] Many human genes

[14] D. Bergsma, ed., "Paris Conference: Standardization in Human Cytogenetics," Birth Defects, Vol. 8, No. 7. Natl. Found., New York, 1972.

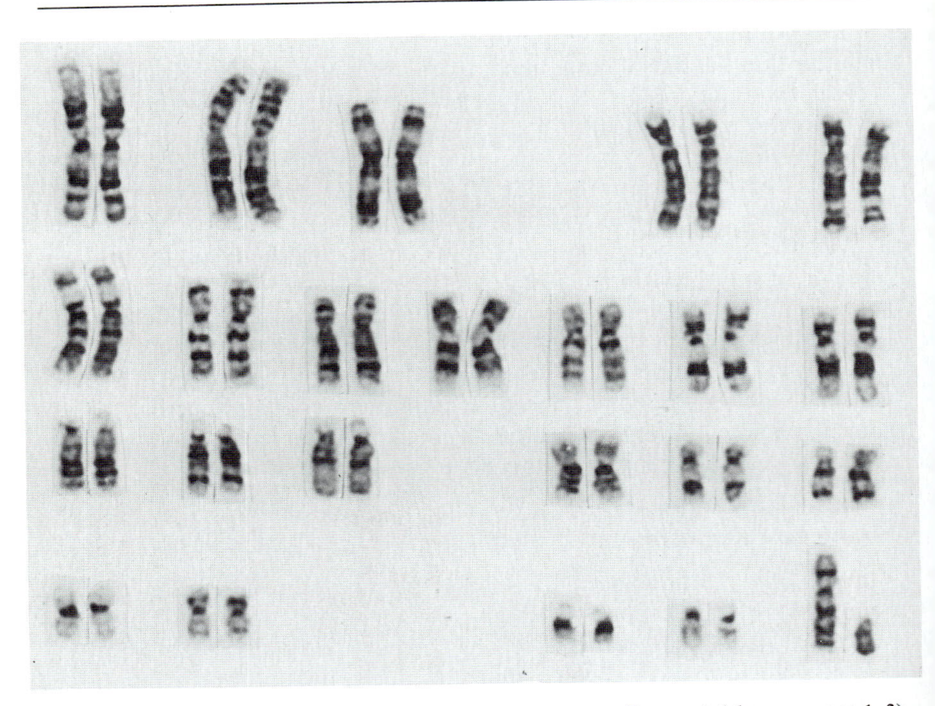

FIG. 4. The human male karyotype, G-banding. Top row: Groups A (chromosomes 1–3) and B (4 and 5). Second row; Group C (chromosomes 6–12). Third row: Groups D (chromosomes 13–15) and E (16–18). Fourth row: Groups F (chromosomes 19 and 20) and G (21 and 22) and the sex chromosomes X and Y.

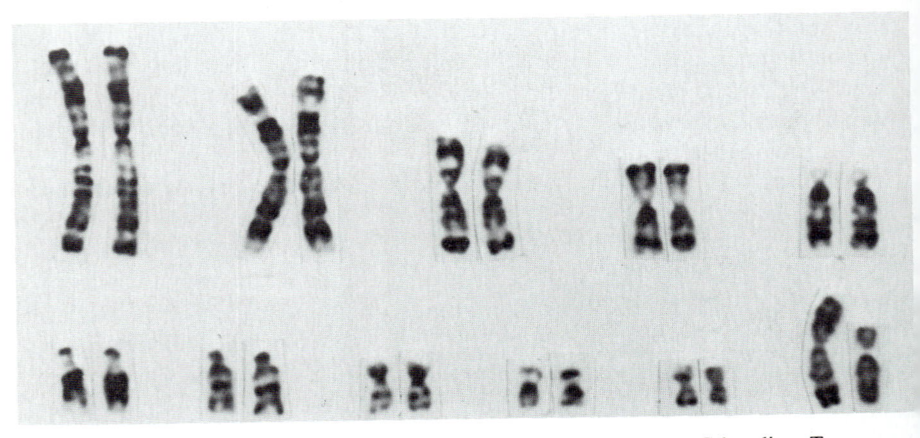

FIG. 5. The Chinese hamster (*Cricetulus griseus*) male karyotype, G-banding. Top row: Chromosomes 1–5. Bottom row: Chromosomes 6–10 plus X and Y.

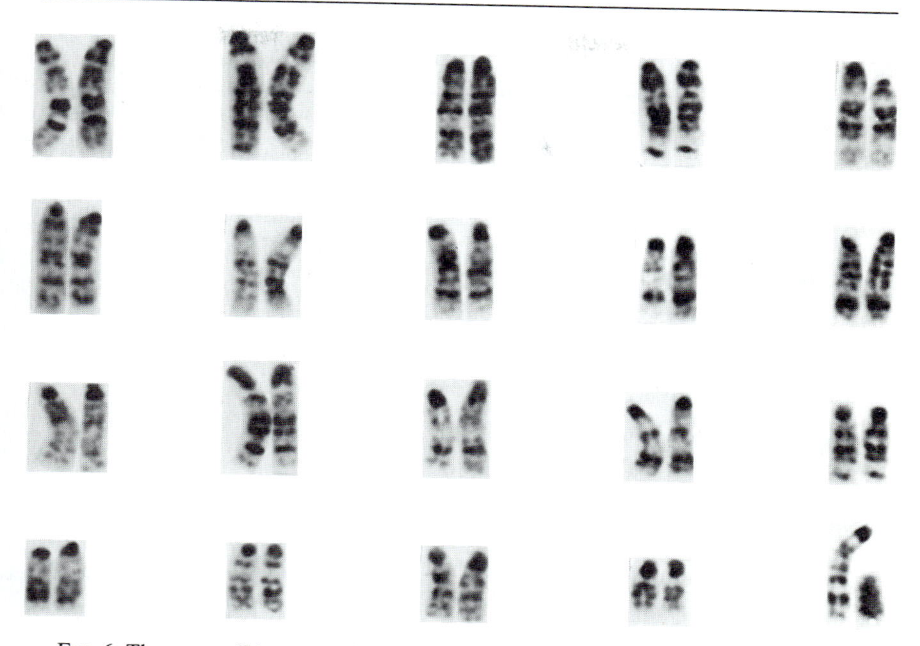

FIG. 6. The mouse (*Mus musculus*) male karyotype, G-banding. Top row: Chromosomes 1–5. Second row: Chromosomes 6–10. Third row: Chromosomes 11–15. Bottom row: Chromosomes 16–19 plus X and Y. Figure prepared by Dr. P. W. Allderdice.

also have been assigned to specific chromosomes or regions of chromosomes.[15]

The Chinese hamster karyotype in Fig. 5 follows the proposed standard karyotype of Ray and Mohandas.[16] While this will likely become the accepted standard, recent publications have appeared using an alternate format in which the #2 chromosomes are placed the other way around (inverted) and the X and Y are referred to as the #3 pair. Thus, pairs 3–10 of Fig. 5 are numbered 4–11 in the alternate scheme.

The mouse karyotype has also been standardized,[17] and the genetic linkage groups that were established several years ago have all been assigned to specific chromosomes.[18]

The karyotypes of a great many other species have been published, but it is beyond the scope of this article to reference them all. There are hundreds of different cell lines in existence, many of which have not been

[15] V. A. McKusick and F. H. Ruddle, *Science* **196,** 390 (1977).

[16] M. Ray and T. Mohandas, *Cytogenet. Cell Genet.* **16,** 83 (1976).

[17] "Standard Karyotype of the Mouse, *Mus musculus,*" *J. Hered.* **63,** 69 (1972).

[18] O. J. Miller and D. A. Miller, *Annu. Rev. Genet.* **9,** 285 (1975).

TABLE III
KARYOTYPE CHARACTERISTICS OF A FEW CELL STRAINS AND ESTABLISHED CELL LINES

| Species | Name of line | Modal no. of chromosomes | No. of marker chromosomes | Reference to karyotype |
|---|---|---|---|---|
| Human ($2n = 46$) | WI-38 (diploid fibroblast) | 46 | 0 | a,b |
| | HeLa | 63–78 | 16–25 | b–e |
| | KB,Hep-2, D98,HEK, etc. | (HeLa contaminants) | | b–e |
| | HT-1080 | 46 | 2 | f |
| Chinese hamster ($2n = 22$) | diploid | 22 | 0 | g |
| | CHO | 20–21 | 10–13 | h,i |
| | CHW | 22 | 4 | j |
| Syrian hamster ($2n = 44$) | diploid | 44 | 0 | k |
| | BHK | 44 | 0 | l |
| Mouse ($2n = 40$) | diploid | 40 | 0 | m |
| | CAK | 41 | 1–3 | n |
| | RAG | approx. 60 | approx 15[o] | p |
| | A9 | approx. 60 | several | q |
| | LMTK$^-$ | 52 | approx. 9[o] | r |
| Rat ($2n = 42$) | diploid | 42 | 0 | s |
| *Drosophila melanogaster* ($2n = 10$) | GM2 and GM3 | Variable with time | | t |

[a] D. Bergsma, ed., "Paris Conference," Birth Defects, Vol. 8, No. 7. Na. Found., New York, 1972.

[b] O. J. Miller, D. A. Miller, P. W. Alderdice, V. G. Dev, and M. S. Grewal, *Cytogenetics* **10,** 338 (1971).

[c] C. C. Lin and S. Goldstein, *J. Natl. Cancer Inst.* **53,** 298 (1974).

[d] W. K. Heneen, *Hereditas* **82,** 217 (1976).

[e] W. A. Nelson-Rees and R. R. Flandermeyer, *Science* **191,** 96 (1976).

karyotyped. Although it is impractical to list all of the lines that have been karyotyped, a few of the more commonly used lines are listed in Table III with references to the detailed karyotypes.

In attempting to karyotype one's own cell line for the first time, there are a few facts to keep in mind. First, chromosomes are continually contracting as the cell proceeds into mitosis and in so doing several fine bands present on a chromosome of a prophase cell or early metaphase cell will coalesce into a smaller number of thicker bands by mid-metaphase. Thus, cells with long thin chromosomes are generally in an early stage of mitosis and have more total bands (and hence different patterns) than cells with shorter thicker chromosomes. With long Colcemid treatment, all cells tend to be at the same stage, that represented in Figs. 4, 5, and 6, thereby considerably reducing this source of variation.

Second, there is subtle variation in chromosomes between the individual members of a species. In humans, for example, this variation can be in the width of certain bands that stain darkly by C-banding, or in the fluorescent intensity of certain bands after staining with quinacrine. However, such inherited chromosomal variants or polymorphisms are not readily apparent by solid staining or G-banding and should not present major difficulties in karyotyping by these techniques.

Third, in comparing the karyotype of one's cell line with published karyotypes, keep in mind that karyotypes evolve with continued culture.

---

[f] S. Rasheed, W. A. Nelson-Rees, E. M. Toth, P. Arnstein, and M. B. Gardner, *Cancer* **33,** 1027 (1974).

[g] M. Ray and T. Mohandas, *Cytogenet. Cell Genet.* **16,** 83 (1976).

[h] L. L. Deaven and D. F. Petersen, *Chromosoma* **41,** 129 (1973).

[i] R. G. Worton, C. C. Ho, and C. Duff, *Somatic Cell Genet* **3,** 27 (1977).

[j] P. A. Gee, M. Ray, T. Mohandas, G. R. Douglas, H. R. Palser, B. J. Richardson, and J. L. Hamerton, *Cytogenet. Cell Genet.* **13,** 437 (1974).

[k] N. C. Popescu and J. A. Di Paolo, *Cytogenetics* **11,** 500 (1972).

[l] R. Marshall, *Chromosoma* **37,** 395 (1972).

[m] "Standard Karyotype of the Mouse," *J. Hered.* **63,** 69 (1972).

[n] R. A. Farber and R. M. Liskay, *Cytogenet. Cell Genet.* **13,** 384 (1974).

[o] The number of marker chromosomes is enhanced in mouse cell lines because the normally telocentric (centromere at one end) mouse chromosomes tend to fuse together at the centromere creating biarmed markers that are not as unique as those created by random breakage and reunion.

[p] P. S. Hashmi, P. W. Allderdice, G. Klein, and O. J. Miller, *Cancer Res.* **34,** 79 (1974).

[q] P. W. Allderdice, O. J. Miller, D. A. Miller, D. Warburton, P. L. Pearson, and H. Harris, *J. Cell Sci.* **12,** 263 (1973).

[r] A. R. Rushton, *Can. J. Genet. Cytol.* **15,** 791 (1973).

[s] "Standard Karyotype of the Norway Rat," *Cytogenet. Cell Genet.* **12,** 199 (1973).

[t] S. F. Dolfini, *Chromosoma* **58,** 73 (1976).

For example, a CHO line may have several normal Chinese hamster chromosomes, several marker chromosomes characteristic of CHO lines already described, and additional new marker chromosomes not previously described that have evolved by structural rearrangements during continuous cell culture. Because of this evolution, it is not valid to assume that one's own CHO line will have the same karyotype as CHO lines in other laboratories. If it is important for the interpretation of biochemical or other data, then the karyotype of the specific subline under study must be determined.

## Comments

Karyotyping can be an extremely valuable research tool for anyone working with cultured cells. In some instances, a quick chromosome count may be all that is required. For this, solid staining would be sufficient, and photographs of chromosomes would not likely be required. The solid-staining technique is easily learned, and for many cell lines one can expect adequate chromosome spreads on the first attempt (also see this volume [27]). Thus, it should take no more than a few hours to obtain a chromosome count on 20–25 cells.

If more detailed information is necessary, then chromosome banding may be required. The G-banding technique presented above will be adequate for most applications, although some situations may require one of the special techniques that have been noted. Since the requirement for well-spread chromosomes is greater for banding than for solid staining, a few days may have to be invested in optimizing the techniques for one's own cell line. Banded chromosomes are also difficult to analyze in the microscope without considerable experience, thus making it necessary to photograph, print, cut out, and arrange the chromosomes in a karyotype. However, once the technique is developed, even a detailed banded analysis can be performed in a day or two.

While several days or even a few weeks invested in developing karyotyping expertise may seem excessive, it is completely inconsequential compared to the many years that already have been spent on HeLa cell research by scientists who thought they were studying something else. It is similarly inconsequential when compared to the time that may be involved in studying gene regulation in heteroploid cell lines where abnormal gene dosage relationships may preclude the operation of normal regulatory mechanisms.

Put in these terms we suggest that karyotyping expertise is not only a valuable tool, but often an essential one for the biochemist who works with cultured cells.

# [28] Cell Fusion

*By* ROGER H. KENNETT

Cell fusion is a biological phenomenon that has been studied and developed from two distinct points of view. Biochemists and cell biologists have been interested in the role that the fusion of biological membranes may play in normal cellular processes such as plasma membrane formation and in the movement of vesicles in the processes of endocytosis and exocytosis.[1] On the other hand much progress in the areas of eukaryotic genetics,[2] virology,[3] and the analysis of malignancy[4] has been made possible by application of cell fusion techniques to the production of heterokaryons and hybrid cells. Heterokaryons result from the fusion of plasma membranes of two different cell types so that one or more nuclei from each type are included within a common cytoplasm. A hybrid cell is formed when growth and division processes take place in a heterokaryon and the genetic materials from the two types of nuclei are included in the single nucleus of a proliferating cell. Such hybrid cells can then be propagated in culture with the growth characteristics usually depending on the cell types fused. Generally, the hybrids can be maintained in culture under the same conditions and for as long as the most actively growing of the two parental cells. Several selective systems have been devised that allow preferential growth of hybrid cells and eventual elimination of the two parental types. These methods will be described below.

There has been interaction among those approaching cell fusion from these two points of view at the level of the various agents used to promote the fusion of cellular membranes. These promoters of membrane fusion fall into two general categories: enveloped viruses and lipophilic or lipolytic reagents such as lysolecithin and polyethyleneglycols.[5]

I will concentrate here on those methods for cell fusion that allow production of a high proportion of viable heterokaryons or hybrids and will discuss the numerous ways in which such fused cells are useful to investigators involved in analysis of biochemical and genetic properties of cultured cells. For those who would like a more detailed account of the

[1] G. Poste, *Int. Rev. Cytol.* **32,** 157 (1972).
[2] V. A. McKusick and F. H. Ruddle, *Science* **196,** 390 (1977).
[3] H. Koprowski, *Fed. Proc., Fed. Am. Soc. Exp. Biol.* **30,** 914 (1971).
[4] G. Barski, *in* "Tissue Culture: Methods and Applications" (P. F. Kruse, Jr. and M. K. Patterson, eds.), p. 469. Academic Press, New York, 1973.
[5] G. Poste and A. C. Allison, *Biochim. Biophys. Acta* **300,** 421 (1973).

METHODS IN ENZYMOLOGY, VOL. LVIII

ISBN 0-12-181958-2

historic development of cell fusion techniques, the two divergent views of Ephrussi[6] and Harris[7] provide interesting reading.

## Heterokaryons and Hybrids: Their Use in Analysis of Eukaryotic Gene Expression

The ability to include genetic material from two cell types within the same cytoplasm makes it possible to pose questions concerning the interactions of two sets of genes. The two types of questions that have been asked are:

1. What positive or negative regulatory effects can be observed when genetic material from two cells in different states of differentiation are included in the same cytoplasm? Do elements from one genome turn on or off genes in the other genome?

2. Is genetic complementation observed when nuclei from cells, each deficient in some function, are included in the same cytoplasm?

In general, it can be concluded that when two cells expressing different states of differentiation are fused, there is usually suppression of differentiated functions.[8] For example, fusion of melanomas with fibroblasts produces hybrids that do not produce melanin as do the melanoma parental cells.[9] Similarly, fusion of a Friend erythroleukemia mouse cell line, which can be induced to synthesize hemoglobin, with a human fibroblast suppresses globin message formation when the mouse globin loci are present as shown by nucleic acid hybridization studies.[10]

A notable exception to this rule is that fusion of a rat liver tumor cell line synthesizing several liver-specific proteins with other types of cells sometimes results in hybrids that synthesize some of these proteins.[11] There are even reports that, when a rat liver tumor cell is hybridized with some types of cells, e.g., human fibroblast, not only are the rat liver proteins synthesized but, in some clones, liver-specific proteins are synthesized from the genes of the other species included in the hybrid.[12] In hybrids in which these liver-specific functions are suppressed, one often finds reexpression of one or more of the liver-specific functions after the

[6] B. Ephrussi, "Hybridization of Somatic Cells," Princeton Univ. Press, Princeton, New Jersey, 1972.

[7] H. Harris, "Cell Fusion," Harvard Univ. Press, Cambridge, Massachusetts, 1970.

[8] R. C. Davidson, *Symp. Soc. Study Dev. Biol.* **31,** 295 (1973).

[9] R. L. Davidson, B. Ephrussi, and K. Yamamoto, *J. Cell. Physiol.* **72,** 115 (1968).

[10] A. Deisseroth, R. Velez, R. D. Burk, J. Minna, W. French Anderson, and A. Nienhuis, *Somatic Cell Genet.* **2,** 373 (1976).

[11] G. J. Darlington and F. H. Ruddle, *Mod. Trends Hum. Genet.* **4,** 111 (1976).

[12] G. J. Darlington, H. P. Bernhard, and F. H. Ruddle, *Science* **185,** 859 (1974).

cells are grown in culture for several passages.[13] This may result from loss of chromosomes and a resulting change in relative gene dosage favoring reexpression of one or more of the liver-specific proteins. These observations are consistent with the finding that when a near tetraploid derivative of the liver cell line is fused with a diploid cell, the liver-specific functions are expressed in most of the hybrids.[14]

The suppression and reexpression of differentiated functions in hybrids, and the possibility that this results from chromosome loss, point out a significant difference between hybrids and heterokaryons. Hybrids start out after the fusion event with at least one full set of chromosomes from each parental cell. As the hybrids are continued in culture, variants arise that have lost chromosomes derived from one or both of the parental cell types.[15] This presumably is a consequence of unequal distribution of chromosomes during mitosis and cell division and the growth advantage that some of the resulting variants have over the original hybrid clone. How rapidly a variant with a reduced chromosome number comes to predominate in the hybrid culture depends on the types of cells fused and the species from which they each originate. An extreme case of chromosome loss exists in the formation of most rodent cell × human cell hybrids. In most cases, the human chromosomes are rapidly lost leaving only a few human chromosomes in each hybrid clone.[16] Since each clone has a different set of a few human chromosomes, mouse × human hybrids have in fact been very useful tools for human gene mapping as indicated below.[17]

One confirmed exception to the rule of extinction in heterokaryons occurs when fibroblasts fused with chick erythroblast nuclei are reactivated and begin to synthesize globin. This also happens when the erythroblasts are fused with cells from which the nuclei have been removed.[18]

In regard to the second question above, genetic complementation has been demonstrated in several systems either by assaying the individual heterokaryons or the mass culture after the fusion for the particular factor(s) missing from the unfused cells.[19,20] Another method is to detect

[13] M. C. Weiss, R. S. Sparkes, and R. Bertolotti, *Somatic Cell Genet.* **1,** 27 (1975).

[14] J. A. Peterson and M. C. Weiss, *Proc. Natl. Acad. Sci. U.S.A.* **69,** 571 (1972).

[15] E. Engel, J. Empson, and H. Harris, *Exp. Res.* **68,** 231 (1971).

[16] M. Weiss and H. Green, *Proc. Natl. Acad. Sci. U.S.A.* **58,** 1104 (1967).

[17] D. Bergsma, ed., "Human Gene Mapping III," Birth Defects: Orig. Art. Ser. XII(7). Karger, Basel, 1976.

[18] T. Ege, J. Zerthen, and N. R. Ringertz, *Somatic Cell Genet.* **1,** 65 (1975).

[19] Reviewed in N. R. Ringertz and R. E. Savage, "Cell Hybrids." Academic Press, New York, 1976.

[20] E. Weed-Kastelein, W. Keijzer, and D. Bootsma, *Nature (London), New Biol.* **238,** 80 (1972).

complementation by using selective conditions under which neither of the unfused cell types will grow but in which the fused cells can grow if complementation takes place. By this procedure, Puck, for instance, has shown that several glycine auxotrophs can be classified into four complementation groups.[21]

## The Use of Hybrid Cells for Gene Mapping

Because most mouse × human hybrids lose human chromosomes, most clones in a collection of such hybrids contain different combinations of human chromosomes. By relating the expression of a human gene product to the presence of a specific human chromosome one can identify the chromosome containing the gene for that particular gene product. This has made possible the chromosomal localization of several hundred human genes. Mouse × human hybrids, made with certain combinations of cell types, result in retention of human chromosomes with an apparently random loss of mouse chromosomes. This, of course, allows the analysis of the mouse genome by similar methods.[2,16,17,22]

## Hybrids and Malignancy

If a malignant cell is fused with another cell type, one can ask if the hybrid retains the malignant phenotype. One may thus inquire as to whether malignancy is a dominant or recessive characteristic and, at this point, it appears that the answer depends on the specific malignant cell that is chosen. In most cases, the malignancy is defined by the ability to grow as a tumor in immunologically deficient "nude" mice or to grow in semisolid medium without attachment to a substrate[4,22a,23] (see also this volume [30]).

## Detection of Viruses in Nonpermissive Cells and Study of Viral Integration Sites

Cell fusion has been used as an effective tool for detecting the presence of viruses in nonpermissive cells. By fusing a cell that permits the replication of a virus with a cell line that contains the virus genome but is

[21] F. T. Kao, L. Chasin, and T. T. Puck, *Proc. Natl. Acad. Sci. U.S.A.* **64,** 1284 (1969).
[22] C. Croce, A. Talavera, C. Basilico, and O. J. Miller, *Proc. Natl. Acad. Sci. U.S.A.* **74,** 694 (1977).
[22a] S. Shin, V. H. Freedman, R. Risser, and R. Pollack, *Proc. Natl. Acad. Sci. U.S.A.* **72,** 4435 (1975). See also this volume [31].
[23] C. O. Poulson and J. Rugaard, *Acta Pathol. Microbiol. Scand.* **79,** 397 (1975).

nonpermissive, virus replication is initiated and the virus is thereby detected.[3]

Croce *et al.* have used cell hybrids to detect the chromosomal location of viral (SV40) integration in human fibroblasts transformed with SV40 virus. The transformed cells were fused with mouse peritoneal macrophages that do not normally divide in culture. Hybrids isolated from a given fusion or series of fusions with the same SV40 transformed line all contained the same human chromosome; some had only that single human chromosome. All the hybrids contained the SV40 genome, expressed SV40 antigens, and were of a malignant phenotype when tested by the criteria stated above.[24]

### Hybrids as a Source of Cell Products for Immunological and Biochemical Analysis

It has recently become clear that somatic cell hybrids will become an important source of specific cellular products that cannot be obtained from short-term primary cultures. The best example of this is the system developed by Milstein and others for the production of hybrid myelomas making monoclonal antibody against an antigen of choice.[25,26] If an antigen is injected into mice the spleen cells can be removed from the mouse and put into culture. These cells will divide and produce antibody but will last for only a short period in culture and then stop dividing.[27] Cell hybridization makes it possible to combine the spleen cells making antibody with mouse plasmacytoma lines. The mouse lines are plasma cell tumors that can be cultured continually or passaged from mouse to mouse as a tumor. They secrete their own myeloma protein, a monoclonal antibody molecule, the binding specificity of which is usually unknown. By hybridizing the continuous cell line with mouse spleen cells and screening the resulting hybrids for production of antibody against the injected antigen, one effectively inserts the genes for the desired antibody into a continuous cell line that has the necessary machinery to express these genes. A continuous supply of the desired gene product is thereby provided which, in this case, is an antibody that binds the antigen with which the mouse was immunized. The procedure can be refined so that a variant of the plasmacytoma line that no longer produces its own myeloma protein is used for hybridization and the resulting hybrid secretes only the antibody mol-

[24] C. Croce, D. Aden, and H. Koprowski, *Proc. Natl. Acad. Sci. U.S.A.* **72,** 1397 (1975).

[25] G. Kohler and C. Milstein, *Nature* (*London*) **256,** 495 (1975).

[26] G. Galfre, S. C. Howe, C. Milstein, G. W. Butcher, and J. C. Howard, *Nature* (*London*) **266,** 550 (1977).

[27] N. Klinman and J. L. Press, *Transplant. Rev.* **24,** 41 (1975).

ecules produced by the antibody genes introduced into the hybrid by the spleen cell nucleus.[28] The culture medium from the hybrids contains microgram per milliliter amounts of the specific antibody. If the "hybrid tumor" is injected into mice that have been pristane primed[29] to encourage ascites fluid formation, the tumor will grow and produce the antibody in milligram per milliliter amounts in the ascites fluid.

Such hybrids will provide a constant supply of monoclonal antibody against antigens of choice. These antibodies can be used as reagents for any procedures for which antibodies were previously used, but with the added advantages of higher levels of discrimination, lower background, and a continuously available supply. The other obvious application of the procedure is that large amounts of homogeneous antibody against specific antigens will be available for sequence analysis; using current techniques, the nucleic acid for these specific antibody genes can be isolated from the hybrid cell lines and characterized.[30] It will soon be possible to compare the genes and gene products of several cell lines that produce antibodies against the same antigenic determinant. One could even choose lines making such antibodies on the basis of those types of antibodies that appear early in mouse neonatal development and those that appear later.[31] The analysis of antibody and gene structure could have significant implications in regard to the genetic mechanism involved in the generation of antibody diversity.

The above system may be but a prototype for others in which a chosen cell line is hybridized with a specific cell type that produces a desired product; the "immortalized" primary cell type thereby provides a continuous hybrid line producing enough of the chosen product to allow its use as a reagent or to make it available for biochemical analysis. Systems of this sort may be applicable to the production of cell lines expressing specific immunological T cell receptors,[32] hormone receptors, or such other cell surface molecules as differentiation antigens or tumor antigens. It may also provide the means of producing significant quantities of hormones and other cellular products.

Some general principles that may be applicable to other possible systems can be derived from the work with hybrid plasmacytomas. It is important that the cell line used has the capacity to produce the desired

[28] D. Margulies, N. M. Keuhl, and M. D. Scharff, *Cell* **8**, 405 (1976).

[29] M. Potter, *Physiol Rev.* **52**, 631 (1972).

[30] S. Tonegawa, C. Brack, N. Hozumi, and R. Schuller, *Proc. Natl. Acad. Sci. U.S.A.* **74**, 3518 (1977).

[31] N. Klinman and J. L. Press, *Fed. Proc., Fed. Am. Soc. Exp. Biol.* **34**, 47 (1975).

[32] R. Goldsby, B. Osborne, E. Simpson, and L. A. Herzenberg, *Nature (London)* **267**, 709 (1977).

product. As indicated previously, hybridization of an antibody-producing cell with another cell type such as a fibroblast usually results in the suppression or drastic reduction in the amount of differentiated product produced. Also, certain cell lines of the same type may hybridize better than others. We have been working with two mouse plasmacytoma lines for producing hybrids P3X63Ag8[33] and 45.6TGl.7.[34] Which of these cell lines produces the most hybrids in a fusion depends on the source of primary mouse cells used in the fusion. P3X63Ag8 produces more hybrids when fused with adult immunized spleen cells whereas 45.6TGl.7 produces more hybrids when fused with stimulated neonatal spleen cells or cells from *in vitro* stimulated monoclonal spleen fragments.[35] Some other plasmacytoma lines do not produce sufficient hybrids to be of practical use in these experiments.

Another factor to consider is the metabolic state as well as the source of primary cells used for the fusion. In our experience, neonatal spleen cells give very few hybrids when fused with the above cell lines. However, when the spleen cells are stimulated *in vitro* with lipopolysaccharide (LPS), a mouse B cell mitogen, for only a few hours the number of hybrids produced is increased dramatically.[35] This is consistent with a report that mitotic cells are fused preferentially when polyethyleneglycol (PEG) is used as a fusing agent.[36] A similar phenomenon may explain why a large proportion of hybrids produced by fusion of spleen cells with a plasmacytoma line are making antibody against a recently injected antigen.[25,26] Thus the production of the desired hybrids will depend upon selection of the proper combination of cell line and primary cell.

Another factor that must be considered is that there must be a method for selecting the hybrids and eliminating the parental nonhybrid population after cell fusion.

## Hybrid Selection Procedures

Littlefield[37] reported the first use of selective medium allowing the growth of hybrid cells but killing the two parental cell types. This selection is based on the use of HAT medium, which was devised by Szybalşki

[33] G. Kohler, S. C. Howe, and C. Milstein, *Eur. J. Immunol.* **6,** 292 (1976).

[34] D. Margulies, N. Geplinski, B. Dhermgrongartama, M. L. Gefter, S. L. Morrison, T. Kelley, and M. D. Scharff, *Cold Spring Harbor Symp. Quant. Biol.* **41,** 781 (1977).

[35] K. Denis, personal communication.

[36] D. Hansen and J. Stadtler, *Somatic Cell Genet.* **3,** 471 (1977).

[37] J. Littlefield, *Science* **145,** 709 (1964).

*et al.*[38] to kill cells that lack hypoxanthine-guanine phosphoribosyl transferase (HGPRT, EC 2.4.2.8.). The medium used by Littlefield contains hypoxanthine (0.1 m*M*), thymidine (16 $\mu M$), and aminopterin (0.4 $\mu M$) Aminopterin inhibits dihydrofolate reductase. Blocking the reductase inhibits *de novo* synthesis of purines and pyrimidines. The medium also must contain glycine (3 $\mu M$) because, in the presence of aminopterin, the conversion of serine to glycine is blocked (the medium is sometimes designated as GHAT medium).

With *de novo* synthesis of the nucleotides blocked, cells depend on the transferase and thymidine kinase to utilize the hypoxanthine and thymine in the medium. Only cells with both of these enzymes will grow in HAT medium. Littlefield used two cell lines: one lacked thymidine kinase and was selected by growth in the presence of the toxic thymidine analog, 5′-bromodeoxyuridine;[39] the other lacked the transferase and thus was resistant to the toxic analogs, 8-azaquanine or 6-thioquanine.[38] Both of these parental cell types are killed in HAT medium whereas hybrids, i.e., cells containing both enzymes by complementation, will grow in HAT.

A modification of this selective system, the so-called half-selective system, can be used: a cell line that does not grow in HAT is fused with a cell type that grows slowly, e.g., a senescent human fibroblast, or does not grow at all in culture, e.g., peripheral blood cells. Only the hybrids form colonies in the HAT medium while the two parental cells either are selected against or do not form colonies. A further modification allows the fusion of HAT-sensitive cells with cells that normally grow in suspension such as lymphoblastoid lines; hybrids are retained as attached colonies and the parental cells that grow in suspension can be washed off from the culture by frequent medium changes.

A similar medium (AA) contains alanosine, an antibiotic blocking conversion of inosine-5′-phosphate to adenosine-5′-phosphate and adenosine. This medium will inhibit cells lacking adenine phosphoribosyltransferase.[40] The same system selects against cells lacking adenosine kinase.[41] Puck *et al.* have derived a auxotrophic mutant requiring an amino acid or nucleotide that has been used in selective medium without the required component. Such media allows growth of hybrids in which genetic complementation takes place but inhibits the auxotrophic mutants.[42]

[38] W. Szybalski, E. Szybalski, and G. Ragni, *Natl. Cancer Inst., Monogr.* **7,** 75 (1962).
[39] W. J. Rutler, R. L. Pictet, and P. W. Morris, *Annu. Rev. Biochem.* **42,** 607 (1973).
[40] T. Kusano, C. Long, and H. Green, *Proc. Natl. Acad. Sci. U.S.A.* **68,** 82 (1971).
[41] T. Chan, R. Creagen, and M. P. Reardon, *Somatic Cell Genet.* **4,** 1 (1978).
[42] F. T. Kao, R. Johnson, and T. T. Puck, *Science* **164,** 312 (1970).

Another system of selection uses the drug ouabain[43] as a specific inhibitor of the $Na^+$–$K^+$ activated ATPase of the plasma membrane, the enzyme responsible for the active transport of $K^+$ into the cell and the extrusion of $Na^+$. Fusion of a cell that can be selected by HAT medium or by its growth characteristics but that is not sensitive to a chosen concentration of ouabain, with another cell type that is sensitive to the same concentration, allows selection of hybrid cells. Since human cells are killed by low concentrations ($10^{-7}$ *M*) and mouse cell by higher concentrations ($10^{-3}$ *M*) of ouabain this agent has become useful for selecting mouse × human hybrids.

One recently described selective medium, glucose-free medium, results in selective growth of hepatoma cell lines producing gluconeogenic enzymes.[44] Variations of the medium select for hybrid clones reexpressing one or more of these enzymes from hybrid lines in which liver-specific functions were extinguished in the initial hybridization.

It should be clear that many selective systems are potentially available. One is limited only by the ability to derive cell lines that are differentially sensitive to metabolic inhibitors or differing in metabolic substrate requirements.

Selection procedures are usually applied to hybrid cells 24–48 hr after fusion to assure expression of the functions necessary for survival although it is not evident from our experience that this delay is really necessary when HAT medium is used. It is important in selective media such as HAT, media in which cells must grow to be selected against, that the fused cells should have enough surface area to divide 2–3 times after the selective medium is added. Plating of the cells at too high a density will slow the growth and increase the time that the cultures must be handled before hybrid clones are isolated.

## Fusion Protocols

Fusion with Sendai virus or polyethylene glycol can be done with cells in mixed monolayers or with cells in suspension. In both cases it is essential that healthy cells with high viability be used for fusion. When cells are removed from a plastic or glass surface in order to fuse in suspension the high viability should be established by phase microscopy or dye exclusion after the cells are in suspension. During the washing and fusion procedures, care should be taken to avoid extremes in pH (maintain 7.2–7.6). This is particularly important with Sendai virus fusions that involve long

[43] L. H. Thompson and R. M. Baker, *Methods Cell Biol.* **6,** 209 (1973).
[44] R. Bertolotti, *Somatic Cell Genet.* **3,** 579 (1977).

incubation periods with large numbers of metabolizing cells in a small volume of medium.

After cells are fused they should be treated as any cells with which one desires to obtain maximum cloning efficiency. Actually, the hybrids are being cloned at a very low cell density because selective conditions will often quickly kill the parental cell types. Therefore one should use a rich medium with 20% fetal calf serum and avoid fluctuations in the pH of cultures. It is helpful to use organic buffers such as HEPES to maintain the pH and thereby avoid the pH fluctuations of bicarbonate-buffered media.[45,46]

## Fusions with Sendai Virus (Parainfluenza I)

The efficiency with which preparations of Sendai virus promote cell fusion is variable. A given batch should be tested by fusing two cell types that are known to fuse well. Virus can be obtained from Microbiological Associates, Bethesda, Maryland. When a good virus preparation is identified it should be dispensed as concentrated allantoic fluid and stored in liquid nitrogen or at least at $-70°$. It can be thawed as needed and inactivated with ultraviolet light[47] or $\beta$ propiolactone.[48] In our experience inactivated virus can be refrozen at least once without significantly reducing its fusing efficiency. It is clear that Sendai virus will promote the fusion of some cell types and not others. For instance, Sendai virus has not increased the amount of fusion above that of the spontaneous fusion frequency when mouse limploid cell lines are used.[49]

Because polyethylene glycol can promote fusion of a wider variety of cell types and is much easier to obtain than an effective preparation of Sendai virus, it will soon replace Sendai virus as a fusing agent. Therefore, fusion methods with PEG one described in greater detail whereas Sendai virus fusion techniques are presented in outline.

## Sendai Virus Fusion

Klebe *et al.*[50] and Giles and Ruddle[51] describe the procedures for virus preparation and mixed monolayer fusions and for fusion of cells in suspension. I describe here the procedure used in my laboratory for fusion of

[45] C. Ceccarino and H. Eagle, *Proc. Natl. Acad. Sci. U.S.A.* **68,** 229 (1978).

[46] C. Croce, H. Koprowski, and H. Eagle, *Proc. Natl. Acad. Sci. U.S.A.* **69,** 1953 (1972).

[47] G. Yergarian and M. B. Nell, *Proc. Natl. Acad. Sci. U.S.A.* **55,** 1066 (1966).

[48] J. M. Neff and J. F. Enders, *Proc. Soc. Exp. Biol. Med.* **127,** 260 (1968).

[49] D. H. Margulies, N. M. Kuehl, and M. D. Scharff, *Cell* **8,** 405 (1976).

[50] R. J. Klebe, T. Chen, and F. H. Ruddle, *J. Cell Biol.* **45,** 74 (1970).

[51] R. E. Giles and F. H. Ruddle, *in* "Tissue Culture: Methods and Applications" (P. F. Kruse, Jr. and M. K. Patterson, eds.), p. 475. Academic Press, New York, 1973.

cells in suspension, a method that is more generally applicable than monolayer fusion because it can be used with cells that grow in suspension as well as those in monolayers. When the precautions of maintaining proper pH control, using healthy cells, and plating so as to allow rapid selection mentioned above are taken, this method yields more hybrid clones than does a monolayer fusion with the same cell types.[52] We use modified Eagle's medium (MEM) with nonessential amino acids, Dulbecco's MEM with high glucose (4.5 g/liter), or RPMI-1640 medium, depending on which of these media are routinely used to grow the cells to be fused.

1. Cells are grown for at least 1 day in medium containing 20% serum that is buffered with 15 m*M* HEPES at pH 7.6.[45]

2. The fusion is done in Hanks's balanced salt solution without glucose, adjusted to pH 7.6 with Tris · HCl (10 m*M*) (Modified Hanks, MH).[53]

3. Harvest cells and wash twice in MH by centrifugation. Centrifugations are done in IEC MS centrifuge (radius: 15 cm) at 1000 rpm for 5 min at room temperature. The cells may be mixed in appropriate ratio and washed together. Use 1 : 1 ratio unless it is evident that one cell type fuses better than the other. The 1 : 1 ratio may be used for initial experiments, and if this does not provide sufficient hybrid clones, other ratios should be tried.

4. Mix $3 \times 10^6$ of each cell type in a round-bottomed tube (Falcon 2001) in 1 ml of MH.

5. Chill on ice.[54]

6. Add 1000 hemagglutination units (HAU) of inactivated virus in 0.5 ml MH.

7. Maintain on ice at pH 7.6 for 5 min or until clumped.

8. Place tube into 37° water bath for 5 min.

9. Add 0.3 ml of dialyzed fetal calf serum. The pH of this should be adjusted so that addition of 0.3–1.5 ml of MH does not change pH from 7.6.

10. Incubate tube at 37° by rocking gently for 30 min.

11. Add medium with 20% serum and gently dispense into 20 35-mm petri dishes or other appropriate culture vessels. One should consider that separation into several separate plates will assure that colonies isolated from each plate arise from separate fusion events.

12. Change medium to HAT medium on the next day. Littlefield's original concentration[37] of HAT constituents has worked satisfactorily in

[52] R. Kennett, unpublished observations.

[53] Y. Okada and J. Tadokoro, *Exp. Cell Res.* **32,** 117 (1963).

[54] There is an indication that although this is a routine procedure it is really unnecessary [L. Donner and L. Turek, *Somatic Cell Genet.* **2,** 77 (1976)].

all HAT selections we have done. Change to fresh medium 2–3 times per week.

13. Colonies usually appear macroscopically between 10 and 20 days.

14. Pick individual colonies into 35-mm petri dishes and grow large enough quantities so that stocks of several ampules, containing at least $10^6$ cells each, can be frozen in 95% serum–5% dimethyl sulfoxide (DMSO).[53,55]

## Fusion with Polyethylene Glycol (PEG)

Polyethylene glycol, which had been used to promote fusion of plant protoplasts, was applied to fusion of mammalian cells by Pontecorvo.[56] Davidson and Gerald modified conditions to include a shorter exposure to PEG and fusion of cells in suspension.[57,58] O'Malley and Davidson found that the frequency of hybridization of cells that are not attached to monolayers could be increased if the cells to be fused were pelleted onto a coverslip before exposure to PEG.[59] Other investigators have described the effectiveness of fusion of PEGs of different molecular weights, concentrations, temperatures, and fusion conditions.[60–62] Norwood *et al.* have shown that exposure of cells to DMSO increases PEG-mediated fusion.[63] Care must be taken not to expose cells to DMSO in medium containing HEPES since DMSO will allow the buffer to gain entrance into the cells with resultant toxicity.

The amount of exposure that a cell can tolerate depends on the molecular weight and concentration of PEG.[57,58] These factors may have to be adjusted for each type of cell in order to maximize the fusion/toxicity effects of PEG treatment. Davidson *et al.* discuss the effects of these factors.[57,58]

The method used here has been effective for several types of cells. The procedures are similar to those reported by Gefter *et al.*[64] for fusion of mouse plasmacytomas.

For fusion of certain types of cells it may be necessary to increase the PEG concentration to 50%, and the addition of DMSO may increase the

[55] W. S. Sly, G. Sekhon, R. Kennett, W. F. Bodmer, and J. Bodmer, *Tissue Antigens* **7**, 165 (1976).

[56] G. Pontecorvo, *Somatic Cell Genet.* **1**, 397 (1975).

[57] R. L. Davidson and P. S. Gerlad, *Somatic Cell Genet.* **2**, 165 (1976).

[58] R. L. Davidson, K. O'Malley, and T. B. Wheeler, *Somatic Cell Genet.* **2**, 271 (1976).

[59] K. O'Malley and R. L. Davidson, *Somatic Cell Genet.* **3**, 441 (1976).

[60] A. Hales, *Somatic Cell Genet.* **3**, 227 (1977).

[61] V. L. Vaughn, D. Hensen, and J. Stadler, *Somatic Cell Genet.* **2**, 537 (1976).

[62] Z. Steplewski, H. Koprowski, and A. Leibovitz, *Somatic Cell Genet.* **2**, 559 (1976).

[63] T. H. Norwood, C. J. Zeigler, and G. M. Martin, *Somatic Cell Genet.* **2**, 263, (1976).

[64] M. L. Gefter, D. Margulies, and M. D. Scharff, *Somatic Cell Genet.* **3**, 231 (1977).

fusion efficiency.[63] With increased PEG concentration, the time of exposure will probably have to be decreased to avoid toxicity. It is our experience that the centrifugation step does increase the frequency although we have fused several combinations of cells simply by exposing the loosened pellet of cells to 1 ml of 50% PEG 1000 or PEG 6000, mixing the pellet with PEG solution by gently rotating the tube, and, after 1–2 min of exposure, adding 10 ml of medium without serum taking care that the PEG is mixed and diluted with the medium.

The ratio of cells fused may vary from 1 : 1 for two cell lines to 5–10 : 1 when spleen cells or other primary cells that do not divide in culture are fused with a cell line. In the cases in which 5–10 : 1 spleen cells : cell line are used, it is clear that the unfused spleen cells act as a feeder layer and increase the cloning efficiency of the hybrid cells. When we fused human peripheral blood cells with mouse plasmacytoma cell lines the number of hybrid clones derived was increased significantly by adding 1000 irradiated (4500 R) human diploid fibroblasts per microwell (Linbro FB 96TC) as a feeder layer.

The procedure below is used for fusion of plasmacytoma lines, cells that are killed by HAT medium, with spleen cells from immunized mice.[25,65] We have also used it for making mouse × human hybrids with several combinations of cell types. Spleen cells for the fusion are removed 3–4 days after the mouse is immunized. The spleen is placed in a 60-mm petri dish in tissue culture medium with serum and is perfused with medium by injecting with a 26-gauge needle at several sites thereby forcing medium into the spleen. This is continued until most of the cells are removed; the number of spleen lymphocytes recovered is usually 5–10 × $10^7$ per organ. Care must be taken not to repeatedly draw the suspended cells up into the syringe and force them through the needle. Remove the cells to a centrifuge tube, pellet them (1000 rpm, IEC MS), remove the supernatant liquid, and suspend them in 5 ml of cold 0.17 *M* $NH_4$ Cl. Keep the tube on ice for 10 min to lyse erythrocytes and then add 10 ml of cold medium with 20% serum. Pellet the cells, count and check viability, and mix with $10^7$ of the plasmacytoma cell line.

1. Wash the cells in medium free of serum by centrifugation in a round-bottomed tube (Falcon 2001).

2. Remove *all* the supernatant liquid by suction and loosen the pellet by tapping the tube.

3. Add 0.2 ml of 30% PEG in medium without serum. The 30% PEG is made by adding 3 ml of PEG 1000 at 41° to 7 ml of medium without serum at 41°. PEG tends to lower the pH of the medium which should be ad-

[65] R. H. Kennett, K. A. Denis, A. S. Tung, and N. R. Klinman, *Current Topics in Microbiol. and Immunol.* **81,** 77 (1978).

justed to approximately 7.6 with NaOH. After mixing the PEG solution it is kept at 37° until used.

4. The cells are maintained in the 30% PEG for 8 min. During the 8-min period, cells are pelleted in the 30% PEG (1000 rpm, IEC MS, radius 35 cm). The centrifugation time should be 3–6 min. It is our impression that the longer the centrifugation, the more hybrids. The limitation, as indicated above, is the toxic effect on the particular cells used. This may be monitored by phase microscopy or trypan blue exclusion. At the end of 8 min the PEG is diluted with 5 ml of medium without serum and then 5 ml of medium with 20% serum is added.

5. The cells, in diluted PEG, are pelleted and resuspended in 30 ml of culture medium with serum. We have found the following medium optimal for cloning and growth of hybrid plasmacytoma lines:

> Dulbecco's MEM with high glucose (4.5 g/liter) to which are added 10% NCTC 109 medium,[28] 20% fetal calf serum, 0.150 mg/ml oxaloacetate, 0.050 mg/ml pyruvate, 0.200 units/ml bovine insulin (Sigma), and antibiotic of choice.

In addition to the purine and pyrimidine bases present in the NCTC 109 medium, thymidine (16 $\mu M$) and hypoxanthine (0.1 m$M$) are added to the medium in which the fused cells are plated.

6. The 30 ml of cells are evenly suspended and gently distributed into 6 microplates (Linbro FB96TC), 1 drop (about 50 $\mu$l) per well. The next day, an additional drop of the above medium with aminopterin (0.8 $\mu M$, twice Littlefield's concentration) is added to make HAT selective medium.

7. The wells are fed two additional drops of medium 6–7 days later; clones appear macroscopically within 2 weeks.

8. If necessary, the clones may be fed weekly by removing most of the medium in the well and replacing it with fresh medium.

9. In most fusions, clones arise in nearly all the wells of the six microplates. The valuable wells are identified by screening the supernatant liquid for production of the desired antibody, and these clones are transferred to larger wells (Linbro FB16-24-TC) with not more than 0.5 ml of medium.

10. At early stages, the clones must be watched carefully so that they are not allowed to grow to too high a density nor diluted too sparsely.

11. When the cells have been transferred into four of the larger wells, they may be transferred to flasks, grown, and stocks prepared and frozen.

12. The frequency of hybrids usually requires that the cells be cloned in agarose over a feeder layer of human fibroblasts to be sure that the cells

are truly clonally derived. We have often cloned the cells from the original well into two cloning plates successfully.[66]

13. If the reversion rate of the parental lines is not known, control cells should be put under selective conditions. For those cells selected against a HAT medium it is wise to grow the parental cells in the appropriate selective drug, 5-bromodeoxyuridine or 8-azaguanine, before the fusion. The hybrid nature of the clones can of course be assessed by testing for expression of parental enzymes, antigens, or chromosomes.[51]

In addition to being used to fuse cells to form heterokaryons and hybrids, the fusion methods with Sendai virus and PEG also have been used recently to incorporate nuclei into enucleated cells,[18] to transfer a few chromosomes that are enclosed in microcells into other cells,[67,68] and to transfer substances enclosed in lipid vesicles[69] or erythrocyte ghosts[70] into cells. These techniques are likely to become important and useful applications of membrane fusion techniques in the future.

### Acknowledgments

The author is supported by NIH Grant CA 18930, NSF Grant PCM 76-82997, and the Cystic Fibrosis Foundation. I thank Kathleen Denis for her discussions and work on hybridizations of mouse plasmacytomas with mouse neonatal cells and spleen fragments.

[66] K. Sato, R. Slesinski, and J. Littlefield, *Proc. Natl. Acad. Sci. U.S.A.* **69,** 1244 (1972).

[67] R. E. K. Fournier and F. H. Ruddle, *Proc. Natl. Acad. Sci. U.S.A.* **74,** 3937 (1977).

[68] R. E. K. Fournier and F. H. Ruddle, *Proc. Natl. Acad. Sci. U.S.A.* **74,** 319 (1977).

[69] G. Poste, D. Papahadjopoulos, and W. J. Vail, *Methods Cell Biol.* **14,** 34 (1976).

[70] M. Furusawa, M. Yamaizumi, T. Nishimura, T. Uchida, and Y. Okoda, *Methods Cell Biol.* **14,** 73 (1976).

# [29] Fusion of Higher Plant Protoplasts

*By* ALBERT W. RUESINK

For many years the higher plant cell wall prevented any significant research utilizing the fusion of higher plant cells, but the development of techniques for removing the cell walls to leave behind living protoplasts[1] has recently provided exciting opportunities for research on genetics, development, and crop modification. In contrast to animal cells, plant cells are often totipotent, and whole plants can be regenerated from individual cells, from isolated protoplasts, or even from the product of the fusion of two protoplasts,[2] which is called a somatic hybrid. Though the route to

[1] A. W. Ruesink, this series, in press, (1979).

[2] P. S. Carlson, H. H. Smith, and R. D. Dearing, *Proc. Natl. Acad. Sci. U.S.A.* **69,** 2292 (1972).




ISBN 0-12-181958-2

useful genetic engineering by means of protoplast fusion will surely be long and difficult, a tremendous potential exists. Both the problems and possibilities have been reviewed elsewhere.[3-5] Because of the difficulty of controlling the nature of the fusion product when one fuses whole cells, another promising approach to genetic manipulation involves various methods being developed to transfer only a selected part of one cell's genome into another.[6] However, isolated DNA is very often degraded in transit in contrast to the minimal genetic modification that occurs during the process of protoplast fusion. In addition to providing a novel means for studying nuclear genes, the fusion of protoplasts also permits a new approach to the study of cytoplasmic genetic factors.[7]

An appropriate fusion process will yield up to 35% of the cells in a preparation as fusion products. Many different species have been successfully fused. Though the fusion itself usually is readily achieved, the subsequent selection and regeneration of useful progeny are not. This article will emphasize the fusion process and suggest strategies for dealing with the later processes needed for genetic engineering. Two short general articles on fusion already exist, one in a handbook on tissue culture methodology[8] and the other in a symposium volume.[9]

## Protoplast Release and Preparation for Fusion

Protoplasts are readily released from many different plants by an array of commercially available enzymes.[1] Leaf mesophyll cells (the green photosynthesizing cells within the leaf) and suspension-cultured cells release protoplasts most readily, but almost any cell type that lacks thick secondary cell walls has at one time or another been used to prepare protoplasts. Although the osmotic strength of the solution surrounding intact plant cells is usually of small concern, once the cell wall is weakened or removed the osmotic strength of the medium is critical. Usually a solution of mannitol or a similar compound at a concentration

[3] P. S. Carlson and J. C. Polacco, *Science* **188,** 622 (1975).

[4] P. R. Day, *Science* **197,** 1334 (1977).

[5] H. H. Smith, *BioScience* **24,** 269 (1974).

[6] D. Hess, *in* "Microbial and Plant Protoplasts" (J. F. Peberdy *et al.,* eds.), p. 125. Academic Press, New York, 1976.

[7] G. Belliard, G. Pelletier, and M. Ferault, *C. R. Hebd. Seances Acad. Sci., Ser. D.* **284,** 749 (1977).

[8] K. N. Kao, *in* "Plant Tissue Culture Methods" (O. L. Gamborg and L. R. Wetter, eds.), p. 22. Nat. Res. Counc. Can., Ottawa, 1975.

[9] T. Eriksson, H. Bonnett, K. Glimelius, and A. Wallin, *in* "Tissue Culture and Plant Science 1974" (H. E. Street, ed.), p. 213. Academic Press, New York, 1974.

of about 0.5 *M* effectively stabilizes protoplasts. Sometimes protoplast viability is enhanced by the additional presence of a complete nutrient medium.

A clean preparation of healthy protoplasts is required for fusion studies. The health of protoplasts can be monitored by measuring the number showing normal cyclosis,[10] by using vital dyes,[11] or by determining the fraction of the protoplasts that survive subsequent culture. To remove protoplasts from undigested tissue, from the digesting enzymes, and from various debris, the preparation first can be poured gently through a nylon net or stainless-steel screen with a pore size of 75–300 $\mu$m. Several subsequent washings with fresh osmoticum will remove both the enzymes used to release the protoplasts and most cytoplasmic debris from any cells that have broken, although the fusion capability may decrease quickly following removal of the enzymes as will be described. To separate good protoplasts from cell wall fragments and other particulate debris, several kinds of density gradients may be used[1] although such a separation is usually unnecessary for fusion studies. Being decidely fragile, protoplasts must be handled gently throughout the isolation procedures and must never be centrifuged in excess of 500 g; usually about 100 g is better.

Protoplasts are not suitable for fusion if they have already formed a new wall. The best way to prevent wall regeneration is simply to work quickly enough so that a wall has no time to form. In one case, the available working time was as short as 10 min after removing the enzyme, at which time the fusion rate of pea mesophyll and *Vicia* suspension culture protoplasts had started to decrease, a decrease that correlated with the formation of new wall.[12] Though workers apparently do not use them for fusion studies, two methods do exist for reducing the rate of wall regeneration—the use of an ionic osmoticum[13] and the use of high osmotic strength.[9]

## The Fusion Process

Three main methods for promoting protoplast fusion have been reported, but the most recent one is so superior to the earlier ones that the first two will be mentioned only briefly.

The first successful method used $NaNO_3$ as the fusing agent.[14] Proto-

[10] A. W. Ruesink and K. V. Thimann, *Proc. Natl. Acad. Sci. U.S.A.* **54,** 56 (1965).

[11] J. M. Widholm, *Stain Technol.* **47,** 189 (1972).

[12] G. Weber, F. Constabel, F. Williamson, L. Fowke, and O. L. Gamborg, *Z. Pflanzenphysiol.* **79,** 459 (1976).

[13] R. K. Horine and A. W. Ruesink, *Plant Physiol.* **50,** 438 (1972).

[14] J. B. Power, S. E. Cummins, and E. C. Cocking, *Nature (London)* **225,** 1016 (1970).

plasts were prepared and washed in a sucrose osmoticum. As soon as they were placed into 0.25 *M* $NaNO_3$ the fusion process began. Centrifugation of the protoplasts in the $NaNO_3$ promoted fusion, but the number of protoplasts fusing remained low.[2] Moreover, protoplasts treated with this salt are markedly altered in their uptake capabilities.[15]

The second method utilized 0.05 *M* $CaCl_2$ in 0.4 *M* mannitol with the pH at 10.5 (0.05 *M* glycine–NaOH buffer) and the temperature at 37°.[16] When protoplasts of tobacco leaves were centrifuged in this medium, they aggregated at once and started to fuse in 10 min. By 40 min, 20–50% of the protoplasts were involved in fusion events. Many were still viable and could readily be cultured, although very large fusion products became unstable and disintegrated. Magnesium did not substitute for the calcium requirement.

Essentially all present work on plant protoplast fusion is carried out with polyethylene glycol (PEG) as the fusing agent, an agent that will also induce fusions of animal cells even though the other agents that have been used successfully to fuse animal cells seem not to work with plant protoplasts. Two laboratories did the early work with the PEG method.[17,18] The following method is one that gives good results with many kinds of protoplasts.

Release protoplasts from two kinds of tissues in separate containers, planning the wall digestions so that the releases are completed simultaneously. Obviously the two tissues should have some characteristic that permits the detection of hybrid protoplasts either visually or with appropriate selection media. The nature of the stabilizing osmoticum is not too critical as far as fusion itself is concerned, but is important to the stability of the protoplasts. Mix the protoplasts together and pour them through an 80-$\mu$m screen. Centrifuge for 3 min at 100 g to sediment the protoplasts. Remove the supernatant. Wash the protoplasts with 10 ml of a solution such as one of the following:

1. 0.5 *M* glucose, 3 m*M* $CaCl_2$, 0.7 m*M* $KH_2PO_4$ (pH 5.9)
2. nutrient salts with 0.1 *M* sucrose, 0.25 *M* sorbitol

Centrifuge as above, remove the supernatant liquid, and resuspend in the same medium to give a 6% (v/v) suspension of cells. Proceed immediately to the following fusion process:

1. Place 0.2 ml of suspension onto a coverslip in the bottom of a small petri dish and cover the dish.
2. Let the protoplasts settle for 5 min.

[15] A. W. Ruesink, *Physiol. Plant.* **44,** 48 (1978).
[16] W. A. Keller and G. Melchers, *Z. Naturforsch., Teil C* **28,** 737 (1973).
[17] K. N. Kao and M. R. Michayluk, *Plants* **115,** 355 (1974).
[18] A. Wallin, K. Glimelius, and T. Eriksson, *Z. Pflanzenphysiol.* **74,** 64 (1974).

3. Add 0.6 ml of PEG solution drop by drop. (Make the PEG solution by dissolving 1 g of polyethylene glycol (mol wt 1500) in 2 ml of 0.1 *M* glucose, 10 m*M* $CaCl_2$, and 0.7 m*M* $KH_2PO_4$.) PEG sources: PEG 1540, Polysciences, Inc., Warrington, Pennsylvania; MODO-PEG 1500: Modokemi, Stenungssund, Sweden.

4. Incubate at room temperature for 40 min. Occasional rocking of the petri dish should prove helpful if the protoplasts are not touching each other.

5. Gently add 0.5 ml of protoplast culture medium that should contain about 5 m*M* Ca. A pH of 5.7 is usually used, but a higher pH, up to 10, will enhance the number of protoplasts fusing.

6. After 10 min, add another 1.0 ml of medium.

7. Wash the preparation several times with 2-ml aliquots of culture medium, using a Pasteur pipette to remove the old medium each time. Be sure to leave a thin layer of medium over the protoplasts at all times.

8. Culture the protoplasts or study them microscopically while they are still on the coverslip. Alternatively, gently wash them free from the glass and embed them in soft agar (0.6%, procooled before adding protoplasts).[19] An optimal arrangement is to embed the protoplasts in a 1-mm thick layer on top of a previously solidified 5-mm layer.

Occasionally, protoplasts will fuse spontaneously as they are being released from tissue, apparently by expansion of the plasmodesmata linking cells together in the tissue.[20] Such spontaneous fusions are rare and are of little concern to workers interested in interspecific fusions. The method described above will produce up to 35% interspecific hybrids in a preparation. The rest are unfused cells and intraspecific fusion products. Although fusion products resulting from the fusion of just two protoplasts predominate, the products of triple, quadruple, and multiple fusions are also present. Very large cells resulting from the fusion of many protoplasts tend to be unstable and hence are typically present only in small numbers.

Both the molecular weight and the concentration of PEG are critical to inducing successful fusions. Polyethylene glycols of less than 1000 mol wt are not able to produce tight adhesion, and those of up to 6000 are more and more active per mole in inducing fusions but produce too viscous a solution for easy handling. The choice of 1500 is a good compromise. The concentration used above (about 0.33*M*) provides maximal fusions, with significantly less fusion at both lower and higher concentrations.[21]

A distinction must be made between aggregation, where two or more protoplasts just stick to each other, and true fusion, where mixing of

[19] I. Takebe, G. Labib, and G. Melchers, *Naturwissenschaften* **58,** 318 (1971).

[20] L. A. Withers and E. C. Cocking, *J. Cell Sci.* **11,** 59 (1972).

[21] K. N. Kao, F. Constable, M. R. Michayluk, and O. L. Gamborg, *Planta* **120,** 215 (1974).

cytoplasm occurs. Treatment with gelatin,[22] antisera,[23] or Concanavalin A[24] causes only aggregation and does not promote fusion. Only by careful microscopic work can one ascertain whether there are aggregations or fusions.

The nature of the change at the membrane surface that leads to fusions is not totally clear. With sodium nitrate, some electrophoretic evidence suggests that the elimination of negative charges on the protoplast surface is a key aspect.[25,26] The effect of PEG is even less clear. The PEG may provide a bridge by which $Ca^{2+}$ can link membrane surfaces together, or it may somehow lead to a disturbance in the surface charge during PEG washout.[21] On the other hand, certain concentrations of PEG are known to cause a phase separation,[27] and such a separation at the plasma membrane surface could conceivably alter its potential for fusion.

Brief mention must be made of other methods of inducing protoplast fusions. When protoplasts are formed from cultured meiotic cells, they fuse very readily. Merely tapping a depression slide containing them is enough to cause considerable fusion.[28] Treatment with sea water or lysozyme somewhat promotes fusions in petunia.[29]

## Hybrid Fusions

Much of the interest in plant cell fusion results from its ability to bring together new combinations of genetic information. The following tabulation lists the cases where hybrid fusions have occurred and indicates that usually PEG was the fusion-inducing agent. Often, some cell divisions subsequently occurred. In only two cases, however, has it been possible to regenerate a hybrid plant.

Considerable attention has been devoted to means by which hybrid fusions can be detected. One important way is the use of selective growth media as will be described below, but microscopic methods are also used. Workers have looked for: nucleus size differences,[30] nucleus staining

[22] T. Kameya, *Planta* **115,** 77 (1973).
[23] J. Burgess and E. N. Fleming, *Planta* **118,** 183 (1974).
[24] K. Glimelius, A. Wallin, and T. Eriksson, *Physiol. Plant* **31,** 225 (1974).
[25] A. W. Ruesink, *Plant Physiol.* **47,** 192 (1971).
[26] B. W. W. Grout and R. H. A. Coutts, *Plant Sci. Lett.* **2,** 397 (1974).
[27] R. Kanai and G. E. Edwards, *Plant Physiol.* **52,** 484 (1973).
[28] M. Ito, *Plant Cell Physiol.* **14,** 865 (1973).
[29] H. Binding, *Z. Pflanzenphysiol.* **72,** 422 (1974).
[30] C. W. Jones, I. A. Mastrangelo, H. H. Smith, H. Z. Liu, and R. A. Meck, *Science* **193,** 401 (1976).

Interspecific Fusion of Higher Plant Protoplasts

| Species | Fusion method | Fate | Ref. |
|---|---|---|---|
| Maize × oat roots | $NaNO_3$ | Unreported | [a] |
| *Torenia baillonii* × *T. fournieri* petals | $NaNO_3$ | Unreported | [b] |
| *Nicotiana glauca* × *N. langsdorffii* mesophyll | $NaNO_3$ | Hybrid plant | [c] |
| *Vicia* culture × pea mesophyll | PEG & Ca | Some cell divisions | [d] |
| Soybean culture × barley mesophyll | PEG & Ca | Some cell divisions | [d] |
| Soybean culture × *Vicia* mesophyll | PEG & Ca | Some cell divisions | [e] |
| Soybean culture × corn mesophyll | PEG & Ca | Some cell divisions | [e] |
| Soybean culture × rapeseed mesophyll | PEG & Ca | Colonies up to 10 cells | [f] |
| Soybean culture × pea mesophyll (2 vars.) | PEG & Ca | Some cell divisions | [g] |
| Soybean culture × tobacco mesophyll (4 vars.) | PEG & Ca | Some cell divisions | [g] |
| Soybean culture × *Colchium* mesophyll | PEG & Ca | Some cell divisions | [g] |
| Tobacco mesophyll × HeLa culture | PEG & Ca | Nuclei cohabit cells up to 6 days | [h] |
| *Petunia hybrida* mesophyll × *P. parodii* mesophyll | PEG & Ca | Hybrid plant | [i] |
| Carrot × petunia | PEG | Small colonies | [j] |
| Petunia × *Atropa* | PEG | Small colonies | [k] |
| Tobacco mesophyll × chicken red cell | PEG | Mixed cytoplasm; intermediate cell surface structure | [l] |

[a] J. B. Power, S. E. Cummins, and E. C. Cocking, *Nature* (*London*) **225,** 1016 (1970).

[b] I. Potrykus, *Nature* (*London*), *New Biol.* **231,** 57 (1971).

[c] P. S. Carlson, H. H. Smith, and R. D. Dearing, *Proc. Natl. Acad. Sci. U.S.A.* **69,** 2292 (1972).

[d] K. N. Kao and M. R. Michayluk, *Planta* **115,** 355 (1974).

[e] K. N. Kao, F. Constabel, M. R. Michayluk, and O. L. Gamborg, *Planta* **120,** 215 (1974).

[f] K. K. Kartha, O. L. Gamborg, F. Constabel, and K. N. Kao, *Can. J. Bot.* **52,** 2435 (1974).

[g] F. Constabel, G. Weber, J. W. Kirkpatrick, and K. Pahl, *Z. Pflanzenphysiol.* **79,** 1 (1976).

[h] C. W. Jones, I. A. Mastrangelo, H. H. Smith, H. Z. Liu, and R. A. Meck, *Science* **193,** 401 (1976).

[i] J. B. Power, S. F. Berry, E. M. Frearson, and E. C. Cocking, *Plant Sci. Lett.* **10,** 1 (1977).

[j] J. Reinert and G. Gosch, *Naturwissenschaften* **63,** 534 (1976).

[k] G. Gosch and J. Reinert, *Naturwissenschaften* **63,** 534 (1976).

[l] G. E. Willis, J. X. Hartmann, and E. D. deLamater, *Protoplasma* **91,** 1 (1977).

properties,[31] pigment differences,[32] plastid differences,[33] chromosome size distinctions,[34] isozyme production,[35] and the tritiation of one type of nuclei.[30]

Once the cells fuse, nuclei can fuse in either or both of two ways: (1) by a fusion of the nuclear membranes and a direct mixing of the interphase chromatin, and (2) by synchronous division of two nuclei in such a manner that chromosomes become intermingled and the two daughter nuclei each receive a full set of chromosomes from each species. The relative frequency of these two fusion methods is not yet clear.[31] It is possible that better development of hybrid cells may occur where the two parent lines have similar generation times. However, in one report in which two fused species had division times of 1 day in one case and over 5 days in the other, the hybrid cells divided rather uniformly at day 2 or 3.[36] Interspecific hybrid cells have been shown to retain their hybrid character for more than 4 weeks in culture—for five or six generations—but throughout that time the chromosomes of the two species entered metaphase with a slight asynchrony and assembled in groups rather than as a true cell plate.[37]

### Development Following Protoplast Fusion

For most research needs, the limiting factor is not the fusion itself, but rather what is done with that initial fusion product. The demonstration that a somatic hybrid produced by fusing protoplasts of *Nicotiana glauca* and *Nicotiana langsdorfii* could be both selected from other protoplasts and regenerated into an intact plant[2] was convincing proof that significant work with fusion products is possible. In that case, the selection involved plating all protoplasts on a nutrient medium whose hormone content was such that none of the parent protoplasts would multiply, a method that unfortunately does not have general applicability.

Once protoplasts from two species have been fused, there are four subsequent problems to be solved when using protoplasts for development studies or genetic engineering. (1) The desired fusion product must be selected from among cells that are not the desired hybrid. (2) The fusion product must be induced to divide. (3) The desired portion of the

[31] F. Constabel, D. Dudits, O. L. Gamborg, and K. N. Kao, *Can. J. Bot.* **53,** 2092 (1975).

[32] I. Potrykus, *Nature* (*London*), *New Biol.* **231,** 57 (1971).

[33] K. K. Kartha, O. L. Gamborg, F. Constable, and K. N. Kao, *Can. J. Bot.* **52,** 2435 (1974).

[34] G. Gosch and J. Reinert, *Naturwissenschaften* **63,** 534 (1976).

[35] L. R. Wetter and K. N. Kao, *Z. Pflanzenphysiol.* **80,** 455 (1976).

[36] F. Constabel, G. Weber, J. W. Kirkpatrick, and K. Pahl, *Z. Pflanzenphysiol.* **79,** 1 (1976).

[37] F. Constabel, G. Weber, and J. W. Kirkpatrick, *C. R. Hebd. Seances Acad. Sci., Ser. D* **285,** 319 (1977).

genome from each parent plant must be selected and maintained in the culture. (4) An intact plant must be regenerated. Significant progress has been made on all problems but number 3.

To select a desired fusion product, the genetic complementation of auxotrophic mutants has been used with such systems as *Neurospora*[38] and liverworts.[39] So far, it has not been successful with higher plant cell fusions, apparently because of cross-feeding between cells that allows parent cells to grow up.[2] Differential drug sensitivity of parent species has recently shown promise as a selection device in a petunia system,[40] and related work indicates that complementation, bringing green coloration into albino tissue culture cells, may prove useful.[41] An alternative method has been to use semilethal, recessive chlorophyll-deficient mutants that complement in diploids.[42]

To procure some cell divisions from protoplast fusion products is not a major problem as can be seen from the table. However, the appropriate culture conditions for maintainance of continued proliferation of the cells have not been worked out. An equally difficult problem has been the regeneration of whole plants from protoplasts. Whole plants can be regenerated from tobacco protoplasts more readily than from those of most species, although there is no intrinsic reason against regeneration of plants from the protoplasts of many other species. Only six plants have yielded to efforts to obtain whole plants from protoplasts: tobacco,[19] carrot,[43] petunia,[44] asparagus,[45] *Atropa*,[46] and rapeseed.[47] A number of laboratories that want to perform genetic engineering work are now focusing their primary attentions on the elucidation of basic information needed to convert tissue-cultured cells, particularly of crop plants, into differentiated plants; these researchers realize that a much better understanding of that preliminary process is essential before protoplast fusions can be used in genetic engineering.

[38] L. Ferenczy, F. Kevei, and M. Szegedi, *Experientia* **31,** 50 (1975).

[39] O. Schieder, *Z. Pflanzenphysiol.* **74,** 357 (1974).

[40] J. B. Power, S. F. Berry, E. M. Frearson, and E. C. Cocking, *Plant Sci. Lett.* **10,** 1 (1977).

[41] E. C. Cocking, D. George, M. J. Price-Jones, and J. B. Power, *Plant Sci. Lett.* **10,** 7 (1977).

[42] G. Melchers and G. Labib, *Mol. Gen. Genet.* **135,** 277 (1974).

[43] H. J. Grambow, K. N. Kao, R. A. Miller, and O. L. Gamborg, *Planta* **103,** 348 (1972).

[44] E. M. Frearson, J. B. Power, and E. C. Cocking, *Dev. Biol.* **33,** 130 (1973).

[45] D. Bui-Dang-Ha and I. A. Mackenzie, *Protoplasma* **78,** 215 (1973).

[46] G. Gosch, Y. P. S. Bajaj, and J. Reinert, *Protoplasma* **86,** 405 (1975).

[47] K. K. Kartha, M. R. Michayluk, K. N. Kao, O. L. Gamborg, and F. Constabel, *Plant Sci. Lett.* **3,** 265 (1974).

# [30] Cell Transformation

*By* IRA PASTAN

Cultured fibroblastic cells are frequently used in studies of "malignant transformation"; the term "transformation" is also often used by itself. "Malignant transformation" was coined to distinguish it from the "transformation" process by which DNA is introduced into bacteria. A "transformed" cell is one that has undergone a stable heriditable change that enables it to grow into a tumor in an appropriate recipient animal. Since the animal from which a cell is derived may not be available for testing for tumorgenicity *in vivo,* other tests have been developed. Nude mice have an immunological deficiency that interferes with their ability to reject foreign cells, thereby allowing their use for tumorgenicity testing (see this volume [31]).

Since it is cumbersome to test whether a cell has become transformed by injecting it into an animal, other simpler tests are widely used. One that strongly correlates with results of animal tests is the ability of transformed cells to grow in the absence of substratum (agar; methylcellulose; suspension culture).[1] Another prominent metabolic feature of transformed cells that has long been recognized is their rapid metabolism of glucose;[2] cells develop this feature shortly after transformation.[3]

Frequently, transformed cells have a different morphologic appearance than normal cells in culture. Normal fibroblasts are elongated and flattened. Transformed cells often are more rounded. This change in shape is accompanied by changes in the cell surface. When viewed by scanning electron microscopy, cells are usually covered with many microvilli and lamelopodia; at times the surfaces are covered with blebs.[4] Another property of transformed fibroblasts is their increased ability to be agglutinated by plant lectins.[5,6] Several theories have been advanced to explain their enhanced agglutinability. The simplest of these is that agglutinability is enhanced because the microvilli and other irregularities

[1] V. H. Freedman and S. Shin, *in* "The Nude Mouse in Experimental and Clinical Research" (J. Fogh and B. Giovanella, eds.). Academic Press, New York, 1978 (in press).

[2] B. Warburg, "The Metabolism of Tumors." R. N. Smith, New York, 1931.

[3] S. G. Martin, S. Venuta, M. Weber, and H. Rubin, *Proc. Natl. Acad. Sci. U.S.A.* **68,** 2739–2741 (1971).

[4] K. Porter, D. Prescott, and J. Frye, *J. Cell Biol.* **57,** 815–836 (1973).

[5] M. M. Burger, *Biochemistry* **62,** 994 (1969).

[6] B. Inbar, and L. Sachs, *Proc. Natl. Acad. Sci. U.S.A.* **63,** 1418 (1969).




ISBN 0-12-181958-2

on the transformed cells' surface provide large surfaces that are available for the lectins to "glue" the cells together.[7]

A very important factor contributing to the abnormal shape of transformed cells is their decreased adhesion to substratum.[8] Mutant cells have been obtained that have the morphologic phenotype of transformed cells: rounded shape, decreased adhesion to substratum, increase in surface microvilli, and increased agglutinability by lectins. In such cells the primary defect is a decrease in cell adhesion.[9] Apparently decreased adhesion to substratum causes the cell to assume a rounded shape with microvilli or blebs on the cell surface. Although such adhesion-defective mutants have the morphologic phenotype of transformed cells, they are not, in fact, transformed since they neither form tumors in animals nor grow in agar.[9] Thus, in the absence of other information, it is not safe to use morphologic or adhesive features for establishing that the phenomenon of transformation has occurred.

## Selection of Transformed Cell Lines for Study

Many transformed cell lines are available for study and can be conveniently obtained from the American Type Culture Collection or other sources (see this volume [37]). Some of these lines have been in culture for many generations and may have developed characteristics that are not related to the original transformation event. Therefore, it is generally best to use recently transformed cells in which the normal parent is available for comparative purposes.

## High and Low Transformation Systems

It is possible to transform almost all the cells in a culture into cancer cells within a few days with RNA tumor viruses (high-frequency transformation). A widely used system is that of chick embryo fibroblasts (CEF) transformed by various strains of Rous sarcoma virus (RSV). Methods for preparing pure clones of RSV and obtaining RSV-transformed chick embryo cells are described in this volume [32] and [33]. Similarly, murine sarcoma virus (MSV) when combined with a helper virus murine leukemia virus (MuLV) will rapidly transform various mouse and rat cell lines.[10] An advantage of the Rous sarcoma system is the

[7] M. Willingham, and I. Pastan, *Proc. Natl. Acad. Sci. U.S.A.* **72,** 1263–1267 (1975).

[8] K. K. Sanford, B. E. Barker, M. W. Woods, R. Parshad, and L. W. Law, *J. Natl. Cancer Inst.* **39,** 705–718 (1967).

[9] J. Pouyssegur, M. Willingham, and I. Pastan, *Proc. Natl. Acad. Sci. U.S.A.* **74,** 243–247 (1977).

[10] J. W. Hartley and W. P. Rowe, *Proc. Natl. Acad. Sci. U.S.A.* **55,** 780–786 (1966).

availability of viruses with temperature-sensitive mutations in the gene *src* responsible for the transformation process (reviewed in Vogt *et al.*[11]). These viruses should be maintained in the same manner as for wild-type virus, but only at the permissive temperature in order to prevent selection of revertants. They then can be checked at a nonpermissive temperature prior to use. Mutants with deletions in the *src* gene are also available.[11]

In general, the DNA tumor viruses transform only a small fraction of cells they infect (low-frequency transformation). This applies to SV40,[12] polyoma,[12] and adenovirus (see this volume [35]). Therefore, a number of transformed clones should be studied to determine if the event of interest is "transformation-specific." With SV40-infected cells, temperature-sensitive mutants in the gene responsible for transformation are available.[13]

It is also possible to transform cells with chemical carcinogens. A detailed method for obtaining such cells is included in this volume [23].

[11] P. K. Vogt, R. A. Weiss, and H. Hanafusa, *J. Virol.* **13,** 555–554 (1974).

[12] J. Tooze, *in* "The Molecular Biology of Tumor Viruses" (J. Tooze, ed.), Chapter 6, pp. 350–402. Cold Spring Harbor Lab., Cold Spring Harbor, New York, 1973.

[13] T. J. Kelley and D. Nathans, *Adv. Virus Res.* **21,** 86 (1977).

# [31] Use of Nude Mice for Tumorigenicity Testing and Mass Propagation

*By* SEUNG-IL SHIN

The hairless nude phenotype in the mouse results from an inherited developmental defect, which is determined by an autosomal recessive mutation that segregates according to the simple Mendelian laws. In addition to being virtually hairless, homozygous mutant nude mice also suffer from a congenital failure to develop a normal thymus gland. As a result, nude mice are completely deficient in the thymus-dependent (T-cell) immunological functions.[1]

Since immunological rejection of heterologous tissue grafts in an animal depends primarily on the T-cell-mediated immunity of the host, the nude mouse is unable to reject implanted cells or tissues from a genetically nonidentical donor.[2] This discovery led to the suggestion that the nude mouse should be useful as a test system for determination of cellular

[1] J. Rygaard, "Thymus and Self: Immunobiology of the Mouse Mutant Nude." F.A.D.L., Copenhagen, 1973.

[2] J. Rygaard and C. O. Povlsen, *Acta Pathol. Microbiol. Scand.* **77,** 758 (1969).




ISBN 0-12-181958-2

tumorigenicity and as a biological "incubator" for mass propagation of animal cells that can form tumors in this host.[3,4] Subsequent developments confirmed that nude mice are indeed a useful adjunct to the biochemist who needs to grow a large quantity of cells or tumor tissues in a relatively simple and reliable manner. Using a number of cell mutants carrying specific biochemical and genetic markers, Freedman and coworkers demonstrated that mass production of mammalian cells by inoculation of small numbers of cells in nude mice does not cause changes in these markers.[5]

Heteroploid mammalian cell lines, most of which exhibit variable degrees of transformed phenotypes *in vitro* and are tumorigenic in nude mice, can be grown as large tumors in nude mice by injecting $10^6$ or fewer cells. The maximal size of tumors that can be obtained depends upon the cell types used, but large single tumors weighing as much as 20 g have been observed. The usefulness of the nude mouse as an experimental vehicle for production of animal cells is particularly obvious for the cell types that cannot be grown in tissue culture and for the tumors that cannot be transplanted in the usual experimental animals. For example, a number of different human primary tumors removed at surgery have been successfully propagated in nude mice as serial transplants.[6,7]

For genetic and biochemical studies of cellular transformation, the availability of a generally applicable assay for neoplastic growth potential *in vivo* is *sine qua non*. Cells transformed *in vitro* by oncogenic viruses, chemical carcinogens, or other agents usually express tumor-specific cell surface antigens. Established heteroploid cell lines not deliberately exposed to known transforming agents may also express new surface antigens through spontaneous mutations. Therefore, even cells originally derived from inbred strains of experimental animals may elicit immunological rejection when they are tested for tumorigenicity in the strain of origin. For cells derived from noninbred animals, notably human, an immunologically neutral test system is clearly even more essential.

Several studies established that the nude mouse provides a rare and unique advantage in this regard, since cell surface antigens that normally induce graft rejection in normal animals fail to do so in the thymusless

[3] V. H. Freedman and S. Shin, *Cell* **3,** 355 (1974).

[4] S. Shin, V. H. Freedman, R. Risser, and R. Pollack, *Proc. Natl. Acad. Sci. U.S.A.* **72,** 4435 (1975).

[5] V. H. Freedman, A. L. Brown, H. P. Klinger, and S. Shin, *Exp. Cell Res.* **98,** 143 (1976).

[6] M. Schmidt and R. A. Good, *J. Natl. Cancer Inst.* **55,** 81 (1975).

[7] Y. Shimosato, T. Kameya, K. Nagai, S. Hirohashi, T. Koide, H. Hayashi, and T. Nomura, *J. Natl. Cancer Inst.* **56,** 1251 (1976).

nude mice (for a review, see Freedman and Shin[8]). Some recent experiments suggest that nude mice contain elevated levels of certain nonT-cell-mediated immunity, and that they are in fact capable of inhibiting the growth of some tumorigenic cells.[9] It may thus become necessary to employ additional immunosuppressive techniques on nude mice to assess the tumorigenicity of selected cell types. However, for the general purpose of cellular tumorigenicity assay of animal cells in culture, the nude mouse offers the best system presently known. The association between cellular tumorigenicity in nude mice and various *in vitro* parameters of cellular transformation has been reviewed recently.[8] The parameter best correlated with tumor formation *in vivo* is the "anchorage-independent" phenotype, as measured by colony-forming efficiency in soft agar or methylcellulose.

## Source, Maintenance, and Breeding of Nude Mice

Due to their severe immune deficiency, homozygous mutant (*nu*/*nu*) mice are highly susceptible to infectious diseases. During the early days after the original discovery of this mutation, when the nature of the genetic defects was unidentified and therefore conventional animal husbandry techniques were used, nude mice rarely survived beyond the first few weeks. However, if relatively simple precautions are taken it is now possible to maintain nude mice for at least several months in most research laboratory environments.

For intermittent short-term uses of nude mice, the most practical arrangement would be to obtain young adult nude mice from commercial breeders. Once received, the animals should be strictly segregated from other experimental animals, especially rodents; this is a most important factor in maintaining good viability. Nude mice in different inbred backgrounds can now be purchased from commercial suppliers in the United States and several other countries. Preferably, the mouse cages should be placed inside a laminar-flow air filtration system. Conventional laminar-flow hoods designed for cell culture can be used for this purpose. For short-term experiments, the use of clean filterpaper covers that fit over the cage tops has also proven to be adequate. Cages, bedding, cage tops, and water bottles should be autoclaved before use if possible. Drinking water should be boiled beforehand, but normal animal feed can be used without further modification. Access to the animals should be limited to

[8] V. H. Freedman and S. Shin, *in* "The Nude Mouse in Experimental and Clinical Research" (J. Fogh and B. Giovanella, eds.), p. 353. Academic Press, New York, 1978.

[9] L. M. Reid, J. Holland, and C. Jones, submitted for publication.

the investigator or the research assistant as much as possible. The use of disposable surgical gowns, masks, and gloves is recommended for experiments that require longer duration.

With a greater awareness of the nature of immune deficiency in the nude animals, breeding of nude mouse colonies has become increasingly practical for many research laboratories not otherwise engaged in animal breeding programs. In most cases, heterozygous (+/*nu*) females, which are phenotypically normal, are mated with homozygous mutant (*nu*/*nu*) males. However, breeding and maintaining an independent nude mouse colony would still require resources not usually available in a biochemical laboratory. Detailed information and practical guides for breeding of nude mice have been published.[10,11] Information on various aspects of experimentation on nude mice in different countries can be obtained from the Nude Mouse Secretariat, c/o Pathological Anatomical Institute, Kommunenhospitalet, DK 1399 Copenhagen K, Denmark. In the United States, research workers may also receive practical information from the Veterinary Resources Branch, National Institutes of Health, Bethesda, Maryland.

## Procedures

### *Test of Cellular Tumorigenicity*

Harvest cells from the culture vessels in the usual manner, by trypsinization of monolayer cells or by centrifugation of cells grown in suspension. Determine the number of viable cells by the trypan blue exclusion test. (Suspend the cells in phosphate-buffered saline, containing 0.2% trypan blue, for 5 min and count the unstained cells on a hemacytometer.) Take up an appropriate volume of the cell suspension to contain enough viable cells for $2 \times 10^6$ cells per injection point. Wash the cells once with serum-free culture medium, and resuspend the cells in the same medium to a final cell concentration of $10^7$ cells/ml.

Inject 0.2 ml of the suspension for each assay point, using sterile 1-ml-capacity tuberculin syringes fitted with a 21-gauge needle. Injecting the cells into a mouse can be made much easier if a second person holds the mouse, by firmly grasping at the base of the tail and at the back of the

[10] "Guide for the Care and Use of the Nude (Thymus-Deficient) Mouse in Biomedical Research," a Report of the Committee on Care and Use of the "Nude" Mouse. *ILAR News* **19,** No. 2 (1976).

[11] R. Ediger and B. C. Giovanella, *in* "The Nude Mouse in Experimental and Clinical Research" (J. Fogh and B. Giovanella, eds.), p. 16. Academic Press, New York, 1978.

head behind the ears, and keeps the mouse stretched out. The person who holds the syringe can then lift the fold of skin on the flank of the mouse midway between the front and the hind legs, and use the other hand for injection. It is neither necessary nor desirable to anesthetize the mouse for this purpose. The cells should be deposited subcutaneously at a single point, as far away as possible from the point of penetration in order to prevent the cells from oozing out when the needle is withdrawn. Immediately after removal of the needle, hold the needle opening briefly between two fingers. The fact that the cells are injected subcutaneously, and not into the abdominal cavity or the rib cage, can be seen easily during the injection process because the needle is visible throughout due to the translucency of the skin of the hairless mouse.

In order to increase the assay points per nude mouse, injections can be made bilaterally. For the determination of the threshold number of cells required to initiate the tumor growth, etc., the number of cells injected accordingly can be decreased or increased. Up to $10^8$ cells have been injected at a single site.

Check the injected mice at least every 2 days thereafter, by both visual inspection and manual palpation. The latency period between the injection of cells and the first appearance of a palpable nodule at the site of injection varies greatly, depending upon the malignancy of the cells and the inoculum used.[8] A typical tumorigenic cell line, such as HeLa or mouse A9 cells, generates palpable nodules 1–2 weeks following injection of $2 \times 10^6$ viable cells. The minimum threshold number of cells required to initiate tumor growth also varies widely. With the most tumorigenic cells, such as the mouse melanoma cell line PG19, less than 10 cells can initiate tumor growth, while near-diploid Chinese hamster cell lines such

FIG. 1. A phenotypically normal, heterozygous (+/*nu*) BALB/c mouse (left), and its hairless, thymus-deficient homozygous mutant nude littermate (right).

as CHO and WOR-6 form growing tumors only if $10^6$ cells or more are injected. Nontumorigenic cells, such as human diploid cell strains, will fail to form tumors even with $10^8$ cells. A tumorigenic cell line should result in the formation of a progressively growing tumor that, if left alone, would eventually kill the host. Positive results are rarely ambiguous (see Fig. 1).

### *Test of Mycoplasma Infection*

Infection of test cells by mycoplasma can drastically alter their tumorigenicity.[12] It is therefore imperative that the absence of contaminating mycoplasma in the cell lines be established by appropriate tests for mycoplasmas before the cell lines are injected into nude mice. This subject has been reviewed extensively[13] (see this volume [2]).

### *Mass Production of Cells in Nude Mice*

Procedures identical to those described above are used for the generation of large tumors in nude mice, except that greater numbers of cells should be injected in order to shorten the latency period. It is practical to inject up to $4 \times 10^7$ cells in 0.2 ml volume each at both sides of a mouse. In general, tumors induced in nude mice by heterologous malignant cells grow as well-encapsulated masses and do not as a rule metastasize to distant sites. The tumor mass can therefore be removed from the host relatively free of the host tissue. Variable amounts of host blood elements, however, are present in the tumor masses.[5] The final weight of the tumor that is produced in nude mice depends primarily on the cell type used. Many of the commonly used established cell lines such as HeLa, mouse L cells, and Syrian hamster BHK 21 can produce enormous tumors that weigh more than the host mouse itself.[5] Certain other cell lines, such as SV40-transformed 3T3, rarely result in large fast-growing tumors. The factors that determine the final tumor size in nude mice are not well understood, even though the residual tumor immunity of the nude mouse against specific cellular antigens may play a role.

In Tables I,[3,14–17] II,[3–5,15,18] and III,[3,4,16,18–21] representative cell lines that have been shown to be tumorigenic in nude mice are presented. It can be seen that a very wide variety of cell types can be grown in nude mice as tumors. In Table IV, commonly used established cell lines that have been shown to be *non*tumorigenic are listed. Tumorigenicity of virus-transformed cells in nude mice has been discussed elsewhere.[22]

[12] O. P. van Diggelen, S. Shin, and D. M. Phillips, *Cancer Res.* **37,** 2680 (1977).

[13] G. J. McGarrity, D. G. Murphy, and W. W. Nichols, eds., "Mycoplasma Infection of Cell Cultures," Vol. 3. Plenum, New York 1978.

TABLE I
REPRESENTATIVE TUMORIGENIC CELL LINES DERIVED FROM HUMAN TISSUES

| Cell line | Tissue of origin | References |
|---|---|---|
| HeLa | Adult cervical carcinoma | 3 |
| D98/AH2 | 8-azaguanine-resistant mutant of HeLa | 3 |
| HS 0643 | Lung carcinoma | 3 |
| RPMI 2650 | Nasopharyngeal carcinoma | 14 |
| KE-1 | Fibrosarcoma | 15 |
| SH-1 | Melanoma | 15 |
| PA-1 | Ovarian teratoma | 15 |
| NB20 | Lymphoblastoid cell line, established *in vitro* | S. Shin, unpublished |
| WI-L2 | Lymphoblastoid cell line, established *in vitro* | 14 |
| A5/FG/HEK | Embryonal cells, transformed *in vitro* by adenovirus 5 | 16 |
| LN-SV | Lesch–Nyhan fibroblasts, transformed *in vitro* by simian virus (SV) 40 | 17 |

TABLE II
REPRESENTATIVE TUMORIGENIC CELL LINES DERIVED FROM MOUSE TISSUES

| Cell line | Tissue or cell line of origin | References |
|---|---|---|
| L929 | Normal connective tissue, transformed *in vitro* | 3 |
| A9 | 8-azaguanine-resistant mutant of L929 | 3 |
| C1-1-D | BrdUrd-resistant mutant of L929 | 5 |
| Ehrlich/Lettrè | Ascites tumor, culture-adapted | 3 |
| RAG | Renal adenocarcinoma; 8-azaguanine-resistant mutant | 3 |
| B-16 | Spontaneous melanoma | 15 |
| Y-1 | Carcinoma of adrenal cortex | 18 |
| LLC | Lewis lung carcinoma | 15 |
| PCC4.azal | Teratocarcinoma from strain 129; 8-azaguanine-resistant mutant | S. Shin, unpublished |
| MPC 45-6 | Plasmacytoma, a derivative of MPC 11 | S. Shin, unpublished |
| MOPC 315 | Myeloma | S. Shin and H. Eisen, unpublished |
| SV101 | 3T3 (Swiss), transformed *in vitro* by SV40 | 3 |
| Py3T3 | 3T3 (Swiss), tranformed *in vitro* by polyoma virus | 18 |
| KA31 | 3T3 (Balb), transformed *in vitro* by Kirsten murine sarcoma virus | 4 |
| EL4 | Lymphoma, induced by Moloney virus | S. Shin, unpublished |

TABLE III
TUMORIGENIC CELL LINES FROM OTHER SPECIES

| Cell line | Species | Tissue or cell line of origin | References |
|---|---|---|---|
| $GH_3$ | Rat | Pituitary tumor secreting growth hormone | 19 |
| LG | Rat | Clonal line of embryonal myoblast | S. Shin and B. Nadal-Ginard, unpublished |
| C6 | Rat | Chemically induced glioma | 18 |
| F17 | Rat | Embryonal cells, transformed *in vitro* by adenovirus 2 | 16 |
| SVRE-9 | Rat | Embryonal cells, transformed *in vitro* by SV40 | 4 |
| (M-MuSV)NRK | Rat | Rat kidney cells (NRK), transformed *in vitro* by Moloney sarcoma virus | 20 |
| (KSV)NRK | Rat | NRK, transformed *in vitro* by Kirsten sarcoma virus | 20 |
| CHO | Chinese hamster | Normal ovary, spontaneously transformed | H. Klinger, unpublished |
| WOR-6 | Chinese hamster | Normal lung fibroblast, mutant resistant to 8-azaguanine and ouabain | 3 |
| PSN4 | Syrian hamster | Embryonal cells, chemically transformed | 21 |
| BHK 21 | Syrian hamster | Baby kidney, spontaneously transformed | 3 |
| BHK-T6a | Syrian hamster | BHK 21, 8-azaguanine-resistant mutant | 3 |
| DMN6A | Syrian hamster | BHK 21, chemically transformed *in vitro* | P. Canary and N. Bouck, unpublished |
| PySp1A | Syrian hamster | BHK 21, transformed *in vitro* by polyoma virus | P. Canary and N. Bouck, unpublished |
| B(a)P104Cl | Guinea pig | Embryonal cells, transformed *in vitro* by benz(a)pyrene | 21 |
| DENA HM2Cl | Guinea pig | Embryonal cells, transformed *in vitro* by diethylnitrosamine | 21 |
| RKT-TG1 | Rabbit | Normal adult kidney, transformed *in vitro* by SV40 | 3 |
| RSV-CE-Pr | Chick | Embryo fibroblast, transformed *in vitro* by Rous sarcoma virus | S. Shin and R. Junghans, unpublished |

TABLE IV
REPRESENTATIVE ESTABLISHED CELL LINES THAT ARE NONTUMORIGENIC IN NUDE MICE

| Cell line | Species | Tissue or cell line of origin | References |
|---|---|---|---|
| CV-1 | Monkey | Normal adult kidney | 4 |
| MDCK | Dog | Normal kidney | 18 |
| Balb/3T3 | Mouse | Embryonic fibroblasts | 3 |
| 3T3 (Swiss) | Mouse | Embryonic fibroblasts | 3 |
| SVR95 | Mouse | 3T3 (Swiss), partially transformed *in vitro* with SV40 | 4 |
| F1-SV | Mouse | Revertant of SV40-transformed 3T3 (Swiss) | 4 |
| M22 | Mouse | Revertant of KA31, Balb/3T3 transformed *in vitro* by Kirsten murine sarcoma virus | 4 |
| NRK Cl3 | Rat | Normal kidney | 20 |
| BRL | Rat | Normal liver | 18 |
| F2408 | Rat | Normal embryonic fibroblasts | S. Shin, unpublished |
| SVRE 12 | Rat | Embryonic fibroblast, partially transformed *in vitro* by SV40 | 4 |

*Reestablishment of Cell Cultures from Tumors*

Tumors produced in nude mice by established culture lines can easily be reintroduced into culture for further characterization. All steps should be carried out under strict aseptic conditions to prevent contamination of the cultures:

Sacrifice the tumor-bearing mouse, either by placing it in a chamber with a loose-fitting lid that contains Dry Ice or by cervical dislocation. Pin the animal to a cork board and move it into a cell culture hood for the

[14] C. D. Stiles, W. Desmond, G. Sato, and M. H. Saier, Jr., *Proc. Natl. Acad. Sci. U.S.A.* **72,** 4971 (1975).

[15] B. C. Giovanella, J. S. Stehlin, and L. J. Williams, *J. Natl. Cancer Inst.* **52,** 921 (1974).

[16] L. B. Chen, P. H. Gallimore, and J. D. McDougall, *Proc. Natl. Acad. Sci. U.S.A.* **73,** 3570 (1976).

[17] C. M. Croce, D. Aden, and H. Koprowski, *Proc. Natl. Acad. Sci. U.S.A.* **72,** 1397 (1975).

[18] C. D. Stiles, W. Desmond, L. M. Chuman, G. Sato, and M. H. Saier, Jr., *Cancer Res.* **36,** 1353 (1976).

[19] S. Shin, A. L. Brown, and F. C. Bancroft, *Endocrinology* **103,** 223 (1978).

[20] J. A. Bilello, V. H. Freedman, and S. Shin, *J. Natl. Cancer Inst.* **58,** 1691 (1977).

[21] C. H. Evans and J. A. DiPaolo, *Cancer Res.* **36,** 128 (1976).

[22] C. D. Stiles and A. A. Kawahara, *in* "The Nude Mouse in Experimental and Clinical Research" (J. Fogh and B. Giovanella, eds.), p. 385. Academic Press, New York, 1978.

subsequent operations. After spraying the area around the tumor with 95% ethanol, open the tumor with scissors and remove a section of tumor about 5 × 5 × 5 mm from an internal area free of necrotic tissue. Place the tumor tissue in a sterile petri dish and remove the mouse from the hood. Rinse the tumor tissue 3 times with at least 5 ml of phosphate-buffered saline or culture medium, placing the tissue into a new petri dish each time. Mince the tissue into very small fragments of 1 $mm^3$ or less and remove all obvious fatty material. Many tumors can be reduced to mostly single cells and cell clumps in this manner. Tumor tissues that are not easily dissociated into single cells can at this point be passed through a sterilized stainless-steel wire mesh to further disaggregate the cells by rubbing the fragments against the mesh with blunt-ended scissors. Alternatively, stepwise, gentle trypsinization can be used to dissociate single cells from tumor fragments. In any case, it is not necessary to obtain pure single-cell suspensions in order to start a new culture from the tumor. Suspend the cell or minced tissue preparation in 10–25 ml culture medium containing antibiotics, and distribute the contents into a set of 60-mm culture dishes. Change the medium after overnight incubation; most of the unattached lymphocytes and cell debris can be washed off at this time. Red cells, loosely attached to the dish, may last for many days, but will not prevent the growth of the cells. Tumors originally produced by established cell lines almost always give rise to vigorous cultures in a few days.

# [32] Biological Techniques for Avian Sarcoma Viruses

*By* ERIC HUNTER

Rous sarcoma virus (RSV), an avian retrovirus, is capable of inducing rapidly growing solid tumors of the connective tissue in fowl, and morphological transformation of cells in culture. This virus group is therefore of particular interest as a model system for studies on viral oncogenesis. Moreover, it has become clear that the avian sarcoma viruses provide an excellent system for studies of the control of eukaryotic gene expression and of the mechanism of integration of viral nucleic acids into a host cell genome.

Experiments of this nature require methods for obtaining a biologically and biochemically homogenous virus population and quantitating the biological activity of such a population. This article describes a quantitative and reproducible assay of RSV, the focus assay,[1,2] and methods for ob-

[1] R. A. Manaker and V. Groupé *Virology* **2,** 838 (1956).

[2] H. M. Temin and H. Rubin, *Virology* **6,** 669 (1958).




ISBN 0-12-181958-2

taining cloned preparations of RSV, through both the focus and soft-agar-colony assays. Large-scale growth of RSV is considered in this volume [33].

Infection of chick embryo fibroblasts by RSV *in vitro* results in a gradual change in the fibroblastic morphology of the infected cells to that of a rounded, refractile transformed cell. Division of the initially transformed cell, in addition to infection and transformation of adjacent cells, results in the formation of an island or "focus" of morphologically transformed cells within the uninfected monolayer of fibroblasts. A single particle of RSV is sufficient to produce a focus since the relationship between virus concentration and focus count is strictly linear. The shape of the transformed cell as well as the morphology of the focus are largely controlled by the viral genome,[3] so that a focus induced by the Schmidt-Ruppin strain of RSV is normally diffuse with interdispersed cells of normal morphology while those induced by the Bryan high-titer strains of RSV are generally compact and contain only transformed cells. To a lesser extent focus characteristics are also influenced by the physiological state of the cells in culture. For instance, cells that have been transferred several times may tend to respond with the production of transformed cells with a fusiform morphology.

Most strains of RSV are nondefective for replication; that is, infection of a cell with a single virion results in both infectious-virus replication and cell transformation. Viruses of the Schmidt-Ruppin strain of RSV, the Prague strain of RSV, and the B77 strain of avian sarcoma virus (ASV) are of this type. The most notable exception is the Bryan high-titer (BH) strain of RSV, a replication-defective virus that requires a coinfecting helper virus in order to produce infectious progeny. BH-RSV is a deletion mutant, lacking most of the envelope glycoprotein gene.[4] It can be complemented by an accompanying helper virus that provides the missing envelope glycoproteins that are necessary for infectivity.[5] BH-RSV strains exist primarily, therefore, as "pseudotypes" in conjunction with a nontransforming helper virus. Such a mixture may be unsuitable for some biochemical experiments where a single homogenous population of virus is required.

Viruses of the Schmidt-Ruppin and Prague strains of RSV and the B77 strain of ASV are most frequently used for biochemical studies, since they can be readily cloned and yield good titers of virus. The initial isolates of the Schmidt-Ruppin and Prague strains of RSV were mixtures of viruses

[3] H. M. Temin, *Virology* **10**, 182 (1960).

[4] P. H. Duesberg, S. Kawai, L.-H. Wang, P. K. Vogt, H. M. Murphy, and H. Hanafusa, *Proc. Natl. Acad. Sci. U.S.A.* **72**, 1569 (1975).

[5] H. Hanafusa, *Natl. Cancer Inst., Monogr.* **17**, 543 (1964).

with different host range characteristics. The Schmidt-Ruppin strain yielded virus of subgroups A, B, and D while the Prague strain was found to consist of viruses of subgroups A, B, and C.[6] Subgroup D virus of the Schmidt-Ruppin strain of RSV (SR-D) and the B77 strain of ASV (subgroup C) are worth noting since they infect mammalian cells with a relatively high efficiency. Indeed, there is evidence to suggest that B77 virus can infect rat cells as efficiently as it can chicken cells; with mammalian cells, however, infection may not necessarily be accompanied by expression of viral genetic information,[7] even though the cells contain an integrated provirus that can be rescued after fusion with permissive chicken cells.

## Preparation and Culture of Chicken Embryo Fibroblasts

### Materials

#### *1. Avian Embryos*

Fertile chicken eggs, most frequently of the white leghorn breed, may be obtained from commercial flocks, many of which have only a low (less than 5%) incidence of congenital leukosis virus infection. Among those breeders in the United States whose flocks have proven satisfactory are: Hy-Line International, Dallas Center, Iowa; H & N Inc., Redmont, Washington; and Spafas Inc., Norwich, Connecticut. Of these H & N and Spafas also offer eggs from leukosis-free flocks. For crucial experiments, however, tests for leukosis virus in the embryo are worthwhile.

All chicken cells contain, integrated within their chromosomal DNA, the genome of an endogenous leukosis virus, RAV-0. The degree to which the endogenous virus genes are expressed varies with different chicken flocks, and ranges from undetectable expression to assembly and release of infectious RAV-0 virions. In most commercial flocks release of complete virions is rare, but the synthesis of the internal structural or *gs* proteins and envelope glycoproteins of the virus is common. Since the latter cells can complement the BH-RSV strain by providing the missing envelope glycoproteins, they are generally termed chick helper factor (*chf*)-positive cells.

Detectable transcription of the endogenous viral genome may be undesirable, and for such studies, embryos from chicken flocks negative for

[6] P. K. Vogt, personal communication.

[7] D. Boettiger, *Cell* **3**, 71 (1974).

both *chf* and *gs* expression can be obtained from Spafas Inc. and H & N Inc. For crucial experiments, quantification of *chf* expression, as described later in this article, is recommended.

Fertile eggs of other avian species generally can be obtained from local vendors. Cells from the Japanese Quail and Peking Duck are particularly useful in some experiments since they appear to lack any endogenous viral sequences related to RSV.

### 2. *Sera*

*a. Calf Serum.* Not all serum lots offered commercially will support focus formation by RSV in tissue culture, and samples from several different lots should be tested before larger quantities are purchased. Satisfactory sera have been obtained from Biocell Inc., Venice, California; Flow Laboratories, Bethesda, Maryland; and Grand Island Biological Company, Grand Island, New York. Newborn calf serum is usually not required for cultivation of chick cells. In fact, bovine cadet serum, collected from animals up to 6 months of age, has proved highly satisfactory. Calf sera are not inactivated. They may be stored for as much as 1 year at −20°.

*b. Chicken Serum.* Chicken serum also may be purchased from the suppliers mentioned above. It should not contain antibody to RSV. Since these sera are not usually obtained from leukosis-free flocks, they *must* be inactivated at 56° for 2 hr prior to use in order to destroy any infectious-leukosis viruses that may be present. Storage up to 1 year at −20° is permissible.

### 3. *Media and Reagents*

*a. Cell Culture Media.* Several media are satisfactory both for the growth of chick embryo cells and assay of RSV. Nutrient mixture F10[8] is commonly used, but Minimal Essential Medium,[9] and Medium 199[10] may be substituted. Single- and double-strength concentrations are required. Commercially available solutions and powders are acceptable. The latter are particularly suitable where large volumes of media are used, since each batch can be tested in advance for its ability to support cell growth and focus formation. Sterilization of tissue culture media is achieved by pressure filtration through 0.22-μm (average pore size) filters. Penicillin

[8] R. G. Ham, *Exp. Cell Res.* **29,** 515 (1963).
[9] H. Eagle, *Science* **130,** 432 (1959).
[10] J. F. Morgan, H. J. Morton, and R. C. Parker, *Proc. Soc. Exp. Biol. Med.* **73,** 1 (1950).

($1 \times 10^5$ IU/liter) and streptomycin (100 mg/liter) are added prior to sterilization.

*b. Tris-Buffered Saline.* This contains 8 g/liter NaCl, 0.37 g/liter KCl, 0.1 g/liter Na $HPO_4$, 1 g/liter glucose, 3 g/liter Tris, $10^5$ IU/liter penicillin, 100 mg/liter streptomycin and is adjusted to pH 7.2 with 10 *N* HCl. Tris-buffered saline is used for washing cells and as a general diluent for trypsin and other reagents. It is sterilized by pressure filtration.

*c. Trypsin.* This is prepared as a 0.25% solution in Tris-buffered saline from Trypsin 1:300 (Grand Island Biological Company). The powdered trypsin is stirred overnight in $\frac{1}{10}$ the final desired volume, clarified, and filtered under pressure through a 0.22-μm filter. Prefiltration through filters of 0.8 μm and 0.45 μm average pore size can help prevent rapid clogging of the final 0.22-μm filter. The sterile 0.25% solution may be stored at −20°, and distribution in volumes of approximately 100 ml makes it feasible to use freshly thawed trypsin for preparation of primary chick embryo cultures.

*d. Tryptose Phosphate Broth.* Bacto tryptose phosphate broth (Difco Laboratories) is prepared according to the manufacturer's directions and sterilized by autoclaving.

*e. Agar.* Purified agar (Difco Laboratories) is prepared at a concentration of 1.8% in double-distilled water and sterilized by autoclaving. Purified agar is preferable to Bacto agar (Difco Laboratories) for the nutrient overlay of primary cultures, but is unsuitable for use in the RSV focus assay where Bacto agar is used.

*f. Sodium Bicarbonate.* A 2.8% solution of sodium bicarbonate is prepared in distilled water and sterilized by autoclaving.

*g. Folic Acid.* A ×100 stock solution of folic acid is prepared by dissolving 80 mg folic acid in 100 ml of 1 *N* sodium bicarbonate. The solution is sterilized by filtration and stored in small aliquots at −20°.

*h. Vitamins.* A ×100 stock solution of the following nonessential vitamins may be purchased commercially or prepared in the laboratory and sterilized by filtration: biotin, calcium pantothenate, choline chloride, inositol, niacinamide, pyridoxine · HCl, riboflavin, thiamine · HCl.

*i. Dimethyl Sulfoxide.* Analytical grade dimethyl sulfoxide (DMSO) is used as an additive to media and for freezing cells.

### 4. *Complex Media Formulations*

The media solutions and reagents described above are combined to form complex media, such as liquid growth medium for primary cultures (PGM); semisolid "overlay" medium for primary cultures (POL); growth medium for secondary cultures (GM); seeding medium (SM); and agar

TABLE I
COMPLEX MEDIA FOR PRIMARY AND SECONDARY CULTURES OF CHICK EMBRYO CELLS

| Media | PGM | POL | SM | GM | OL |
|---|---|---|---|---|---|
| Nutrient mixture F10,[a] ×1 concentration | 80[b] | 40 | 80 | 80 | — |
| Nutrient mixture F10,[a] ×2 concentration | — | 20 | — | — | 40 |
| Calf serum | 8 | 8 | 1 | 5 | 5 |
| Chicken serum | 2 | 2 | — | — | — |
| Tryptose phosphate broth | 10 | 10 | 10 | 10 | 10 |
| Sodium bicarbonate (2.8%) | 2.0 | 2.0 | 2.0 | 2.0 | 2.0 |
| Purified agar[c] | — | 20 | — | — | — |
| Bacto agar | — | — | — | — | 40 |
| Dimethyl sulfoxide | — | — | — | 1.0[d] | 1.0 |

[a] Other culture media may be substituted for nutrient mixture F10; see text.
[b] Expressed in parts of total volume.
[c] Purified and Bacto agar (1.8%) are melted in a boiling-water bath or microwave oven. Both agar and media components are then equilibrated in a 45° water bath prior to mixing.
[d] GM is supplemented with 1% DMSO when virus is to be harvested from transformed cells; see text.

overlay medium for the RSV focus assay (OL) (Table I). The formulations for the bottom agar (BA) and top agar (TA) media of the soft-agar-colony assay and for cloning medium (CM) used in conjunction with this assay are presented in Table II.

TABLE II
COMPLEX MEDIA FOR SOFT-AGAR CLONING OF RSV

| Media | BA | TA | CM |
|---|---|---|---|
| Nutrient mixture F10, ×1 concentration | — | — | 70 |
| Nutrient mixture F10, ×2 concentration | 35[a] | 20 | — |
| Calf serum | 10 | 6 | 10 |
| Chicken serum | 3 | 2 | 3 |
| Tryptose phosphate | 15 | 10 | 15 |
| ×100 vitamins | 1.0 | 1.0 | 1.0 |
| ×100 folic acid | 1.0 | 1.0 | 1.0 |
| Quail conditioned medium[b] | — | 40 | — |
| Bacto agar | 35 | 20 | — |
| Dimethyl sulfoxide | — | — | 0.5 |

[a] Expressed in parts of total volume.
[b] PGM harvested from quail primary culture plates 72 hr after seeding.

## Cell Culture Methods

### *1. Preparation of Primary Chick Embryo Cell Cultures*

Fertile chicken eggs, incubated for 9–11 days, are candled and the position of the air sac marked. The shell above the air sac is swabbed with 70% alcohol and cracked with an egg punch (Tri R Instrument Co., Jamaica, New Jersey), which allows this piece of the shell to be removed essentially in one piece with a pair of sterile forceps. From this point, it is important to use separate instruments for each embryo in order to avoid possible cross-contamination. Using sterile curved forceps for each operation, the air sac membrane is peeled away carefully and the embryo removed to a sterile petri dish. The embryo is decapitated with the aid of another pair of sterile forceps, and then it is transferred to a 50-ml conical centrifuge tube (Corning or Falcon Plastics). Contaminating red blood cells are removed by addition and removal of 10 ml of Tris-buffered saline, and the embryo is minced to 1–2 mm square pieces using the spoon-shaped end of a sterile stainless-steel spatula. Ten milliliters of warm, freshly thawed trypsin (0.25%) are added, and the tissue suspension is pipetted up and down *gently* 10–12 times in a 25 ml wide-mouth serological pipette in order to free cells from the tissue matrix. Pieces of tissue and cell clumps are allowed to settle, and the supernatant liquid is decanted into 20 ml of ice-cold PGM. Gentle swirling ensures that further digestion of the cells in suspension is inhibited by the high levels of calf serum in the PGM. Five milliliters of warm Tris-buffered saline are added to the remaining tissue, and cells that are freed by pipetting gently as before are again decanted into ice-cold PGM after any tissue pieces have been allowed to settle. This washing procedure with Tris-buffered saline is carried out of a total of 4–5 times and is usually sufficient to free most cells from the embryo tissue. If significant amounts of tissue remain at the end of this procedure, a further 5 ml of warm trypsin may be added, and the washing procedure repeated.

The cell suspension obtained is pelleted by low-speed centrifugation in an PR2 refrigerated centrifuge I.E.C. or equivalent (600–800 rpm, 10 min), and *gently* resuspended in 20 ml of PGM by means of a wide-mouth pipette or a Vortex mixer. Since some tissue pieces and cell clumps are unavoidably transferred during decanting, these are allowed to settle and the supernatant liquid is carefully decanted once more to a sterile 50-ml centrifuge tube. A small aliquot is removed, diluted 1 : 10 in PGM, and the number of cells determined by counting the suspension in a hemocytometer. Care should be taken to avoid counting red blood cells (elipse-shaped, nucleated cells) and badly damaged embryo cells. The yield from a 10-day-old embryo should be between 1.0 and $1.5 \times 10^8$ viable cells.

Approximately 0.8–1.2 $\times$ $10^7$ cells are required for tests of susceptibility to RSV infection and endogenous virus gene expression; the remainder may be seeded in 100-mm petri dishes in 10 ml PGM (5–7 $\times$ $10^6$ cells/plate) for later use as secondary cultures. Plates containing PGM are equilibrated in a 37–38° incubator, gassed with 95% air–5% $CO_2$, for 1–2 hr prior to seeding, and returned to it immediately after addition of the cells. The primary cultures are overlayed with POL (12.5 ml/100-mm plate) 3 days after seeding. This stimulates the growth of the cells and maintains them in a viable, healthy condition for up to 9 days after seeding.

The general procedure described above can be used to establish primary embryo cultures from most avian species, with only minor modifications to the method described here.

### 2. *Assay of Primary Cells for Susceptibility to Virus Infection*

In general, chicken embryos obtained from the larger commercial breeders, particularly Spafas and H & N, will show uniform susceptibility to specific virus subgroups. For example, H & N offer embryos susceptible to all subgroups (C/O) or susceptible to only A and C subgroups (C/BDE). However, it is worthwhile to test the cells from each embryo in order to confirm the susceptibility of the cells to the particular virus under study and, in the case of less well defined flocks, to detect possible congenital infection with avian leukosis virus.

A portion of the cells from each embryo is seeded into 35-mm petri dishes containing 2 ml PGM (1.0 $\times$ $10^6$ cells/plate), and within 4 hr of seeding these cultures are infected with both a high and a low concentration of the RSV under study. In general, approximately 1 $\times$ $10^4$ and 1 $\times$ $10^2$ focus-forming units are added to the respective plates. Viruses of subgroups A, B, and C are routinely added in this fashion to each embryo to ascertain its susceptibility to RSV. The inoculum is left on the cells overnight, and the following day the liquid medium is replaced with 3 ml of OL which is allowed to harden at room temperature for 10–20 min before the plates are returned to the incubator.

A qualitative estimate of susceptibility to the various virus subgroups can generally be obtained on the third or fourth day after seeding by observing the morphology of the cells on those plates receiving the most virus. Since the large uninfected plates are generally ready to be used for secondary cultures at this time, cells with the desired susceptibility can be assured.

The number of foci developing on those plates receiving the lower quantity of virus can be determined 7 days after infection and used as a quantitative measure of the susceptibility of the cells.

Cellular resistance of primary chicken cultures to a particular RSV can result from an inheritable characteristic in the cells or from congenital infection of the embryo with an avian leukosis virus. The latter type of resistance can be transmitted with cell-free supernatant fluids to susceptible cultures and is due to viral interference, whereas the former cannot. Without carrying out this type of test it is not possible to distinguish between the two types of resistance. However, as mentioned previously, the flocks of the large commercial breeders have well-defined susceptibility to virus infection. Furthermore, Vogt[11] has pointed out that in commercial white leghorn flocks most resistance against subgroup B is of genetic origin, whereas practically all resistance to subgroup A is caused by congenital infection with a subgroup A leukosis virus.

### 3. *Assay of Primary Cells for Chick Helper Factor (chf) Expression*

For some experiments it is necessary to know the extent to which the endogenous virus genes are being expressed. The most convenient method for assaying such expression is to determine the ability of the cells to complement the defective Bryan high-titer (BH) strain of RSV in the chick helper factor test.[12,13] For this test, cells are inoculated with RSV(RAV-7), a virus mixture obtained after dual infection of cells with BH-RSV and the leukosis virus, RAV-7. The BH-RSV in this mixture is referred to as a pseudotype since, having incorporated the envelope glycoproteins of RAV-7, it possesses the host range and antigenic properties of the leukosis virus. Infection of embryo cells that are not expressing *chf* will result in the replication of more RSV(RAV-7) with the host range of RAV-7, whereas infection of cells expressing *chf* results in the synthesis of both RSV(RAV-7) and RSV(RAV-0); the latter is a BH-RSV pseudotype that has incorporated the envelope glycoproteins of the endogenous leukosis virus, RAV-0. The virus produced from "*chf*-positive" cells thus has the host-range of both RAV-7 (subgroup C) and RAV-0 (subgroup E) and can be distinguished from RSV(RAV-7) by its ability to form foci on quail embryo cells. RSV(RAV-7) is a particularly useful pseudotype virus for the *chf* test since, having the subgroup C host-range, it can infect most commercially available chicken embryo cells; it differs from most subgroup C viruses, however, in that it is com-

[11] P. K. Vogt, *in* "Fundamental Techniques in Virology" (K. Habel and N. P. Salzman, eds.), p. 198. Academic Press, New York, 1969.

[12] T. Hanafusa, H. Hanafusa, and T. Miyamoto, *Proc. Natl. Acad. Sci. U.S.A.* **67,** 1797 (1970).

[13] R. A. Weiss, W. S. Mason, and P. K. Vogt, *Virology* **52,** 535 (1973).

pletely excluded from quail embryo cells, and thus provides a very sensitive system for the detection of RSV(RAV-0) pseudotypes.

For the test, a portion of the embryo cells ($2 \times 10^6$/plate) are seeded into a 35-mm petri dish containing 2 ml of PGM + polybrene (2 μg/ml—see below) and inoculated with approximately $10^5$ focus-forming units of RSV(RAV-7) shortly after seeding. On the second day after seeding, 1 ml of the culture medium is removed and assayed on secondary cultures of quail embryo cells for infectious virus as described in the following section. Control plates are inoculated with RSV(RAV-7) and appropriate dilutions of a stock of RSV(RAV-0). Culture medium from embryos that are *chf*-positive will generally induce $10^2$ to $10^3$ foci in this quail focus assay.

## Quantitation of Transforming Virus: The Focus Assay

### *1. Preparation of Secondary Chicken Embryo Cultures*

Quantitative assays of RSV are generally carried out on secondary chick embryo cultures since more cells are available and the population is more uniform. Two methods for making secondary cultures have been employed in this laboratory and are described below.

#### A. Method 1

The soft-agar overlay is aspirated or poured gently from the primary cultures, and the cell monolayer is washed with warm Tris-buffered saline. Warm trypsin (0.05% in Tris-buffered saline, 5 ml/100-mm dish) is added, and the plate is rocked gently for 1 min after which the trypsin is aspirated from the dish. When the monolayer takes on a granular appearance, the cells are removed from the surface of the dish in 5 ml of GM by means of a 5-ml hand pipette and rubber bulb of corresponding size. The cell suspension, which should consist predominantly of single cells, is transferred to a sterile tube; the plate is rinsed with another 5 ml of GM to remove any remaining cells. This washing is added to the tube, and after gentle mixing, the total number of cells obtained is determined with a hemocytometer or an electronic cell counter; between $10–20 \times 10^6$ cells/dish can be expected from one 100-mm primary plate, depending on the number of days the cells have been in culture. Cells are seeded in 60-mm tissue culture dishes containing 4 ml of GM ($0.9–1.2 \times 10^6$ cells/dish; other sizes of dishes receive cells in proportion to their surface area, e.g., $4–5 \times 10^5$ cells/35-mm dish). The density at which cells are seeded is particularly important for the RSV focus assay since overseeding can lead to suppression of focus formation.

B. METHOD 2

The medium is removed from the primary cultures and the monolayers washed with Tris-buffered saline as described above. Only 2.5 ml of 0.05% trypsin are added to a 100-mm plate, and this is left on the cells until the monolayer takes on a granular appearance. The cells are suspended with a hand pipette and rubber bulb and are added to 3 ml of transfer buffer (TB: 10% calf serum in Tris-buffered saline). After determining the cell concentration, the suspension is seeded into 60-mm tissue culture dishes containing 3 ml SM, at $1.2 \times 10^6$ cells/plate. The cells settle and spread rapidly in this medium, which is changed, 0.5–2 hr after seeding, to 5 ml GM.

Each method has its advantages. The first generally avoids any possibility of overtrypsinizing the cell monolayer since excess trypsin is removed from the plates. It also avoids the obligatory medium change of method 2 that can be excessively time consuming in large assays. In general, assays of RSV set up by method 2 have somewhat cleaner background monolayers since unhealthy or dying cells are removed with the SM.

### 2. *Inoculation and Agar Overlay*

Secondary chick embryo cultures may be inoculated immediately after seeding if prepared according to method 1 or after the medium change in method 2. In general, virus is added to the cells within 2–4 hr after seeding. The cells remain susceptible to virus infection for about 1 day after seeding, but inoculation with virus beyond this time reduces the efficiency of the assay since one round of cell division appears necessary for the expression of viral genes.[14,15] Suitable dilutions (usually $10^{-1}$, $10^{-2}$, $10^{-3}$) of the virus are inoculated directly into the GM and left on the cells overnight. The next day, the GM is replaced by OL (7 ml/60-mm plate), which is allowed to harden for 10 min at room temperature before the plates are returned to the incubator. The extended period of incubation of virus and cells prior to addition of OL is acceptable since under conditions of low multiplicity infection progeny virus is not detected until 16–18 hr after infection. Secondary infection by progeny virus is thereby avoided.

The efficiency with which viruses of subgroups B, C, D, and E infect cells is increased significantly (up to a 40-fold enhancement for focus formation) if polycations are added to the infecting medium.[16] Of these,

[14] E. H. Humphries and H. M. Temin, *J. Virol.* **10,** 82 (1972).
[15] E. H. Humphries and H. M. Temin, *J. Virol.* **14,** 531 (1974).
[16] K. Toyoshima and P. K. Vogt, *Virology* **38,** 414 (1969).

polybrene (Aldrich Chemical Co.) is the most efficient and least toxic to cells. It may be added to the GM (2 μg/ml) on a routine basis unless viruses of subgroup A are to be assayed, since infection by these viruses is inhibited to a limited extent by polycations.

Focus assays on secondary cells of avian species other than chicken are carried out in essentially the same manner. However, the OL added in focus assays on quail embryo cells should always contain 1% chick serum since this increases the efficiency of the assay severalfold.

### 3. *Quantitation of Foci*

The focus assay is usually counted 6–8 days after infection, although foci may be visible from the fifth day. In order to quantitate the number of foci a glass plate with a 2-$mm^2$ grid (Technical Instrument Co., San Francisco, California) is attached to the slide holder on the mechanical stage of an inverted microscope. The optimum total magnification for counting foci is between ×30 and ×40 since this allows a 4 mm wide section of the plate to be counted at one time. A ×10 objective may be useful for determining fine details of focus morphology. Cell clumps may resemble foci, but generally can be distinguished from them by the presence of cells radiating from the clump and by the absence of the rounded, refractile cell type that characterizes RSV foci. Counting foci is usually facilitated if 1–2 ml of GM are added on top of the agar 2 hr prior to counting since this appears to increase the refractile nature of the transformed cell and distinguishes it further from the background monolayer.

## Cloning of RSV

The use of a homogenous virus population is crucial in many biochemical experiments. This is particularly true for RSV since in addition to the natural variation that might be expected in a virus population, the growth of nondefective sarcoma viruses in cells is accompanied by the generation of transformation-defective (*td*) mutants of RSV.[17,18] In fact, continued passage of a virus stock can result in a population that is predominantly transformation-defective virus. Therefore, it is important to produce cloned virus stocks with which to perform experiments with RSV. Two methods are currently used to clone RSV: in the first, the virus is cloned in the focus assay described above; in the second, the virus is cloned by means of the soft-agar-colony assay.

[17] P. K. Vogt, *Virology* **46,** 939 (1971).

[18] S. Kawai and H. Hanafusa, *Virology* **49,** 37 (1972).

*1. Focus Assay Cloned RSV*

Appropriate dilutions of the virus to be cloned are inoculated onto secondary chick cells so that between 10 and 20 foci will develop on each plate. The focus assay is carried out in the normal manner, except that an intermediate reoverlay with 2 ml OL can be useful on the fifth day. A well-isolated individual focus is chosen, and the transformed cells together with a small plug of overlying agar are withdrawn with a finely drawn-out sterile Pasteur pipette. This material is transferred to a tube containing 1 ml of TB, sonicated gently to destroy live cells, and used as a fresh inoculum in a second focus assay. Several dilutions are again inoculated onto secondary chick cells in order to obtain widely separated foci. A total of three such clonings are generally carried out prior to growing up stocks of the virus. Usually, a focus from the third serial focus assay is transferred directly onto a monolayer of secondary chick embryo cells ($1.2 \times 10^6$/60-mm plate). The transfer of these cells, after 4–5 days, directly to a 100-mm plate results in complete transformation of the cell population.

Virus can be harvested at this stage for future experiments. Significantly higher titers of cloned virus are obtained if the GM is supplemented with 1% DMSO, and supernatant fluids are harvested on a daily basis. Variations on this method are used by others (see this volume [33]).

Preparation of cloned virus stocks in this manner is sufficient for most experiments. However, in this method the virus in the final cloned stock is the result of several cycles of infection and will, therefore, even at this stage, contain a small *td* virus component. In those experiments in which any *td* virus is unacceptable the following method should be used.

*2. Soft-Agar-Colony Assay Cloned RSV*

The generation of td virus appears to be associated with a stage in the replication of the nondefective sarcoma virus and may in fact occur during reverse transcription.[19] Cloned sarcoma virus lacking a *td* component can thus be obtained if a single RSV transformed cell is propagated into a virus-producing clone of cells. Since cells transformed by RSV acquire the ability to grow as colonies when suspended in a soft-agar medium,[20] it is possible to obtain clones of cells that have been transformed by a single nondefective sarcoma virus.

The method for the agar-colony assay[21] employs a bottom layer of

[19] J. Hillova, D. Dantchev, R. Mariage, M. P. Plichon, and M. Hill, *Virology* **62**, 197 (1974).
[20] H. Rubin, *Exp. Cell Res.* **41**, 149 (1966).
[21] J. A. Wyke and M. Linial, *Virology* **53**, 152 (1973).

0.6% agar medium (BA) containing Japanese quail feeder cells and a top layer of 0.36% agar medium (TA) into which infected cells are suspended. Bacteriological dishes (60 mm) are used in preference to tissue culture dishes in order to prevent adherence and overgrowth of the quail feeder cells. A volume of 3.0 ml of bottom agar containing $5 \times 10^5$ quail cells is added to each dish and allowed to harden at room temperature.

Chicken cells are infected in suspension by mixing $1–2 \times 10^6$ cells in 1 ml GM with virus at a multiplicity of infection of between 0.1 and 0.01 focus-forming units per cell. The mixture is incubated at room temperature for 1–2 hr, after which aliquots of dilutions of the infected cells are added to 3 ml of TA in tubes at 45°. The agar suspension of cells is then poured immediately onto the bottom agar medium. After the top agar has hardened (20–30 min), the dishes are moved carefully to incubate at 37°. At day 4 and day 7, dishes are overlaid with 2 ml of OL medium. Colonies of transformed cells are generally visible to the naked eye by 7–12 days.

In some instances clumping of cells can occur during the 1–2 hr incubation of cells and virus. To ensure that only single cells are suspended in the top agar layer, the cell–virus mixture may be seeded into a 35-mm plate and retrypsinized from the plate 4–6 hr later.

In order to establish transformed cell clones, colonies (about 1 mm in diameter) are aspirated into a finely drawn-out Pasteur pipette and placed into 16-mm wells of a 24-well cluster plate (Costar, Cambridge, Massachusetts) containing 1 ml of cloning medium. In general, no more than 12 colonies should be grown up in each 24-well plate in order to reduce the possibility of cross-contamination. Plates are incubated at 37° and checked every second day to determine the stage of growth. When the cells are confluent they can be transferred to a 35-mm culture dish and from there, via a 60-mm dish, to a 100-mm plate, from which cloned virus can be harvested. It is important when transferring transformed cell clones not to underseed the tissue culture dish; in general, seeding cells onto twice the original area results in rapidly growing healthy cells. Once established, transformed cell clones can be grown in GM containing 1% DMSO.

## Frozen Storage of Chick Embryo Fibroblasts

Chick embryo cells can be frozen and stored in a medium containing high concentrations of serum and DMSO. The cells are trypsinized and suspended in GM as described previously. After centrifugation (600–800 rpm, 5 min), the cells are resuspended in GM at a concentration of about $20 \times 10^6$ cells/ml. An equal volume of freezing medium (11 parts GM, 5 parts calf serum, 4 parts dimethyl sulphoxide) is added drop by drop, and the cell suspension is added in 0.5-ml aliquots to 2-ml freezing vials (Vangard International, Neptune, New Jersey). The suspension is allowed to freeze slowly (1°/min is optimal) to −70° before being transferred to a liquid nitrogen freezer. The cells can be stored at −70°, with reduced survival.

Acknowledgments

Many of the techniques described in this article were developed in the laboratory of Dr. P. K. Vogt, University of Southern California. The work of the author is supported by Grant VC-215 from the American Cancer Society.

# [33] Large-Scale Growth of Rous Sarcoma Virus

*By* RALPH E. SMITH

Biochemical investigations frequently require large quantities of cells and virus. For example, the investigator may wish to isolate a specific subcellular component that comprises a small fraction of the total material present in the cell. In addition, biochemical investigations of viral structural components frequently require milligram quantities of purified virus. Finally, the investigator may need to prepare large quantities of highly infectious virus to take advantage of the unique ability of nondefective Rous sarcoma virus (RSV) to rapidly and uniformly transform all cells in a susceptible population. This article describes techniques for the growth of transformed cells. It also outlines procedures for the preparation of purified RSV. Certain of the biological techniques that are useful for working with the avian sarcoma viruses are presented separately in this volume [32].

## Media Reagents, Viruses, and Cells

### *Media and Reagents*

*Growth Medium (GM).* Ham's F10 nutrient mixture (Flow Laboratories) is supplemented with 10% (v/v) tryptose phosphate broth (Difco), 5% calf serum (Gibco), streptomycin (50 μg/ml), penicillin (50 U/ml), Fungizone (2 μg/ml), and dimethylsulfoxide (1%). Vitamin $B_{12}$ (Gibco; crystalline) is added to a concentration of 0.1% (w/v).

*Maintenance Medium (MM).* Ham's F10 is supplemented as above for growth medium, except that the calf serum concentration is reduced to 1%.




ISBN 0-12-181958-2

*Primary Growth Medium (PGM).* Ham's F10 is supplemented as above for growth medium, except that the serum concentration is 8% calf serum and 2% chicken serum (heat inactivated at 56° for 1 hr).

*Overlay Medium.* Part A: 2% Bacto agar (Difco) is prepared in water, autoclaved, and stored at room temperature. Just before use, the agar is melted in a boiling-water bath, cooled to 45°, and mixed 40 : 60 with part B at 45°. Part B: Double-strength Ham's F10 (prepared by reconstituting powdered medium with half the usual amount of water), 10% calf serum, 20% tryptose phosphate broth, 100 $\mu$g/ml streptomycin, 100 U/ml penicillin, and 4 $\mu$g/ml Fungizone (Squibb No. 43760, E. R. Squibb and Sons, Inc., Princeton, New Jersey).

*Phosphate-Free Maintenance Medium.* Maintenance medium is made from powdered Ham's F10 that has no phosphate (available as a special order from most media suppliers), made without tryptose phosphate broth, and buffered with 5 m*M* Tricene at pH 7.2.

*Polybrene.* Available as desiccated powder from Aldrich Chemical Co. Prepare as a 2 mg/ml stock in water, autoclave, and store at 4°. Dilute in Tris-buffered saline 1 : 10 to make a working stock of 0.2 mg/ml; add 0.05 ml (one drop from a Pastuer pipette is sufficiently accurate) to 5 ml medium to make a 2 $\mu$g/ml solution.

*Dilution Buffer.* Ten percent calf serum in Tris-buffered saline.

*Tris-Buffered Saline.* 0.14 *M* NaCl, 25 m*M* Trizma base, 5 m*M* glucose, 5 m*M* KCl, 0.7 m*M* $Na_2HPO_4$, 100 Ug penicillin/ml, 100 $\mu$g streptomycin/ml. The pH is adjusted to 7.4 with 1 *N* HCl (final concentration of HCl is approximately 0.018 *N*), and the material is sterilized by filtration.

*Trypsin in Tris-Buffered Saline.* Trypsin (Gibco catalog no. 70730), at a final concentration of 0.25%, is dissolved by constant stirring for 3–4 hr in Tris-buffered saline and sterilized by filtration.

*TE.* 5 m*M* Tris · HCl (pH 8.6) and 1 m*M* sodium ethylenediaminetetraacetate.

*STE.* 0.1 *M* NaCl, 10 m*M* Tris · HCl (pH 7.5), and 0.1 m*M* sodium ethylenediaminetetraacetate.

*Viruses*

Any of the sarcoma virus strains can be used for large-scale culture. The most commonly used strains are the Schmidt-Ruppin strain of RSV, the Prague strain of RSV, and the B77 strain of avian sarcoma virus. Although most are used interchangeably, differences exist which merit discussion. Both the Schmidt-Ruppin and Prague strains were originally mixtures of viruses that have been cloned to reveal three viruses with different antigenic and host-range properties. The Schmidt-Ruppin RSV strain was cloned into members of the A, B, and D subgroups, while the Prague RSV strain was purified into members of the A, B, and C subgroups.[1]

Members of the A subgroup are difficult to work with because extracellular virus does not accumulate reproducibly to a high level. The reason for this is not clear, but seems to be related to the tendency of viruses of subgroup A to accumulate at the cell surface in large aggregates, which are difficult to remove and purify. An occasional preparation will show a high titer of virus, but this result is not reproducible. Viruses of subgroups B and D are slightly cytopathogenic, and prolonged passage of infected cells is usually difficult. In contrast, viruses of subgroup C (the B77 avian sarcoma virus and the C subgroup of Prague RSV) are attractive, since they are not cytopathogenic and extracellular virus accumulates readily.

*Cells*

Chicken cells are used for most biochemical procedures. Methods for the preparation of chicken embryo fibroblasts (CEF) will not be given, since detailed descriptions are available[2] (see also this volume [32]). However, a brief discussion of the appropriate cell for study is useful. All chicken cells possess an endogenous virus (RAV-0), which may be either completely latent or expressed as infectious particles. For certain biochemical applications, it may be important to know whether the endogenous virus is being expressed in infected cells. Procedures have been published for determining the level of RAV-0 production.[3] Breeding programs have been initiated to provide fertile eggs from flocks that are free of RAV-0 expression. Such fertile eggs are available from commercial sources (SPAFAS, Inc., Storrs, Connecticut; Heisdorf and Nelson, Redmon, Washington).

Fibroblasts that are free of the endogenous virus of the chicken may be obtained from other avian species, but these cells may have endogenous viruses of their own. Further, other avian species often have unusual susceptibility patterns, and only certain subgroups of virus may be used efficiently. The most common nonchicken avian species are the Japanese quail and the Pekin duck. Both are readily available from commercial breeders (Truslow Farms, Inc., Chestertown, Maryland) and can be purchased throughout the year. Other more exotic avian species, such as the

[1] J. Tooze, "The Molecular Biology of Tumor Viruses." Cold Spring Harbor Lab., Cold Spring Harbor, New York, 1973.

[2] P. K. Vogt, *in* "Fundamental Techniques in Virology" (K. Habel and N. P. Salzman, eds.), p. 198. Academic Press, New York, 1969.

[3] See this volume [32].

pheasant, are usually more difficult to obtain, and fertile eggs may be available only once per year.

Mammalian cells can be transformed by avian sarcoma viruses, but replication of the virus is blocked. Mammalian cells can therefore be used to grow virus-transformed cells in the absence of virus replication. Infection of mammalian cells is an inefficient process, and large quantities of highly infectious virus are usually required to obtain a few transformed cells. Furthermore, one should keep in mind that the mammalian cells should be frequently recloned, and cells selected for the particular attribute desired.

## Preparation of Virus Stocks

The preparation of high-titer, well-characterized stocks of virus is a key ingredient for the successful use of RSV in biochemical investigations. Stock virus preparation is usually carried out in two stages: *cloning* of the virus stock and preparation of an *intermediate* stock.

*Cloning*. RSV cloning is started by performing a focus assay, in which a monolayer ($1.2 \times 10^6$ CEF in 5 ml GM/60-mm tissue culture dish) is infected with 10-fold dilutions of RSV (dilutions are prepared in dilution buffer). Infection with subgroups B, C, and D RSV are performed in the presence of 2 $\mu$g/ml polybrene. After 18–24 hr, the growth medium is replaced with overlay medium, and the cells are incubated for an additional 7 days (an intermediate overlay of 3 ml of overlay medium at day 4 is usually helpful). Individual well-isolated foci are identified by visual inspection and removed by drawing up a plug of agar directly over the focus with a drawn-out sterile capillary pipette. The small amount of material in the capillary pipette is transferred to 1 ml of dilution buffer. After brief sonication in an ultrasonic bath, this material may be used to start a second round of cloning, since sufficient virus is usually present to obtain foci at a $10^{-1}$, $10^{-2}$, or $10^{-3}$ dilution. Using this procedure, a virus can be consecutively cloned 3 times in as many weeks.[4]

*Intermediate Stock Preparation*. The relatively low virus titer of the clone necessitates the preparation of a consistent, high-titer intermediate virus stock for routine use. After investment of a large amount of time and effort in the cloning of the seed virus, it is desirable to preserve the clone in a pristine condition. This is accomplished by leaving the clone in its milliliter of dilution buffer and entering the tube infrequently. To prepare the intermediate stock, 0.1 ml of fluid from the isolated clone is put onto a monolayer ($1.2 \times 10^6$ CEF/60-mm dish) in the presence of 2 $\mu$g/ml of

[4] T. Graf, H. Bauer, H. Gelderblom, and D. Bolognesi, *Virology* **43**, 427 (1971).

polybrene, and cells are allowed to grow without overlay until all the cells are transformed (one to two transfers). If other viruses are being grown at the same time, it is a good practice to set aside a bottle of medium specifically for the feeding of this stock in order to minimize the possibility of cross-virus contamination. The size of the intermediate stock is contingent upon the anticipated use and available storage space. One should attempt to harvest sufficient virus to last for 1–2 years; otherwise, the frequent removal of fluid from the original clone will necessitate recloning at an early date. Storage should always be at −70° or lower. Aliquots of the intermediate stock should be made in sufficiently small volume that repeated freezing and thawing are not necessary, since the titer of the virus will then drop below a useful level. An effective method is to prepare several large containers of the virus stock (150 ml in a screw-cap 200-ml polyallomer Sorvall centrifuge bottle works well), and then thaw and dispense the virus into smaller samples (10 ml or less) as the need arises.

### Large-Scale Culture

Large amounts of virus are usually prepared by culture of virus-producing cells in roller culture bottles. Roller culture bottles are generally maintained at 37°, and since the bottles are sealed with screw caps, humidification and carbon dioxide addition are unnecessary. Roller culture bottles are rotated at the rate of 6–10 revolutions per hour.

#### *Preparation of Cells*

There are two common methods for the preparation of virus-producing cells in roller culture bottles. In method A, a small number of cells are infected with virus in plastic tissue culture dishes, and these cells are transferred until sufficient numbers are available to seed roller culture bottles. The advantage of this method is that little intermediate stock virus is used, and few cells are required for the initial seeding. Further, the cells transform during passage in plastic dishes, and transformed cells are easier to grow than nontransformed cells.

Specifically, method A involves the following procedure. Secondary CEF are seeded into 150-mm plastic tissue culture dishes at $20 \times 10^6$ cells/dish and infected with 0.2 ml virus (undiluted) per dish. Cells are subsequently trypsinized and replated into plastic tissue culture dishes at $20 \times 10^6$ cells/dish every 2–3 days until sufficient number of cells are available for a roller culture bottle. In order to estimate the number of cells required for a roller culture bottle, approximately $4 \times 10^7$ cells can be obtained from a single 150-mm dish of confluent CEF. A single roller culture bottle (1330 $cm^2$; Bellco) requires inoculum from about five tissue culture dishes. Cells are seeded into the roller culture bottle ($2 \times 10^8$ cells in 200 ml growth medium per bottle), and the medium is replaced the next day. After an intervening day without feeding, the growth medium in the roller culture bottle is replaced daily. After about 4–7 days, the cells can be further subdivided by trypsinization and transfer to roller culture bottles. The following procedure is used to trypsinize confluent CEF in roller culture bottles. The medium is poured off, and 50 ml of warm 0.25% trypsin in Tris-buffered saline are added to the bottle. The bottle is tilted to distribute the trypsin solution over the entire surface of the bottle, and then the trypsin solution is poured off. When visual inspection indicates that the cell sheet has started to become granular (a few seconds is generally sufficient), 100 ml of growth medium are added, and the bottle is vigorously rotated by hand to strip the cells from the surface of the bottle. The cell suspension is transferred to a glass bottle and set aside momentarily. Since the roller culture bottle that has been trypsinized almost always has cells remaining on the glass surface, it is fed with 200 ml of growth medium and returned to the roller rack. The trypsinized cells are counted and $2 \times 10^8$ cells are seeded into each new roller culture bottle in 200 ml of growth medium. This procedure can be repeated as frequently as necessary to obtain the desired number of roller culture bottles. Four to five days after seeding, sufficient numbers of cells can be recovered to seed at least one new bottle (which amounts to a 2 : 1 split, since the trypsinized bottle is retained).

One disadvantage of method A is that 2 weeks may be required to obtain sufficient cells for seeding a single roller culture bottle. Although this time interval makes it difficult to respond to fluctuations in experimental needs, one can compensate by having several large-scale stocks in various stages of preparation. For example, it may be advisable to start new large-scale preparations weekly.

In method B, cells are obtained for seeding roller culture flasks by trypsinizing chick embryo fibroblasts, and the cells are seeded directly into roller culture bottles without an intervening period of growth in plastic tissue culture dishes. One advantage of method B is that the investigator can rapidly respond to experimental needs by preparing large numbers of roller culture bottles in a short time. Another advantage is that no time is spent in plastic tissue culture dishes, and the risk of contamination is consequently diminished because less handling of the cells is involved.

Specifically, method B involves the following procedures. Eleven- to twelve-day-old embryos are pooled, so that two embryos are used to prepare cells for one roller culture bottle. The embryos are decapitated, minced, and cells are dissociated with 0.25% trypsin in Tris-buffered saline. A fluted trypsinization flask (Bellco) is helpful for this step. After two consecutive trypsinizations (using 50 ml trypsin per embryo), the cells are centrifuged at 1500 rpm for 15 min (800 g) in an International refrigerated centrifuge, and the trypsin is decanted from the cell pellet. The cell pellet is resuspended at the rate of 100 ml PGM per embryo, and 25 ml of a high-titer virus stock are added per embryo. Roller culture bottles are seeded in 250 ml PGM per bottle. It is important to note that polybrene cannot be used in this procedure, since it appears to change the surface charge of the glass and prevents adherence of the cells. The primary cells will form a monolayer within 4–5 days, at which time the roller culture bottle is trypsinized (see method A) and cells are transferred to a new roller culture bottle. Transformation is obtained within 4–5 days.

A disadvantage of method B is the requirement for large quantities of intermediate virus stocks. It is impractical to infect a roller culture bottle with less than 50 ml of virus-containing supernatant fluid, because transformation and high-titer production are considerably delayed when small amounts of virus are used. The reason for the delayed appearance of transformation probably stems from the large number of chicken cell fragments and nonviable cells present in primary embryo fibroblast preparations that may absorb virus and prevent infection of the minority of cells that eventually emerge to form the monolayer. Another disadvantage of method B is that the cell monolayer is usually less uniform than monolayers prepared as outlined in method A. However, the rough appearance of the cell sheet is rapidly eliminated once trypsinization and transfer of the cells in the roller culture bottles are initiated. Cells in roller culture bottles are harvested, trypsinized, and transferred in the same manner, whether prepared by method A or B.

### *Harvest of Virus and Cells*

Once cells are confluent, virus can be harvested from roller culture bottles for an extended period of time.[5] The amount of medium used, and the frequency and duration of harvest, are influenced by the intended use of the virus.

When the investigator is interested in harvesting virus for examination of structural polypeptides, virus can be collected daily or every other day, using 200 ml of growth medium per bottle for the first 7–10 days of culture, then maintenance medium for harvest periods of up to several

[5] R. E. Smith and E. H. Bernstein, *Appl. Microbiol.* **25**, 346 (1973).

months. Prolonged harvest periods are only possible with RSV-transformed cells, since they will grow back into areas of the roller culture bottle that have been exposed due to overgrowth of the cell monolayer and loss of the cell sheet. The investigator determines when to stop harvesting by monitoring the virus yield. A good rule of thumb is that a liter of supernatant fluid should yield 1 mg of purified virus. When production falls below this level, it is hard to justify the time and expense required for continued harvest.

When the investigator is interested in harvesting virus for the isolation of 70 S RNA, daily harvesting is adequate. The duration of harvest is usually not as long, and 1 month of production is generally all that should be attempted. When the investigator is interested in harvesting virus for the isolation of 35 S RNA, the roller culture bottles should be harvested at intervals of 2 hr, but can usually be harvested for only 3 weeks. An automated system is available for this type of harvest, and this subject will be covered in the next section.

Virus-producing cells can be harvested at any time during the preparation or harvest of large-scale cultures, as long as the cells have not begun to decline due to prolonged harvest. Cells in roller culture bottles are washed with 50 ml of Tris-buffered saline and detached from the surface of the roller culture bottle with the aid of a rubber-bladed scraper (Bellco). The cells are rinsed into a centrifuge bottle with the aid of 200 ml of Tris-buffered saline and centrifuged at 800 g for 15 min in a refrigerated centrifuge. The cell pellet is then processed further for experimental purposes, or stored frozen at $-70°$.

### *Mechanization of Virus Production*

The requirement for frequent harvest intervals for the production of 35 S RNA leads to considerable expenditure of time and effort when large quantities of RNA are needed. For example, it is difficult to manually harvest several roller culture bottles more than 3 times in a working day, especially when one is also required to process the supernatant fluid. The automated collection of roller culture bottle supernatant fluids[6] is therefore helpful. The frequent collection of supernatant fluid also has the additional benefit of providing virus that has a high specific infectivity.[7]

Roller culture bottles are prepared for automated harvesting in much the same way as outlined previously. Cells are seeded into roller culture bottles and grown until nearly confluent; then the roller culture bottles are fitted with special caps and placed onto a harvesting device (Bellco Glass,

[6] R. E. Smith and F. Kozoman, *Appl. Microbiol.* **25,** 1008 (1973).

[7] R. E. Smith, *Virology* **60,** 543 (1974).

Inc.). It is helpful to start with subconfluent roller culture bottles, since cell growth is usually vigorous as a consequence of frequent feeding.

Roller culture bottles are usually fed at the rate of 200 ml of maintenance medium per bottle per day, or about 17 ml per bottle every 2 hr. This small amount of medium must evenly cover the bottom of the bottle, so an accurate leveling of the harvesting device is required.

Bottles are harvested for 3 weeks, at which time virus production declines.[8] When necessary, the monolayers in the roller culture bottles may be inspected by placing an inverted microscope on a laboratory cart and manipulating the bottles to view the monolayers without removing the attached tubing.

We have experimented recently with the addition of 5% $CO_2$ to the roller culture bottles while they are being harvested. Two types of $CO_2$ addition are satisfactory. In the first method, bottles are continuously flushed with 5% $CO_2$ in air at the rate of 30 ml/min for five bottles. The second method consists of adding a 70-sec pulse of 5% $CO_2$ at the same flow rate (30 ml/min for five bottles) immediately after each harvest. The first method gives faster cell growth, and thus high virus yields are achieved sooner. However, cells overgrow rapidly, and harvest is terminated sooner. When the second method is employed, roller culture bottles require more time to achieve maximal virus yield, but high production is maintained for a longer period of time, and less $CO_2$ is consumed.

A special application of automated harvesting is the collection of $^{32}P$-labeled virus for RNA.[9] Cells are grown in roller culture bottles until confluent; then phosphate-free maintenance medium is added (50 ml/bottle) for 4–6 hr. Ten to twelve milliCuries of [$^{32}P$]orthophosphate (carrier-free; New England Nuclear) are added in 50 ml of fresh phosphate-free maintenance medium, and the cells are labeled overnight (12–18 hr). The supernatant fluid is removed, the roller culture bottle is fitted with a special cap, and the bottle is harvested every 2 hr for up to 8 days after labeling.

## Purification of Virus

The purification of RSV is divided into two steps, concentration and purification.

### *Concentration*

The large quantities of fluid generated by roller culture bottles require the investigator to reduce the volume of the sample prior to processing the

[8] R. E. Smith, S. Nebes, and J. Leis, *Anal. Biochem.* **77**, 226 (1977).
[9] R. E. Smith and K. Quade, *Anal. Biochem.* **70**, 354 (1976).

virus. The most convenient method is to pellet the virus with ultracentrifugational forces. This is done in three stages. The first stage is to remove cells and large debris by centrifugation at 800 g for 15 min. The second stage is to centrifuge the supernatant fluid at 8000 g for 15 min to remove subcellular debris and improve the subsequent purification steps. The third stage is to pellet the virus. Three fluid-volume capacities are used. For up to 210 ml, a Beckman SW27 rotor is used, and centrifuged 30 min at 24,000 rpm (80,000 g). For up to 420 ml, a Beckman 35 rotor is used, and centrifuged 45 min at 24,000 rpm (46,000 g). For up to 1200 ml, a Beckman 19 rotor is used, and centrifuged 105 min at 18,000 rpm (21,000 g). For even larger fluid-volume capacities, zonal rotors are available that process several liters per hour.

The virus pellet is resuspended (at a volume $\frac{1}{100}$ to $\frac{1}{200}$ of the original) in a buffer appropriate for the use of the virus. For most applications, TE (1 m*M* EDTA, 5 m*M* Tris · HCl, pH 8.6) is a suitable buffer. Two special cases deserve mention. The first is that the virus pellet can usually be extracted for 70 S RNA, in which case it is appropriate to resuspend the virus in STE (0.1 *M* NaCl, 1 m*M* EDTA, 10 m*M* Tris, pH 7.4). The second is that virus pellets may be used to make complementary DNA, in which case the investigator resuspends the virus pellet in a buffer appropriate for DNA synthesis.

*Purification*

Highly purified virus is obtained in a two-step process: equilibrium (isopycnic) density gradient sedimentation, followed by velocity (rate zonal) density gradient sedimentation. The concentrated virus pellet is resuspended thoroughly (a brief sonication in an ultrasonic cleaning bath is helpful) and layered onto a 15–60% (w/v) sucrose (in TE) gradient. The Beckman SW27 rotor is used most frequently in our laboratory for this application. The virus suspension is layered onto the sucrose gradients, and the rotor is centrifuged at 24,000 rpm (80,000 g) for 3–6 hr, at which time the virus is visualized near the center of the gradient. The virus band is removed with a Pasteur pipette, diluted 4-fold with TE, and the virus is pelleted at 24,000 rpm (80,000 g) for 35 min in a Beckman SW27 rotor. The virus pellet is resuspended in TE, layered onto a 20–40% sucrose gradient, and centrifuged 90 min at 24,000 rpm (80,000 g) in a Beckman SW27 rotor, or 40 min at 36,000 rpm (175,000 g) in a Beckman SW41 rotor. The virus band is diluted with TE, and the virus is pelleted as before. This virus can be stored at −20° indefinitely. Virus concentration can be measured by the method of Lowry *et al.*[10] or estimated by an optical density measurement[11] in which 1 optical density unit equals 158 $\mu$g protein.

### Large-Scale Growth of Nontransforming Viruses

Growth of nontransforming viruses in large-scale cultures is performed in essentially the same fashion as outlined previously for RSV. Virus cloning is more difficult, since the viruses do not form foci in chick embryo fibroblasts. However, viruses of subgroups B, D, and F and certain members of subgroup A form plaques on CEF,[12,13] and virus can be recovered from a plaque by aspiration into a Pasteur pipette. The preparation of roller culture bottles of cells infected with nontransforming viruses is performed in the same manner as outlined for RSV, except that growth medium is used for a longer period of time. Nontransformed cells produce less organic acids than transformed cells, and therefore they seem less susceptible to overgrowth or toxic effects from feeding on an every-other-day schedule. The production of virus is followed by the size of the virus pellet obtained upon concentration of the virus by ultracentrifugation. Subgroups B, D, and F viruses are somewhat cytopathogenic, and healthy roller culture bottles can decline due to the replication of the virus. This effect can be overcome by periodically adding $1\text{–}2 \times 10^8$ uninfected cells to the roller culture bottle. Once the cells are confluent, cell transfer by trypsinization is performed in the manner outlined for RSV. Cell yields are lower than with transformed cells, so a longer period of time is required to obtain a large number of virus-producing cells. Virus yields are also lower, a consequence of a smaller number of cells per roller culture bottle.

### Acknowledgments

The author wishes to express his thanks to Jann Bodie, Christine Chastain, Kay Izard, and Sue Nebes, each of whom have substantially contributed to the development of the procedures outlined in this review. Work performed in the author's laboratory was supported by National Institutes of Health Grant RO1-CA12323.

[10] O. H. Lowry, N. J. Rosebrough, A. L. Farr, and R. J. Randall, *J. Biol. Chem.* **193,** 265 (1951).

[11] R. E. Smith and E. H. Bernstein, *Appl. Microbiol.* **25,** 346 (1973).

[12] T. Graf, *Virology* **50,** 567 (1972).

[13] C. Moscovici, D. Chi, L. Gazzolo, and M. G. Moscovici, *Virology* **73,** 181 (1976).

# [34] Preparation of Simian Virus 40 and Its DNA

*by* GEORGE KHOURY and CHING-JUH LAI

Simian virus 40 (SV40), a small (45-nm diameter) protein-encapsidated DNA virus, is a member of the papovavirus B genus (polyomavirus) of the papovaviruses.[1] Its nucleoprotein core consists of a double-stranded superhelical closed circular DNA genome with a molecular weight of $3.6 \times 10^6$, complexed with host-cell-derived histones. The genome contains approximately 5200 base pairs, which is enough genetic information to code for only five or six proteins. SV40 was discovered in 1960 by Sweet and Hilleman[2] as an inapparent infectious agent in rhesus monkeys. It was capable of producing a lytic infection in African green monkey kidney (AGMK) cells characterized by cytoplasmic vacuolization. Initially, SV40 was of medical interest because it was present as a contaminant of certain batches of poliovirus vaccine that had been administered to humans. In addition, SV40 could produce tumors in suitable laboratory animals and was able to transform human cells in tissue culture. More recently, investigators have focused on the ability of SV40 to undergo different interactions with various host cells. While SV40 replication in permissive cells such as AGMK leads to amplification of the virus and cell death, quite a different response occurs in nonpermissive cells. The inoculation of mouse cells with SV40, for example, results in an "abortive infection" with only partial expression of the viral genome. SV40 DNA does not replicate in mouse cells, and progeny virions are not produced. A subpopulation of these abortively infected cells eventually become stably transformed by the virus. The transformed state results from stable integration of the viral DNA into the host chromosome, continued expression of the viral genes, and an altered cellular morphology and growth pattern reflecting the presence of the viral function.

The simplicity of the SV40 genome and its ability to undergo transforming as well as lytic interactions with various cells has made it a model system for virologists, geneticists, and molecular biologists who are interested in the study of gene regulation in eukaryotic cells. SV40 is probably the best understood of all DNA animal viruses and has provided insight into the mechanisms of DNA replication, transcription, and posttranscriptional controls, translational regulation, and DNA–protein interactions. The use of purified viral DNA has been crucial to most of these

[1] J. L. Melnick, A. C. Allison, J. S. Butel, W. Eckhart, B. E. Eddy, S. Kit, A. J. Levine, J. A. R. Miles, J. S. Pagano, L. Sachs, and V. Vonka, *Intervirology* **3,** 106 (1974).

[2] B. H. Sweet and M. R. Hilleman, *Proc. Soc. Exp. Biol. Med.* **105,** 420 (1960).




ISBN 0-12-181958-2

studies. In addition, SV40 DNA has been used to study the physical and chemical properties of supercoiled molecules, to assay restriction enzymes, and to evaluate DNA–protein interactions. The recent determination of the complete nucleotide sequence for SV40[3,4] will certainly enhance its value for numerous future biological and biochemical investigations. Simple methods for the production of SV40 and SV40 DNA are described below.

## A. Growth of SV40

### *Cell Cultures*

SV40 produces a lytic infection in monolayer cultures of African green monkey kidney cells. Several continuous cell lines are available including BSC-1,[5] CV-1,[6] and Vero[7] as well as derivative clones of these lines. Titers between $10^8$ and $10^9$ plaque-forming units (PFU) per milliliter of cell lysate usually can be obtained. Primary or secondary AGMK cells apparently yield 5–10 times this amount of virus. However, primary cells are expensive and tedious to prepare, which makes them a less satisfactory choice for most laboratories.

### *Stock Virus*

A number of strains of SV40 are commonly used including 776 (SVS), 777, and Va 4554. Many strains differ slightly in their DNA sequence and, in some cases, this difference is reflected in variations in the restriction enzyme cleavage patterns or in the morphology of the plaques that the virus induces.[8] Stocks of virus are generally prepared by infecting cells (in Eagle's minimal essential medium, for example, supplemented with antibiotics, 0.03% glutamine, and 2% fetal bovine serum; abbreviated, MEM-2) with less than 0.01 PFU/cell of a seed virus preparation. The low multiplicity avoids the generation of defective virions that have been

[3] V. B. Reddy, B. Thimmappaya, R. Dhar, K. N. Subramanian, B. S. Zain, J. Pan, C. L. Celma, and S. M. Weissman, *Science* (in press).

[4] W. Fiers, R. Contreras, G. Haegeman, R. Rogiers, A. Van de Voorde, H. Van Heuverswyn, J. Van Herreweghe, G. Volckaert, and M. Ysekaert, *Nature (London)* (in press).

[5] H. E. Hopps, B. C. Bernheim, A. Nisalak, J. H. Tjio, and J. E. Smadel, *J. Immunol.* **91,** 416 (1963).

[6] F. C. Jensen, A. J. Girardi, R. V. Gilden, and H. Koprowski, *Proc. Natl. Acad. Sci. U.S.A.* **52,** 53 (1964).

[7] E. Earley, P. H. Peralta, and K. M. Johnson, *Proc. Soc. Exp. Biol. Med.* **125,** 741 (1967).

[8] T. J. Kelly and D. Nathans, *Adv. Virus Res.* **21,** 85 (1977).

shown to arise after high multiplicity infections or during undiluted serial passage of virions.[8]

For inoculation, a small volume of virus, sufficient to cover the cell sheet, is added to confluent monolayer cultures. After adsorption at 37° for 1–2 hr, cultures are fed with a suitable maintenance medium, such as MEM-2. Cultures infected by wild-type SV40 are incubated at 37°, whereas cultures infected by temperature-sensitive mutants of SV40 are usually grown at 32°. A suitable stock of virus is prepared by simply freeze-thawing monolayer cultures of SV40-infected cells 3 times in their own maintenance medium. This is generally done at a time when the cell monolayer shows extensive cytopathic effects (rounding or detachment of greater than 80% of the cell sheet), an event that occurs, for example, 10–14 days after incubation at 37° of cells infected at a multiplicity of 0.01 PFU per cell. The stock virus can be titered by the plaque assay of serial dilutions on monolayers of primary or continuous African green monkey kidney cultures.[9] SV40 is quite stable and retains its titer for long periods of time when frozen in portions at −20 to −70°. Since a significant amount of virus remains associated with cell membranes, clarification of viral stocks by pelleting and discarding cellular debris will result in the loss of large fractions of the virions. If it is important to clarify or concentrate stocks of SV40, this can be achieved without significant losses by purifying virions. A convenient method for virus purification, which employs the equilibrium banding of virus in CsCl, is described below in the second procedure for preparation of viral DNA.

## B. Preparation of SV40 DNA

Since defective virions do not generally arise in significant quantities before several serial passages at high multiplicities of infection, it is convenient to infect confluent monolayers of kidney cells at a multiplicity of approximately 10 PFU/cell for the preparation of SV40 DNA. There are two basic methods of viral DNA preparation; the choice of technique depends on the intended use for the SV40 DNA as is explained below.

### *Preparation of SV40 DNA by the Selective Extraction Method of Hirt*

In 1967, Bernard Hirt developed a convenient method for selective extraction of polyoma virus DNA (and other low-molecular-weight DNAs including SV40 DNA) from infected cells.[10] This method, which is based

[9] R. Dulbecco and M. Vogt, *J. Exp. Med.* **99,** 167 (1954).
[10] B. Hirt, *J. Mol. Biol.* **26,** 365 (1967).

on the removal of high-molecular-weight cellular DNA by precipitation in the presence of high salt concentrations and sodium dodecyl sulfate (SDS), has a number of advantages. It is both rapid and efficient, producing as much as 2–6 times the yield of SV40 DNA per equivalent of infected cells as the alternative method mentioned below (e.g., 10–20 μg per 150 cm$^2$ flask of cells). Since the Hirt extraction procedure is based on the isolation of intracellular DNA, rather than encapsidated viral DNA, cells are harvested earlier than if DNA were to be prepared from purified virions. A disadvantage of the Hirt method is the presence of small amounts, less than 1%, of contaminating cellular DNA in most preparations. While this level of contamination is acceptable for most purposes, there are certain circumstances under which essentially pure SV40 DNA is required and for which the second method is preferable, e.g., *in vitro* labeling of viral DNA (see below). A slight modification of the Hirt extraction procedure in which the supercoiled SV40 DNA is further purified by dye-bouyant-density centrifugation[11] is described below.

*Reagents and Materials*

Hirt lysing solution (0.6% sodium dodecyl sulfate, 10 m*M* EDTA at pH 7.4)
Sodium chloride, 5*M*
RNase (Worthington Biochemicals, Freehold, New Jersey; preheat a 1 mg/ml solution to 80° for 10 min to inactivate traces of DNase)
Potassium acetate solution (20%) at pH 6.0
Ethidium bromide (aqueous solution, 2 mg/ml)
Standard saline citrate (SSC solution = 0.15 *M* sodium chloride, 15 m*M* sodium citrate)
Isopropyl alcohol saturated with CsCl by adding a few crystals to about 10 ml isopropyl alcohol
Phenol, saturated with a solution of 10 m*M* Tris · HCl at pH 8.0, and 10 m*M* NaCl
Sucrose solutions: 5% and 30% (w/v) in 0.1 *M* NaCl, 10 m*M* Tris · HCl at pH 7.5, and 2.5 m*M* EDTA

Cell monolayers are infected with SV40 at 1–10 PFU/cell. Between 3 and 5 days after infection when cells show early cytopathic effects (some cells are rounded or vacuolated, but all cells remain attached to surface), cultures are harvested as follows. The cell sheet is gently washed twice with cold PBS. After draining, about 1–2 ml of Hirt lysing solution is added per 150-cm$^2$ flask and allowed to remain on the surface for 20 min at room temperature. The cell lysate is then scraped into a Sorvall centrifuge

[11] R. Radloff, W. Bauer, and J. Vinograd, *Proc. Natl. Acad. Sci. U.S.A.* **57,** 1514 (1967).

tube, the volume measured, and 5 $M$ NaCl added to a final concentration of 1 $M$. The tube is covered, inverted gently (to reduce potential shearing forces) 10 times, and stored at 4° overnight. After centrifugation for 40 min at 25,000 g and at 0°, the supernatant fraction is gently removed into another glass centrifuge tube. RNase is added to a concentration of 10 $\mu$g/ml followed by incubation at 37° for 30 min. After the addition of an equal volume of saturated phenol, the mixture is shaken for 10–15 min at room temperature, and the phases are separated by centrifugation at 3000–5000 rpm for 10 min at room temperature. The aqueous (upper) phase is removed into a centrifuge tube and to it is added 0.1 volume of 20% potassium acetate and 2–2.5 volumes of cold ethanol. After mixing, this material is stored at −10 to −20° for at least 8 hr. The DNA that precipitates in ethanol is pelleted by centrifugation at 10,000 rpm for 20 min at 0° in a Sorvall SS-34 rotor. After the supernatant liquid is carefully decanted, the DNA precipitate is dissolved in 1–2 ml of SSC solution at pH 7. For subsequent dye-density equilibrium centrifugation, the DNA solution should contain ethidium bromide at approximately 200 $\mu$g/ml and CsCl with a density of 1.56 g/ml. This is conveniently accomplished by first weighing an empty nitrocellulose or polyallomer (see below) centrifuge tube, and then adding the DNA solution that has been increased to 6.5 ml with SSC solution. After the addition of 0.75 ml of ethidium bromide stock solution (2 mg/ml), the tube is again weighed and the content weight increased to 7.5 g by the further addition of SSC. After dissolving 6.5 g of CsCl into each tube, the density should be approximately 1.56. This may be checked by weighing 1 ml of the solution or by measuring its refractive index $n$ ($n$ = 1.3880). Tubes are filled with mineral oil, capped, and centrifuged to equilibrium, e.g., 40,000 rpm for 48 hr at 10° in a Beckman 50.1 rotor. Other rotors, e.g., SW41 and type 40, also can be used by appropriately adjusting the speed and time of centrifugation.

Nitrocellulose centrifuge tubes are preferable to polyallomer tubes since they are more transparent, i.e., bands can be more easily visualized, and are easier to puncture. However, DNA adheres more readily to nitrocellulose. Thus, for the preparation of very small amounts of DNA, some investigators prefer to use polyallomer ultracentrifuge tubes and siliconized glass centrifuge tubes.

After centrifugation to equilibrium, two bands of DNA should be detectable near the center of the tube separated by approximately 1 cm. The lower of these bands contains the supercoiled DNA (higher density) and can be removed by tapping from the bottom of the tube or by pipetting from the top. Small amounts of DNA are more easily visualized with an ultraviolet (UV) lamp in the dark. However, exposure to UV should be limited to avoid thymidine dimer formation or nicking of DNA. Ethidium bromide is frequently removed by passing the DNA through a column of cationic resin, e.g., Dowex-50. Perhaps the easiest way to remove the dye is by extracting several times with equal volumes of isopropyl alcohol saturated with CsCl. The pink, intercalating dye is fractionated into the organic (upper) phase and removed. The CsCl can then be removed from the aqueous phase by dialysis against an appropriate buffer (e.g., 10 m*M* NaCl and 10 m*M* Tris-chloride at pH 8.0). Since supercoiled DNA molecules other than SV40 may still be present, i.e., mitochondrial DNA, SV40 DNA can be further purified by sedimentation in 5–30% sucrose gradients at 30,000 rpm for 9 hr at 4° with an SW41 rotor. These conditions will place supercoiled SV40 DNA near the center of the gradient.

*Preparation of SV40 DNA from Purified Virions*

As an alternative to the Hirt procedure, SV40 DNA can be isolated from purified SV40 virions. The principle advantage of this method is the high degree of purity of the DNA released from SV40 capsids. This purity is especially important when viral DNA is subsequently labeled to very high specific activities by *in vitro* methods (see "*In vitro* labeling" below) in order to detect small amounts of SV40 DNA in transformed cells. As mentioned, the disadvantages of this method are that it requires more time, both for the growth of virions and for the additional steps in the purification procedure, and that the yield of DNA is generally lower.

*Reagents and Materials*

CsCl solution, density = 1.4 ($n$ = 1.3723)
CsCl solution, density = 1.34 ($n$ = 1.3670)
Tris-NaCl buffer (10 m*M* Tris · HCl, pH 8.0, 10 m*M* NaCl)
Sarkosyl: 10% solution of sodium sarcosinate

Cell monolayers are infected at about 10 PFU/cell. At 6–10 days, when monolayers show extensive cytopathic effects (see above), the remaining cells in infected flasks are scraped from the surface. Media and cells are poured into large centrifuge tubes, and the cells are pelleted at about 5000 g for 10 min at 4°. The supernatant fluid is decanted and saved; the cell pellet from 5–10 flasks is resuspended in 16 ml of the supernatant fluid. After the addition of 2 ml of 0.25% trypsin (0.025% final concentration) and 2 ml of 13% desoxycholate (1.3% final concentration), the resuspended cells are incubated at 37° for 20 min. Cellular debris is removed by pelleting at 10,000 g for 10 min at 4°, and the supernatant fluid is gently layered onto a 10-ml cushion of CsCl ($\rho$ = 1.4 g/ml) in an SW27 nitrocellulose tube. This tube and additional tubes are filled to capacity with the medium saved from the infected cells; centrifugation is carried out at 24,000 rpm for 3 hr (or 12,000 rpm overnight) at 4° in an SW27 rotor. The virions, which under these conditions enter about 1 cm into the CsCl cushion, can be visualized best with a high-intensity lamp in a dark room as a blue translucent band. The band is removed in 1–2 ml of CsCl, preferably by dripping through the bottom of a punctured tube. The density of this virus-containing CsCl should be 1.33–1.34 g/ml ($n$ = 1.3660), and this material can be diluted directly into 7–8 ml of CsCl of the same density (1.34 g/ml) in a nitrocellulose tube for either an SW40 or a 50Ti rotor. The virus is centrifuged to equilibrium, e.g., at 35,000 rpm for 16 hr at 4° in a 50Ti rotor. After visualization and removal of the SV40 band (which should be near the center of the tubes; $\rho = 1.34$), the virus is dialyzed for 24 hr against the Tris-NaCl buffer at 4° with 3–4 changes of the dialysate. The protein capsids can be removed from virions by incubation at 50° for 30 min in 1% Sarkosyl. After two extractions with phenol that is saturated with Tris-NaCl buffer, the viral DNA, which remains in the aqueous phase, can be either dialyzed at room temperature or precipitated in 2.5 volumes of cold ethanol prior to use. A further purification of supercoiled SV40 molecules from relaxed molecules can be obtained as described in the previous purification method by dye-bouyant-density centrifugation and/or sucrose gradient sedimentation. However, when SV40 DNA is extracted from virions these steps are frequently unnecessary.

## C. Radiolabeling and Quantitation of SV40 DNA

### *Radiolabeling of SV40 DNA*

The intended use for SV40 DNA frequently requires that it be radiolabeled. Addition of [$^3$H]thymidine (10–100$\mu$Ci/ml) or [$^{14}$C]thymidine (0.01–0.1 $\mu$Ci/ml) to the medium of infected cells generally provides labeled DNA with a specific activity of $10^5$ cpm/$\mu$g and $10^2$–$10^3$ cpm/$\mu$g, respectively. Radiolabeling with carrier-free [$^{32}$P]orthophosphate can produce SV40 DNA with specific activities of $1–2 \times 10^6$ cpm/$\mu$g if the cellular phosphate pools are lowered. This can be achieved by the following protocol.

Cells are washed with phosphate-free medium several times prior to infection and fed with phosphate-free medium supplemented with 2% dialyzed fetal calf serum, antibiotics, and glutamine. Twenty-four hours after infection, the monolayers are again washed three times with phosphate-free medium. Cells are fed again with phosphate-free medium that contains approximately 100 μCi/ml of carrier-free [$^{32}$P]orthophosphate. The time for harvesting radiolabeled viral DNA depends on the method to be used for purification (see above).

*In vitro Labeling of SV40 DNA*

A number of laboratories now are routinely labeling DNA preparations *in vitro* using a method referred to as nick-translation. There are at least two recently published procedures that describe in detail the methodology involved.[12,13] Both procedures require DNase I and DNA polymerase I and involve the nicking and removal of nucleotides from the unlabeled input DNA followed by repair synthesis with radiolabeled nucleoside triphosphates. The advantages of *in vitro* labeling compared with *in vivo* labeling include the ability to obtain radiolabeled DNA within hours by a relatively simple procedure and the production of DNA with extremely high specific activities, e.g., $10^8$ cpm/μg using deoxynucleoside triphosphates, two of which are $\alpha^{32}$P-labeled. Lower specific activities also can be achieved by limiting the DNase concentration, the specific activity of the radiolabeled precursors, or the time of incubation.

*In vitro* labeled SV40 DNA recently has been used to obtain qualitative and quantitative information relating to the integrated viral genomes in transformed cells. These high-specific-activity probes have been successfully used for reassociation kinetic analyses[13] and for detection of viral DNA transferred to nitrocellulose filters[14] by the Southern blotting procedure.[15]

*Quantitation and Assay of SV40 DNA*

The yield of SV40 DNA is most easily determined spectrophotometrically. The concentration of a sample in milligrams per milliliter is approximately equal to the optical density at 260 nm divided by 20. Small amounts of viral DNA are quantitated more accurately by reassociation kinetics[16] or by gel electrophoresis. All three forms of SV40 DNA, i.e., form I (supercoiled molecules), form II (relaxed circles), and form III (linear duplexes), are easily separated by agarose gel electrophoresis. In order to quantitate small amounts of SV40 DNA (approximately 0.05–1.0 μg), a DNA sample of unknown concentration is electrophoresed in parallel with serial dilutions of a DNA preparation of known concentration.

[12] T. Maniatis, S. G. Kee, A. Efstratiadis, and F. C. Kafatos, *Cell* **8,** 163 (1976).
[13] P. W. J. Rigby, M. Dieckmann, C. Rhodes, and P. Berg, *J. Mol. Biol.* **113,** 237 (1977).
[14] M. Botchan, W. Topp, and J. Sambrook, *Cell* **9,** 269 (1976).
[15] E. Southern, *J. Mol. Biol.* **98,** 503 (1975).
[16] L. D. Gelb, D. E. Kohn, and M. A. Martin, *J. Mol. Biol.* **57,** 129 (1971).

The gel is stained with ethidium bromide solution (1 μg/ml in the electrophoresis buffer) for 15–30 min and visualized under UV light. By comparing the intensity of staining of the unknown sample with a similar standard, one can rather accurately assess the DNA concentration.

Perhaps the best criterion of purity for a viral DNA preparation is its restriction enzyme cleavage pattern. Restriction cleavage maps of the SV40 genome are known for numerous restriction enzymes[17] (also see Kelly and Nathans[8]). In a typical assay, a small quantity of DNA is cleaved with a restriction endonuclease, and the product is subjected to agarose or polyacrylamide gel electrophoresis. If the DNA is $^{32}$P-labeled, the gel can be examined by autoradiography; otherwise, the restriction pattern of the cleaved DNA can be analyzed by ethidium bromide staining as described above.

A number of experiments have recently been carried out to directly test the biological activity of either native SV40 DNA or segments of the viral genome. In the first approach, permissive cells can be effectively inoculated with the viral DNA in the presence of DEAE-dextran.[18] This technique has made it possible to assay the relative infectivity of various viral DNA preparations. In addition, the technique has enabled investigators to propagate altered forms of viral genomes, either naturally occurring molecules or those constructed *in vitro*. An alternative procedure for introducing viral DNA into nonpermissive cells involves coprecipitation of the DNA with calcium phosphate[19]; this method had proven more reliable for obtaining morphologically transformed cells, and it has been employed successfully to determine the limits of the viral DNA regions required for establishment and maintenance of cell transformations. Finally, direct microinjection of viral DNA into cells has been valuable for studying several biological interactions between viruses and cells.[20] While this technique is limited by the number of cells that can be microinjected, the ability to inoculate efficiently large numbers of molecules into single cells has led to a number of important observations.

[17] K. J. Danna and D. Nathans, *Proc. Natl. Acad. Sci. U.S.A.* **69,** 3097 (1969).
[18] J. McCutcheon and J. M. Pagano, *J. Natl. Cancer Inst.* **41,** 351 (1968).
[19] F. L. Graham and A. J. van der Eb, *Virology* **52,** 456 (1973).
[20] A. Graessman, *Exp. Cell Res.* **60,** 273 (1970).

# [35] Purification and Assay of Murine Leukemia Viruses

*By* CHARLES J. SHERR and GEORGE J. TODARO

Murine type C viruses have been classified into two major groups based on their potential for producing neoplastic disease. The mouse



ISBN 0-12-181958-2

leukemia viruses (MuLVs) can replicate in cultures of fibroblastic and epithelioid cells and generally do not produce cytopathic effects or morphological transformation. By contrast, the mouse sarcoma viruses (MSVs) are defective in their ability to replicate, require "helper" MuLVs for their continued propagation in cell culture, and produce morphological transformation of susceptible cells *in vitro*. This article describes techniques for growing murine leukemia viruses in tissue culture and discusses methods for purifying virus in sufficient quantities for subsequent biochemical studies. Since the success of these techniques involves determinations of viral titer, several of the commonly used, quantitative *in vitro* assays.for viral replication are also described.

## Growth and Purification of Murine Leukemia Virus

### *Sources of Cells and Viruses*

MuLVs have been classified into several different host-range types. These include ecotropic viruses that replicate predominantly or exclusively in cells from the same or closely related species; xenotropic viruses, restricted from replicating in mouse cells, but able to grow in cells from heterologous species; and amphotropic viruses that grow in both mouse and nonmouse cells. Ecotropic viruses from laboratory mice (*Mus musculus*) can be further distinguished by their ability to grow preferentially in cells derived from certain prototype mouse strains. The N-tropic viruses replicate preferentially in cells derived from NIH Swiss mice (N-type cells) while B-tropic viruses grow best in cells derived from BALB/c mice (B-type cells); NB-tropic viruses grow readily in either N- or B-type cells.[1] The restriction of N-tropic viruses in B-type cells and vice versa is controlled by the *Fv*-1 locus.[2,3]

Successful propagation of MuLVs in culture depends on the host range of the virus chosen for study as well as the cell line chosen to support its replication. For routine propagation of ecotropic virus stocks, the feral mouse SC-1 cell line is widely used since it lacks *Fv*-1 restriction and can be infected with N-, B-, or NB-tropic viruses.[4] Alternatively, cell lines such as NIH/3T3[5] or BALB/3T3[6] established from the prototypic N- and B-mouse strains, respectively, can be employed. Xenotropic viruses can

[1] J. W. Hartley, W. P. Rowe, and R. J. Huebner, *J. Virol.* **5,** 221 (1970).
[2] T. Pincus, J. W. Hartley, and W. P. Rowe, *J. Exp. Med.* **133,** 1219 (1971).
[3] T. Pincus, W. P. Rowe, and F. Lilly, *J. Exp. Med.* **133,** 1234 (1971).
[4] J. W. Hartley and W. P. Rowe, *J. Virol.* **19,** 19 (1976).
[5] J. L. Jainchill, S. A. Aaronson, and G. J. Todaro, *J. Virol.* **4,** 549 (1969).
[6] S. A. Aaronson and G. J. Todaro, *J. Cell. Physiol.* **72,** 141 (1968).

be propagated to high titer in the SIRC rabbit corneal cell line (CCL 60, American Type Culture Collection [ATCC], Rockville, Maryland), in MvlLu mink lung cells (ATCC line CCL 64), or in Cf2Th canine thymus cells (Naval Biomedical Research Laboratories, Oakland, California). Viably frozen virus seed stocks can be obtained from the ATCC, from Pfizer Laboratories (Maywood, New Jersey), from ElectroNucleonics, Inc. (Bethesda, Maryland), or from the Office of Resources and Logistics of the National Cancer Institute (Bethesda, Maryland). Seed cultures of cells infected with virus also can be obtained from numerous investigators working in the field.

*Infection of Cultured Cells*

Cells used for virus production are grown at 37° on the surfaces of 75-$cm^2$ (250-ml) plastic tissue culture flasks (Falcon Labware, Cockeysville, Maryland) or in 490-$cm^2$ polystyrene roller bottles (Corning Glass Works, Corning, New York) at 1.0–1.5 rpm using Dulbecco's modified Eagle's medium supplemented with 4 m*M* L-glutamine, 4.5 mg/ml glucose, 10% heat-inactivated fetal calf serum, 100 units/ml penicillin, and 100 $\mu$g/ml streptomycin (Grand Island Biological Co., Grand Island, New York). All cell lines should be periodically monitored for opportunistic infections by fungi, adventitious viruses, mycoplasma, and resistant bacteria.

Prior to infection, cells are seeded at $1 \times 10^6$ cells per 75 $cm^2$ tissue culture flask in 12 ml of complete medium containing 2 $\mu$g/ml polybrene (Aldrich Chemical Company, Milwaukee, Wisconsin). After 24 hr, the medium is removed and the cells are exposed at 37° to 1.5 ml of undiluted, filtered (0.45 $\mu$m) medium from a virus-producing culture, or with a 10-fold dilution of concentrated virus seed stock obtained as described above. After 1 hr, 12 ml of complete medium are added to the culture, and the medium is changed after 24 hr and subsequently at 3-day intervals. Cells are subcultured when they reach confluence using 0.1% trypsin in phosphate-buffered saline to dissociate the cells from the plastic substrate and from each other. Supernatants from the cultures are tested at weekly intervals for the production of viral RNA-dependent DNA polymerase (see "Supernatant Reverse Transcriptase Assay" below). In general, cultures become highly positive for viral replication from 1–2 weeks after infection.

*Virus Purification*

Infected cells are grown and seeded into plastic 490-$cm^2$ roller bottles (0.5 rpm) each containing 50–70 ml of medium. After cells attach, the roller belt speed is increased to 1.0–1.5 rpm. Cells should be maintained at 60–100% confluency to insure high levels of virus production, but should not be allowed to remain at confluency for long periods as virus production will diminish, and cells will tend to slough from the surfaces of the bottles. Medium containing virus is sterilely decanted at 24- to 48-hr intervals and is rapidly cooled and kept at 4°. If maximum infectivity is desired, roller bottles should contain only 20–40 ml of medium and should be "rapidly harvested" at 3- to 6-hr intervals. In general, $10^{12}$ viral particles (approximately 1–2 mg of protein) can be obtained from approximately 5 liters of medium.

All further steps are performed at 4°. Virus-containing fluids are clarified of cells and debris by centrifugation at 12,000 g for 10 min at 4°. The supernatant fluid is then centrifuged over a cushion of 20% glycerol containing 0.05 *M* Tris · HCl, pH 7.8, and 0.1 *M* KCl (105,000 g for 90 min) to pellet the virus. If the supernatant fluid is large, virus can be concentrated by ultrafiltration prior to pelleting (30 psi, XM 300 membrane; Amicon Corp., Lexington, Massachusetts). Virus pellets are resuspended in 0.05 *M* Tris · HCl, pH 7.8, containing 10 m*M* KCl and 0.1 m*M* EDTA using gentle passage through a #18 gauge syringe needle. The suspension is then loaded over 20–60% (w/w) linear sucrose (ribonuclease-free) gradients and banded isopycnically at a density of 1.16 g/cc. Visible gradient bands can be harvested from the top of the tube using a Pasteur pipette, or gradients can be collected by needle puncture from the bottom, and the density of the band determined by refractometry. If the latter technique is employed, aliquots of each tube can be assayed for viral reverse transcriptase activity (as described below) to confirm the position of the viral band. Pooled, isopycnically banded virus is then diluted in the same buffer used for banding and is repelleted over a glycerol cushion as described above. For studies of viral structural proteins, viral pellets can be directly frozen at −70° or can be suspended in 0.05 *M* Tris · HCl, pH 7.8, containing 0.1 m*M* EDTA prior to freezing. For maximum viral infectivity, preparation of viral RNA, or for synthesis of complementary DNA transcripts, suspended virus particles should not be frozen but should be stored at 4° and used as soon as possible.

With very large volumes of virus-containing medium, continuous-flow centrifugation is the most efficient method for purifying viral particles. Virus fluids are clarified of cells and debris by a single pass through a Beckman J-CF continuous-flow rotor (15 liters/hr at 12,000 rpm; calculated to retain particles with a sedimentation coefficient of ≥10,000 S). Clarified virus fluids (20 liters maximum volume) are then passed through a CF-32 rotor at 5 liters/hr, 32,000 rpm, and sedimented into a 20–60% (w/w) linear sucrose gradient. Particulate and protein bands are allowed to equilibrate for an additional 0.5 hr after cessation of virus fluid flow.

The rotor contents are unloaded at 2000 rpm using a dense "chase" solution (62% sucrose) and are fractionated on the basis of optical density (254 nm) and buoyant density (1.120–1.180 g/cc). Virus-rich fractions (concentrated 500–1000 times at $\rho \sim 1.16$ g/cc) are normalized to $\leq 15\%$ sucrose by dilution with 0.05 *M* Tris · HCl, pH 7.8, and centrifuged in a Beckman type 35 rotor at 35,000 rpm for 1.5 hr (143,000 g). The virus pellets are then suspended and stored as described above.

## Assays for Murine Leukemia Virus

### *I. Enzymatic and Immunological Techniques*

#### A. Supernatant Reverse Transcriptase Assay

This test depends upon the packaging of RNA-dependent DNA polymerase activity in extracellular type C viral particles.[7] The extremely sensitive assay is useful for recognizing cultures infected by any type C virus, including all classes of MuLVs.[8,9] In general, a linear correlation exists between the amount of reverse transcriptase activity in a viral stock and the number of infectious viral particles.[10] Supernatant fluids can be serially collected and assayed from a single cell culture allowing the study of the time course of viral infection in a single flask.

*Procedure*

1. Five to fifteen milliliters of medium from an infected culture are clarified by centrifugation (12,000 *g* for 10 min at 4°).
2. Virus in the supernatant is pelleted by centrifugation through a cushion of 20% glycerol containing 0.05 *M* Tris · HCl, pH 7.8, and 0.1 *M* KCl (105,000 *g* for 90 min at 4°).
3. Pellets are resuspended in 0.1 ml of 0.05 *M* Tris · HCl, pH 7.8, containing 0.1 *M* KCl, $10^{-3}$ *M* dithiothreitol (DTT), and 0.1% Triton X-100.
4. An aliquot of the suspension (0.01 ml) is added to a 0.060 ml total reaction mixture containing 0.05 *M* Tris · HCl, pH 7.8, 0.06 *M* KCl, $2 \times 10^{-3}$ *M* DTT, $6 \times 10^{-4}$ *M* manganese acetate, 0.02 $A_{260}$ units poly rA (Miles Laboratories, Kankakee, Illinois), 0.02 $A_{260}$ units oligo $dT_{12-18}$ (Collaborative Research, Waltham, Massachusetts), and $3 \times 10^{-6}$ *M* [$^3$H] Thymidine 5′-triphosphate, tetrasodium salt ([$^3$H]TTP, 60,000 dpm/pmol; New England Nuclear Corporation, Boston, Massachusetts).

[7] H. M. Temin and D. Baltimore, *Adv. Virus. Res.* **17,** 129 (1972).

[8] J. Ross, E. M. Scolnick, G. J. Todaro, and S. A. Aaronson, *Nature* (*London*), *New Biol.* **231,** 163 (1971).

[9] M. M. Lieber, C. J. Sherr, and G. J. Todaro, *Int. J. Cancer* **13,** 587 (1974).

[10] J. R. Stephenson, R. K. Reynolds, and S. A. Aaronson, *Virology* **48,** 749 (1972).

5. The reaction mixture is incubated for 60 min at 37° and terminated by the addition of 3.0 ml 10% trichloroacetic acid (TCA).

6. Acid-precipitated mixtures are collected on Millipore filters (type HA, white plain), washed 5 times with 5% TCA, dried, and counted in a toluene-based scintillation cocktail.

*Remarks*. This test can be used for assaying not only type C viruses but also the other classes of reverse transcriptase-containing viruses.[11] In particular, certain mouse cell lines or cultured tissues produce type B viruses like the mouse mammary tumor virus (MTV) as well as another newly described group of "B-like" retroviridae (M432[12]). The latter two groups of viruses contain reverse transcriptases that prefer magnesium ($10^{-2}$ *M*) to manganese as the divalent cation in reactions performed with synthetic templates.[13,14]

Some cells also release nonviral coded polymerases into the culture medium. Some of these enzymes can incorporate [$^3$H]TTP into acid-precipitable poly(dT) product. The cellular polymerases have different template-primer requirements,[8,14] and control assays using different template primers [e.g., poly (rC)-oligo (dG), poly (rA)-poly (dT)] can be performed in conjunction with the reverse transcriptase test. Inhibition of reverse transcriptase activity with specific antibodies also can be used to confirm the nature of the polymerase being measured.[15]

In the above assay, confluent uninfected host cells in one 75-cm$^2$ (250-ml) flask generally produce less than $3 \times 10^3$ cpm of acid-precipitable poly (dT) product. Polymerase activities greater than $1 \times 10^4$ cpm/flask are scored as positive for viral replication, although activities in positive cultures are routinely much higher than this minimum level. Although the assay can be performed in the absence of synthetic templates, thus allowing only the endogenous viral RNA to be transcribed, the endogenous reaction is considerably less sensitive.

### B. Radioimmunoassays for Viral Structural Proteins

The genomes of murine type C viruses code only for a limited number of viral proteins. These include the reverse transcriptase, the major envelope glycoprotein (gp70), and at least four additional low-molecular-weight polypeptides (p30, p15, pp12, and p10). By convention, the virion

[11] A. J. Dalton, J. L. Melnick, H. Bauer, G. Beaudreau, P. Bentvelzen, D. Bolognesi, R. Gallo, A. Graffi, F. Haguenau, W. Heston, R. Huebner, G. J. Todaro, and U. I. Heine, *Intervirology* **4**, 201 (1974).

[12] R. Callahan, C. J. Sherr, and G. J. Todaro, *Virology* **80**, 401 (1977).

[13] E. M. Scolnick, E. Rand, S. A. Aaronson, and G. J. Todaro, *Proc. Natl. Acad. Sci. U.S.A.* **67**, 1789 (1970).

[14] M. Green and G. F. Gerhard, *Prog. Nucleic Acid Res.* **14**, 187 (1974).

[15] W. P. Parks, E. M. Scolnick, J. Ross, G. J. Todaro, and S. A. Aaronson, *J. Virol.* **9**, 110 (1972).

structural proteins are designated "p" for protein, "gp" for glycoprotein, and "pp" for phosphoprotein, followed by their apparent molecular weight in thousands of daltons.[16] Although standard immunological techniques such as immunofluorescence, immunodiffusion, and complement fixation can be employed to assay for type C viral structural proteins, the ability to purify viral proteins to homogeneity, label them to high specific activity with $^{125}$I, and generate antisera that react specifically with them has led to the development of extremely sensitive radioimmunoassays (RIAs) for each of the above virion components.

Viral antigenic determinants can be assigned operationally to one of three major classes of immunological reactivities, although in actuality there exists a broad spectrum of different antigenic reactivities for each of the viral proteins. Antigenic determinants that are shared in common by type C viruses isolated from different mammalian species are designated "interspecies-specific," while antigenic determinants shared in common among a group of isolates from a single species (e.g., all MuLVs from *M. musculus*) are classified as "species-specific" or "group-specific" determinants. In addition, "type-specific" antigenic determinants are those that distinguish different isolates within a group (e.g., xenotropic and ecotropic MuLVs).[16]

Four virion proteins (reverse transcriptase, p30, p10, and gp70) possess predominantly group-specific and interspecies-specific reactivities while p15 and pp12 appear to exhibit predominantly type-specific reactivities. For general assays of MuLVs isolated from laboratory mice, p30 and gp70 are the most useful proteins for radioimmunoassay studies since they are relatively easy to purify, are present in higher concentrations than other proteins in virions, and contain readily detectable group- and interspecies-specific antigenic determinants. Techniques for the purification of antigenically reactive MuLV p30,[17,18] gp70,[19] p15, p12,[20] and p10[21] structural proteins as well as MuLV reverse transcriptase[22] have been described in detail. Each of these proteins can be labeled with $^{125}$I to high specific activity (5 $\mu$Ci/$\mu$g) using the chloramine T method[23] and can be used as "tracer antigens" in competitive RIAs that can detect subnanogram quantities of competing viral protein.

[16] J. T. August, D. P. Bolognesi, E. Fleissner, R. V. Gilden, and R. C. Nowinski, *Virology* **60**, 595 (1974).
[17] S. Oroszlan, C. L. Fisher, T. B. Stanley, and R. V. Gilden, *J. Gen. Virol.* **8**, 1 (1970).
[18] E. M. Scolnick, W. P. Parks, and D. M. Livingston, *J. Immunol.* **109**, 570 (1972).
[19] M. Strand and J. T. August, *J. Biol. Chem.* **248**, 5627 (1973).
[20] R. C. Nowinski, E. Fleissner, N. H. Sarkar, and T. Aoki, *J. Virol.* **9**, 359 (1972).
[21] M. Barbacid, J. R. Stephenson, and S. A. Aaronson, *J. Biol. Chem.* **251**, 4859 (1976).
[22] J. M. Krakower, M. Barbacid, and S. A. Aaronson, *J. Virol.* **22**, 331 (1977).
[23] F. C. Greenwood, W. M. Hunter, and J. S. Glover, *Biochem. J.* **39**, 114 (1963).

A representative RIA protocol for MuLV p30 proteins is described below. This assay utilizes an antiserum prepared to the feline leukemia virus (FeLV) and detects those antigenic determinants shared by the p30 proteins of MuLVs and FeLV. For screening infected cells in tissue culture, assays for interspecies determinants are generally most useful, since detectable immunological differences between various MuLV isolates are minimized. Competition assays are initiated using conditions in which 50% of the $^{125}$I-labeled MuLV tracer p30 protein is bound in soluble immune complexes formed with antibodies to FeLV p30 protein. Following incubation with various dilutions of competing protein, immune complexes are precipitated with a second antiserum to 7 S immunoglobulins, and residual radioactivity in the precipitates is determined. The dilution of antiviral serum required to bind 50% of the labeled tracer antigen must be determined by titration as described below. Although the outlined procedure utilizes a rabbit antiserum to viral p30 protein and a goat antiserum to rabbit IgG, antiviral sera produced in other species are equally suitable as long as appropriate "second antibodies" are employed. The procedure can be adapted for assaying the different viral proteins and depends only on the availability of tracer antigens and antisera.

*Reagents*

1. RIA buffer: 0.01 *M* potassium phosphate, pH 7.5, containing 0.1 *M* NaCl, 0.01 *M* ethylenediaminetetraacetate (EDTA), and 0.5% normal rabbit serum.
2. $^{125}$I-labeled p30 protein from MuLV (Rauscher strain or some high-titer equivalent). Specific activity should be approximately 5 $\mu$Ci/$\mu$g.
3. Rabbit antiserum to p30 protein of FeLV.
4. Goat antiserum to rabbit 7 S globulin ("second antibody").

*1. Procedure for Titration of Antiviral Antibodies*

1. Serial dilutions of rabbit antiserum to viral p30 protein are prepared in RIA buffer. The dilutions required will depend upon the titer of antibodies in a particular serum but, for most sera, 2-fold dilutions from $10^{-2}$–$10^{-6}$ will suffice.
2. Titration reaction mixtures (0.5 ml) include: 0.3 ml RIA buffer, 0.1 ml of $^{125}$I antigen (diluted in RIA buffer to give $10^4$ cpm/0.1 ml), and 0.1 ml of rabbit antiserum at various dilutions. Two "background" control tubes without antiserum should be included and should contain 0.4 ml of RIA buffer and 0.1 ml of $^{125}$I antigen.
3. Reactions are incubated for 2 hr at 37° and for 18 hr at 4°.
4. A titered excess of second antibody is added (i.e., enough to precipitate all rabbit IgG in a 0.5 ml reaction) to each tube.

5. Incubation is continued for 1 hr at 37° and for 3 hr at 4°.

6. Immune precipitates are collected by centrifugation (5000 g for 10 min at 4°), washed with phosphate-buffered saline, and radioactivity in the pellets is determined in a gamma counter.

7. The dilution of rabbit antiserum that binds 50% (5000 cpm) of $^{125}$I-labeled tracer antigen is chosen as the antiserum titer for competition assays.

*2. Procedure for Competitive RIA*

1. Competing antigen extracts of infected, cultured cells are prepared by scraping the cells from the culture flask with a rubber policeman, washing the cells in serum-free medium, and suspending the washed cells in 0.5 ml phosphate-buffered saline. The cells are lysed by two cycles of rapid freezing and thawing in acetone–Dry Ice, followed by the addition of 10–15 volumes of ether. The ether is evaporated under a nitrogen stream, debris is removed by centrifugation (5000 g for 10 min), and the supernatant is used as the competing antigen. Protein is determined by the Lowry method[24] and should ideally be at least 2 mg/ml.

2. Two-fold serial dilutions of competing antigen are prepared in RIA buffer.

3. Two background control tubes contain 0.4 ml RIA buffer without antiserum or competing antigen. Two "binding" control tubes contain 0.3 ml RIA buffer and 0.1 ml titered rabbit antiserum at appropriate dilution (see procedure 1 above).

4. Competition reaction mixtures contain 0.25 ml RIA buffer, 0.1 ml titered rabbit antiserum at appropriate dilution, and 0.05 ml diluted competing antigen extract.

5. Reaction mixtures are incubated for 2 hr at 37°.

6. $^{125}$I-labeled tracer p30 protein (diluted in RIA buffer to give $10^4$ cpm/0.1 ml) is added to all tubes.

7. Reaction mixtures are incubated for 1 hr at 37° and for 18 hr at 4°.

8. Immune complexes are precipitated using the second antibody as described for steps 4–6 in procedure 1 above.

9. Calculations: The "working range" ($R$) of the assay is determined by subtracting [average cpm in binding control tubes ($Bi$)] – [average cpm in background control tubes ($Ba$)]. The degree of competition ($\%C$) for each competition reaction mixture is determined as:

$$\%C = \frac{[\text{cpm in competition reaction mixture}] - Ba}{R} \times 100$$

[24] O. H. Lowry, N. J. Rosebrough, A. L. Farr, and R. J. Randall, *J. Biol. Chem.* **193,** 265 (1951).

*Remarks.* All assays should include a standard competition curve generated using known concentrations of purified, unlabeled viral p30 protein as the competing antigen. This permits quantitative estimation of the amount of p30 antigen in any heterogeneous competing protein mixture, if the total quantity of competing protein at each dilution is known.[25] Technical problems in RIAs for type C viral proteins are similar to those encountered using general RIA techniques and are reviewed elsewhere.[26]

Antisera to viral proteins can be prepared by immunization with detergent-disrupted, isopycnically banded virions.[25] While such sera contain antibodies to many different viral structural proteins, the specificity of the RIA depends only on the purity of the $^{125}$I-labeled tracer antigen. Second antibodies from several species are commercially available, but can be prepared at minimal cost using purified IgG as the immunizing antigen.

Assays for group-specific determinants of p30 proteins can be designed using antiserum to the homologous p30 protein (i.e., anti-MuLV p30) instead of antiserum to FeLV.[18] In such tests, minor type-specific differences between different MuLV p30 proteins will be reflected in the slopes of the competition curves as well as in the final percent competition at "saturating" levels of competing protein.[25,26]

While RIAs can be used as a general screen for MuLVs from laboratory mice (*Mus musculus*), they also can be used for detecting MuLVs from several other *Mus* species (e.g., *M. caroli*[27] and *M. cervicolor*[28]). The immunological properties of some of the latter isolates are, however, distinctly different from those of laboratory mouse MuLVs.

## *II. Virological Assays*

### A. Mixed Culture Cytopathogenicity (The "XC Test")

The XC cell line was established from a Wistar rat tumor induced with the Prague strain of Rous (avian) sarcoma virus.[29] These cells contain the Rous sarcoma virus genome but do not release infectious viral particles. When infected with certain MuLVs, XC cells undergo syncytium forma-

[25] C. J. Sherr and G. J. Todaro, *Virology* **61,** 168 (1974).

[26] W. H. Hunter, *in* "The Handbook of Experimental Immunology" (D. M. Weir, ed.), p. 608. Davis, Philadelphia, Pennsylvania, 1967.

[27] M. M. Lieber, C. J. Sherr, G. J. Todaro, R. E. Benveniste, R. Callahan, and H. G. Coon, *Proc. Natl. Acad. Sci. U.S.A.* **72,** 2315 (1975).

[28] R. E. Benveniste, R. Callahan, C. J. Sherr, V. Chapman, and G. J. Todaro, *J. Virol.* **21,** 849 (1977).

[29] V. Klement, W. P. Rowe, J. W. Hartley, and W. E. Pugh, *Proc. Natl. Acad. Sci. U.S.A.* **63,** 753 (1969).

tion resulting in the appearance of plaques. The number of plaques produced is a direct function of the number of infectious viral particles.[30]

*Procedure*

1. Mouse cells (see below) are seeded sparsely into 60-mm plastic petri dishes ($\sim 1 \times 10^5$ cells/plate) and are grown at 37° in Dulbecco's modified Eagle's medium containing 10% fetal calf serum and antibiotics.
2. One day later, the cells are exposed to medium containing 25 $\mu$g/ml DEAE dextran (Sigma Chemicals, St. Louis, Missouri) or 2 $\mu$g/ml Polybrene (Aldrich Chemical Company) for 1 hr at 37°.
3. The medium is removed, and cultures are infected with dilutions of virus-containing tissue culture medium ($\sim$1.0 ml) for 2 hr on a rocking platform. The medium is changed after infection, and cultures are kept at 37° for 5 additional days until the cells reach confluence.
4. The medium is removed, and the cell layers are exposed for 10–30 sec to two GE germicidal bulbs at $\sim$10 cm (60 erg/mm$^2$/sec).
5. XC cells in fresh medium ($10^6$ cells/plate) are added to the irradiated monolayer, and cultures are incubated at 37° until plaque formation occurs (generally about 4 days after the addition of XC cells). Medium is changed 2 days after the addition of XC cells.
6. Cultures are washed with phosphate-buffered saline, fixed with methanol, and stained with Giemsa (Grand Island Biological Corp., Grand Island, New York). The plaques appear as holes in the XC cell sheet which contain some multinucleated giant cells.

*Remarks.* The XC test is useful for detecting most ecotropic MuLVs that replicate efficiently in culture, but it cannot be used for either xenotropic or amphotropic viruses, or for certain genetically transmitted ectropic viruses that replicate poorly. Thus the test detects only a subset of MuLVs, and a negative XC test does not mean that virus is not present. For routine assays, the mouse SC-1 cell line is ideal since it lacks *Fv*-1 restriction and can be infected with N-, B-, or NB-tropic viruses.[4] Mouse secondary embryo cultures[30] may be used to type N- and B-tropic isolates. the NIH/3T3[5] and BALB/3T3[6] cell lines can also be used as prototype N- and B-cells, respectively.

Ultraviolet (UV) irradiation of the mouse cells prior to the addition of XC cells kills the mouse cells at a faster rate than it inactivates their capacity to produce virus. The dose of irradiation is approximate and should be determined empirically for each mouse cell line. In addition, the time after infection when UV irradiation is performed must be long enough to allow plaque development but short enough to prevent the

[30] W. P. Rowe, W. E. Pugh, and J. W. Hartley, *Virology* **42,** 1136 (1970).

development of secondary, "satellite" plaques.[30] This will depend on the relative "infectivity" of the viral stock and the susceptibility of the mouse cells employed.

B. SECONDARY FOCUS-FORMATION ("$S^+L^-$ TESTS")

The Moloney sarcoma-positive, leukemia-negative ($S^+L^-$) strain of mouse sarcoma virus can transform susceptible cells from a variety of mammalian species. Nonproducer transformed clones containing the $S^+L^-$ genome do not produce infectious viral particles but do express certain viral coded antigens in addition to the function(s) required for transformation.[31] Several nonproducer cell lines undergo further morphological alterations when infected by MuLVs. These changes (focus or plaque formation) form the basis for several quantitative assays for infectious MuLVs.

Nonproducer derivants used in these assays include clones derived from mouse 3T3 cells (line "C116"[32]), from feline CCC cells (line "8C"[33]), or from mink MvlLu cells (line $MiCl_1$[34]). Mouse C116 cells support the replication of ecotropic MuLVs while 8C and $MiCl_1$ cells are employed in assays for xenotropic MuLVs. In general, the procedures employed are similar for each cell line. The number of foci or plaques produced is directly proportional to infectious virus concentration.[32-34]

*Procedure*

1. C116 cells are grown in Dulbecco's modified Eagle's medium with 10% calf serum, $MiCl_1$ cells in RPMI-1640 medium with 10% fetal calf serum, and 8C cells in McCoy's 5A medium with 15% fetal calf serum. All cultures are supplemented with antibiotics.

2. Cells are seeded in 60-mm plastic petri dishes at $\sim 1 \times 10^5$ cells/plate.

3. On the next day, the medium is withdrawn and cells are treated with medium containing 25 $\mu$g/ml DEAE dextran (8C cells) or with 1 $\mu$g/ml Polybrene ($MiCl_1$ or C116 cells) for 1 hr.

4. Serial dilutions of virus-containing medium (0.5–1.0 ml/plate) are adsorbed onto the cells for 1 hr at 37° with occasional rocking of the plates.

5. Four milliliters of medium are then added, and plates are incubated at 37° in a 5% $CO_2$ atmosphere.

[31] R. H. Bassin, L. A. Phillips, M. J. Kramer, D. K. Haapala, P. T. Peebles, S. Nomura, and P. J. Fischinger, *Proc. Natl. Acad. Sci. U.S.A.* **68,** 1520 (1971).

[32] R. H. Bassin, N. Tuttle, and P. J. Fischinger, *Nature (London)* **229,** 564 (1971).

[33] P. J. Fischinger, C. S. Blevins, and S. Nomura, *J. Virol.* **14,** 177 (1974).

[34] P. T. Peebles, *Virology* **67,** 288 (1975).

6. The medium is changed 24 hr after infection, and again at 3-day intervals until foci can be scored (generally 5–12 days after infection).

*Remarks.* The foci induced in C116 or 8C cells appear as lytic areas containing a few grape-like clusters of round, refractile cells. This occurs because cells undergoing morphological change tend to round up and detach from the plate. By contrast, foci produced in $MiCl_1$ cells appear as clustered, refractile cells that overgrow the monolayer and tend to remain attached to the plate. As a result $MiCl_1$ foci are most readily scored by direct microscopy while C116 or 8C "plaques" can be enumerated most readily after fixation and staining of the cells. For quantitative assessment of MuLV titers, foci should be read as early as possible after infection to avoid the possibility of satellite focus-formation resulting from further rounds of infection.[32–34] Use of early-passage $MiCl_1$ cells is necessary since, at late passages, these cells spontaneously revert to multilayer growth and tend to round up and float in the culture medium.

## Concluding Remarks

Investigators working with murine type C viruses should be aware that genetic information coding for these viruses is vertically transmitted in the germ line of all mouse species, and that cells in tissue culture can begin to produce "endogenous," genetically transmitted viruses. Moreover, mouse cells have the capacity to produce at least four distinct classes of retroviridae with different host-range, antigenic, and genomic properties.[28] Activation of endogenous viruses may occur after infection of mouse cells with standard MuLVs; therefore, viruses purified for biochemical studies should be tested for the acquisition of new properties not detected in the original infecting stock. The mouse cell lines recommended here for virus propagation (SC-1, BALB/3T3 and NIH/3T3) all have endogenous viral information that can be activated by the infectious leukemia virus and that may be able to recombine with the infecting virus. For example, stocks of NB-tropic Rauscher MuLV propagated in cells derived from Balb/c mice are frequently contaminated by endogenous xenotropic viruses that were activated in the host cells. These viral stocks contain mixtures of virions with different antigens and genomes, as well as phenotypically mixed particles and potential recombinant viruses. Such events would not be detected by assaying viral particles for reverse transcriptase activity but could be scored by appropriate immunological or virological tests. The prevalence of many different retroviral classes among different species of *Mus* also suggests that currently employed assays for viral replication may be insufficient in screening for new isolates. Thus, the failure to detect murine type C viruses in certain assays for viral replicative functions may not mean that viruses are absent.

# [36] Human Adenoviruses: Growth, Purification, and Transfection Assay

*By* MAURICE GREEN and WILLIAM S. M. WOLD

Human adenoviruses (Ads) provide some of the most powerful eukaryotic systems for the experimental analysis of DNA tumor virus replication, cell transformation, and the molecular biology of human cells (reviewed in Green *et al.*[1-3]). Thirty-one well-defined human adenovirus serotypes (Ad1-31) have been identified. Most types fall into five distinct groups based upon several properties including DNA–DNA homology, immunological cross-reactivity of "early" viral proteins (tumor antigens), and tumorigenicity in newborn hamsters. The groupings are as follows: group A (Ad12, 18, 31), group B (Ad3, 7, 11, 14, 16, 21), group C (Ad1, 2, 5, 6), group D (Ad8–10, 13, 15, 17, 19, 20, 22–30), and group E (Ad4).

Human Ads have a duplex, linear, noncircularly permuted DNA genome of 20–25 $\times 10^6$ daltons.[4] The DNA molecules have inverted terminal repetitions and a protein linked, probably covalently, to both 5'-termini.[3] The Ad virion (virus particle) is assembled from about 13 viral polypeptides and the viral DNA genome into an icosahedron structure, consisting of an external capsid and an inner core containing viral DNA and at least two basic proteins.

The productive infection of human KB cells by Ad2 has been the most widely used model for Ad replication.[5] Recently, Ad5 (very closely related to Ad2) and HeLa cells have been used as well. The genomes of these viruses have been dissected with a variety of restriction endonucleases, and detailed physical and genetic maps constructed.[6-9] Productive

[1] M. Green, *Annu. Rev. Biochem.* **39,** 701 (1970).
[2] M. Green, J. T. Parson, M. Pina, K. Fujinaga, H. Caffier, and I. Landgraf-Leurs, *Cold Spring Harbor Symp. Quant. Biol.* **35,** 803 (1970).
[3] W. S. M. Wold, M. Green, and W. Büttner, *in* "Adenoviruses" (D. P. Nayak, ed.). Dekker, New York, 673–768 (1978).
[4] M. Green, M. Pina, R. Kimes, P. C. Wensink, L. A. MacHattie, and C. A. Thomas, Jr., *Proc. Natl. Acad. Sci. U.S.A.* **57,** 1032 (1967).
[5] M. Green and G. E. Daesch, *Virology* **13,** 169 (1961).
[6] J. Sambrook, M. Botchan, P. Gallimore, B. Ozanne, U. Pettersson, and J. Williams, *Cold Spring Harbor Symp. Quant. Biol.* **39,** 615 (1974).
[7] P. A. Sharp, P. H. Gallimore, and S. J. Flint, *Cold Spring Harbor Symp. Quant. Biol.* **39,** 457 (1974).
[8] J. F. Williams, C. S. H. Young, and P. E. Austin, *Cold Spring Harbor Symp. Quant. Biol.* **39,** 427 (1974).
[9] H. S. Ginsberg, M. J. Ensinger, R. S. Kauffman, A. J. Mayer, and U. Lundholm, *Cold Spring Harbor Symp. Quant. Biol.* **39,** 419 (1974).



ISBN 0-12-181958-2

infection of "permissive" human cells results in the production of approximately 200,000 virus particles per cell, and the eventual death of the infected cell.[10] The viral genome is transcribed and replicated in the cell nucleus. There are at least two stages of expression of the Ad genome, "early," before initiation of viral DNA replication at 6–7 hr postinfection, and "late."[2] Early genes, which are transcribed by host cell enzymes, are arranged in four noncontiguous gene blocks, two on the *r*-strand and two on the *l*-strand, and represent about 30% of the single-stranded genome. Proteins encoded by early genes (as many as 12 or 14 early polypeptides have been identified) probably function in viral DNA replication, regulation of transcription, and cell transformation. Late genes, which code virion structural proteins, are mainly on the *r*-strand, and represent most of the remainder of the genome. Most exciting, late mRNA consists of "spliced" mRNA molecules (reviewed in Wold *et al.*[3]), wherein the main coding body of the RNA is covalently linked to short "leader" RNA sequences coded by DNA regions far upstream from each gene.

Ad infection of nonpermissive or semipermissive rodent cells may result in the stable transformation of a small fraction ($10^{-5}$–$10^{-6}$) of the cells. The transforming information [presumably gene(s)] has been localized on the left-hand end of the *r*-strand within map position 1 to 7.5 (the entire genome consists of 100 map units), in that cells can be transformed by transfection with DNA restriction fragments containing this region,[11] and all virion-transformed cells retain this region[6] and express it as RNA[7] and protein. The transforming "genes" of Ad12 (group A) and Ad7 (group B) are in a similar region. The transforming region lies within an early gene block.

In this article, we describe in detail methods used in our laboratory for growth of cells for the propagation and plaque assay of all human Ad serotypes. We also describe the isolation of viral DNA, and the transfection assay to determine Ad DNA infectivity and to localize Ad-transforming genes.

## Growth of Cultured KB Cells Used for the Propagation of Adenoviruses

The 31 human Ads (prototype strains are available from the American Type Culture Collection) are grown in suspension cultures of human KB cells for large-scale virus propagation, and in monolayer culture (petri dish) for plaque assay of Ad infectivity. HeLa cells are used in some laboratories with comparable results. There is some evidence that HeLa

[10] M. Green, *Cold Spring Harbor Symp. Quant. Biol.* **27**, 219 (1962).

[11] F. L. Graham, P. J. Abrahams, C. Mulder, H. L. Heijneker, S. O. Warnaar, F. A. J. de Vries, W. Fiers, and A. J. van der Eb, *Cold Spring Harbor Symp. Quant. Biol.* **39**, 601 (1974).

and KB cell lines are derived from the same HeLa cell parental line. Both lines are available from the American Type Culture Collection. Below are described the preparation of growth media and the cultivation of KB cells for large-scale production of virus and for infectivity assay.

## Preparation of Cell Culture Media

Triple-distilled water is used to prepare all media. Monolayer cultures are grown in serum-supplemented Eagle's minimal essential medium (MEM),[12] a synthetic medium consisting of a balanced salt solution and amino acids and vitamins essential for growth of cultured human cells; it is often supplemented with nonessential amino acids. Joklik-modified MEM which lacks calcium (to prevent cell clumping) and contains a 10-fold higher phosphate concentration (increased buffering capacity) is used for suspension cultures. There are several ways to prepare media: by assembly of laboratory-prepared or commercially available concentrates ($\times 10$ to $\times 200$) of balanced salts, amino acids, and vitamins; by purchase of complete $\times 1$ or $\times 10$ media; and by hydration of powdered media (complete except for sodium bicarbonate). Most commercial powdered media appear to be as satisfactory as laboratory-prepared media for the growth of cells, and they cost about the same.

As a precaution against contamination and errors in media components, we routinely maintain cells on two independent media (A and B), using sera, media, phosphate-buffer saline (PBS), and trypsin-EDTA, from two different sources or lots.

*Preparation of 20 Liters of MEM from Powdered Medium.* Two packages of powdered media [for 10 liters each, Grand Island Biological Co. (GIBCO) or KC Biologicals] are dissolved in 18 liters of water in a stainless-steel pressure vessel (Millipore Corp., 5-gallon capacity). Twenty milliliters of a stock $\times$ 1000 solution of antibiotics are added (either penicillin and streptomycin, or $\times 1000$ Gentamycin), followed by 44 g of $NaHCO_3$ for monolayer MEM or 40 g of $NaHCO_3$ for suspension MEM. The pH is adjusted to 7.2 with 1 *N* NaOH or 1 *N* HCl, made up to 20 liters with water, and sterilized with a pressure pump by filtration through a 127-mm disc prefilter (Millipore AP2512750) placed on top of a 142-mm, 0.20-$\mu$m metricel filter (Gelman, from Fisher Scientific Co.) or a 0.22-$\mu$m GS filter (Millipore). Media are sterility-tested for at least a week prior to use, using both thioglycolate broth (Difco) and blood agar plates [tryptic soy agar containing 5% sheep blood (Gibco Diagnostics)], and stored at 4°. Before use, 5% or 10% sera (donor calf or donor horse obtained from KC Biologicals, Flow Laboratories, Inc., or Gibco) are added.

[12] H. Eagle, *Science* **130**, 432 (1959).

*Preparation of Minimal Essential Medium from Individual Components.* Assembling medium from individual components provides control of quality and the opportunity to alter the concentration of specific substances for experiments, e.g., reduction in phosphate concentration in order to label nucleic acids with $^{32}PO_4$.

*Minimum Essential Medium for Suspension Culture.* The following materials are assembled and sterilized by filtration as described above: 15 liters of water, 2000 ml of ×10 suspension salt A (the preparation of concentrates is described below), 200 ml of ×100 suspension salt B, 1000 ml of ×20 MEM essential amino acids, 200 ml of ×100 MEM nonessential amino acids, 200 ml of ×100 MEM vitamins, 1380 ml of water containing 12.5 g of L-glutamine (Sigma Chemical Co.) and 100 mg of phenol red (sodium salt, Fisher Scientific Co.), and 20 ml of either ×1000 penicillin and streptomycin or ×1000 Gentamycin.

*Minimal Essential Medium for Monolayer Culture.* The procedure described above for suspension medium is used except that monolayer salt A and B concentrates (see below) are substituted for monolayer salt concentrates.

*Balanced Salt Concentrates* (stored at 4°, nonsterile):

SUSPENSION SALTS. Salt A (×10) is 68 g NaCl, 4 g KCl, 15 g $NaH_2PO_4 \cdot H_2O$, 10 g dextrose, and 20 g $NaHCO_3$ (added slowly) per liter. Salt B (×100) is 20 g $MgCl_2 \cdot 6\ H_2O$ per liter.

MONOLAYER SALTS. Salt A (×10) is 68 g NaCl, 4 g KCl, 1.4 g $NaH_2PO_4 \cdot H_2O$, 10 g dextrose, and 20 g NaHCO per liter. Salt B (×100) is 20 g $MgCl_2 \cdot 6\ H_2O$ and 25 g $CaCl_2 \cdot 2\ H_2O$ per liter.

*Essential Amino Acids* (×20). The following amino acids (Sigma) are dissolved in 4 liters of water and 80 ml of concentrated HCl: 25.3 g L-arginine · HCl, 4.8 g L-cystine, 8.2 g L-histindine · HCl, 10.4 g L-isoleucine, 10.4 g L-leucine, 14.6 g L-lysine, 3.0 g L-methionine, 6.4 g L-phenylalanine, 9.6 g L-threonine, 2.0 g L-tryptophan, 7.2 g L-tyrosine, and 9.2 g L-valine. Then, 5920 ml of water are added and the solution is sterilized by pressure filtration and stored at 4°.

*Nonessential Amino Acids (×100).* The following amino acids (Sigma) are dissolved in 2 liters of water and stored frozen in 100-ml portions: 1.78 g L-alanine, 3.00 g L-asparagine, 2.66 g L-aspartic acid, 2.94 g L-glutamic acid, 1.5 g glycine, 2.1 g L-serine, and 2.3 g L-proline.

*Vitamins (×100).* The following vitamins (Sigma) are dissolved in 1500 ml of water: 24 mg niacinamide, 41 mg pyridoxal · HCl, 64 mg thiamine · HCl, 8 mg riboflavin, 48 mg D-pantothenic acid (hemi-calcium salt), 280 mg choline chloride, and 108 mg myo-inositol. Then 5.4 ml of 0.1 *N* NaOH containing 5 mg d-biotin and 88 mg folic acid are added, and the solution is adjusted to 2 liters. Aliquots (200 ml) are stored at −20°.

*Penicillin and Streptomycin ( ×1000).* Twenty million units of penicillin G (Sigma) and 20 g of streptomycin sulfate (Sigma) are dissolved in 200 ml of water. The solution is sterilized by suction filtration through a 47-mm, 0.22-$\mu$m filter (Millipore) and stored at 4°.

*Gentamycin ( ×1000).* A sterile solution containing 50 mg/ml of Gentamycin is purchased from the Schering Corporation and stored at 4°.

*Phosphate-Buffered Saline (PBS).* For washing cell sheets prior to plaque assay, PBS containing magnesium and calcium is used; it is usually supplemented with 0.1% bovine serum albumin (BSA) (Sigma, crystalline). To wash cell sheets before trypsinization and for the preparation of trypsin-EDTA, PBS lacking magnesium and calcium is used. PBS is made by dissolving with stirring 80 g NaCl, 2 g KCl, 1 g $CaCl_2 \cdot 2\,H_2O$, and 1 g $MgCl_2 \cdot 6\,H_2O$ in 8 liters of water. Then, 11.5 g $Na_2HPO_4$ and 2 g $KH_2PO_4$ are dissolved in 2 liters of water and added to give 10 liters of PBS. The solution is sterilized by pressure filtration and stored at 4° in 100-ml portions. PBS lacking magnesium and calcium is prepared as described above except that $CaCl_2 \cdot 2\,H_2O$ and $MgCl_2 \cdot 6\,H_2O$ are deleted. BSA is prepared as a 5% stock solution, sterilized by suction filtration, and stored at 4°. To prepare PBS–0.1% BSA, 2 ml of stock BSA are added to 100 ml of PBS.

*Trypsin-EDTA.* Four grams of trypsin (Difco Laboratories, 1:250), 2 g of disodium EDTA, 4 g of dextrose, 20 mg of phenol red (sodium salt, Fisher), and 4 ml of ×1000 stock solution of penicillin and streptomycin are dissolved in PBS without magnesium and calcium and adjusted to 4 liters. The solution is filtered and stored at 4°.

*Tris-Saline-Glycerol.* This is used to maintain the infectivity of purified Ad stocks at −35°. A ×5 stock of Tris-saline is 40 g of NaCl, 1.9 g of KCl, 0.5 g of $Na_2HPO_4$, 5 g of dextrose, and 150 ml of 1 *M* Tris · HCl buffer (pH 8.1 at 25°), made up to 1 liter, and sterilized by filtration. Tris-saline-30% glycerol is made by adding 100 ml of ×5 Tris-saline to 150 ml of sterile glycerol (autoclaved at 120° for 15 min) and adjusting the volume to 500 ml with sterile water.

*Agar Overlay Medium for Plaque Assay.* The following sterile solutions are combined to prepare 100 ml of ×2 overlay medium: 18.0 ml of ×10 Earle's balanced salt solution, 2.0 ml of ×100 MEM essential amino acids, 2.0 ml of ×100 MEM vitamins, 2.0 ml of ×100 L-glutamine, 0.7 ml of 2.1% L-arginine, 0.4 ml of 1.0 *N* NaOH, 12 ml of horse serum, 12 ml of chicken serum (KC Biological), ( recently, we have used 12 ml of fetal calf serum plus 12 ml of water instead of horse and chicken serum ) 2.0 ml of antibiotic mixture (see below), and 43 ml of sterile water. The solution is stored at −20°. (1) *Earle's salt solution (×10).* Solution A is 136 g of NaCl, 8 g of KCl, 2.88 g of $KH_2PO_4$, and 2 g of $MgSO_4$ in 800 ml of water. Solution B is

4.0 g of $CaCl_2 \cdot 2\ H_2O$ in 200 ml of water. Solution C is 40 g of dextrose and 100 ml of 0.2% phenol red in 200 ml of water. A, B, and C are autoclaved separately and cooled. A and B are mixed, and then C is added. The solution is adjusted to 2 liters with sterile water and stored at 4°. (2) *Glutamine (200 mM, ×100).* L-glutamine, 29.2 g, is dissolved in 1 liter of water, sterilized by filtration, and 100 ml amounts stored at −20°. (3) *Phenol red (0.2%).* Phenol red (2.0 g) is dissolved in 50 ml of 95% ethanol and 950 ml of water and stored at 4°. (4) *Arginine.* L-arginine (Sigma), 2.1 g, is dissolved in 100 ml of water, sterilized by autoclaving, and stored at 4°. (5) *Antibiotic mixture.* During plaque assay, petri dish cultures are incubated at 37° for up to 20 days in a humidified $CO_2$ incubator. Fungal and bacterial contamination are often a problem, and therefore an antibiotic mixture containing neomycin and Fungizone in addition to penicillin and streptomycin is used. Six million units of penicillin, 2 g of streptomycin, 0.1 g of amphotericin B (Fungizone, Squibb Pharmaceuticals), and 1 g of neomycin sulfate (Upjohn) are dissolved aseptically in 200 ml of sterile water and stored in 10-ml amounts at −20°.

*Growth of Monolayer and Suspension Cultures.* Cells are incubated at 36–37°. Monolayers are cultured in 32-ounce Saniglass prescription bottles or plastic cell culture flasks with growth surface areas of 25–500 $cm^2$ (Nunc, from Vanguard International; Falcon Plastics, from Fisher Scientific; or Corning Glassworks, from Fisher Scientific). Cells are grown in suspension in screw-cap spinner flasks, ranging from 50 ml to 8 liters, with adjustable hanging magnetic bars (Bellco Glass, Inc.). The suspension is stirred with a magnetic stirrer. Cells are subcultured and media changed using aseptic techniques in an isolation cubical or, preferably, in a laminar-flow hood. Glassware is cleaned with nontoxic detergents designed for cell culture, and sterilized by autoclaving, or by heating in an oven at 180° for 4–8 hr.

A KB cell clonal line, grown in our laboratory for the past 19 years, is readily converted from monolayer to suspension by scraping monolayers with a rubber policeman and culturing in suspension. KB cells grow in suspension indefinitely and readily form monolayers. Suspension cultures are propagated in MEM supplemented with 5% horse serum for the routine growth of Ads. Cells are counted each day with a hemocytometer (American Optical Corp., Bright-Line) under a microscope (×100 magnification); the number of cells deposited on 10 large squares (five on each side of the hemocytometer) × 1000 is the cell density (cells/ml). The cell density is adjusted to $2 \times 10^5$ cells/ml by the addition of fresh medium each day. The cell doubling time ranges from 18–24 hr.

Monolayers maintained in Eagle's MEM with 10% calf serum are used mainly to seed petri dishes for plaque assay. Cultures are observed every other day with an inverted microscope for cell culture (×40 to ×100 magnification). Confluent monolayers are subcultured about twice a week as follows. After removal of the medium by aspiration or pouring, the cells are scraped into 8 ml of fresh medium with a rubber policeman, dispersed gently with a 10-ml pipette (fitted with Propipet), and seeded into four or five new flasks (about $2 \times 10^6$ cells per 32-ounce flask) containing 40 ml of medium.

## Large-Scale Growth and Purification of Adenoviruses

*Infection of KB Cell Suspension Cultures.*[13-15] Exponentially growing KB cells ($3\text{–}6 \times 10^5$ cells/ml, 2–20 liters total) are collected by centrifugation at 180 g for 15 min at room temperature in the International PR-J centrifuge (No. 276 rotor) in sterile 1-liter polypropylene screw-capped bottles (International No. 2939). Cells are suspended in 1/20th volume of MEM (no serum), infected by the addition of 20–100 plaque-forming units (PFU) per cell of a sterile Ad stock, and stirred for 1 hr at 37°. The suspension is diluted with warm MEM (Joklik-modified) containing 5% horse serum to give $3\text{–}4 \times 10^5$ cells/ml. The infected cells are incubated with stirring at 37° until 30–40 hr after infection (48 hr for Ad12 strain Huie), chilled in ice, and harvested by centrifugation at 230 g for 20 min at 4°. The cell pellet (over 90% of virus remain cell-associated) is resuspended in 10 ml of 10 m*M* Tris · HCl, pH 8.1, per 3 liters of infected cells, and frozen until purified (usually within 3 days after harvesting).

*Purification of Adenoviruses.*[13,14] Infected cell-pellet suspensions are thawed briefly in a 37° water bath. All subsequent operations are at 0–4°. Cell pellets are suspended in an additional 40 ml of 10 m*M* Tris · HCl, pH 8.1, and the cells disrupted with a Raytheon DF-101 sonic oscillator at full power for 5 min. The suspension is homogenized with 40 ml of trichlorotrifluoroethane (Taylor Chemical) in a 200-ml stainless-steel cup of a Sorvall Omnimixer for 1 min at about 10,000 rpm, and then centrifuged in a 250-ml polycarbonate screw-capped bottle (Nalgene No. 3123), or in a disposable sterile 250-ml screw-capped centrifuge tube (Corning No. 25350), using the PR-J centrifuge (No. 259 rotor) at 1000 g for 10 min. The upper aqueous layer is removed with a pipette, and the lower layer is homogenized with 20 ml of 10 m*M* Tris · HCl, pH 8.1. The aqueous layers are combined and carefully layered over 10 ml of CsCl at 1.43 g/cm$^3$ (e.g., Schwarz-Mann, biological grade, 43 g of CsCl in 60 ml of 10 m*M* Tris · HCl, pH 8.1) in a cellulose nitrate tube ($1 \times 3.5$ inch). After centrifugation in a SW-27 rotor for 1 hr at 20,000 rpm, the supernatant liquid above and most of the CsCl solution below the opalescent virus band are removed. The virus band (3–5 ml) is adjusted to 1.34 g/cm$^3$ with powdered CsCl, placed in a $0.5 \times 2.5$ inch (6.5 ml) cellulose nitrate tube (which fits

[13] M. Green and M. Pina, *Virology* **20,** 199 (1963).
[14] M. Green and M. Pina, *Proc. Natl. Acad. Sci. U.S.A.* **51,** 1251 (1964).
[15] M. Pina and M. Green, *Virology* **38,** 573 (1969).

the Ti 50 rotor with an adaptor), filled with CsCl solution (density = 1.34 g/cm$^3$, 56 g of CsCl in 116 ml of 10 m$M$ Tris · HCl, pH 8.1), and centrifuged for 16–24 hr at 30,000 rpm. The visible band in the middle of the gradient, at a buoyant density of 1.34 g/cm$^3$, is collected from the side by tube puncture, adjusted to 6.5 ml by the addition of CsCl solution at 1.34 g/cm$^3$, and recentrifuged as described above. The twice-banded virus is collected and stored at 4°.

*Preparation of Purified Adenovirus Stocks and Plaque Assay of Infectivity.* It is essential for biochemical studies that the amount of infecting virus be known, because all cells must be infected, and because the course of the viral infection and the quantity of viral macromolecules synthesized varies somewhat with the multiplicity of infection. Virus is purified as described above, with the use of aseptic procedures, for the preparation of Ad stock viruses for infection. All materials are either autoclaved or soaked for at least 30 min in 70% ethanol (e.g., cellulose nitrate tubes). Virus, banded once, is diluted in 10 ml of sterile Tris-saline-30% glycerol per liter of initial infected cells, and stored in 5-ml aliquots in plastic snap-cap tubes at −35°. The infectivity of Ad stocks remains constant for years. Stocks range in titer (assayed on KB cells) from 2–5 × $10^{11}$ PFU/ml for Ad2, 6 × $10^{10}$ to 2 × $10^{11}$ PFU/ml for Ad7, and 2 × $10^8$ to 1 × $10^9$ PFU/ml for Ad12 (Huie).

A plaque assay[16] developed for Ad2 has been used for all human Ads.[17] A plaque is a cluster of infected, nonstained, nonviable cells, produced by infection of a single cell by one virus particle with secondary infection of contiguous cells by progeny virus, and is surrounded by uninfected, viable cells stained by neutral red. The plaquing efficiency of purified virus ranged from virus particle:PFU ratios of 11 : 1 for Ad3 to 2300 : 1 for Ad25; the size and time of appearance of plaques varied with the Ad serotype.[17] Assays are performed in 60-mm plastic petri dishes seeded usually the day before with 1–1.5 × $10^6$ KB cells. Cells are collected from suspension cultures by centrifugation at 180 g for 7 min, and resuspended in MEM with 10% calf serum to give a density of 2–3 × $10^5$ cells/ml; 5-ml amounts are seeded in petri dishes. Alternatively, KB cell monolayers are harvested with trypsin-EDTA, resuspended in MEM with 10% calf serum, and petri dishes seeded as described above. Dishes are incubated in a $CO_2$ incubator (flushed with 5% $CO_2$ and maintained at 90–100% humidity, e.g., Forma Hydro-Jac, Forma Scientific) until cell sheets are 70–80% confluent. The medium is removed by aspiration with a Pasteur pipette, and cells are washed by gentle addition and aspiration of 5 ml of PBS–0.1% BSA at 37°. Serial 10-fold dilutions of virus samples are

[16] H. C. Rouse, V. H. Bonifas, and R. W. Schlesinger, *Virology* **20**, 357 (1963).

[17] M. Green, M. Pina, and R. C. Kimes, *Virology* **31**, 562 (1967).

prepared in cold PBS–0.1% BSA (or alternatively in MEM with 2% calf serum), and 0.5-ml amounts (incubated briefly in a 37° water bath) are added to triplicate dishes. Dilutions are chosen to contain about 20–200 PFU/0.5 ml. Ad2 plaques are about 5 mm in diameter, and therefore over 50–100 plaques per plate will produce confluent, noncountable areas, whereas as many as 500 Ad12 plaques (0.5 mm) per plate can be scored.[17] Dishes are rocked gently and placed in a $CO_2$ incubator for 90 min to permit virus adsorption. Five milliliters of agar overlay medium are slowly added at the edge of each dish (to avoid dislodging cells) with a 10-ml pipette. Overlay medium is prepared at 46° just before use by mixing 94 ml of ×2 overlay medium (containing 6% horse serum and 6% chicken serum or 6% fetal calf serum) with 100 ml of melted 1.8% Bacto-agar and 6 ml of sterile 7.5% sodium bicarbonate. After the agar overlay has hardened at room temperature (15–30 min), dishes are incubated in a humidified $CO_2$ incubator. Five milliliters of agar overlay medium are added again after 5 days. An additional 5 ml of agar overlay medium, containing 0.001 or 0.002% neutral red (Fisher), are added after 10 days. Plaques are visible after 12–14 days with most Ads (13–15 days with Ad12) and are counted macroscopically against a lighted background. Small plaques of Ad2, Ad1, and Ad5 can be seen as early as 8 days after infection if the neutral red overlay is added on the 7th day.[17]

## Isolation and Purification of Adenovirus DNA[14,18]

DNA is isolated from virus by digestion with a protease, treatment with SDS, and phenol extraction. Protease treatment is necessary to remove a tightly bound protein[18] that is covalently linked to the termini of Ad DNA molecules.[19] Twice-banded virus (in CsCl solution) is dialyzed against two changes of 200 volumes of 10 m*M* Tris · HCl, pH 8.1, 1 m*M* EDTA at 4° for 2 hr or longer. To each milliliter of virus are added 150 μl of 50 m*M* EDTA (pH 6.9), 150 μl of 1 *M* sodium phosphate (pH 6.0), 15 μl of 1 *M* cysteine-HCl, and 30 μl of papain (Sigma, twice crystallized, 25 mg/ml in 50 m*M* sodium acetate, pH 4.5). After 1 hr at 37°, 150 μl of 5% SDS are added per milliliter of virus, the solution is incubated at room temperature for 30 min, and it is extracted with an equal volume of phenol (saturated with 10 m*M* Tris · HCl, pH 8.1, 1 m*M* EDTA) by gentle rotation (to avoid shear degradation of Ad DNA molecules) in a multipurpose rotator (Scientific Industries, Model 150 V) for 15 min at 4°. After centrifugation at 7800 g at 4° for 10 min, the upper aqueous phase is removed and extracted twice more with phenol. The aqueous phase is extracted 3

[18] M. Green and M. Pina, *Proc. Natl. Acad. Sci. U.S.A.* **50,** 44 (1963).

[19] D. M. K. Rekosh, W. C. Russell, A. J. D. Bellet, and A. J. Robinson, *Cell* **11,** 283 (1977).

times with water-saturated ether, and then dialyzed against 200 volumes of 10 m*M* Tris · HCl, pH 8.1, containing 1 m*M* EDTA for 24 hr. The absorbance at 250, 260, and 280 nm is determined. The amount of DNA is calculated based on an $A_{260}$ unit equivalent to 50 $\mu$g DNA/ml. The 260 nm/280 nm and 260 nm/250 nm ratios are about 1.9 and 1.1, respectively. Yields of 60–80% homogeneous, intact duplex DNA molecules, sedimenting at 30–31 S, are obtained.[4] Approximately 0.5 mg of Ad2 DNA is isolated from virus produced in 1 liter of cells.

## Transfection-Infectivity Assay for Viral DNA and DNA Restriction Fragments

Transfection assays with purified viral DNA (rather than virions) are useful for several kinds of studies. For example, recent recombinant DNA procedures, and related procedures, are beginning to be exploited for the construction of Ad mutants *in vitro*.[20] In addition, transfection procedures using Ad DNA and DNA restriction endonuclease fragments have been used to map the Ad-transforming genes. The infectivity of Ad DNA and the transforming activity of Ad DNA and restriction DNA fragments are assayed with the calcium phosphate procedure,[11,21] modified to include treatment of cells with dimethylsulfoxide (DMSO)[20,22] to enhance the transfection activity. Cells used for transfection-infectivity assay are permissive human KB cells in petri dishes as described in preceding sections. Cells (nonpermissive) used for transfection-transformation assay are prepared from 16-day rat (Wistar) whole embryos. Cells prepared from baby rat (6- to 7-day-old) kidneys are used by Graham *et al.*[11] A permanent rat cell line for transfection-transformation experiments has also been described.[23]

Cell procedures are performed under aseptic conditions. Several embryos are minced with scissors, washed 2 or 3 times with PBS lacking calcium and magnesium, and digested with warmed (37°) trypsin-EDTA for 10–20 min at room temperature. Cells are centrifuged for 10 min at 230 g, chilled, suspended in MEM with 10% calf serum, and filtered through two layers of gauze. Cells are collected from the filtrate by centrifugation, a 0.5% suspension is prepared in MEM containing 10% calf serum, and 60-ml amounts are seeded in 32-ounce culture flasks; usually one embryo provides sufficient cells for one flask. When monolayers are nearly confluent (3–4 days), cells are subcultured by trypsin-EDTA treat-

[20] G. Chinnadurai, S. Chinnadurai, and M. Green, *J. Virol.* **26,** 195 (1978).

[21] F. L. Graham, J. Smiley, W. C. Russell, and R. Nairn, *J. Gen. Virol.* **36,** 59 (1977).

[22] N. D. Stow and N. M. Wilkie, *J. Gen. Virol.* **33,** 447 (1976).

[23] G. Kimura, A. Itagaki, and J. Summers, *Int. J. Cancer* **15,** 694 (1975).

ment into 60-mm dishes in MEM with 10% fetal calf serum for transfection ($5 \times 10^5$ cells/dish). When the cell sheets are 80–90% confluent, 0.5 ml of solution, containing either Ad DNA or DNA restriction fragments, is added to each dish without removing the medium. Ad DNA is prepared as described above. Preparative isolation of Ad DNA restriction endonuclease fragments is described in detail in Wold *et al.*[24] The viral DNA solution is prepared as follows.

(1) Ad DNA (0.1 $\mu$g/ml to 0.5 $\mu$g/ml for DNA fragments smaller than $2–3 \times 10^6$ daltons, or 1.0 $\mu$g/ml to 10 $\mu$g/ml for larger fragments) is diluted into HEPES-buffered saline (a $\times 2$ solution is prepared from 8 g of NaCl, 370 mg of KCl, 125 mg of $Na_2HPO_4 \cdot 2\ H_2O$, 1 g of glucose, and 5 g of *N*-2-hydroxyethyl-piperazine-*N*′-2-ethanesulfonic acid, made up to 500 ml and pH 7.05, and sterilized by filtration) containing 10 $\mu$g/ml of salmon sperm DNA.

(2) $CaCl_2$ is added (from a 2.5 *M* stock solution sterilized by autoclaving) to a final concentration of 125 m*M*. After 15–20 min at room temperature, the mixture becomes slightly turbid and is added to petri dishes containing rat embryo cells. After 4 hr of incubation at 37° in a $CO_2$ incubator, the medium is removed, the cultures washed once with medium, and 1 ml of 25% DMSO in HEPES-buffered saline is added to each plate. After 2–4 min, the DMSO is removed, the cultures washed with medium, and 5 ml of fresh MEM with 10% fetal calf serum are added. Three or four days later the medium is replaced with calcium-free MEM with 5% fetal calf serum. Incubation is continued (the medium is changed twice weekly) until transformed cell foci appear (positive controls are always included in assays). Generally, about 2–4 weeks are required for foci to develop. Foci appear macroscopically as raised colonies, and microscopically as multiple layers of densely packed, criss-crossed cells. Transformed colonies can be quantitated under a dissecting microscope at a magnification of $\times 30$ to $\times 40$. Alternatively foci can be counted after the cultures have been fixed by a treatment with absolute methanol for 5 min, air dried, and stained with Giemsa. The efficiency of transfection of an Ad2 DNA HindIII digest is approximately 10 foci per microgram of DNA. As final proof that foci represent cells transformed by viral DNA, clonal lines must be isolated and shown to contain integrated viral DNA and to synthesize viral RNA. For this purpose, individual foci are taken up into a Pasteur pipette by gentle aspiration with a rubber bulb and transferred into a new petri dish containing calcium-free MEM with 10% fetal calf serum. Clonal lines are then developed by selection and subculture and used for subsequent biochemical analyses.

[24] W. S. M. Wold, M. Green, and J. K. Mackey, *Methods Cancer Res.* **15,** 69–161 (1978).

Section III

# Specific Cell Lines

*Articles 37 through 54*

# [37] Sources of Stable Cell Lines

By GERARD J. MCGARRITY

## I. Introduction

T. C. Hsu summarized the advantages of cell culture systems:[1]

1. They provide a continuous supply of homogenous cellular material for biochemical experiments as well as for practical use in medical and public health work.

2. The cells *in vitro* can be manipulated advantageously in many ways. This cannot be done with cells *in vivo*.

3. They can be stored in a deep-frozen state without changing their growth rate and genetic composition, and they can be revived at will.

4. Using cell cultures is more economical than rearing animals and performing experiments with intact animals.

5. They save lives of animals.

To obtain full benefit of these advantages, cell cultures used in research and diagnosis should be standardized as fully as possible and subjected to extensive quality-control procedures. (For detailed methods on this topic, see this volume [2] and [13].) Quality control should include selection of the cell culture to be used and the source of the culture. Many cell cultures contain undetected adventitious agents such as bacteria, yeast, viruses, and mycoplasmas. In this laboratory the annual mean infection rate of cell cultures with mycoplasmas has ranged from 3.4–14.9%. This is based on testing of more than 6000 cell cultures. Mycoplasma-infected cell cultures are useless in controlled standardized procedures. Introduction of a mycoplasma-infected culture into a laboratory generally results in spread of the infection to all cultures in the laboratory.[2]

A significant number of cell cultures are contaminated with other cells, especially HeLa. Gartler concluded that many of the permanent human cell lines may in fact be HeLa, based on electrophoretic variant forms of isoenzymes as genetic markers;[3] this observation has been reinforced.[4,5] More recently, Nelson-Rees and Flandermeyer have reported that 41 of

[1] T. C. Hsu, *in* "The Future of Animals, Cells, Models, and Systems in Research Development, Education and Testing," p. 180. Natl. Acad. Sci., Washington, D.C., 1977.

[2] G. J. McGarrity, *In Vitro* **12,** 643 (1976).

[3] S. M. Gartler, *Natl. Cancer Inst., Monogr.* **26,** 167 (1967).

[4] O. J. Miller, D. A. Miller, P. W. Allderdice, V. G. Dev, and M.S. Grewal, *Cytogenetics* **10,** 338 (1971).

[5] W. Nelson-Rees, R. R. Flandermeyer, and P. K. Hawthorne, *Science* **184,** 1093 (1974).




ISBN 0-12-181958-2

253 cell cultures examined (16%) were not as purported; HeLa cell contamination and wrong species accounted for 36 of these 41 errors.[6]

To protect against mycoplasma infection, interspecies and intraspecies contamination, and other problems, cell cultures should be acquired only from sources that have performed reliable quality-control checks. The common practice of acquiring cultures from colleagues is risky and is discouraged; several episodes of mycoplasma infection have been traced to this practice. Even reassurances of negative mycoplasma assays made months, even weeks, before is no guarantee of a culture's present status.

The purpose of this article is to list the major sources of cell cultures and information pertinent to each source.

## II. Cell Repositories

### *American Type Culture Collection (ATCC)*

The ATCC was founded in 1925 for the collection, preservation, and distribution of authentic cultures of living microorganisms and animal cells. Its 1975 catalogue lists 153 characterized reference animal cell lines derived from 40 different species. An additional 170 human skin fibroblast lines are derived from apparently normal individuals and from patients with various diseases, including genetic disorders. Animal species represented include: amphibian, aves, bat, buffalo, cattle, dog, dolphin, gerbil, goat, hamster, horse, human, insect, minipig, mink, monkey, mouse, muntjac, pig, pisces, potoroo, rabbit, raccoon, rat, and reptile. Widely used human cultures include Hela and several HeLa derivatives; EB-3 and RAJI from Burkitt's lymphoma; and WI-38 and MRC-5, both human diploid lung. IMR-90, a human diploid lung, although not listed in the 1975 catalogue, is available from ATCC.

A copy of the latest catalogue can be purchased (approx. cost: $10, includes postage). The catalogue lists a history of each cell line and a description of the repository reference seed stock that includes freeze medium, viability, growth medium, growth characteristics, plating efficiency, age of culture since origin, morphology, karyology, sterility tests, species confirmation, virus susceptibility, and isoenzyme patterns, if available. All ATCC stocks of questionable human origin (possible HeLa contaminants) are indicated in the catalogue.

Frozen ampules of cell cultures can be purchased from ATCC. Current prices are $28.00 per ampule to nonprofit institutions and $43.00 per ampule to commercial firms. Information regarding purchase of cultures or

[6] W. Nelson-Rees and R. R. Flandermeyer, *Science* **195,** 1343 (1977).

catalogues can be obtained directly from the ATCC, 12301 Parklawn Drive, Rockville, Maryland 20852.

Additional information regarding handling of frozen ampules and guidelines for selection of cells and deposition of new cell lines are also listed in the catalogue.

Special arrangements can be made with ATCC for safe-deposit of individual cell cultures. These cultures are not accessioned into the general stock. The depositor retains all proprietary rights. Other special services are also available on a for-fee basis.

### *Human Genetic Mutant Cell Repository*

This repository was established at the Institute for Medical Research, Camden, New Jersey in 1972 by the National Institute of General Medical Sciences. The objective of this repository is to store cultures in low passage with single and multiple gene defects, both defined and undefined at the molecular level; chromosome abnormalities, including translocations, inversions, and deletions; polymorphisms (antigens, isoenzymes) as well as carrier, sibling, and control cultures.[7] Fibroblast, lymphocyte, and amnion cultures are included. Most are human cells.

The purposes of this repository are:

1. The study of cells from a number of persons with the same inherited disorder which may reveal subtle variations in the underlying defect that would be important to its diagnosis and treatment.

2. Use of identical cell lines for study in different laboratories should facilitate the comparison of data and interpretation of research results.

3. Cell lines from some rare diseases which otherwise would not be readily available to all interested investigators.

4. Investigators in different laboratories can develop and maintain numerous cell lines.

5. The availability of the skills for characterization and identification of genetic mutant cells will assure uniformity and reliability of cells from the repository.

Cell cultures in the repository are tested for sterility, including mycoplasma, species of origin, karyotype, viability, and expression of the biochemical or chromosomal defect.

The Fourth Edition of the Catalogue of the Human Genetic Mutant Cell Repository was published in October 1978 and is available free of charge. The catalogue lists over 1700 cultures, including approximately 200 biochemical mutants and 250 chromosome aberrations. Where

[7] L. L. Coriell, *Science* **180,** 427 (1973).

known, McKusick numbers and HL-A antigens are listed after appropriate cultures.

Starter cultures are supplied. These are obtained by inoculating a frozen ampule into a T25 flask and growing for several days prior to shipping. The cost per T25 flask is $20 plus shipping. The catalogue can be obtained by writing to the Human Mutant Cell Repository, Institute for Medical Research, Copewood Street, Camden, New Jersey 08103.

*Aging Cultured Cell Repository*

The repository was established at the Institue for Medical Research by the National Institute on Aging. The purpose of this repository is to collect, store, and distribute cultures of interest to aging research.[8] The repository contains fibroblast cultures from apparently normal individuals including large quantities of well-characterized early passage cultures from fetal lung; cultures from clinical conditions associated with cellular aging and growth disorders; cultures from tumor patients; and transformed cell lines. The human diploid fibroblast cell line IMR-90 was established from lung tissue of a female fetus by this repository and is available.[9] A limited supply of early passage IMR-90 is also available to qualified investigators by special action of the repository's advisory committee.

Another human diploid fibroblast, IMR-91, has been established from lung tissue of a male fetus. IMR-91 will be available from this repository when characterization studies have been completed. Two human diploid fibroblast lines will be established from adult tissue and will be available when characterization studies have been completed.

The list of available cultures is contained in the Catalogue of the Human Genetic Mutant Cell Repository described above. Currently, 157 cell lines are stored in the Aging Cultured Cell Repository. Cultures are tested for sterility, including mycoplasma, species of origin, karyotype, viability, and, where applicable, expression of the specific characteristic of interest.

Shipping and prices are the same as listed for the Human Genetic Mutant Cell Repository. Cultures are free to laboratories engaged in aging research. Further information can be obtained from the Aging Cultured Cell Repository, Institute for Medical Research, Copewood Street, Camden, New Jersey 08103.

[8] D. G. Murphy and W. W. Nichols, *Cytogenet. Cell Genet.* **15,** 30 (1975).

[9] W. W. Nichols, D. G. Murphy, V. J. Cristofalo, L. H. Toji, A. E. Greene, and S. A. Dwight, *Science* **196,** 60 (1977).

## III. National Science Foundation Cell Culture Centers

The NSF has established Cell Culture Centers at Massachusetts Institute of Technology and the University of Alabama at Birmingham. These are intended to serve as a facility and research resource for scientists who wish to obtain large-scale cell and virus production for highly meritorious research projects.

The MIT Center is designed for large-scale monolayer and suspension cell cultures and viral production. The University of Alabama, Birmingham, has been participating in large-scale production of lymphocyte cell cultures. The objective of both Centers is to produce cells and viruses on a scale sufficient to permit investigators to perform experiments that could not otherwise be performed.

Proposals to use these facilities should be made to the individual Centers. Each proposal is evaluated by a review committee. Proposals should list the amount of culture required and a description of the objectives and significance of the research and supporting materials. Further information is available from:

NSF Cell Culture Center
E17-21
Massachusetts Institute of Technology
77 Massachusetts Avenue
Cambridge, MA 02139

NSF Cell Culture Center
University of Alabama Hospitals & Clinics
1808 7th Avenue, South
Birmingham, AL 35294

## IV. Other Sources

Certain commercial firms sell various types of cell cultures. Some of these cultures are established by the firms and others are obtained from other sources, including the American Type Culture Collection. Further information can be obtained by consulting the catalogues of suppliers of cell culture reagents for types of cultures available and types of quality-control tests performed on these cultures.

## V. Comments

Selection of the proper cell culture system is an integral component of any *in vitro* experimental procedure. The source of the cell culture can be just as critical. Where possible, cultures should be obtained from sources that have performed adequate quality-control tests. If such tests have not been performed, recipient laboratories should quarantine the culture until tests (especially for mycoplasma) are complete. Procedures for sterility testing of cell cultures and media are described elsewhere in this text.

Quality control should be performed periodically to detect any significant alteration in the cell culture that could influence results of experimental procedures.

## [38] Isolation of Fibroblasts from Patients

*By* WILLIAM S. SLY and JEFFREY GRUBB

Over the past 10 years, fibroblasts cultured from biopsies of human patients have provided unique opportunities for exciting biochemical studies. Studies of Lesch-Nyhan syndrome fibroblasts made important contributions to knowledge of purine metabolism and gout, to X-chromosome inactivation, and to somatic cell hybridization.[1] Studies of fibroblasts from patients with mucopolysaccharide storage diseases defined the enzymes involved in mucopolysaccharide degradation and led to discovery of secretion and receptor-mediated uptake of lysosomal enzymes by fibroblasts.[2] Studies of fibroblasts from patients with familial hypercholesterolemia have revolutionized our thinking about the uptake and metabolism of cholesterol and of diseases involving defective cell-surface receptors.[3] These three examples provide dramatic evidence that fibroblasts, despite their reputation as cells in which many differentiated cell functions are not expressed, can contribute enormously to solution of certain biochemical problems.

In this article, we will describe how one obtains a skin biopsy and propagates cultured fibroblasts for biochemical studies from this source. Before doing so, it is worth emphasizing that cell lines from patients with many human genetic disorders are available on request from the Human Genetic Mutant Cell Repository, Institute for Medical Research, Camden, New Jersey. This is a federally financed collection of human mutant cell lines developed under contract for the National Institute of General Medical Sciences. Its purpose is to provide material from patients with human genetic diseases to investigators who wish to study these diseases, but who may have difficulty gaining access to cells from affected patients. Mutant strains are established from skin biopsies solicited from around the world, grown up, tested for contamination, and stored frozen in liquid

[1] J. E. Seegmiller, *Adv. Hum. Genet.* **6,** 75 (1976).
[2] E. F. Neufeld, T. W. Lim, and L. J. Shapiro, *Annu. Rev. Biochem.* **44,** 357 (1975).
[3] M. S. Brown and J. L. Goldstein, *Science,* **191,** 150 (1976).



ISBN 0-12-181958-2

nitrogen. For a small charge, samples are thawed, started in culture, and shipped to investigators on request. Investigators contemplating initiating a cell culture laboratory for studies on normal or mutant human fibroblasts would be well advised to obtain the catalogue of existing strains.

1. Obtaining a Skin Biopsy

*Materials*

Biopsy tray with 2 4-mm punch biopsy instruments, 2 curved scissors, and 2 forceps. A second set of instruments is provided in case one gets contaminated accidentally while attempting the biopsy.
Sterile gauze
Isopropyl alcohol, 70%, 30 ml
Lidocaine without epinephrine, 1%
1 tuberculin syringe
1 screw-capped tube containing 10 ml of medium
*Note:* All glassware items—dishes, tubes, coverslips—are acid washed with concentrated $HNO_3$ and rinsed extensively with deionized water before sterilizing by autoclave.

The site of removal of tissue depends on the purpose. Studies of testosterone binding and/or metabolism should be done on fibroblasts from genital skin.[4] For nearly all other general purposes, we take a biopsy from the subdeltoid area of the upper arm (the lateral aspect of the arm $\frac{1}{3}$ the distance from the shoulder to the elbow). Others use the anterior aspect of the forearm, 2 inches below the elbow crease.[5] The area is first scrubbed twice thoroughly with a sterile gauze flat moistened with 70% isopropyl alcohol. Then an area about 1–1.5 $cm^2$ is anesthetized by injection of 0.2 ml of 1% Lidocaine without epinephrine intradermally with a tuberculin syringe. After 2–3 min, a 4-mm sterile punch biopsy instrument is rotated with pressure to cut a circular flap through the epidermis from the center of the anesthetized area. The edge of the flap is raised with a sterile forceps, and the flap is removed by cutting under it with a small curved scissors. The removed flap is dropped into a sterile 125-mm screw-capped tube containing 10 ml of tissue culture medium (MEM–Earle's containing 15% heat-inactivated fetal calf serum, penicillin, and streptomycin, see below).

Biopsies are usually "planted" immediately, but they may be planted up to 4 days after the biopsy, if stored at room temperature in cell culture

[4] J. E. Griffin, K. Punyashthiti, and J. D. Wilson, *J. Clin. Invest.* **57**, 1342 (1976).
[5] J. T. Cooper and S. Goldstein, *Lancet* **2**, 673 (1973).

medium. If the biopsy is to be mailed, it may be transferred to a sterile 5-ml polypropylene cryotube (Vangard International), which should be filled to the top, sealed, and mailed in a well-insulated shipping carton.

## 2. "Planting" the Skin Biopsy and Initial Propagation of Fibroblasts

*Materials*

1 100-mm sterile glass petri plate
2 Bard-Parker disposable scalpels with No. 15 blades (may be autoclaved and reused)
2 60 × 15 cm tissue culture dishes
4 25-mm round coverslips, Corning 2915 "circles," thickness No. 2; these have been acid washed, and distributed between layers of filter paper in 100-mm glass petri dishes and autoclaved
Dow Corning silicone stopcock grease, placed in 13 × 75 screw-cap vial and autoclaved
Flat-ended forceps for handling coverslips (sterilized by immersion of operational end in 95% ethanol, and flame-dried prior to use)
Growth medium: Eagles Modified Minimal Essential Medium containing penicillin, 100 U/ml; streptomycin, 100 $\mu$ g/ml; and sodium pyruvate 0.10 g/liter (all components are commercially available in solution or may be made up from commercially available powdered medium)
Heat-inactivated fetal calf serum (75 ml) is added to each 500-ml bottle of medium (fetal calf serum is heat-inactivated prior to use by placing in a 60° water bath for 30 min)
Tissue culture trypsin 0.25% (obtained commercially as 2.5% trypsin and diluted to 0.25% in 0.01 *M*, Tris · HCl, pH 7.0, and 0.015*M* NaCl)

This operation is carried out in a laminar-flow hood to avoid contamination of the sample. The aim is to cut the sample into small pieces, distribute it into two dishes that will be fed from separate media sources to reduce chances of loss by contamination, and to hold the tiny pieces in place by placing them under a coverslip which in turn is held in place by silicone grease (see Fig. 1).[5,6]

Pour the biopsy and medium from the tube to a sterile glass petri dish. Use two round-bladed scalpels held opposing each other like scissors blades. Bisect the skin biopsy by pressing the two blades down against the glass and rolling the edges as if closing them in opposite directions (as a

[6] R. J. Warren and C. De La Cruz, *Exp. Cell Res.* **71**, 238 (1972).

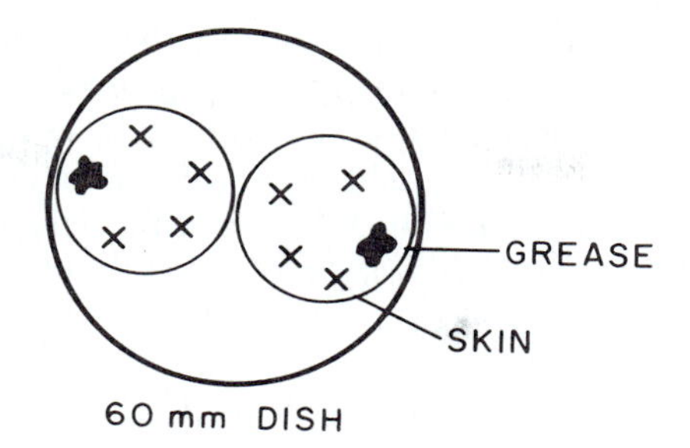


FIG. 1. View of tissue culture dish in which explants (fragments from a skin biopsy) have been "planted" beneath 25-mm coverslips that are held in place with silicone grease.

closing scissors). Then each piece is further divided in two and subdivided until one has 12–16 pieces of 1 $mm^3$ or less.

Transfer the pieces to two dry 60-mm tissue culture dishes. The samples can easily be transferred with the tip of the blade of one of the scalpels and will adhere to the dry dish. Arrange samples to fit under a round 25-mm coverslip as shown. Use a sterile wooden applicator or a disposable needle to place a spot of sterile silicone grease for one quadrant of the coverslip. Lower a coverslip into place with a forceps so that it covers the skin pieces and the silicone grease. If skin pieces lie beyond it, tuck them under the coverslip with a scalpel blade before pressing the coverslip down tightly enough to secure them. Add about 0.2 ml medium with a small pipette (1 ml) at the margin of the coverslip and allow it to move across by capillarity to displace the air beneath the coverslip. Once the coverslips have medium beneath them (and no large air bubbles), 5 ml of medium are added to the dishes directly to the top of the coverslips. The dishes are placed in trays, incubated in a humidified $CO_2$ incubator with 5% $CO_2$ in air, and not touched for 10 days. At 10 days, the dishes are examined with an inverted microscope. By this time an epithelial outgrowth may be evident at the margin of the skin pieces. From this point, dishes are "fed" 3 times a week. Medium is aspirated from the edge of the tilted dishes and replaced with 5 ml of fresh medium. By 2 weeks, fibroblasts are evident at the margins of the epithelial layer and moving out from the explant. Four to six weeks from the time the biopsy was taken, the cells will have grown well beyond the coverslip and be ready for transfer (splitting).

At this point, the medium is removed, and the dishes are washed with 2 ml of 0.25% trypsin, following which 1.0 ml of 0.25% trypsin is added and the dishes placed at 37° for 5–7 min. The cell layer will be seen detaching from the dish. Then the coverslips are dislodged to be certain the trypsin has had access to cells beneath them, and the cell layer is dispersed by adding 4 ml of serum-containing medium and pipetting up and down 3–4 times. The cells are sedimented in a sterile tube by centrifuging at 1000 rpm, resuspended in 5 ml of serum-containing medium, and added to a 60-mm dish. Within 2–4 days, this dish should be confluent and ready to split further. The trypsinization procedure outlined in this paragraph is repeated. Resuspended cells from each 60-mm dish are distributed into 2 100-mm dishes, i.e., a 1:5 split, which are fed 3 times weekly with 10 ml of medium. Within 7 days, the dishes should be confluent and may be propagated further by splitting 1:5 to 1:8 (see Section 4) or prepared for storage by freezing (Section 3).

### 3. Freezing Cells for Storage and Recovering Them from Frozen Storage

*Materials*

Dimethylsulfoxide (DMSO), not sterilized
Fetal calf serum: heat inactivated as in Section 2
Freeze medium is prepared by adding 5 ml of DMSO to 95 ml of heat-inactivated fetal calf serum.
Pro Vials (Cooke), also called sterile serum freeze-vials (Fisher), 2-ml size (2-ml sterile cryotubes, Vangard International are equivalent)
−70° freezer

It is generally advisable to freeze cells in early passage to guarantee a source of material for future investigation. We have found 5% dimethylsulfoxide (DMSO): 95% heat-inactivated (60° for 30 min) fetal calf serum a good freeze medium for every type of fibroblast examined. However, the exposure to DMSO prior to freezing should be as brief as possible, and the DMSO should be diluted out promptly on thawing. For many cell lines (but not mucolipidosis II cell lines), serum-containing medium containing 10% glycerol is also a satisfactory freeze medium. (See this volume [3].)

The following explains the method to prepare five freeze vials from two 100-mm dishes. This can be done as soon as one has two subconfluent to confluent 100-mm dishes from the primary explant. The freezing may be delayed one or two passages if one wants to prepare a considerably larger number of vials of a line that he expects to use for a long time. Cells are detached from two 100-mm dishes at a time by trypsinization. Aspirate medium, wash with 2.0 ml of 0.25% trypsin, and then allow 1.0 ml of trypsin to bathe the cells for 5 min at 37°. Disaggregate the detached cells by pipetting cells in serum-containing medium (5 ml/dish) and sediment. Resuspend in 5.2 ml of freeze medium (which should be at 4°) and transfer 1 ml into each of five freeze vials. Cap and move immediately to a

−70° freezer to allow them to freeze. The vials of frozen cells are transferred to a liquid nitrogen storage tank the next working day. We routinely thaw one vial and initiate the culture to be certain that the cells are recoverable from frozen storage and not contaminated. An alternate method to freezing in a -70° freezer is use of a BF-5 cap attachment for a Linde Liquid nitrogen tank which allows slow freezing of nine vials at a time in the vapor phase. They are held here for 90 min before transfer to a cannister for storage.

To recover cells from frozen storage, one warms the bottom of the vial to 37° in a water bath immediately on removal from the freezer. As soon as the sample has melted (which is obvious as the sample turns from ice to liquid) the vial is wiped with 95% ethanol to sterilize the outside, air dried, and the cells removed from the opened vial with a sterile 3-cc syringe with a 19-gauge needle. The cells are diluted with 10 ml of serum-containing medium to dilute out the DMSO; cells are sedimented, resuspended in 5 ml of serum-containing medium, and placed in a 60-mm dish. Within 1–2 days at 37°, this sample is usually ready to propagate and is "split" 1:5 into two 100-mm dishes.

From that point, most cell lines can be split every 7–8 days from 1:5 to 1:8 to produce large amounts of cells.

## 4. Propagation of Cultured Fibroblasts

Fibroblasts can be propagated by passaging, or splitting, every 7–8 days. Splitting 1:5 means detaching cells by trypsinization and distributing them into vessels with 5 times the original surface area. Note that the actual surface area available for growth is smaller than one would estimate from the outside diameter of plastic dishes (see the table). Plating human fibroblasts at low density (such as produced by 1:15–1:20 splits) is followed by a lag in initial growth rate with a disproportionately long time required to achieve confluence. To generate large amounts of cells, one can easily split 1:5–1:8 every week for most early passage cell lines. After 20–30 passages, growth of cell lines will slow progressively. Unless one is

| Commonly used plastic disposable dishes | Diameter of growth surface (mm) | Area of growth surface (cm) | Media per feeding (ml) |
|---|---|---|---|
| 100 mm × 20 mm | 81 | 51.5 | 10 |
| 60 mm × 15 mm | 51 | 20.4 | 5 |
| 35 mm × 10 mm | 32 | 8.5 | 2 |

interested in studying the phenomenon of senescence in culture, it is best at this point to thaw early passage cells to produce more material.

Growth of cells in roller bottles has considerable advantage when large amounts of cells are desired. Single bottles can provide the surface area of 10–30 100-mm dishes. Once growth is established in roller bottles, feeding each bottle requires far less labor and somewhat less material than feeding 10–30 plates with the corresponding surface area.

The Corning disposable tissue culture bottles (490 $cm^2$) provide a convenient amount of material for many diagnostic studies, such as surveys of lysosomal enzyme levels. HEPES buffer, 15 m*M*, is added to the tissue culture medium. Bottles are started from two 100-mm dishes (a 1:5 split) and fed twice weekly with 100 ml of media for 2 weeks. Cells may be harvested by trypsinization, or without trypsin by washing the cell layer with phosphate-buffered saline, detaching the cells by scraping the surface with a special rubber policeman (Bellco), and sedimenting in the phosphate-buffered saline, after which the cell pellet may be frozen, lyophilized, or disrupted to produce a cell extract.

# [39] Cell Lines from Invertebrates

*By* W. FRED HINK

Cell lines from about 70 different species of invertebrates have been reported. Most are from insects, but several lines have been established from ticks and snails. The lines have developed from primary cultures of embryos, hemocytes, ovaries, imaginal discs, fat bodies, and macerated larvae, pupae, or adults. This article presents general information that is applicable to the culturing of most of these lines. It does not deal with primary cultures or details about each specific cell line.

## Maintenance of Cell Lines

### *Incubation Temperature and Atmosphere*

Invertebrate cell lines grow most rapidly within the range of 25–30°. The *Trichoplusia ni* (TN-368) moth line is a typical example and has maximum growth rates at 26–30°. At 37° the cells are in poor morphological condition and at 20° they grow slower but remain viable for longer periods of time than when grown at 26° or 30°. An incubator, set at 28°, is an adequate environment for culturing all invertebrate cell lines.



ISBN 0-12-181958-2

Many vertebrate cell lines are grown in an atmosphere of 5% $CO_2$ in air to reduce the loss of $CO_2$ from the bicarbonate buffers. Since invertebrate cell culture media are not primarily buffered with bicarbonate, the cells are incubated under a normal atmosphere.

*Subculturing*

While almost any cell culture vessel is adequate (TC petri dish, glass T-flask, polystyrene disposable TC flask, etc.), most lines are maintained in TC flasks with 5-ml volumes of media and 25-$cm^2$ growth surfaces, and this discussion will center around the use of this size flask. Subculturing is done by transferring an aliquot of cells from a parent culture to a new flask containing fresh medium. New cultures are initiated with $1–3 \times 10^5$ cells/ml. The ratio of the volume transferred from the parent culture to the volume of the new culture is termed the split ratio. A 0.5-ml aliquot of suspended cells from a 5.0-ml parent culture transferred to 4.5 ml fresh medium is a 1:10 split ratio. Split ratios for different lines vary between 1:2 and 1:25, and subculture intervals range from 2 days to several weeks. When one first receives a specific line and attempts initial maintenance, it is advisable to follow the split ratio and subculture interval as used by the parent laboratory.

If difficulty is encountered in keeping cells in good condition, the subculture techniques may be modified. Cells should be subcultured while in the exponential growth phase or just as they are entering the stationary phase. To determine when to subculture, growth curves should be done soon after receiving cells. The single cell suspensions are counted with a hemocytometer. Growth curves from representative insect cell lines show that after initiation of new cultures, there are lag periods of several hours to 3–4 days. Exponential growth may last for 2–7 days after which the cell population numbers level off or decline. Population doubling times during exponential growth range from 16 hr for *T. ni* to 48 hr for *Antheraea eucalypti*, and maximum cell densities are from $1 \times 10^6$ to $1 \times 10^7$ cells/ml.

Many insect lines grow loosely attached or in suspension, and these are gently agitated to obtain a homogenous suspension prior to subculturing.

Attached cell lines must be released from the culture flask and dispersed as single cells before subculturing. For most attached lines, the cells are removed by scraping with a rubber policeman and resultant clumps dispersed by gentle pipetting.[1,2] Other techniques employ a small

[1] P. E. Eide, J. M. Caldwell, and E. P. Marks, *In Vitro* **11,** 395 (1975).

[2] J. L. Vaughn, R. H. Goodwin, G. J. Tompkins, and P. McCawley, *In Vitro* **13,** 213 (1977).

TABLE I
RINALDINI'S SOLUTION[a]

| Compound | mg/100 ml solution |
|---|---|
| NaCl | 800 |
| KCl | 20 |
| $Na_2PO_4 \cdot H_2O$ | 5 |
| D-glucose | 100 |
| $NaHCO_3$ | 100 |
| $Na_3C_6H_5O_7 \cdot 2\ H_2O$ | 67.6 |
| Demineralized water | to 100 ml |

[a] Rinaldini.[3]

magnetic spin bar followed by pipetting or simply pipetting medium over the cell layer to flush them off. Insect cells are fragile, and vigorous agitation or shearing must be avoided.

If cells are damaged by physical techniques, they are released from the culture flask with enzymes. Since $Ca^{2+}$ and $Mg^{2+}$ ions apparently function in stabilizing cellular adhesiveness, enzymes are dissolved in $Ca^{2+}$- and $Mg^{2+}$-free saline solutions. Rinaldini's solution (Table I),[3] i.e., Tyrode's solution[4] with Ca and Mg salts replaced with sodium citrate, is one of the most often used.

Trypsin is usually used at 0.1 or 0.2%.[5,6] To prepare, add 100 mg or 200 mg to about 1.0 ml Rinaldini's solution and stir to make a paste. With continuous stirring, add the additional 99 ml of the salt solution. If necessary, the pH should be adjusted to 7.8 and the solution sterilized by filtration. Since enzymic activity is lost over a period of several weeks when stored at 4°, the preparation should be divided into convenient volumes and stored frozen at −20°. Prepared trypsin solutions (2.5%) in salines without Mg and Ca may be purchased from commercial sources.

To subculture attached cells, all medium is removed from the flask and 2 ml of trypsin solution are added. The trypsin and cells are incubated for 2–5 min (depending on cell line), and the solution is gently pipetted to flush cells from the flask. Cells are transferred to sterile centrifuge tubes containing 1 ml of heat-inactivated fetal bovine serum (FBS) to stop enzyme activity. After centrifugation at 380 *g* for 10 min, the serum is removed by pipette and fresh medium added. Aliquots are transferred to new culture flasks to obtain the appropriate split ratio.

[3] L. M. Rinaldini, *J. Physiol.* (*London*) **123,** 20P (1954).

[4] M. V. Tyrode, *Arch. Int. Pharmacodyn. Ther.* **20,** 205 (1910).

[5] I. Schneider, *J. Cell Biol.* **42,** 603 (1969).

[6] J. Mitsuhashi, *Annot. Zool. Jpn.* **48,** 139 (1975).

*Storage*

If cells are not transferred to fresh medium when they reach maximum density, they begin to die and, depending on the cell line, all will be dead in 10–20 days. The longevity of cell viability can be increased by transferring newly subcultured cells to 5°. In my laboratory, the following lines were stored at 5° in fresh media and samples taken every week to determine if they would recover: *T. ni* (TN-368) can be stored 14 days; *A. eucalypti* for 21 days; Grace's *Aedes aegypti* for 40 days; and *Lasperesia pomonella* (CP-1268 and CP-169) for 90 days. In all cases, cell growth was slower after storage, requiring about five subcultures for the growth rate to return to normal.

As with vertebrate cells, insect cells also may be frozen in liquid nitrogen or ultra-low temperature freezers at −70° to −90°. The cell culture media are supplemented with either sterile 10% glycerol or 8% dimethysulfoxide. Cells from a culture in late stage of exponential growth are centrifuged at 380 g for 10 min and resuspended in glycerol-supplemented medium at one-half the original volume. The cell suspension is dispensed in 1.0-ml volumes into glass ampules and immediately sealed by use of a propane torch and glass rod. Animal cells should be slowly frozen at rates of 1–3° per minute to about −30°; programmable freezers may be used for this purpose. We obtain slow freezing by wrapping the ampules in insulation before placing them in a freezer and, after 2 hr, transferring them to liquid nitrogen. Cells are thawed quickly by removing them from the nitrogen and plunging the ampules in a 30° water bath where they are agitated to accelerate melting. The content of one ampule is added to 4 ml of fresh medium in a TC flask. For most lines, the cells will attach within 2–4 hr at which time the medium with the glycerol supplement is removed by pipetting and 5 ml of fresh medium added. If cells do not attach, the cryoprotective agent may be removed by centrifuging freshly thawed cells, discarding the supernatant fluid, and resuspending cells in fresh medium.

*Quality Control*

Occasionally, a batch of basal medium, serum, hydrolysate, or other ingredient will be toxic or will not support normally expected growth rates. For this reason, the lot number of all ingredients in the complete medium should be recorded. When an ingredient with a new lot number is to be employed and before supply of the previously used batch is depleted, it should be incorporated in a small batch of complete medium that contains no other new ingredients. This should be used to culture cells for at least 10 subcultures; the new ingredient is judged satisfactory if cells and growth are normal.

## Suspension Culture

Large numbers of single cells may be obtained by growing cells in vessels where they are kept in suspension. Spin flasks, with magnetically driven impellers designed to avoid shear damage, flasks on gyratory shakers, or fermentors are employed. These cells may be used in molecular and cellular biology studies involving such techniques as extraction of macromolecules or isolation of cellular organelles.

Mosquito (*Aedes albopictus*) cells,[7] established and normally grown in Mitsuhashi and Maramorosch medium (Table VI) have been adapted to grow in vertebrate culture medium consisting of Eagle's medium (Joklik-modified) with 1% seven nonessential amino acids, 10% FBS, and 1% lactalbumin hydrolysate.[8] The salt concentrations and osmotic pressures of these two media are similar. Upon transfer of cells to the new medium, there was a brief lag in growth before multiplication commenced. Cells were cultured in stationary flasks for several months before attempting to adapt them to suspension culture. Thereafter, cells were alternately passed in stationary and suspension cultures until a rapidly growing suspended subline was obtained. Cells in suspension clump with several hundred cells per clump. To quantify cells accurately, they were treated with the detergent, NP40, and nuclei were counted with a hemocytometer. The population doubling time was 21 hr at 25°. After adaptation to suspension culture, the lactalbumin hydrolysate supplement was removed and cells could then be labeled with radioactive leucine.

The same *A. albopictus* cell line was grown in 150-ml spin flasks in Mitsuhashi and Maramorosch medium with no medium modifications.[9] Singh's *A. aegypti* and Hsu's *Culex quinquefasciatus*[10] and *Culex tritaeniorhynchus*[11] were also cultured under the same conditions.[9]

Fruit fly (*Drosophila melanogaster,* line 2) cells,[12] established and normally grown in Schneider medium (Table VIII), were also adapted to a vertebrate cell culture medium.[13] The medium was Dulbecco's modified Eagle's supplemented with 10% FBS, 0.5% lactalbumin hydrolysate, and

[7] K. R. P. Singh, *Curr. Sci.* **36,** 506 (1967).

[8] A. Spradling, R. H. Singer, J. Lengyel, and S. Penman, *Methods Cell Biol.* **10,** 185 (1975).

[9] T. K. Yang, E. McMeans, L. E. Anderson, and H. M. Jenkin, *Lipids* **9,** 1009 (1974).

[10] S. H. Hsu, W. H. Mao, and J. H. Cross, *J. Med. Entomol.* **7,** 703 (1970).

[11] S. H. Hsu, S. Y. Li, and J. H. Cross, *J. Med. Entomol.* **9,** 86 (1972).

[12] I. Schneider, *J. Embryol. Exp. Morphol.* **27,** 353 (1972).

[13] J. Lengyel, A. Spradling, and S. Penman, *Methods Cell Biol.* **10,** 195 (1975).

Gibco MEM nonessential amino acids. The pH was adjusted to 6.9 and cultures gassed with 5% $CO_2$ in air. A series of manipulations consisting of addition of various supplements and culturing in roller bottles were required to obtain suspension cultures. Cells in suspension reach densities of $1–4 \times 10^6$ cells/ml with a generation time of 30 hr. Another *D. melanogaster* line, ($GM_2$),[14] was grown in insect cell culture media in Erlenmeyer flasks on a gyratory shaker at 180 rpm with a 1-inch radial stroke.[15]

The moth cell line (*T. ni*, TN-368)[16] grew in suspension in 100-ml spin flasks and 4-liter fermentors.[17] The culture medium (TNM-FH, Table IV) was modified by addition of 0.1% methylcellulose (50 cps) to prevent cell clumping. Aeration of cultures produced higher growth rates and final cell densities. Periodic adjustment of medium pH also resulted in higher maximum cell densities.

## Media

According to my most recent tabulation, there are 34 different formulations of media for culturing invertebrate cell lines. It is impractical to consider all of these here, and I will therefore discuss only the most common media. Before dealing with specific media, some general comments will be made.

Insect cell culture media differ from media designed for vertebrate cells by having different ion balances, elevated amino acid concentrations, generally phosphates as buffers rather than carbonates, usually lower pH, and higher osmotic pressures. There is some question as to the significance of these differences since a few insect cell lines grow in vertebrate media.[8,13,18,19] Established insect cell lines appear to be rather insensitive to Na/K ratios altered from 0.67–1.38 and calcium and magnesium levels reduced by one-half and three-fourths, respectively.[20] However, as with vertebrate cells, excess potassium is toxic.[21] The increased amounts of amino acids are present in some formulas because insect blood has high concentrations of these compounds. Requirements for specific amino acids vary with different lines, and patterns of utilization are different

[14] G. Mosna and S. Dolfini, *Chromosoma* **38,** 1 (1972).

[15] T. Miyake, K. Saigo, T. Marunouchi, and T. Shiba, *In Vitro* **13,** 245 (1977).

[16] W. F. Hink, *Nature (London)* **226,** (1970).

[17] W. F. Hink and E. Strauss, *in* "Invertebrate Tissue Culture" (E. Kurstak and K. Maramorosch, eds.), p. 297. Academic Press, New York, 1976.

[18] M. Pudney, personal communication.

[19] A. H. McIntosh, K. Maramorosch, and C. Rechtoris, *In Vitro* **8,** 375 (1973).

[20] J. L. Vaughn, *In Vitro* **9,** 122 (1973).

[21] T. J. Kurtti, S. P. S. Chaudhary, and M. A. Brooks, *In Vitro* **11,** 274 (1975).

even between cell lines from the same insect species.[22] In fact, some amino acids essential for growth of living intact insects are not essential for cells *in vitro*.[23]

Osmotic pressure and pH are important parameters in media formulation. Studies with moth cell lines indicate that growth is reduced when osmotic pressure varies more than about 40 msM/kg on either side of the osmotic pressure (316 msM/kg) of the "normal" medium.[24] Our policy is to have less than ±10 msM/kg variation in batches of media. The pH of most media range from 6.2–7.0 with media for some cockroach lines being as high as 7.4. The optimum initial pH for moth (*Heliothis zea*) cells is between 6.5–7.0[25] and for two leafhopper cell lines it is 6.3–6.4.[26] When the pH of the medium was held at preset levels of 5.8, 6.0, 6.3, 6.5, and 6.7 for the life of suspension cultures of *T. ni* cells, the optimum was between pH 6.0–6.5, whereas pH 6.7 was clearly detrimental.

### *Supplements and Additives to Basal Media*

For purposes of discussion, the basal medium is that portion of the complete medium that contains synthetic chemically defined compounds. There are no chemically defined media for invertebrate cell lines, and all basal media are supplemented with undefined natural ingredients. Fetal bovine serum, at concentrations of 5–20%, is used in nearly all media, and it is usually heat inactivated at 56° for 30 min. Other complex ingredients such as lactalbumin hydrolysate furnish amino acids, and yeast hydrolysate provides both amino acids and vitamins.

Antibiotics are often added, but their use should be discouraged because they can mask microbial contamination. If contamination is present, it is better to know it immediately instead of carrying undetected low-level contaminants and having them affect experimental results. It appears that dependence on the presence of antibiotics leads to the development of rather careless nonsterile techniques. The two most common antibiotics and concentrations are penicillin G at 100 U/ml and streptomycin sulfate at 100 $\mu$g/ml. A $\times 100$ stock solution is prepared by dissolving 10,000 U penicillin and 10 mg streptomycin/ml PBS, filter-sterilized; 1.0 ml is added to 99 ml of complete medium just before use.

[22] W. F. Hink, B. L. Richardson, D. K. Schenk, and B. J. Ellis, *Proc. Int. Colloq. Invertebr. Tissue Cult., 3rd, 1971* p. 195 (1973).

[23] J. Mitsuhashi, *J. Insect Physiol.* **22,** 397 (1976).

[24] T. J. Kurtti, S. P. S. Chaudhary, and M. A. Brooks, *In Vitro* **10,** 149 (1974).

[25] T. J. Kurtti and M. A. Brooks, *in* "Insect and Mite Nutrition" (J. G. Rodriguez, ed.), p. 387. North-Holland Publ., Amsterdam, 1972.

[26] G. Martinez-Lopez and L. M. Black, *In Vitro* **13,** 777 (1977).

*Media Sterilization*

Media are usually sterilized by pressure filtration through membranes with 0.22–0.20 μm pore sizes. A common procedure is to filter basal media and then add sterile sera, as purchased from commercial suppliers, aseptically to the basal media. Since sera may contain microbial contaminants, we make up complete media and then filter. It is advisable to check media for sterility before using them to culture cells. One may take a small aliquot from storage bottles of media and incubate at 28° for at least a week or leave all the media at room temperature for 7 days. We use the latter approach as it assures us that the entire contents of the storage bottles are free of viable microorganisms.

*Formulas and Preparation of Media*

GRACE MEDIUM

This medium (Table II)[27] is designed to resemble the chemical composition of silkworm, *Bombyx mori*, hemolymph and is a modification of Wyatt's medium.[28] The basal medium is supplemented with various combinations and amounts of FBS, chicken egg ultrafiltrate, yeast extract, lactalbumin hydrolysate, and bovine serum albumin prior to use. It is used to culture cell lines from the Australian emperor gum moth, *A. eucalypti*;[27] silkworm, *B. Mori*;[29] spruce budworm, *Choristoneura fumiferana*;[30] cotton bollworm, *H. zea*;[31] codling moth, *L. pomonella*;[32] gypsy moth, *Lymantria dispar*;[33] forest tent caterpillar, *Malacosoma disstria*;[34] tobacco hornworm, *Manduca sexta*;[1] cynthia moth, *Samia cynthia*;[35] cabbage looper, *T. ni*;[16] mosquito, *A. aegypti*;[36] mosquito, *Aedes vexans*;[37] and mosquito, *Culiseta inornata*.[37]

PROCEDURE FOR MAKING 10 LITERS OF GRACE MEDIUM[38]

1. Weigh out amino acids (B) and combine them in a 400-ml beaker.
2. Weigh out sugars (C) and combine them in a 400-ml beaker.

[27] T. D. C. Grace, *Nature (London)* **195**, 788 (1962).
[28] S. S. Wyatt, *J. Gen. Physiol.* **39**, 841 (1956).
[29] T. D. C. Grace, *Nature (London)* **216**, 613 (1967).
[30] S. S. Sohi, *Proc. Int. Colloq. Invertebr. Tissue Cult., 3rd, 1971* p. 75 (1973).
[31] W. F. Hink and C. M. Ignoffo, *Exp. Cell Res.* **60**, 307 (1970).
[32] W. F. Hink and B. J. Ellis, *Curr. Top. Microbiol. Immunol.* **55**, 19 (1971).
[33] J. M. Quiot, personal communication.
[34] S. S. Sohi, *Proc. Int. Colloq. Invertebr. Tissue Cult., 3rd, 1971* p. 27 (1973).
[35] J. Chao and G. H. Ball, *Curr. Top. Microbiol. Immunol.* **55**, 28 (1971).
[36] T. D. C. Grace, *Nature (London)* **211**, 366 (1966).
[37] B. H. Sweet and J. S. McHale, *Exp. Cell. Res.* **61**, 51 (1970).
[38] W. F. Hink, unpublished data.

TABLE II
GRACE INSECT TISSUE CULTURE MEDIUM (g/10 liters)[a]

| Group | Component | g/10 liters | Component | g/10 liters | Group |
|---|---|---|---|---|---|
| | *Salts* | | *Sugars* | | |
| (A) | $NaH_2PO_4 \cdot H_2O$ | 10.08 | Sucrose | 266.8 | (C) |
| (A) | $NaHCO_3$ | 3.50 | Fructose | 4.0 | (C) |
| (A) | KCl | 22.40 | Glucose | 7.0 | (C) |
| (A) | $MgCl_2 \cdot 6\ H_2O$ | 22.80 | | | |
| (A) | $MgSO_4 \cdot 7\ H_2O$ | 27.80 | *Organic acids* | | |
| (I) | $CaCl_2 \cdot 2\ H_2O$ (sep.) | 13.25 | Malic | 6.7 | (D) |
| | | | Alpha-ketoglutaric | 3.7 | (D) |
| | *Amino acids* | | Succinic | 0.6 | (D) |
| (B) | L-arginine HCl | 7.0 | Fumaric | 0.55 | (D) |
| (B) | L-aspartic acid | 3.5 | | | |
| (B) | L-asparagine | 3.5 | *Vitamins* | | |
| (B) | L-alanine | 2.25 | Thiamine HCl | 0.0002 | (E) |
| (B) | B-alanine | 2.0 | Riboflavin | 0.0002 | (E) |
| (B) | L-glutamic acid | 6.0 | Ca pantothenate | 0.0002 | (E) |
| (B) | L-glutamine | 6.0 | Pyridoxine HCl | 0.0002 | (E) |
| (B) | L-glycine | 6.5 | *p*-Aminobenzoic acid | 0.0002 | (E) |
| (B) | L-histidine | 25.0 | Folic acid | 0.0002 | (E) |
| (B) | L-isoleucine | 0.5 | Niacine | 0.0002 | (E) |
| (B) | L-leucine | 0.75 | Isoinositol | 0.0002 | (E) |
| (B) | L-lysine HCl | 6.25 | Biotin | 0.0001 | (E) |
| (B) | L-methionine | 0.5 | Choline chloride | 0.002 | (E) |
| (B) | L-proline | 3.5 | | | |
| (B) | L-phenylalanine | 1.5 | *Antibiotics* | | |
| (B) | DL-serine | 11.0 | Penicillin G, Na salt | 0.3 | (F) |
| (B) | L-tryptophan | 1.0 | Streptomycin sulphate | 1.0 | (F) |
| (B) | L-threonine | 1.75 | | | |
| (B) | L-valine | 1.0 | | | |
| (G) | L-cystine HCl (sep.) | 0.25 | | | |
| (H) | L-tyrosine (sep.) | 0.5 | | | |

[a] Grace.[27]

3. Weigh out organic acids (D) and combine them in a 100-ml beaker.

4. Weigh out L-cystine (G) and L-tyrosine (H) and combine them in a 50-ml beaker.

5. Weigh out calcium chloride (I) and place in a 100-ml beaker.

6. A ×1000 stock solution of vitamins (E) is made up by adding 2 mg thiamine, 2 mg riboflavin, 2 mg Ca pantothenate, 2 mg pyridoxine, 2 mg *p*-aminobenzoic acid, 2 mg folic acid, 2 mg niacin, 2 mg isoinositol, 1 mg biotin, and 20 mg choline chloride to a 100-ml volumetric flask and bringing up to volume with water. Filter-sterilize, dispense in 10-ml portions, and store at −20°.

7. If antibiotics (F) are to be used, they are made up in a ×100 stock

solution in Hanks's PBS, filter-sterilized, and added to complete medium just before use.

8. Put (C) into a 15-liter beaker. Add 4 liters water using some of this to rinse the beaker that held the dry sugars. Stir with a magnetic stirrer.

9. Weigh out salts (A) starting with $NaH_2PO_4 \cdot H_2O$, transfer to 15-liter beaker, and proceed in order down through the list in Table II. To prevent precipitation, each salt must be added separately and dissolved before adding the next salt.

10. Put (B) into a separate 4-liter beaker, add 2 liters water, and stir until dissolved.

11. Put (D) into a separate 4-liter beaker, add 2 liters water, and stir until dissolved.

12. Make up 250 ml 1 *N* KOH by placing 13.03 g KOH in a 250-ml volumetric flask and bring up to volume with water.

13. Add the 1 *N* KOH to (D) with continuous stirring to bring pH to 5.9. Note volume of KOH used (usually 170–200 ml).

14. Make up 1 *N* HCl by adding 6 ml concentrated HCl (37%) to 54.0 ml with water.

15. Add 15 ml of the 1 *N* HCl to (GH) and dissolve these amino acids.

16. Add 50 ml water to (I) and dissolve.

17. Add (B) and pH-adjusted (D) to 15-liter beaker.

18. Add (GH) to 15-liter beaker.

19. Add (I) to 15-liter beaker.

20. Adjust pH of the contents of 15-liter beaker to 6.2 with 1 *N* HCl. Note volume of HCl added.

21. Add 10 ml of ×1000 stock vitamins (E).

22. While combining ingredients, the volumes are recorded in order to know how much additional water is needed to bring final volume to 10 liters. Here is an example of our calculations for a typical batch:

| Fraction | Volume occupied by chemicals when dissolved (ml) | | Volume of liquid (ml) |
|---|---|---|---|
| A | 37.0 | | 2000.0 |
| B | 53.0 | | 2000.0 |
| C | 170.0 | | 2000.0 |
| D | 5.0 | | 2000.0 |
| KOH | — | | 175.0 |
| GH | 0.5 | | 15.0 |
| I | 3.0 | | 50.0 |
| HCl | — | | 27.0 |
| E | — | | 10.0 |
| *Totals:* | 268.5 | + | 8277.0 = 8545.5 ml |

23. Add water to bring final volume to 10 liters.

24. Filter through 0.22-$\mu$m membrane and store at 5°.

25. As part of our quality-control program, we take the osmotic pressure of all batches, and variation from normal indicates that errors were made during formulation. We also make up 100 ml complete medium from a new batch of Grace basal medium. This is evaluated through 10 subcultures before concluding that it is satisfactory.

Tables III[38a] and IV give two examples of complete media that use Grace basal medium.

### Goodwin IPL-52 Medium[39]

This medium (Table V) supports growth of cell lines from moths, *Spodoptera frugiperda*, *H. zea*, and *T. ni*.[39] It also has been used to obtain quickly developing monolayers from primary cultures of pupal moth ovaries.

### Preparation of Goodwin IPL-52 Medium[40]

1. Prepare stock solution (A) by dissolving the following in 100 ml water:

| | |
|---|---|
| $(NH_4)Mo_7O_{24} \cdot 4\ H_2O$ | 4.0 mg |
| $CoCl_2 \cdot 6\ H_2O$ | 5.0 mg |
| $CuCl_2 \cdot 2\ H_2O$ | 19.5 mg |
| $MnCl_2 \cdot 4\ H_2O$ | 2.0 mg |
| $ZnCl_2$ | 4.0 mg |

2. Prepare stock solution (B) by dissolving the following in 100 ml water:

| | |
|---|---|
| $FeSO_4 \cdot 7\ H_2O$ | 82.3 mg |
| Aspartic acid | 53.2 mg |

3. Dissolve all ingredients in Table V from, and including, *L*-arginine to, and including, folic acid in order listed in 900 ml $H_2O$.

4. While stirring, add 1 ml stock solution (A) and 0.67 ml stock solution (B).

5. Adjust pH to 6.3 with 10% NaOH (approximately 5 ml).

6. Dissolve $CaCl_2$ separately in 50 ml $H_2O$ and add slowly while stirring.

7. Bring total volume to 1000 ml with $H_2O$.

8. Sterilize by filtration.

[38a] C. E. Yunker, J. L. Vaughn, and J. Cory, *Science* **155**, 1565 (1967).

[39] R. H. Goodwin, *In Vitro* **11**, 369 (1975).

[40] R. H. Goodwin, personal communication (1977).

TABLE III
YUNKER, VAUGHN, AND CORY MEDIUM[a]

| |
|---|
| 90.0 ml Grace medium |
| 10.0 ml FBS |
| 10.0 ml Egg ultrafiltrate |
| 1.0 g Bovine plasma albumin |

[a] Yunker *et al.*[38a]

TABLE IV
HINK (TNM-FH) MEDIUM[a]

| |
|---|
| 90.0 ml Grace medium |
| 8.0 ml FBS |
| 0.3 g Lactalbumin hydrolysate |
| 0.3 g TC Yeastolate |

[a] Hink and Strauss.[17]

TABLE V
GOODWIN IPL-52 MEDIUM (mg/liter)[a,b]

| Component | mg/liter | Component | mg/liter |
|---|---|---|---|
| A: $(NH_4)Mo_7O_{24} \cdot 4\ H_2O$ | 0.040 | DL-Serine | 600 |
| A: $CoCl_2 \cdot 6\ H_2O$ | 0.050 | L-Threonine | 200 |
| A: $CuCl_2 \cdot 2\ H_2O$ | 0.195 | L-Tryptophan | 100 |
| A: $MnCl_2 \cdot 4\ H_2O$ | 0.020 | L-Tyrosine (dissolve in 10% NaOH) | 250 |
| A: $ZnCl_2$ | 0.040 | L-Valine | 500 |
| B: $FeSO_4 \cdot 7\ H_2O$ | 0.0551 | Dextrose | 5000 |
| B: Aspartic acid | 0.0356 | Maltose | 1000 |
| L-Arginine hydrochloride | 800 | $MgSO_4 \cdot 7\ H_2O$ | 1880 |
| L-Aspartic acid | 1000 | KCl | 2600 |
| L-Asparagine | 1300 | $NaHCO_3$ | 350 |
| L-Cystine (dissolve in 10% NaOH) | 100 | $NaH_2PO_4 \cdot H_2O$ | 1160 |
| L-Glutamic acid | 1300 | TC Yeastolate | 5000 |
| L-Glutamine | 1000 | Acetyl β-methyl choline chloride | 250 |
| L-Glycine | 400 | Cyanocobalamin ($B_{12}$) | 1 |
| L-Histidine | 200 | Isoinositol | 10 |
| Hydroxy-L-proline | 800 | Folic acid | 1.2 |
| L-Isoleucine | 500 | $CaCl_2$ | 500 |
| L-Leucine | 400 | Turkey serum | 3% |
| L-Lysine hydrochloride | 700 | Chicken serum | 3% |
| L-Methionine | 1000 | Calf serum | 3% |
| L-Proline | 600 | | |
| L-Phenylalanine | 1000 | | |

[a] Goodwin.[39]
[b] Goodwin.[40]

TABLE VI
MITSUHASHI AND MARAMOROSCH MEDIUM (g/liter)[a]

| | | | |
|---|---|---|---|
| $NaH_2PO_4 \cdot H_2O$ | 0.20 | D-glucose | 4.00 |
| $MgCl_2 \cdot 6\ H_2O$ | 0.10 | Lactalbumin hydrolysate | 6.50 |
| KCl | 0.20 | Yeastolate | 5.0 |
| $CaCl_2 \cdot 2\ H_2O$ | 0.20 | FBS | 20% |
| NaCl | 7.00 | Penicillin | 100 U/ml |
| $NaHCO_3$ | 0.12 | Streptomycin | 100 μg/ml |

[a] Mitsuhashi and Maramorosch.[41]

9. Before use, the above basic medium is supplemented with heat-treated (60° for 30 min) sera as given in Table V.

MITSUHASHI AND MARAMOROSCH MEDIUM[41]

Many mosquito cell lines are grown in this medium (Table VI) which contains relatively few ingredients. It supports growth of cell lines from mosquitoes *A. aegypti,*[7] *A. albopictus,*[7] *Aedes novalbopictus,*[42] *A. taeniorhynchus,*[43] *Aedes w-albus,*[44] and *Armigeries subalbatus.*[43] This medium also is used for cell lines from the leafhopper, *Agallia constricta,*[45] and the potato tuber moth, *Gnorimoschema operculella.*[46]

PREPARATION OF MITSUHASHI AND MARAMOROSCH MEDIUM[43,47]

1. The lactalbumin hydrolysate is dissolved separately in 300 ml $H_2O$ by stirring with a magnetic stirring bar.
2. The remainder of the components (except FBS and antibiotics) are dissolved one at a time in 400 ml water.
3. The two solutions are combined, and the pH is adjusted to 6.5 with 0.1 *N* KOH. The pH of 6.5 is used for leafhopper cells whereas a pH of 6.8–7.0 is used for mosquito cells.
4. Bring volume to 800 ml and filter-sterilize.
5. FBS (200 ml) is added aseptically before use.
6. If antibiotics are to be incorporated, add 20 ml of sterile stock antibiotics consisting of 5000 U penicillin and 5000 μg streptomycin/ml.

[41] J. Mitsuhashi and K. Maramorosch, *Contrib. Boyce Thompson Inst.* **22,** 435 (1964).
[42] U. K. M. Bhat and P. Y. Guru, *Expl. Parasitol.* **33,** 105 (1973).
[43] I. Schneider, personal communication (1977).
[44] K. R. P. Singh and U. K. M. Bhat, *Experientia* **27,** 142 (1971).
[45] R. Chiu and L. M. Black, *Nature (London)* **215,** 1076 (1967).
[46] U. Pant, A. F. Mascarenhas, and V. Jagannathan, *Indian J. Expl. Biol.* **15,** 244 (1977).
[47] D. E. Lynn, personal communication (1977).

TABLE VII
LEIBOVITZ L-15 MEDIUM (mg/liter)[a]

| Group | Component | mg/liter | Component | mg/liter | Group |
|---|---|---|---|---|---|
| | *Amino acids* | | | | |
| (A) | DL-alpha alanine | 450.0 | $CaCl_2$ | 140.0 | (E) |
| (A) | L-Arginine (free base) | 500.0 | $KH_2PO_4$ | 60.0 | (F) |
| (A) | L-Asparagine | 250.0 | $Na_2HPO_4$ | 50.0 | (F) |
| (A) | L-Cysteine (free base) | 120.0 | Phenol red | 10.0 | (F) |
| (A) | Glycine | 200.0 | Sodium pyruvate | 550.0 | (G) |
| (A) | L-Histidine (free base) | 250.0 | | | |
| (A) | DL-Isoleucine | 250.0 | *Vitamins* | | |
| (A) | L-Leucine | 125.0 | DL-Ca pantothenate | 1.0 | (H) |
| (A) | L-Lysine | 75.0 | Choline chloride | 1.0 | (H) |
| (A) | DL-Methionine | 150.0 | Folic acid | 1.0 | (H) |
| (A) | DL-Phenylalanine | 250.0 | Inositol | 2.0 | (H) |
| (A) | L-Serine | 200.0 | Nicotinamide | 1.0 | (H) |
| (A) | DL-Threonine | 600.0 | Pyridoxine HCl | 1.0 | (H) |
| (A) | L-Tryptophane | 20.0 | Riboflavin-5′ phosphate | 0.1 | (H) |
| (A) | DL-Valine | 200.0 | Thiamine monophosphate | 1.0 | (H) |
| (B) | L-Glutamine | 300.0 | $Na_2HPO_4$ | 140.0 | (H) |
| (C) | L-Tyrosine | 300.0 | | | |
| | | | *Sugar* | | |
| | *Salts* | | Galactose | 900.0 | |
| (D) | NaCl | 8000.0 | | | |
| (D) | KCl | 400.0 | *FBS* | 10% | |
| (D) | $MgCl_2 \cdot 6\ H_2O$ | 200.0 | | | |
| (D) | $MgSO_4 \cdot 7\ H_2O$ | 200.0 | | | |

[a] Leibovitz.[48]

LEIBOWITZ L-15 MEDIUM[48]

This medium (Table VII), which was originally developed for vertebrate cells, supports cell lines from the hard tick, *Rhipicephalus appendiculatus*,[49] three other tick species,[50] and the blood-sucking bug, *Triatoma infestans*.[18] The medium is usually supplemented with 10% tryptose phosphate broth and 10% FBS. The final pH of L-15 is 7.6 and must be adjusted to pH 7.0 before use for tick cell lines.

PREPARATION OF L-15 MEDIUM[48]

1. Amino acids (A) are prepared as ×10 solutions, filtered, divided into 100-ml aliquots, and stored at 4°.

[48] A. Leibovitz, *Am. J. Hyg.* **78,** 173 (1963).
[49] M. G. R. Varma, M. Pudney, and C. J. Leake, *J. Med. Entomol.* **11,** 698 (1975).
[50] P. Y. Guru, V. Dhanda, and N. P. Gupta, *Indian J. Med. Res.* **64,** 1041 (1976).

2. Vitamins (H), L-glutamine (B), sodium pyruvate (G), and galactose (I) are prepared as ×100 solutions, filtered, divided into 10-ml aliquots, and stored at −20°.

3. L-tyrosine (C) is prepared by dissolving 400 mg/liter of water with heat and continuous stirring. Then 750 ml are added to a 1-liter flask, autoclaved, and stored at room temperature.

4. Salts (D), (E), and (F) are made up so that when combined they are a ×10 solution. (D) is dissolved in 600 ml water, (E) in 100 ml water, and (F) in 275 ml water plus 25 ml of 0.4% phenol red (see step 5). Each solution is autoclaved separately and cooled; solution (E) is added to (D), and (F) is added to combined (E) and (D). This sequence must be followed to prevent formation of calcium phosphates.

5. Make up 0.4% phenol red (pH indicator) by adding 1 g of the dye into a 500-ml flask and slowly adding 0.05 *N* NaOH until the phenol red is almost in solution (about 60.0 ml). Then add NaOH, drop by drop, until solution is attained; the solution should be deep red. Add water to 250 ml.

6. The solutions are aseptically combined to make 1 liter:

| | |
|---|---|
| Tyrosine (×1) | 750.0 ml |
| Vitamins (×100) | 10.0 ml |
| Pyruvate (×100) | 10.0 ml |
| Galactose (×100) | 10.0 ml |
| Salts (D, E, and F) (×10) | 100.0 ml |
| Amino acids (×10) | 100.0 ml |
| Antibiotics (×100) | 10.0 ml |

7. The combined ingredients are divided in 99-ml portions to which 1.0 ml of glutamine and 10% FBS are added immediately before use.

SCHNEIDER MEDIUM[43,51]

This medium (Table VIII) is for culturing cell lines from fruit flies *D. melanogaster*,[12] *Drosophila immigrans*,[52] and *Drosophila virilis*.[52] There are several other media, not discussed here, that are used for other *D. melanogaster* lines.

PREPARATION OF SCHNEIDER MEDIUM[43]

1. Salts and organic acids (A) are dissolved in 30 ml water. The pH will be about 3.5 and should be adjusted to 6.0 with KOH.

2. Sugars (B) are dissolved in 5 ml water.

3. Yeastolate (C) is dissolved in 5 ml water.

[51] I. Schneider, *J. Embryol. Expl. Morphol.* **15,** 271 (1966).

[52] I. Schneider and A. B. Boumenthal, *in* "Genetics and Biology of Drosophila," (M. Ashburner and T. R. F. Wright, eds.) Vol. 2a. Academic Press, New York, 1978.

TABLE VIII
SCHNEIDER MEDIUM (mg/100 ml)[a,b]

| Group | Component | mg/100 ml | Component | mg/100 ml | Group |
|---|---|---|---|---|---|
| | *Salts and organic acids* | | | | |
| (A) | NaCl | 210 | L-Cystine | 10 | (D) |
| (A) | $Na_2HPO_4$ | 70 | L-Glutamic acid | 80 | (D) |
| (A) | $KH_2PO_4$ | 45 | L-Glutamine | 180 | (D) |
| (A) | KCl | 160 | Glycine | 25 | (D) |
| (A) | $MgSO_4 \cdot 7\ H_2O$ | 370 | L-Histidine | 40 | (D) |
| (A) | $\alpha$-Ketoglutaric acid | 20 | L-Isoleucine | 15 | (D) |
| (A) | Succinic acid | 10 | L-Leucine | 15 | (D) |
| (A) | Fumaric acid | 10 | L-Lysine HCl | 165 | (D) |
| (A) | Malic acid | 10 | L-Methionine | 80 | (D) |
| | | | L-Phenylalanine | 15 | (D) |
| | *Sugars* | | L-Proline | 170 | (D) |
| (B) | Glucose | 200 | L-Serine | 25 | (D) |
| (B) | Trehalose | 200 | L-Threonine | 35 | (D) |
| (C) | TC Yeastolate | 200 | L-Tryptophan | 10 | (D) |
| | | | L-Tyrosine | 50 | (D) |
| | *Amino acids* | | L-Valine | 30 | (D) |
| (D) | $\beta$-Alanine | 50 | $CaCl_2$ | 60 | (E) |
| (D) | L-Arginine | 40 | $NaHCO_3$ | 40 | (F) |
| (D) | L-Aspartic acid | 40 | | | |
| (D) | L-Cysteine | 6 | | | |

[a] Schneider.[43]
[b] Schneider.[51]

4. All amino acids (D) except cystine and tyrosine are dissolved in 40 ml water. Dissolve cystine in 5.0 ml hot acidic water and tyrosine in 5.0 ml alkaline water. After these two amino acids are in solution, they are added slowly to the rest of the amino acid solution.

5. $CaCl_2$ (E) is dissolved in 5.0 ml water.

6. $NaHCO_3$ (F) is dissolved in 4.0 ml water.

7. The components in (B), (C), (E), and (F) dissolve quickly. This is not so with (A) and (D) which require stirring for about 30 min with a magnetic stirrer.

TABLE IX
HANSEN S-301 MEDIUM[a]

|  |  |
|---|---|
| Schneider's medium diluted to 22% | |
| Galactose | 1.3 g/liter |
| Lactalbumin hydrolysate | 4.5 g/liter |
| Fetal bovine serum | 13% |

[a] Hansen.[53]

8. All solutions, except (F), are added to (A) in alphabetical order.

9. Adjust pH to 6.45 and add $NaHCO_3$ (F). This brings the final pH to 6.68 without further adjustment.

10. Filter-sterilize.

HANSEN MEDIUM[53]

This medium (Table IX) is for culturing a cell line (Bge) from the snail, *Biomphalaria glabrata*. This is the first, established, rapidly growing cell line from an invertebrate other than the arthropods. The pH is 7.1–7.3, and osmotic pressure is much lower than insect cell culture media.

[53] E. Hansen, *in* "Invertebrate Tissue Culture: Research Applications" (K. Maramorosch, ed.), p. 75. Academic Press, New York, 1976.

# [40] Cold-Blooded Vertebrate Cell and Tissue Culture

*By* KEN WOLF

## Introduction

Methods of cell and tissue culture[1] range from the very simple to the highly complex and demanding. In general, many of the methods can be routine, but cultures are biological systems that can and at times do evade absolute control. The biochemist planning to use cell cultures will find that cultures at time show greater variability than the biochemical determinations that are performed.

Most of today's animal cell and tissue culture involves materials of mammalian and secondarily, of avian origin. Reptiles, amphibians, and fishes—the poikilotherm vertebrates—are used the least. Without doubt the present status of homeotherm cell and tissue culture reflects health-related research on man and his domestic livestock. Lower vertebrate cell and tissue culture has much to offer the researcher; it extends the phylogeny of vertebrates, allows work to be done through a wide range of temperatures, and is less demanding than culture of homeotherm materials.

Fortunately for all concerned, the physiology and functional systems of the foregoing five classes of vertebrates have many similarities. Within the limits to be described, the methods can be considered realistically as simply being *vertebrate cell and tissue culture*.

[1] This article employs the standard terminology of the Tissue Culture Association as set forth by S. Federoff, TCA Manual **1,** 53 (1975).




ISBN 0-12-181958-2

Although commonly referred to as *fishes*, the elasmobranchs (sharks, skates, rays, and allied forms) and the cyclostomes (lampreys and hagfishes) differ considerably from teleosts or bony fishes. Culture of cells and tissues from the last two classes of vertebrates is best described as being in the exploratory or developmental stage.

One reason for the wide use of cell and tissue culture today is that quality equipment, media, and supplies are readily available from scientific and biological supply houses. Commonly used biologicals were originally formulated for application with homeotherm cells and tissues; with minor modification or, in most cases, without alteration these products also can be used for reptilian, amphibian, and teleostean cells. In other words, the biochemist can buy ready-made tissue culture supplies; virtually everything he will need is readily available.

The intent of this article is to enable the reader to culture lower vertebrate cells or tissues successfully. However, the reader first should be familiar with aseptic technique and should have had at least some tissue culture experience. Obviously there are space limitations here, and the unitiated will probably benefit from Part I of this volume and from cell culture textbooks. In addition, there are review type references on culture of reptile materials by Clark,[2] on that of amphibians by Freed and Mezger-Freed[3] and Rounds,[4] and on fishes by McKenzie and Stephenson,[5] Sigel and Beasley,[6] Sigel *et al.*,[7] and Wolf.[8] The most comprehensive work on fish cell and tissue culture is that of Wolf and Quimby.[9]

Some differences between homeotherm and poikilotherm vertebrates are worth noting, namely their environments and health histories. Microbial contamination is a hazard in initiating almost any primary culture. However, a great many of the mammals and avian embryos used as tissue donors are essentially healthy and usually laboratory-reared. Therefore,

[2] H. F. Clark, *in* "Tissue Culture: Methods and Applications" (P. F. Kruse, Jr. and M. K. Patterson, Jr., eds.), p. 147. Academic Press, New York, 1973.

[3] J. J. Freed and L. Mezger-Freed, *in* "Tissue Culture: Methods and Applications" (P. F. Kruse, Jr. and M. K. Patterson, Jr., eds.), p. 123. Academic Press, New York, 1973.

[4] D. E. Rounds, *in* "Tissue Culture: Methods and Applications" (P. F. Kruse, Jr. and M. K. Patterson, Jr., eds.), p. 129. Academic Press, New York, 1973.

[5] L. S. McKenzie and N. G. Stephenson, *in* "Tissue Culture: Methods and Applications" (P. F. Kruse, Jr. and M. K. Patterson, Jr., eds.), p. 143. Academic Press, New York, 1973.

[6] M. M. Sigel and A. R. Beasley, *in* "Tissue Culture: Methods and Applications" (P. F. Kruse, Jr. and M. K. Patterson, Jr., eds.), p. 133. Academic Press, New York, 1973.

[7] M. M. Sigel, E. C. McKinney, and J. C. Lee, *in* "Tissue Culture: Methods and Applications" (P. F. Kruse, Jr. and M. K. Patterson, Jr., eds.), p. 135. Academic Press, New York, 1973.

[8] K. Wolf, *in* "Tissue Culture: Methods and Applications" (P. F. Kruse, Jr. and M. K. Patterson, Jr., eds.), p. 139. Academic Press, New York, 1973.

[9] K. Wolf and M. C. Quimby, *Fish Physiol.* **3**, 253 (1969).

the risk of contamination in such material is low. In contrast and for all practical purposes, the reptiles, amphibians, and all but aquarium or hatchery-reared fishes are wild animals, and their health history is unknown or uncertain. Such feral animals can harbor systemic parasites, microorganisms, and viruses that can contaminate primary cell cultures.

On the other hand, there are differences between homeotherms and poikilotherms that are to the investigator's advantage. For example, cell cultures from lower vertebrates usually require less frequent handling and attention than homeotherm cultures, and they are far more tolerant of neglect. Intrinsically, the coldblooded vertebrates are adapted to and have a range of body temperatures that cannot be matched by homeotherms. Some cold-water fish cells metabolize at temperatures from near 0–25° whereas many warm-water fishes, reptiles, and amphibians span the range of 10–37°. If rigid temperature control need not be maintained, many kinds of cells are grown on the bench at room temperature, i.e., 20–25°.

## Physiological Salines

The usual composition of physiological salines or balanced salt solutions (BSS) for vertebrates is an isotonic salt solution containing the essential physiologic ions, a system for maintaining pH in the physiological range, an energy source (usually glucose), and, unless contraindicated, phenol red as a pH indicator. With appropriate adjustment of osmolarity the common BSS are appropriate for cell cultures of mammals down through bony fishes (Tables I).

Virtually all routine coldblooded vertebrate cell culture needs will be met with Earle's or Hanks' BSS and Dulbecco and Vogt's phosphate-buffered saline (PBS). All are available commercially. Without their sodium bicarbonate, Earle's and Hanks' BSS are also available as nonsterile powders that can be dissolved in high-purity water and decontaminated by membrane filtration (0.22 μm mean pore diameter). Alternatively, these solutions are easily prepared and decontaminated or sterilized in one's own laboratory.

Physiological salt solutions find many applications in cell culture. Earle's BSS with 2.2 g/liter of sodium bicarbonate is intended to equilibrate in the physiological range in an atmosphere containing about 5% $CO_2$. In normal atmosphere, the pH of Earle's BSS rises comparatively rapidly—the quickest of the three. Hanks' BSS has a phosphate buffer and much less sodium bicarbonate; therefore its pH changes less rapidly in normal atmosphere. For the easiest maintenance of pH, PBS is recommended. Physiological salines may be buffered with about 15 m*M* Tris,

TABLE I
OSMOLARITY ADJUSTMENTS SUGGESTED FOR PHYSIOLOGICAL SALINES AND MEDIA USED FOR CULTURING LOWER VERTEBRATE CELLS AND TISSUES

| Vertebrate class | Habitat | Adjustment |
|---|---|---|
| Reptile | Terrestrial or freshwater | None |
| | Marine | Add 0.06 *M* NaCl[a] |
| Amphibian | Terrestrial or freshwater | Dilute 20% with water or NaCl-free BSS |
| Teleost | Freshwater | None |
| | Marine | Add 0.07 *M* NaCl[b] |
| Elasmobranch | Freshwater | Add urea to isotonic level |
| | Marine | Add 2 *M* urea |
| Cyclostome | Freshwater | None |
| | Marine | Not determined |

[a] Add 17.6 ml of 3.4 *M* stock NaCl solution per liter.
[b] Add 20.6 ml of 3.4 *M* stock NaCl solution per liter.

HEPES, TES, TRICINE, or BES buffer providing that about 10 m*M* $NaHCO_3$, an essential constituent of media for vertebrate cell cultures, is also present.

## Culture Media

Based as they are upon one or another of the aforementioned balanced salt solutions, culture media for poikilotherm vertebrate cells are off-the-shelf media that were originally designed for mammalian cell and tissue culture. At this time lower vertebrate cells have been only maintained and not truly grown, in chemically defined medium. If the reader requires a chemically defined medium, it should be kept in mind that, while the cells may not grow, their viability can be retained and they will metabolize for at least several days in such medium.

In most situations, animal cell cultures are expected to divide and the biological mass to increase. Literature on teleost fishes, amphibia, and reptilia shows that investigators use one or another of four common media for growing cells and that the media are usually supplemented with 10% serum; in a few instances 20% serum is used. Without hesitation, fetal bovine serum is recommended as the growth stimulant of choice. Fetal bovine serum is rather expensive but, if cost is a factor and time permits, comparative testing of agamma calf, horse, bovine, or porcine serum is recommended since they may do as well. Probably because of innate resistance to change and the generally excellent results with fetal bovine serum, it is the single supplement that is recommended for culture of all vertebrate cells.

While it is likely that most of the common commercially available media will be satisfactory for poikilotherm vertebrate cell or tissue cultures, the most widely used are Eagle's minimal essential medium (MEM), Eagle's basal medium (BME), Medium 199, and Leibovitz L-15. The latter has been specifically designed to maintain a pH in the physiological range under conditions of normal atmosphere.

Contrary to what might appear to be logical, serum from the donor species more frequently than not proves to be unsatisfactory for the usual cell cultures. In addition, a sustained supply is usually difficult to obtain, and quality control is less easily achieved than in commercial sources.

For most work with poikilotherm vertebrate cultures a 10% level of serum is standard, but when initiating primary cultures or when vigor appears to be low, an additional 5% serum or 5% whole egg ultrafiltrate can sometimes be used to advantage. For reasons of economy, some workers substitute part of the serum with 0.5% lactalbumin hydrolysate, but the advantage is only marginal.

Leukocyte culture differs somewhat from other cell culture in that a 20% level of serum is usually employed, and some have found it advantageous to use homologous serum. This appears to be particularly true for elasmobranch leukocyte culture; sharks, skates, and rays have little or no serum albumin, and isotonicity with the marine environment is maintained by retention of very high levels of urea.

### pH Control

As a general guideline and in the absence of specific information to the contrary, an initial pH of 7.3–7.4 is suggested for starting lower vertebrate cell cultures. At this range, phenol red indicator is reddish orange. Until one is familiar with the color changes of phenol red, a set of pH standards will be useful. Some cells tolerate a pH well above or below the optimal physiological level, but values much above or below will result in an extension of the lag phase of growth.

### Temperature of Incubation

On the premise that near-optimal growth is desired, the choice of incubation temperature can be approached with the following rule of thumb: use a temperature slightly above that preferred by the intact ani-

TABLE II
PROXIMATE GUIDELINES FOR INCUBATION TEMPERATURES FOR LOWER VERTEBRATE CELL AND TISSUE CULTURES

| Vertebrate class | Environment | Temperature range (°C) | | |
|---|---|---|---|---|
| | | Low | Near-optimal | High |
| Reptile | Temperate | 20 | 23–25 | 37 |
| | Tropical | not known | 30 | 37 |
| Amphibian | Temperate | 15 | 25–26 | 37 |
| | Tropical | not known | 28 | 37 |
| Teleost | Temperate | | | |
| | Cold water | 4 | 20 | 26 |
| | Warm water | 13 | 25 | 30–37 |
| | Tropical | 20 | 25 | 37 |
| Elasmobranch | Temperate | | | |
| | Cold water | 4 | 10–15 | not known |
| | Warm water | not known | not known | not known |
| Cyclostome | Temperate | 4 | 15 | 20 |

mal. Once the cultures are growing, experimentation can determine an optimal or near-optimal temperature. The advantage to be found with poikilotherms is that their cells can be chosen for growth and metabolism in the range of near freezing through 37° (Table II).

Cold-water fishes such as salmon and trout prefer temperatures of about 8–12°, but the optimum for salmonid cell cultures is about 20°. Salmonid cells will grow, however, at 4°, and their thermal death point is about 26 or 27°. Interestingly, the growth rates near their maximum temperature tolerance are less than at 20°. Warm-water fish cells seldom grow at 5–10°; their optimum is generally in the range of 22–25°, and their upper limit is about 30° or higher. With increments of about 1° at each subculture, warm-water fish cells may be adapted to grow at 37°.

Amphibian cells can be handled much like those of warm-water fishes and grown at about 22–25°. They too can be kept at 10–15° to slow metabolic rates. Accordingly, they can be stored at 15° for weeks and in some cases for months without attention. Since most plastic cell culture vessels allow diffusion of $CO_2$, long-term storage of active cultures is better done in glassware.

Reptiles are designated as being terrestrial, aquatic (fresh water), or marine; they are further classified as temperate climate or tropical animals. Reptilian cell cultures therefore should be grown in medium of proper tonicity and also incubated at an appropriate temperature (Table

II). Cells from donors of a temperate climate are incubated at room temperature and those from tropical latitudes at about 30°.

## Leukocyte Culture

Poikilotherm vertebrate leukocytes lend themselves to the usual *in vitro* applications: karyology, migration inhibition studies, investigations of antibody production, and other immunologic research. Viable leukocytes are readily obtained from sterile whole blood drawn with either plastic or siliconized glass syringes containing cold heparin solution in physiological saline at a final concentration of 10 IU per milliliter of blood. If one is unfamiliar with the anatomy of a particular animal, it will be advantageous to dissect a specimen and locate the heart and major blood vessels and to orient them with external anatomy. Direct cardiac puncture with a needle of appropriate gauge works well with normally scaled reptiles. Easy access to the heart of terrapins and turtles, of the size conveniently used in laboratories, is gained by first drilling a small hole in the plastron immediately beneath the heart. Amphibians can be bled from the heart, and the frogs and toads also can be bled from the femoral artery. For best results in culturing amphibian leukocytes, the donors should be kept at normal temperature prior to bleeding; in that way the lymphocyte population is maintained.

Fishes also can be bled from the heart, but if the animal is to remain alive cardiac puncture does run the risk of tamponade formation. Fishes may be bled by opening the mouth and using a shallow approach to enter the dorsal aorta immediately behind the last gill arch. Fishes are perhaps bled most safely, most easily, and repeatedly by lateral or ventral approach in the region behind the vent, the caudal peduncle. By the last method, either the dorsal aorta or the caudal vein will be entered.

When drawn, blood is mixed with the heparin solution and centrifuged in the cold at 200–500 g or until the "buffy coat" of leukocytes is clearly visible above the packed erythrocytes. It is convenient to centrifuge the blood in an inverted but locked syringe. Locks are easily made from split plastic tubing that fits around the plunger; during centrifugation the tubing prevents the barrel from moving down and expressing blood. After centrifugation, a needle is refitted, bent to a 90° angle, and the plasma or "buffy coat", or both, are gently expressed into culture medium.

## Obtaining Tissues

Either external or internal tissues can be cultured, but the latter have the lower risk of contamination. Almost universally, ovarian tissues do well in culture. Ovaries may be used at almost any stage of development and typically provide a sizable mass of usable tissue. Ovaries are highly recommended for beginning exercises in lower vertebrate cell and tissue culture. Embryonic tissues also do well because they are typically clean and usually grow luxuriantly; their only limitations are volume and the fact that they are typically available only during a limited portion of the year. Some viviparous laboratory fishes are an exception, and they produce young throughout the year. Other internal tissues that usually lend themselves well to culture are heart, trachea, gas bladder, mesenteries, spleen, kidneys, and lungs. However, the last two frequently harbor metazoan parasites that are esthetically undersirable but which probably will not grow in vertebrate culture systems. Liver from juvenile animals is an additional candidate tissue but, because of the great mass of differentiated cells, liver can yield active cultures somewhat more slowly than other tissues. Portions of the gastrointestinal tract also may be cultured, but with them the risk of contamination exceeds that of external tissues; thorough washing and decontamination for an hour or more in a mixture of the following bactericidal antibiotics is suggested: 500 IU polymyxin B, 500 $\mu$g neomycin, and 40 IU of bacitracin per milliliter of sterile water.

Although available in only limited quantity, corneal tissue is excellent for culture as are fish fins and amphibian skin. Because of its cornified outer layers the skin of reptiles presents difficulties because of its inert component and greater-than-normal risk of contamination.

Whether internal or external tissues are to be used, the donor animals should be deprived of food for several days prior to use. This precaution will minimize risks of gross contamination by food or feces during handling.

Immobilization during surgical opening is desirable, if not essential. Animals can be pithed or killed with a blow to the head; fishes and amphibians can be euthanized in 1 : 5000 solution of tricaine methanesulfonate and amphibians by immersion in a plastic bag of the anesthetic. Reptiles and elongate amphibians can have muscular contraction for considerable time after death. To reduce problems of vigorous bodily contractions during operations, the carcass can be refrigerated for an hour or more prior to use. One need not fear tissue or cell death; sterile internal tissues may be refrigerated in PBS for at least a day and in some instances as long as 3 days and still yield cultivable cells. If mobility is a problem, the head, limbs, and tail from the vent rearward also can be restrained with hooks or stout pins. The plastron of terrapins and turtles must be removed to expose internal organs.

Decontamination of the integument should be carried out if sterile internal tissues are to be obtained from any lower vertebrate. Soaking for

5 min in a 1 : 10 dilution of household bleach or any hypochlorite solution containing at least 500 ppm available chlorine is adequate. For reptiles, thorough scrubbing with disinfectant is further recommended since they often have external fauna and flora. Strong hypochlorite solutions denature protein and kill living cells; therefore these solutions should not be used if external tissues of amphibia and fishes are to be cultured. Fins, skin, tongue, and gills should be washed well with cold, preferably chlorinated, tap water and further decontaminated for an hour or more at 5–10° in a solution of the three bactericidal antibiotics listed earlier.

Strict aseptic technique should be used to remove internal tissues. A tray of wax or soft wood is used to pin or otherwise immobilize the animal. Additional pins hold back the abdominal flaps and the gastrointestinal tract so that there is unimpaired access to the visceral cavity and organs. Fish, of course, are positioned laterally and also opened laterally. Amphibia and reptiles are best fixed in the supine position and opened midventrally.

As organs or tissues are removed, they should be transferred to a sterile petri dish or other covered container maintained on ice. Materials from lower vertebrates of temperate or tropical origin need not be chilled, but holding in chilled vessels is not inhibitory. If the amount of tissue removed is greater than can be used immediately, it may be kept safely at 4°. Protected from dehydration, the tissue can be expected to yield viable cells for several days.

There is a great deal of evidence that the emphasis of lower vertebrate cell and tissue culture has been on proliferation of undifferentiated or dedifferentiated cells. However, if obtained aseptically as indicated in the foregoing, differentiated tissues or cells, including hepatocytes and endrocrine tissues, among others, can be maintained for days or weeks and remain capable of at least some of their functions.

## Primary Cultures

Primary cultures can be of the explant type, or they can be monolayers that are derived either from tissues that are simply minced and planted or minced and enzymically dispersed before planting. Procedures for initiating primary cultures of poikilotherm cells are essentially the same as those for homeotherms. For details, see this volume [9] and [10].

### *Monolayer Cultures from Minced Tissues*

Monolayer cell cultures can be started simply and quickly from minced tissues that are washed, drained, and transferred to culture vessels. Bent-tip tissue culture type Pasteur pipettes having about a 2-mm orifice or spatulas are convenient for planting the tissues. Fragments are uniformly distributed over the growth surface at a rate of two to four pieces per $cm^2$. The culture vessels are then closed and placed on edge for an hour or so at a favorable incubation temperature. After fluids drain from the tissue and natural adhesion occurs, the liquid is aspirated, the vessels inverted, medium is added, and the cultures are moved to an incubator. The vessels are then oriented so that the medium covers the tissue. Most fragments adhere and many will send out cellular processes or extend areas of epithelial-like cell sheet within a day or two. This initial growth helps the fragments adhere more tightly. After the first growth occurs the cultures can be examined safely microscopically. Confluent sheets of cells may be formed within a few days, but it may take as long as 2 weeks; it depends on donor age, the organ used, pH, temperature, and other variables. When confluency is attained and growth is obviously heavy, the cultures may be divided. Requiring only 2 or 3 hr, this method is undoubtedly the simplest and least time-consuming means of setting up primary monolayer cultures.[10] The cultures thus developed are typically less homogeneous than monolayers that have been initiated from trypsinized tissues.

*Monolayer Cultures from Trypsinized Tissues*

The most commonly used procedure for starting primary monolayer cultures of vertebrate cells is that of "trypsinization." Details of the procedure are described in this volume [10], and the general methods are applicable to lower vertebrate as well as to homeotherm tissues. The following are important comments and exceptions as they apply to reptile, amphibian, and fish tissue trypsinization.[11]

No lower vertebrate tissues should be trypsinized at 37° because the cells will be killed. Instead, there are two alternative lower temperatures that can be used. Tissues from poikilotherm vertebrates of warm environments can be trypsinized at 20–25° with harvest being made at intervals of 15–30 min. Tissues from cold-water fishes may be trypsinized at 15° or lower. Regardless of the environmental temperature preferred by the animal itself, tissues from all of the lower vertebrates can be trypsinized successfully at 5°. Duration of enzymic treatment will depend upon the tissue; embryonic tissues may disperse in several hours but digestion of from 12–16 hr is more common. The choice of room tempera-

[10] For additional reading and a listing by source and catalogue numbers of specific materials needed to carry out the foregoing methods, see K. Wolf and M. C. Quimby, TCA Manual **2,** 445 (1976).

[11] Additional readings and a listing of specific materials by source and catalogue number are to be found in K. Wolf and M. C. Quimby, TCA Manual **2,** 453 (1976).

ture or cold overnight digestion may be dictated by the work situation, or one can compare the two temperatures for dispersing tissues. If that is the case, ciliated cells that occur in various tissues may be used to monitor the digestion and subsequent culture conditions for at least several days. Under favorable trypsinization and culture regimes, cilia will continue to beat actively and provide instant assessment of viability.

Penicillin and streptomycin, a bacteriostatic combination, are almost traditional in animal cell and tissue culture, and some workers include them in the digestion mixture. Gentamicin is a bacteriocidal antibiotic and much more effective in controlling the gram-negative organisms that are common in lower vertebrates. Levels of 10–100 $\mu$g/ml are suggested.

Trypsinization of homeotherm tissues commonly is allowed to proceed until a monodispersed suspension of cells results. That degree of dispersion can be excessive for some lower vertebrate tissues; the resulting suspension may contain too many dead cells. Viability is usually excellent if digestion is allowed to proceed only to the point at which there is a mixture of individual cells and tissue fragments consisting of 10–100 or so cells. That kind of determination may be made with 0.5% aqueous trypan blue; live cells exclude the dye.

Residual trypsin can be inhibitory to cells that are obtained by digestion. Accordingly, the harvest should be washed in PBS or in medium before planting.

Because of the presence of small fragments of nondispersed tissue in the suspension of lower vertebrates after the optimal digestion period, cell counting is pointless. Instead, the harvest is resuspended in growth medium on the basis of its packed volume. For embryonic or soft tissues, the ratio of cells to medium can be 1:2000 to 1:4000 or more. For initial trials, especially when adult tissues are used, a ratio of 1:1000 or less is suggested; prompt success is much preferred to slow results or to failure that can follow if too few cells are seeded.

While vortex mixers are common in most laboratories, they should not be used to suspend living animal cells; the vigorous mixing will likely kill many cells. Cells should be resuspended by repeated gentle pipetting.

### Subculturing or Transferring Cell Lines

Coldblooded and homeotherm vertebrate cell cultures are transferred or divided in essentially the same way. There are advantages, however, in working with lower vertebrate cells since it is seldom, if, ever, necessary to feed cultures between transfers, i.e., to provide them with fresh medium. When not used, cell cultures from the poikilotherms retain viability for months without attention; they typically have a high tolerance for neglect. Virtually all vertebrate cell cultures can be divided or transferred at room temperature, but some workers prefer to handle mammalian and avian cells at 37°, a temperature probably excessive for many poikilotherm cells.

The most widely used method of dispersing lower vertebrate cells for subculturing is with a solution of EDTA (200 μg/ml) and trypsin (1–2.5 mg/ml). A listing of equipment, solutions, and procedures is available.[12] Cultures first should be examined microscopically for quality, cell morphology, and possible contamination. The medium is then removed and the EDTA–trypsin solution added at about 5 ml/75 $cm^2$ of growth area or 2 ml/25 $cm^2$. After several minutes the cell sheet develops marked opacity. The EDTA–trypsin solution is removed and replaced with about half as much fresh dispersant. Cell lines vary considerably in their response. Many will have been freed after several minutes, but some may require gentle scraping. Total treatment time should not exceed 10–12 min, after which fresh medium should be added to stop the action of trypsin. The cells are uniformly dispersed by pipetting, diluted as necessary in fresh medium, and planted. Although most lower vertebrate cell lines can be routinely subcultured with this procedure, some are usually sensitive to trypsin. If cells lyse or otherwise fail to prosper, they should be centrifuged out of the EDTA–trypsin solution, washed in fresh medium or BSS, recentrifuged, and resuspended in fresh medium for planting. Cells grown in the absence of serum or other protein are particularly vulnerable to trypsin and will probably require greatly reduced trypsin or even alternative methods that omit trypsin. Inexperienced investigators are cautioned to use low split ratios for their early trials of subculturing.

At times, cell growth may be more rapid than anticipated or transfers may be unavoidably delayed, so that heavy or dense cell sheets result. In such cases, the usual 10–12 min treatment time may not be sufficient to disperse the cells. The incompletely dispersed population can be resuspended, planted, and allowed to grow for several days. During that time, cells will migrate out of the clumps, and a second EDTA–trypsin treatment can be used to achieve complete dispersion.

In the absence of specific information on density of cells to be cultured, i.e., number per milliliter, it is suggested that the initial density should be at least several hundred thousand cells or more per milliliter. It is usually safe to seed at a high population density although cultures will become confluent quickly and require frequent handling. If cells are seeded too thinly, the lag time will be lengthy or the population too sparse to establish itself, so that the culture may decline and die. Adverse affects are particularly prevalent in media of marginal quality.

[12] K. Wolf and M. C. Quimby, *Tissue Cult. Assoc. Man.* **2,** 471 (1976).

# [41] Plant Cell Lines[1]

*By* JOHN F. REYNOLDS and TOSHIO MURASHIGE

As experimental materials, aseptically cultured cells or tissues have certain advantages over organs and plants that are allowed to develop in the more natural state. Correlative influences can be minimized, and artifacts attributable to bacteria, fungi, and other small organisms can be avoided. It also is possible to regulate more precisely the provisions with respect to nutrients, light, and temperature. Sometimes cultured cells can be treated and interpreted experimentally like simple microorganisms.

Plant cell cultures may have economic significance as well. When totipotentiality is manipulatable the cultures can serve as intermediary in rapid clonal multiplication of plants[1] and in elimination of viruses and related pathogens.[2] Some are potential sources of medicinals and other important plant constituents.[3]

In the simplest and perhaps most widely practiced method of maintaining plant cell lines the callus, or the unorganized tissue that results on wounding, is cultured. For biochemical studies, however, liquid suspension cultures of free-living cells are probably preferred, even though they demand more labor. Indefinitely subculturable callus or cell suspensions are now attainable with virtually any plant, bryophytes through angiosperms, and from nearly every plant organ, including stem, root, leaf, fruit, flower, and seed. Callus is easily initiated by simply placing freshly cut sections of disinfested organs on the surface of an agar-gelled medium.

The nutrient medium of plant callus and cell cultures should contain a balanced salt mixture, sucrose (3–5%) as carbon source,[4] thiamine · HCl (0.1–10 mg/liter), and usually the growth regulators, auxin and cytokinin. These substances should be tested for their effectiveness as auxin in the concentration range 0.1–10 mg/liter: 3-indoleacetic acid (IAA), 1-naphthaleneacetic acid (NAA), and 2,4-dichlorophenoxyacetic acid (2,4-D). Kinetin, $N^6$-benzyladenine (BA), and $N^6$-isopentenyladenine (2iP) are the more readily available cytokinins; they should be examined in the range 0.03–3 mg/liter. *Myo*-inositol in a concentration of 100 mg/liter and other supplements, e.g., citric acid (2 g/liter)[5] and casein hydrolysate (1–3 g/liter), also may be helpful.

[1] T. Murashige, *Annu. Rev. Plant Physiol.* **25,** 135 (1974).

[2] T. Murashige, *A. W. Dimock Lect., Cornell Univ.* No. 2 (1974).

[3] W. Barz, E. Reinhard, and M. H. Zenk, eds., "Plant Tissue Culture and Its Biotechnological Application." Springer-Verlag, Berlin and New York, 1977.

[4] Glucose is used occasionally instead of sucrose.

[5] Y. Erner, O. Reuveni, and E. Goldschmidt, *Plant Physiol.* **56,** 279 (1975).




ISBN 0-12-181958-2

Most callus cultures grow best in darkness. However, free-living cells in liquid suspension may benefit from low levels of illumination, perhaps 1000 lx, provided daily by cool-white fluorescent lamps. The standard practice with callus and liquid suspension cultures has been to incubate at a constant temperature of about 27°.

## Establishing the Callus Culture

Cell and tissue cultures of the tobacco, *Nicotiana tabacum* L. "Wisconsin 38," have been chosen to illustrate all principles because details of their requirements are established, the pattern of their development *in vitro* is readily manipulated, and totipotentiality can be demonstrated.

Plants should be grown under greenhouse conditions to ensure a high percentage of clean and responsive explants and used before flowering begins. The better explants are obtained from the upper or younger regions of the stem. Cell cultures originating from older tissues often manifest retarded growth and organogenic rates as well as high frequencies of polyploid cells.

Sever the plant at its base and remove all leaves. Trim stem of all leaf bases and axillary buds, using a surgeon's scalpel fitted with No. 10 blade. Swab stem with cheesecloth, dampened with 95% ethanol, to remove some of the stickiness and to achieve preliminary disinfestation. Obtain 1-cm-long stem segments from the region 10–30 cm from the shoot apex region. Plance segments in a small Erlenmyer flask and add disinfestant solution.[6] Cover flask loosely with beaker and place under gentle vacuum for 10 min. Decant disinfestant and rinse stem segments 3 times with distilled water that has been previously autoclaved and cooled. Working within a laminar air-flow hood or other protective device, transfer stem segments to sterile Petri dish. Remove the bark from each of the segments. This is accomplished by grasping the segment at its cut ends with a pair of forceps, making a shallow incision longitudinally with a surgeon's scalpel fitted with No. 10 blade, and paring away the bark with the dull edge of the scalpel. Stand the peeled stem sections on end and divide into three or four tangential slabs. Transfer slabs to nutrient agar and plant with the cambium side exposed. If desired, the pith portion that remains also can be cut into explants. It should be prepared as cubes about 2 mm thick.

Presterilized, disposable Petri dishes are convenient. In their absence, the Pyrex type can be used, after autoclaving 30 min at 121° or dry-heating

[6] Laundry bleach diluted to contain 0.5% sodium hypochlorite is used most widely; a few drops of detergent might be added to enhance wetting.

in an oven for 4 hr at 160°. Several dishes can be wrapped together in aluminum foil or placed in a prescribed cannister for sterilization.

Satisfactory callus from the above explants can be expected by employing a medium of the following composition, in mg/l: Murashige and Skoog inorganic salts;[7] sucrose, 30,000; *myo*-inositol, 100; thiamine · HCl, 0.4; kinetin, 0.1; 2,4-D, 0.3; and Phytagar,[8] 6000. Adjust pH of the solution to 5.7 with 1 *N* HCl or NaOH prior to addition of agar. Wash agar with distilled water 3 times before use and dissolve by heating. If a hotplate or burner is employed, stir medium constantly to avoid charring of agar. A simpler way to dissolve agar is by autoclaving the medium at 121° for 3–7 min, depending on the quantity of medium. Mix medium well and dispense 25-ml portions into 25 × 150 mm glass culture tubes. Cap tubes with polypropylene closures.[9] Autoclave 15 min at 121° to sterilize medium, then cool tubes at 45° slants. Inoculate each tube with one tobacco stem explant, and incubate in darkness at 27°. Within 2 weeks, callus will appear as a white growth on the exposed surface of the explant. After 4 weeks it should have developed sufficiently for subculturing and establishing of callus stock.

In subculturing the callus, prepare nutrient tubes as above. Select cultures that show large growths of white callus. Examine tubes carefully and discard those that show bacteria, fungi, or other microbial contaminants. Contaminated cultures should be autoclaved before disposal to minimize spread of fungi and bacteria in the laboratory. Using a pair of long forceps,[10] transfer callused section to sterile Petri dish. Pare away a 2-mm-thick slice of callus from the stem, using a surgeon's scalpel fitted with No. 10 blade. It may be helpful to use a pair of small forceps to hold the stem section during the cutting process. Divide the callus slice into 3–5 mm squares and transfer to nutrient tubes, at a rate one tissue section per tube. The callus is maintained as stock by subdividing established cultures at 4-week intervals and transferring 2 × 4 × 5 mm subsections to freshly prepared medium. Incubation of established callus is also carried out in constant darkness and at 27°.

[7] T. Murashige and F. Skoog, *Physiol. Plant.* **15,** 473 (1962). The salts, in mg/l are: $NH_4NO_3$, 1650; $KNO_3$, 1900; $CaCl_2 \cdot 2\ H_2O$, 440; $MgSO_4 \cdot 7\ H_2O$, 370; $KH_2PO_4$, 170; $Na_2EDTA$, 37.3; $FeSO_4 \cdot 7\ H_2O$, 27.8; $H_3BO_3$, 6.2; $MnSO_4 \cdot H_2O$, 16.9; $ZnSO_4 \cdot 7\ H_2O$, 8.6; KI, 0.83; $Na_2MoO_4 \cdot 2\ H_2O$, 0.25; $CuSO_4 \cdot 5\ H_2O$, 0.025; and $CoCl_2 \cdot 6\ H_2O$, 0.025.

[8] This is agar specially prepared for plant tissue culture by Grand Island Biological Co., Grand Island, New York.

[9] Examples of plastic caps available are Kimble Kimkap, Bellco Kaput, and Bacti Capalls.

[10] Use 9-½ inch dressing forceps, Model MX6-158, obtainable through Miltex Instrument Co., New York, New York.

Cells in Liquid Suspension Culture

Whereas it is possible to obtain free-living cells in liquid media by starting directly with a freshly excised plant section, the easier method is to begin with a suitably prepared callus as inoculum. Cell dissociation in liquid culture can be improved if the callus is first rendered friable. The friable texture can be attained sometimes by culturing the callus on an agar medium that is enriched in salts and auxin. Gibberellin $A_3$ ($GA_3$) may further enhance the friability. Cytokinin, if necessary, should be included at only very low levels. Some natural complexes, e.g., protein hydrolysates and yeast extract, may also increase friability. The ease of altering tissue texture may be an inherent quality, and occasionally cell dissociation may not be possible under any condition. These relationships between nutrient addenda and callus friability are perhaps applicable to freshly excised tissues and might be considered when callus is not available.

To obtain readily dissociating tobacco callus, use a medium whose composition is the same as that employed in establishing the tobacco callus cultures, but with the 2,4-D increased to 2 mg/liter and with an additional supplement of 1 g/liter casein hydrolysate;[11] the vitamin component might be altered by raising thiamine · HCl to 10 mg/liter and by adding 5 mg/liter nicotinic acid and 10 mg/liter pyridoxine · HCl. Adjust the medium pH to 5.7, dissolve the agar in the medium by heating, and dispense medium into 25 × 150 mm culture tubes with 25 ml per tube. Use polypropylene closures and autoclave the medium to achieve sterility. Obtain callus that has been grown for 4 weeks in a given subculture passage of an established stock. Transfer the tissue to a sterile Petri dish and cut into 2-mm-thick slices; then cut each slice further into 4- or 5-mm squares. Inoculate nutrient tubes with callus slices and incubate in darkness at constant 27°.

After 4 weeks a translucent, soft, and relatively watery tissue should result. The callus will have lost its earlier firmness and white appearance. A substantial proportion of the cells will be easily separable from one another by gentle teasing; the cells will be substantially larger than those of the firm, white callus. Transfer the entire new growth to liquid nutrient. A micro-spatula may be better than forceps for handling the tissue.

The composition of the liquid medium can be the same as that employed in rendering the callus friable, except for the omission of agar. Use 125-ml Delong flasks, each with 25 ml of nutrient solution. Cap flasks with Morton stainless-steel closures and sterilize by autoclaving at 121° for 15 min.

[11] Enzymic digest of casein is obtainable from ICN Life Sciences Group, Cleveland, Ohio.

Place the inoculated flasks on shaker[12] and agitate constantly at 150 rpm. Provide 1000 lx illumination, from cool-white fluorescent lamps, 16 hr daily and at a constant 27° temperature.

After 10–14 days a dense suspension composed of free cells and cell aggregates should have developed. Separate free-living cells and small aggregates from larger aggregates and subculture them in liquid medium to establish a suspension culture stock. To achieve separation, first, drain the nutrient solution by filtering the suspension through two layers of sterile Kimwipe tissue; use a 70-mm-diameter powder funnel to support the Kimwipe tissue. The funnel may be held in a small beaker or flask. Sterilization of Kimwipe tissue is accomplished easily by folding and enclosing several sheets in Petri dishes and autoclaving. Complete drainage of the nutrient solution by gathering together the corners of the Kimwipe tissue and gently wringing the contents. Carefully unfurl the Kimwipe tissue and weigh out 1-g samples of cells for subculture. Avoid including the larger cell clusters. Weigh cells in a sterilized Petri dish.

The medium for subcultures is the same in composition as that used in initiating the suspension culture. A stock of vigorously dividing and growing cells can be attained by repeatedly subculturing in fresh medium 1-g portions at 10- to 14-day intervals. Growth of plant cells in liquid suspension cultures can be measured as drained weight, dried weight,[13] sedimented volume of cells after gentle centrifugation,[14] or as cell counts. Accuracy of cell counting may be improved if the mixture of cells and aggregates is first dispersed by treating drained samples with 5% chromic acid or 0.25% pectinase[15] for 24 hr at room temperature. Protein content might also be used to estimate plant cell culture growth.[16]

## Plating to Obtain Strains of Single-Cell Origin

Tissue strains of single-cell origin are obtainable in large numbers by allowing the free-living cells of liquid suspension cultures to regenerate callus in agar plates. The procedure enables establishment of pure lines from an otherwise heterogeneous population of cells. Heterogeneity among cultured cells may be attributable to natural and induced variations.[17] Since suspension cultures are usually comprised of a mixture of

[12] Satisfactory results can be obtained with a Model G-10 gyratory shaker, made by New Brunswick Scientific Co., Edison, New Jersey.

[13] Dry in 70° oven until constant weight is attained.

[14] 2880 *g* for 5 min.

[15] Sigma pectinase, for example.

[16] J.-P. Jouanneau and C. Peaud-Lenoel, *Physiol. Plant.* **20,** 834 (1967).

[17] T. Murashige and R. Nakano, *Am. J. Bot.* **54,** 963 (1967).

free-living cells and cell aggregates of varying dimensions, it is important to fractionate the cultures to obtain a component of predominantly free-living cells. In practice, even the best free-cell preparations will be contaminated by a significant fraction of small aggregates. Thus, to ensure a single-cell origin of developing tissues, each newly prepared plate should be examined under a microscope and the single-cell units located and circled with indelible ink. The tissue arising within the circled area is more likely to be of single-cell derivation. Further assurance of single-cell origin is possible by plating protoplasts instead of intact cells. Since protoplasts have no cell wall, separation into individuals is virtually guaranteed.

The nutrient medium for plating of tobacco cells is the same in composition as the agar medium used in obtaining friable callus. The cells to be plated are mixed with the nutrient agar while the agar is still fluid, but sufficiently cooled to be noninjurious. In actual practice, prepare the complete nutrient medium, but dilute to only 90% of its ultimately desired volume. After pH adjustment and agar dissolution, dispense medium in 9-ml aliquots into 25 × 150 mm culture tubes. Cap tubes with polypropylene closures and autoclave 15 min at 121°. Allow the agar to cool to about 50° and place nutrient tubes in a water bath adjusted to 40°. This prevents premature gelling, but enables maintenance of a favorable temperature. The nutrient tubes should be held in a suitable rack and emersed so that the water in the bath is level with that of the medium in the tubes. Only clean distilled water should be used in the bath. Obtain desired cells by filtering a 10-day liquid suspension culture through a layer of sterile cheesecloth. Collect cells that pass into the filtrate by draining the filtrate through sterile Kimwipe tissue. After gentle wringing, weigh out 1-g quantities and resuspend in 9 ml of sterile liquid nutrient, the composition of which is the same as that used for plating, but without the agar. Transfer 1-ml portions of the suspension to tubes containing the agar nutrient. Mix cells and nutrient agar gently, but thoroughly, and pour contents quickly into sterile 100-mm petri dishes. Spread cells uniformly in the dish. As soon as the medium gels, seal the petri dish with Parafilm. The seal minimizes evaporation of medium and contamination by airborne organisms. Observe plates with an inverted microscope and circle single-cell units with a fine-tipped felt pen. Place cultures in darkness and at constant 27°. Within 2–4 weeks, callus growths of varying sizes, sometimes also diversely textured and pigmented, will become apparent. Remove growths that have attained 2–3 mm in diameter and reculture or subculture, following the procedure described earlier for callus, and establish callus stocks of single-cell origin.

Protoplasts for plating are best obtained from tobacco cells that are in the early exponential phase of cell division in liquid suspension culture, i.e., after 4–5 days of a given passage.[18] Collect cells on sterile Kimwipe tissue. Transfer 0.5-g portions to 50-ml Delong flasks, each containing 5 ml of protoplast-release solution.[19] Place the suspension on a gyratory shaker and agitate continuously for 2–3 hr at 50–70 rpm. Filter through nylon cloth of 50–60 $\mu$m. Centrifuge the filtrate at 100 g for 2 min. Decant and resuspend in 0.7 *M* mannitol and recentrifuge. Repeat the process of washing 3 times, by suspending in mannitol and centrifuging. Finally, suspend the rinsed protoplasts in 2 ml of nutrient medium that has been prepared earlier and held in a 40° water bath. This medium should contain, in mg/l: Murashige and Skoog salts; sucrose, 15,000; mannitol, 110,000; NAA, 0.6; kinetin, 0.1; thiamine · HCl, 10; pyridoxine · HCl, 10; nicotinic acid, 5; *myo*-inositol, 100; glycine, 2; and prewashed Phytagar, 8000. Mix protoplasts and medium gently, but thoroughly, and pour into 40-mm petri dish by following a procedure similar to cell plating. Incubate plated protoplasts as done with cells. Cell-wall regeneration will occur within several hours, and cell division will be observable within a few days. Transferable cultures will be obtained in 2–4 weeks.

### Comments

Some serious misconceptions regarding cultured plant cells need correction. Cultured plant cells are not undifferentiated as often claimed. They are larger, more vacuolated, lower in cytoplasm content, thicker walled, and smaller of nuclei than those of the embryo or apical meristem, where the more truly undifferentiated cells can be found. Very often phloem and xylem elements may be apparent among them. The cells in culture may lack organization and are therefore identifiable as being unorganized.

Uniformity also is not the usual trait among cells of callus or liquid suspension. Obvious variations include shape and size, cytoplasm and organelle contents, wall thickness, and ergastic deposits. More important, the cells within a culture may be genetically diverse. For example, variations in chromosome number have not been uncommon.[20] Subculturing, furthermore, has tended to accentuate the incidence of variants. Retention over long periods of a cell line's characteristics will require employment of cryogenic procedures.[21]

[18] H. Uchimiya and T. Murashige, *Plant Physiol.* **54,** 936 (1974).

[19] Solution containing 1% cellulysin, 0.2% macerase, and 0.7 *M* mannitol. The enzyme preparations are obtainable through Calbiochem, La Jolla, California. The pH of the solution is set at 5–7, and the solution should be filter-sterilized.

[20] F. D'Amato, *Int. Biol. Programme* **2,** 333 (1975).

[21] Y. P. S. Bajaj, *Physiol. Plant.* **37,** 263 (1976).

It is hazardous to assume that the behavior of cultured cells invariably reflects that of their progenitor plant. Frequently, the progenitor's characteristics are dependent on its organized state. For example, tobacco roots have been observed to carry on high rates of synthesis of the alkaloids, nicotine and anabasine, but cultured cells do not.[22] Epigenetic modifications also occur frequently and contribute to aberrant behavior of cultured plant cells and plants.[23]

When inoculating freshly prepared liquid media of suspension cultures it is important that cell density requirements are satisfied. Cell division will not occur below critical concentrations. For tobacco cultures, an initial count of 60,000 cells/ml might be satisfactory. If it is necessary to use cell concentrations below critical levels, nurse cells might be employed. The nurse cells first must have been rendered incapable of proliferation by irradiation or other means. Excessive cell numbers in the inoculum are also undesirable. They necessitate more frequent subculturing of liquid suspension cultures and present difficulty in distinguishing individual callus growth that arises in plate cultures.

A variety of nutrient compositions can be used successfully with plant cell cultures. But maximum advantage is attained only by assessing specific nutritional needs systematically. A first step might be the comparison of the more popular basal salt formulations. Subsequent tests may include vitamins, amino acids, carbohydrates, hormonal substances, and natural complexes. Data regarding salt formulations, vitamin and amino acid mixtures, hormones, and other nutrient constituents have been compiled recently.[24] Agar is still the best gelling agent, but its quality varies according to commercial source, grade, and lot. Regardless of the information contained on the product label, it is good practice to wash the agar before adding it to the nutrient medium. Often, rinsing 3 times with distilled water may be sufficient. This excludes much of the readily dissolved and undesirable substances.

Agitation of liquid suspension cultures can be accomplished by various devices. Reciprocating shakers and rotators may be used instead of gyratory shakers, and agitation rates can be varied from less than 1 to over 400 rpm. Successful liquid suspension cultures have also been achieved without agitation, by continuously bubbling air through the nutrient solution.

The question of pH of culture media has often been raised. Unfortunately, there have been few systematic investigations directed at the reso-

[22] M. L. Solt, R. F. Dawson, and D. R. Christman, *Plant Physiol.* **35,** 887 (1960).
[23] F. Meins and A. Binns, *Proc. Natl. Acad. Sci. U.S.A.* **74,** 2925 (1977).
[24] L. C. Huang and T. Murashige, *Tissue Cult. Assoc. Man.* **3,** 539 (1976).

lution of this problem. The common practice has been simply to set a nutrient medium's starting pH at between 5 and 6, assuming that this is within optimum range, and maintaining this pH range during the course of culture.

### Acknowledgments

This work was supported in part by the Elvenia J. Slosson Fellowship in Ornamental Horticulture and NSF Grant OIP 75-10390 awarded to T. M. We thank S. Hamman and S. Kearns-Sharp for typing the manuscript.

## [42] Lymphocytes as Resting Cells

*By* Shelby L. Berger

The human peripheral blood lymphocyte is an ideal system for the study of both resting and growing cell populations. As they are isolated from the blood, lymphocytes are quiescent. DNA synthesis measured in such cells reflects mainly ongoing repair processes since less than one cell in 5000 is engaged in mitosis.[1] *In vivo,* the cells are capable of surviving in the nondividing state for months or years.[2] In culture, resting lymphocytes can be maintained in excellent condition for several days after which they deteriorate and die. It should be emphasized that these are *physiologically* nongrowing cells; despite the presence of autologous plasma and a complete medium such as Eagle's minimal essential medium (MEM) or RPMI-1640, the cells remain in $G_0$, outside the mitotic cycle. Since most of our knowledge of cellular processes derives either from continuously dividing cells adapted for cell culture, from density-inhibited cells, or from arrested cells deprived of essential nutrients, it is essential to expand our understanding by including the great bulk of normal cells which resemble none of these examples. The lymphocyte provides such an opportunity.

When confronted with a stimulus, the resting lymphocyte undergoes changes in virtually every aspect of cellular metabolism culminating in DNA synthesis and mitosis. *In vivo,* the challenge is an immunological one, and the responding cells are those having specific mechanisms for antigenic recognition. *In vitro,* growth induction of a large number of cells

[1] Reviewed by H. L. Cooper, *in* "Drugs and the Cell Cycle" (A. M. Zimmerman, G. M. Padilla, and I. L. Cameron, eds.), pp. 137–194. Academic Press, New York, 1973.

[2] N. B. Everett and R. W. Tyler, *in* "Formation and Destruction of Blood Cells" (T. Greenwalt and G. Jamieson, eds.), pp. 264–283. Lippincott, Philadelphia, Pennsylvania, 1970.





ISBN 0-12-181958-2

can be elicited by the addition of any one of a number of nonspecific mitogenic materials such as phytohemagglutinin or concanavalin A to the culture medium. Although the latter compounds do not depend on any known prior sensitization of the cells, the mitogenic response *in vitro* is believed to resemble that *in vivo* in all other particulars.[1]

Lymphocytes are also activated when cells from more than one donor are mixed.[3,4] This mixed lymphocyte reaction is of interest to immunologists but presents an impediment to cellular biochemists desiring large numbers of resting cells for metabolic studies. Moreover, metabolic activity differs in newly isolated cells from different donors so that, in practice, an experiment and its control must be performed with lymphocytes from the same unit of blood. Therefore, optimizing the cell yield from a single donor is essential.

## Preparation of Lymphocytes

The method of purification presented here is a modification of Cooper's procedure[5] incorporating improvements and simplifications developed during the past few years. Changes in the catalog numbers of essential apparatus also have been included for the convenience of those desiring to reproduce this scheme exactly.

### *Preliminary Isolation of Leukocytes*

The initial collection of the blood and the preliminary separation of lymphocytes from erythrocytes require equipment available from Fenwall Laboratories.[6] Items and their catalog numbers are as follows: "Blood-Pack," 500 ml, containing heparin for the collection of whole blood (4R0601); "Transfer Pack Unit," 150 ml with coupler (4R2001); "Plasma Transfer Set" with 2 couplers (4C2243); plasma extractor (4R4414). For centrifuging large buckets capable of holding the 500-ml Blood-Pack, a low-speed centrifuge, such as the PR-6000 (International) fitted with a No. 276 or No. 981 rotor, or a RC-3 (Sorvall) with an HG-4L rotor, is also necessary.

Blood from a fasting donor is collected in the 4R0601 pack and after thorough mixing is centrifuged at 1500 rpm (approximately 400 *g*[7]) for 5

[3] B. Bain, M. R. Vas, and L. Lowenstein, *Blood* **23,** 108–116 (1964).

[4] B. Bain and L. Lowenstein, *Science* **145,** 1315–1316 (1964).

[5] H. L. Cooper, this series, Vol. 32, [62].

[6] Fenwall Laboratories, One Baxter Parkway, Deerfield, Illinois.

[7] The *g* values are rounded to one significant figure and refer to the force at the center of the centrifuge tube.

min at room temperature. On removing the blood pack from the bucket, a dense layer of erythrocytes can be seen with overlying plasma. The lymphocytes are found as a diffuse band, suspended in the lower region of the plasma layer. Without disturbing the bands, the unit is carefully hung in an open plasma extractor by the holes provided in the blood pack and connected to the 4R2001 Transfer Pack by plugging the attached coupler into the center port of the blood pack. The blood pack is also connected by the side port to a receiving vessel, usually a sterile 32-ounce prescription bottle with a foil cover, by means of the 4C2243 Transfer Set. The foil can be crimped around the coupler to hold it in place near the rim of the bottle. Sterile technique is used throughout, and all steps are performed at room temperature unless noted otherwise. With the tubing of the transfer set closed, 50 g of plasma are expressed from the unit into the transfer pack by releasing the plasma extractor. Then, with the transfer pack tubing squeezed shut, the remaining plasma and the region of the interphase are expressed into the bottle. To insure complete recovery of the lymphocyte-rich layer, several milliliters of red cells are expressed into the bottle as well. Afterwards, 90–100 g of the remaining red cell layer in the blood pack are added to the transfer pack. As a result of this procedure, most of the lymphocytes are confined to the prescription bottle while the remainder, which were either trapped in the red cell layer or suspended in the plasma, are in the transfer pack. Both the blood pack and the transfer pack are mixed well and centrifuged again under the same conditions to reextract the plasma–erythrocyte interphase. This step necessitates removing the transfer set coupler from the receiving bottle, but not from the side port of the blood pack, and capping it with the original covering to maintain sterility. Other tubing connections are clamped but left in place, and the tubing itself is tucked between the two packs in the bucket during sedimentation. After the centrifugation step the aim is to combine the lymphocytes in a single container. If plasma is visible in the blood pack when subsequently squeezed by the plasma extractor, it is collected into the receiving bottle. The plasma layer and a few milliliters of erythrocytes from the transfer pack are also expressed into the prescription bottle. As a result, virtually all the lymphocytes, in partially purified form, are recovered. To insure against clotting, 5000 U of heparin (1 ml of Liquaemin Sodium "50"[8]) are added at this time.

Many investigators preparing lymphocytes have observed the formation of a "buffy coat" in the discarded blood pack after several hours of standing. When examined microscopically, this band was found to be rich in granylocytes and almost devoid of lymphocytes. Thus the preliminary

[8] Available from Organon Inc., West Orange, New Jersey.

separation described above removes some, but not all, of the contaminating granulocytic leukocytes.

*Nylon Column Filtration*

Separation of lymphocytes from polymorphonuclear leukocytes is accomplished by adsorbing the granulocytes on nylon fibers. The nylon fiber, 3 denier, 1.5 inch, type 200 nylon staple, is obtained from Dupont. The purification is carried out in a waterjacketed column, 60 cm in length by 2.5 cm ID, maintained at 37°. In order to retain sterility of the fractionated blood, it is helpful if the bottom of the column is made to fit the needle adapter of a No. 4C2240 Plasma Transfer Set with coupler and needle adapter (Fenwall). The column effluent can then be conducted to a 32-ounce sterile prescription bottle with the sterile tubing; the coupler end can be crimped in place with foil as before.

Nylon, as obtained from Dupont, is compressed and clumped, and contains a fabric finish that must be stripped. The fiber mass is first teased apart by hand into a fluffy puff of single strands. A column is packed with 3.5 g of the nylon so that the height is approximately 13 cm. This is best accomplished by rinsing the inside of the column with water to wet the surface and pushing the nylon to the bottom with a stick. The fabric finish then can be removed by soaking the teased nylon *in situ* for 1 hr in 1 *N* HCl and washing it for at least 3 hr with running distilled water. Excess water is suctioned out of the nylon, and the column, with its ends covered with foil, is autoclaved for 20 min. Prolonged drying is not recommended because the nylon is destroyed by dry heat. If the column is to be used immediately, it is cooled and made ready by equilibration at 37°, connection to the receiving vessel with the aforementioned transfer set, and application of approximately 100 ml of Eagle's MEM to wash out condensate. Traces of the wash solution adhering to the nylon fiber are not detrimental, but soggy nylon should be avoided. It is advantageous to prepare sterile nylon columns in advance.

The lymphocyte-rich fraction is warmed to 37°, mixed thoroughly, and poured into a waiting nylon column of the type described above. With the bottom outlet clamped, the cell suspension is incubated for 5 min before passing the entire sample through the nylon at a flow rate of 4 ml/min. With the apparatus suggested, the flow rate is 20–25 drops every 15 sec. In this laboratory, the rate of flow has been controlled by first drawing a partial vacuum above the surface of the blood in the column with a Pharmacia P-3 peristaltic pump during the 5-min incubation, reversing the pump, opening the outlet, and finally adjusting the speed of the pump to effect the desired flow rate. A tight-fitting sterile stopper with a sterile cotton plug is required to couple the pump to the column aseptically.

When the cell suspension has been thoroughly drained or pumped from the column, it is held at 37° while the nylon is rinsed. Approximately 100 ml of warmed MEM are passed through the column at the same rate of flow to release trapped cells. The wash solution is reserved in a separate vessel.

*Removal of Platelets*

The partially purified cell suspension is freed of platelets by differential centrifugation at 1000 rpm (200 *g*) for 10 min. Sterile, plastic conical centrifuge tubes with caps, holding 50 ml, such as Corning (No. 25335), Falcon (No. 2074), Kimble (No. 58331), or the equivalent are satisfactory and can be centrifuged in an HL-8 (Sorvall) or a No. 269 (International) rotor. The plasma layer, containing platelets, is carefully decanted or removed with a sterile pipette without mixing the two layers. The deep red pellet contains the lymphocytes. At a later time the platelets are separated from the plasma centrifugally (2000–3000 *g*) in 50-ml sterile tubes at 20° for 15–20 min. Although room temperature should be acceptable for this process, in practice, the plastic tubes become deformed at high speeds in a warm centrifuge. The supernatant fraction, composed of cell-free plasma, is filtered through a Nalge, 0.45-$\mu$m filter[9] to insure sterility and retained for use as a supplement in the final culture medium.

*Separation of Lymphocytes from Erythrocytes*

The cell pellets, in their original tubes, are fractionated immediately by sedimentation through a dense medium. Toward this end, they are each diluted to 35 ml with the reserved wash solution and, if necessary, with fresh MEM. Each tube is then underlaid with 13–14 ml of a mixture of Ficoll and diatrizoate salts (density 1.077 $\pm$ 0.001 g/ml) (LSM solution[10]), which is kept refrigerated until required. This is best accomplished by filling a 10-ml pipette nearly to the sterile cotton plug and slowly expelling the liquid with the pipette tip held near the bottom of the tube. A Pipet-aid[11] simplifies this task. The resultant two-phase system is centrifuged at 1500 rpm (500 *g*) for 20 min at room temperature to effect cell separation.

[9] Available from Nalgene Labware Division, Rochester, New York.

[10] LSM stands for Lymphocyte Separation Medium and can be purchased from Litton Bionetics, Inc., 5510 Nicholson Lane, Kensington, Maryland.

[11] Pipet-aid is an automatic pipetting device sold by Drummond Scientific Co., Broomall, Pennsylvania.

The dense red cells penetrate the LSM layer and sediment to the bottom. Lymphocytes band at the interphase while granulocytes, if present, are found as a whitish layer immediately above the erythrocytes. Contaminating platelets band with lymphocytes.

The lymphocyte layer is recovered with a pipette and separated from LSM by centrifugation at 1500 rpm for 15 min. The final purified preparation is resuspended in 30 ml of MEM and combined in a single tube. The entire procedure takes 4 hr, once the blood is drawn.

### *Yield*

For counting, approximately 0.2 ml are withdrawn to a spot plate, and an aliquot is diluted 20-fold with 0.02% crystal violet in 1% acetic acid. A white blood cell diluting pipette is convenient for this purpose, and a hemocytometer is required for quantitation.

The yield from one unit of blood (1 pint; 500 ml) is approximately $4-12 \times 10^8$ small lymphocytes. Immediately after purification polymorphonuclear leukocytes account for 0–2% of the total purified leukocyte population. However, these contaminating cells do not survive well in culture and are gradually eliminated during incubation. An occasional monocyte is evident on microscopic examination; such cells are essential for mitogen stimulation of lymph node lymphocytes from guinea pigs[12] and are believed to be necessary for activating other types of lymphocytes as well.

### *Culture Conditions for Human Lymphocytes*

In this laboratory the purified cells are usually cultured at $2 \times 10^6$ per milliliter in MEM modified for suspension cultures; higher concentrations, e.g., $10^7$ per milliliter, can be used for short incubations. The MEM is supplemented with 10% autologous plasma, 100 $\mu$g/ml streptomycin, 100 U/ml penicillin, 4 m*M* glutamine, 0.01 *M* HEPES buffer,[13] and 0.4 m*M* nonessential amino acids. The cells are incubated without stirring at 37° in sealed recumbent prescription bottles. A 32-ounce bottle with a volume of approximately 950 ml holds 200–275 ml of cell suspension. It is important to use a smaller bottle for smaller volumes in order to maintain the depth of the cell suspension at about 1.5–2 cm. Incubation for 15–18 hr is recommended before commencing biochemical studies of resting cells.

[12] D. L. Rosenstreich, *in* "Mitogens in Immunobiology" (J. J. Oppenheim and D. L. Rosenstreich, eds.), pp. 385–398. Academic Press, New York, 1975.

[13] HEPES (*N*-2-hydroxyethylpiperazine-*N*-2-ethanesulfonic acid) buffer is available as a 1 *M* solution in saline from Microbiological Associates, Walkersville, Maryland.

If the cells are to be stimulated, they can be incubated with mitogens immediately. A dose of 5 $\mu$g/ml phytohemagglutinin or 10 $\mu$g/ml concanavalin A is optimal for human peripheral blood lymphocytes.[14] Since the cells, purified by the procedure reported here, undergo vigorous mitogen-induced proliferation, the method obviously does not completely deplete monocytes and macrophages. A lower cell density of $2-3 \times 10^5$ per milliliter has been recommended for growing lymphocytes.[15] In this laboratory a starting concentration of $2 \times 10^6$ per milliliter has been used routinely, and the activated lymphocytes have been cultured for as long as 10 days. During this interval they are resuspended in fresh medium every 48 hr or diluted with 50% more fresh medium on the same schedule. Nevertheless, the cultures tend to become acid and survive in better condition if maintained with loose caps in an environment equilibrated with 8.5% $CO_2$.

The response to mitogens is asynchronous. For example, phytohemagglutinin stimulates one, two, or three rounds of cell division in some cells and is toxic to others.[1] Thus the total number of cells rarely increases. However, direct determinations of cell number are complicated by a somewhat variable tendency to adhere to glass between 6 and 10 hr after incubation with phytohemagglutinin, and by cell aggregation later in the activation process.

*Impurities*

The major contaminants in lymphocyte preparations are platelets and erythrocytes. Both are destroyed by the dilute acetic acid in the staining solution used for counting the cells but can be viewed microscopically in unstained samples. On the average, one erythrocyte is seen per 20 lymphocytes after incubating the cells overnight.

Platelet contamination is more difficult to assess. In this laboratory, platelets account for 8–25% of the packed cell volume. However, such contamination can be reduced by repeating the centrifugation step in 100% plasma. For best results the cell pellets, containing predominantly lymphocytes and erythrocytes, should be resuspended in a small amount of medium while the recovered plasma is cleared of platelets by sedimentation. When the plasma is ready, the impure lymphocytes are mixed with

[14] For a more complete list of mitogens and their properties, see B. A. Cunningham, B.-A. Sela, I. Yahara, and G. M. Edelman, *in* "Mitogens in Immunobiology" (J. J. Oppenheim and D. L. Rosenstreich, eds.), pp. 13–30. Academic Press, New York, 1975, or see N. Sharon, *ibid.* pp. 31–41.

[15] J. L. Bernheim and J. Mendelsohn, *in* "Regulatory Mechanisms in Lymphocyte Activation" (D. O. Lucas, ed.), pp. 479–505. Academic Press, New York, 1977.

the plasma and separated from remaining platelets by differential centrifugation as indicated above.

In the procedure detailed by Cooper,[5] platelets are removed by agglutination with adenosine diphosphate.[16] This method is capricious in its usefulness; it ranges from highly effective to completely ineffective depending on the donor. If platelet-free cultures are essential, defibrination with glass beads is recommended. Although some loss of lymphocytes from clotted blood cannot be avoided, the fibrin clots trap and bind virtually all of the platelets.[17]

### Comments

The inconvenient and tedious preparation of nylon can be circumvented by purchasing this material from Fenwall. The Leuko-Pak (No. 4C2401) contains sufficient teased nylon for 3–4 units of blood. There are two disadvantages: processed nylon is considerably more expensive than the cruder product from Dupont, and one becomes dependent on the availability of Leuko-Paks, which can be out of stock for months.

The separation of erythrocytes by LSM solution has been adopted as a convenience. The Ficoll-Hypaque method as described by Perper *et al.*[18] and used by Cooper[5] is equally satisfactory.

The composition of the lymphocytes produced by nylon-column filtration has been a matter of conjecture. It has been suggested that nylon may deplete the population of B cells (bone marrow-derived lymphocytes) leaving predominantly T cells (thymus-derived lymphocytes). When the distribution of the two types of lymphocytes was measured by rosetting with sheep erythrocytes treated with S-2-aminoethylisothiouronium bromide hydrobromide,[19] approximately 85% were nominally T cells in our preparations. Studies of the composition of lymphocytes purified without recourse to nylon gave similar results.[19] There is therefore no evidence for depletion of B cells during the brief exposure to nylon which forms part of this procedure for purifying lymphocytes.

A major disadvantage of the human peripheral blood lymphocyte in research is the difficulty of obtaining large amounts of material. For most purposes, pooled cells are unsatisfactory even if mitogen treatment is anticipated. This problem is solved in part by leukophoresis, an alternate

[16] A. Gaarder, J. Johnsen, S. Laland, A. Hellem, and P. A. Owren, *Nature (London)* **192,** 531–532 (1961).

[17] S. Niewiarowski, E. Regoeczi, G. J. Stewart, A. Senyi, and J. F. Mustard, *J. Clin. Invest.* **51,** 685–700 (1972).

[18] R. J. Perper, T. W. Zee, and M. M. Mickelson, *J. Lab. Clin. Med.* **72,** 842–848 (1968).

[19] M. E. Kaplan and C. Clark, *J. Immunol. Methods* **5,** 131–135 (1974).

yielding large numbers of white cells which must then be purified as described. However, autologous plasma sufficient for culturing the entire yield is not provided so that pooled AB serum must be used as a substitute. The effect of foreign substances in serum on the cellular biochemistry of cultured cells has not been fully investigated.

# [43] Macrophages

*By* DOLPH O. ADAMS

Macrophages are currently studied *in vitro* in almost all fields of biology. The mononuclear phagocyte system, which includes macrophages, is a host-wide system of phagocytic cells with similar properties.[1] Mononuclear phagocytes arise in the marrow, circulate briefly in the blood as monocytes, and immigrate into the tissues and inflammatory foci where they mature into macrophages. Macrophages can be further stimulated to develop altered function and metabolic characteristics—a state often termed activation. Activation in this broad sense can be induced by many stimuli and is characterized by the presence of certain markers[2] as well as by increases in size, adherence, secretory capacity, content of lysosomes, and various functions such as phagocytosis and chemotaxis.[1] Since most activated states are not identical functionally, activated macrophages should be identified by both the eliciting stimulant and the altered capacity tested.

Macrophages as cultivated cells offer the following advantages: They are easy to obtain, can be cultured in relatively pure form, are primary cultures, and are available in numbers sufficient for analytical biochemical manipulations. On the other hand, macrophages do not generally replicate in culture, are relatively short-lived, and may be difficult to obtain in numbers sufficient for preparative biochemistry. It is important to note that macrophages also are very sensitive to small changes in their environment and are thereby modified considerably from their native state *in vivo,* even when delicately handled and observed after very short periods of culture.

[1] D. O. Adams, *Am. J. Pathol.* **84,** 163 (1976).

[2] P. Edelson and Z. A. Cohn, *in* "In Vitro Methods in Cell Mediated and Tumor Immunity" (B. R. Bloom and J. R. David, eds.), pp. 333–340. Academic Press, New York, 1976.




ISBN 0-12-181958-2

## Sources of Macrophages

### *General*

Macrophages can be obtained from blood, lung, spleen, liver, and the peritoneal cavity.[(2–5)] However, mononuclear phagocytes from most of these sources must be used in short-term experiments since they do not survive much more than 24–48 hr in culture. Alveolar macrophages from all sources and peritoneal macrophages from small rodents other than mice are particularly difficult to maintain. The spleen does not yield many macrophages, and those obtained are often contaminated by other adherent cells. Hepatic macrophages are considerably altered by the techniques necessary for their isolation. Consequently, peripheral blood monocytes and peritoneal macrophages are the commonly employed sources of mononuclear phagocytes.

### *Monocytes*

Monocytes are generally obtained from humans, but equine monocytes have been studied.[3] Monocytes are separated from the blood by differential centrifugation and subsequent adherence to culture vessels. Approximately $10^8$ monocytes can be obtained from 500 ml of blood. This represents the most readily accessible source of human mononuclear phagocytes. However, these cells are extensively manipulated before plating, are difficult to culture, and may be contaminated with large numbers of platelets. A detailed protocol for the culture for human monocytes is available.[2]

### *Peritoneal Macrophages*

The peritoneal cavity offers a ready source of mononuclear phagocytes. The unstimulated peritoneal cavity of mice contains usable numbers ($2–3 \times 10^6$) of resident macrophages. These may be washed out directly and placed into culture. The resident cells should not possess any of the characteristics of inflammatory or activated macrophages.[2] If they do, the likelihood that the mice are infected should be strongly considered.

[3] Z. A. Cohn, this series, Vol. 32, pp. 758–765.

[4] G. D. Wasley and R. John, *in* "Animal Tissue Cultures" (G. D. Wasley, ed.), pp. 101–137. Butterworth, London, 1972.

[5] A. E. Stuart, J. Habeshaw, and A. E. Davison, *in* "Handbook of Experimental Immunology" (D. M. Weir, ed.), pp. 24.1–24.36. Blackwell, Oxford, 1973.

tributing to the purity of the macrophages. The resultant macrophages should be used directly in the separation vessel if possible, since they are extremely difficult to remove. Despite these disadvantages, the adherence technique yields highly purified macrophages that can be used within hours of removal and that have been subjected to little manipulation. Most subsequent experimental manipulations can be conducted without removing the macrophages by appropriate selection of the vessel used for purification.

## Identification of Macrophages

Precise identification of a given population of cells as mononuclear phagocytes is vital because of the number of other cells that resemble them morphologically, particularly lymphocytes. Mononuclear phagocytes are defined as adherent cells that have a characteristic morphology, the capacity for extensive phagocytosis, and bear receptors for the activated third component of complement (3) and for the activated C-terminal portion of immunoglobulin.

### *Morphology and Histochemistry*

The morphology of mononuclear phagocytes is well defined.[10] A population of spread adherent cells in the size range of 20–50 $\mu$m and having characteristic nuclei and ruffled cytoplasm can be tentatively identified as macrophages. Resistance to removal with trypsin further strengthens this tentative identification.[11] Mononuclear phagocytes contain abundant quantities of a nonspecific esterase, the presence of which distinguishes these cells from other leukocytes.[12] It is worth emphasizing that differences in the nonspecific esterase, as with most histochemical procedures, represent a quantitative rather than a qualitative difference between macrophages and other cells.

### *Phagocytic Uptake*

Extensive phagocytic uptake by a mononuclear cell population is clear evidence of a lineage in the mononuclear phagocyte system. Procedures for determining phagocytic uptake are well detailed.[7] A test particle such as starch, latex beads, or erythrocytes plus an opsonin (fresh serum or specific antibody) are incubated for 30–60 min at 37° with the mac-

[12] L. T. Yam, C. Y. Li, and W. H. Crosby, *Am. J. Clin. Pathol.* **55,** 283 (1971).

rophages. The cells are then washed and the percentage of *mononuclear* cells containing five or more particles *within the cytoplasm* (not adherent to the cells) is determined. Lab-Tek chambers (Lab-Tek Products, Napierville, Illinois) or coverslips suspended over slides with a depressed well (Arthur H. Thomas, cat. no. 6688-D20) are particularly convenient for this purpose.

### *Markers*

Mononuclear phagocytes are distinguished by the presence of surface receptors for both $C3_b$ and Fc.[13,14] These may be demonstrated by use of appropriately coated red cells. Detailed protocols for these procedures are available.[13,14]

## Culture Techniques

The selection of appropriate culture conditions for macrophages depends upon the intent of the particular experiment and usually represents a balance between multiple factors. First, basal media are less expensive and will support viability of the macrophages but may not permit full expression of all functional capabilities. Second, macrophages are easily stimulated by alterations in their environment such as enriched media, high concentrations of serum, trace concentrations of endotoxin, or phagocytosis of other cells. Such endogenous stimuli can obscure differences between experimental and control cultures. Third, the inherent inconsistencies of tissue culture must be considered. In our hands, for example, Dulbecco's minimum essential medium (MEM) supports accumulation of acid hydrolases much better than does Eagle's MEM, whereas the reverse is true for the expression of nonspecific cytotoxicity.

### *General Requirements*

Macrophages usually are cultured in moist air containing 5% $CO_2$. Precise regulation of $CO_2$ content with a controller may be advantageous for critical experiments. Acidity is common with cultured macrophages, and pH should be checked daily. The pH can be adjusted to 7.2 by addition of sterile sodium bicarbonate (sterilized by filtration).

Plastic vessels coated for tissue culture are generally suitable for

[13] J. Michl, D. J. Ohlbaum, and S. C. Silverstein, *J. Exp. Med.* **144,** 1465 (1977).

[14] E. M. Shevach, E. S. Jaffe, and I. Green, *Transplant. Rev.* **16,** 3 (1973).

growth of macrophages. We have found clusters of four 60-mm wells useful for large numbers ($10 \times 10^6$/well) of macrophages and plates containing twenty-four 16-mm wells useful for small numbers ($5 \times 10^5$/well). These plates must be maintained in a moist atmosphere to prevent evaporation. T-flasks do not have this disadvantage, but are difficult to wash vigorously for purification and difficult to scrape completely when removing the macrophages. Small culture chambers placed over glass slides (Lab-Tek) are extremely useful for assessing morphology and for autoradiography.

Macrophages generally do best in culture at densities of from $2\text{–}4 \times 10^5$ adherent macrophages per $cm^2$. The number of suspended leukocytes to be added to a culture to achieve this density can be roughly calculated by multiplying the number of macrophages per milliliter of suspension by the proportion of macrophages expected to adhere, i.e., by $\frac{1}{2}$ to $\frac{2}{3}$ if the macrophages are activated and by $\frac{1}{4}$ to $\frac{1}{3}$ if they are not.

Cultures are washed free of nonadherent cells after 3 or 4 hr, except in the case of human monocytes, which are washed after 1 hr. Leaving the cultures intact for 24 hr generally improves the number of adherent cells and the purity of the resulting population.

*Media*

Medium 199 and Eagle's minimal essential medium are useful basal media. These are generally supplemented with streptomycin, penicillin, and *fresh* glutamine. HEPES buffer, 5–10 m*M*, aids in stabilizing pH but may be toxic to some cultures.

NCTC-135, MEM alpha, RPMI-1640, and Ham's F12 are richer media that have been successfully employed in culturing macrophages.

Basal media can be selectively enriched by addition of a variety of substances including sodium pyruvate (0.11 mg/ml), ascorbic acid (50 $\mu$g/ml), other vitamins, and nonessential amino acids.[15]

Serum-free medium may be necessary for some experiments. A variety of supplements have been tried; we have had the most success with lactalbumin hydrolysate (Grand Island Biological Co.). The cell-free medium of Newman-Tytell is particularly useful, especially when supplemented with 10 m*M* HEPES. We have recently begun using the supplement described by Guilbert and Iscove of selenite, transferrin, lecithin, and albumin[16] and have found that it improves the viability and maturation of macrophages when added to serum-free cultures.

[15] C. Waymouth, *Int. Rev. Cytol.* **3**, 1 (1954).

[16] L. J. Guilbert and N. N. Iscove, *Nature (London)* **263**, 594 (1976).

*Serum*

Most cultures of macrophages are supported by 10–40% serum. Fetal calf serum is routinely employed, but equine and newborn calf serum also have been used. Newborn calf serum may contain antibodies that stimulate maturation of the macrophages.[3] Human macrophages usually are cultured with human serum, preferably autologous. For murine macrophages, we routinely used 10% heat-inactivated fetal calf serum.

Serum can vary considerably from lot to lot in its ability to support both survival and function of macrophages. For critical experiments, serum may have to be screened by testing the ability of various lots to support the function being tested. Furthermore, certain components of serum are apparently labile. In critical experiments, we obtain frozen serum shipped in Dry Ice from a previously screened lot. All is gently thawed at 37°, heat-inactivated with constant swirling for 30 min at 56°, apportioned into small aliquots, and stored at −20°. On the day an experiment begins, one container of serum is gently thawed and used. Any left over serum is not reused for an experiment, not even the next day, but is saved for routine tissue culture.

*Detachment of Macrophages*

Detaching macrophages that have already adhered to a culture vessel particularly if the macrophages are activated, is extremely difficult. Neither trypsin, cold, nor chelating agents remove macrophages well, and it is generally best to conduct experimental manipulations in the culture vessel used for purification. Gentle scraping of macrophage cultures with a rubber policeman will remove the cells, approximately ½ of which may be viable. Macrophages may be detached partially by use of a local anesthetic.[17] Purified Lidocaine, without preservatives (Astra Pharmaceuticals, Worchester, Massachusetts) is made to 360 m*M* in phosphate-buffered saline (PBS) and adjusted to pH 6.6 with 1 *M* NaOH. The stock is diluted in medium to 12 m*M*, and macrophages are incubated in the medium containing Lidocaine for 5 min at 37°. The macrophages should then appear rounded and are easier to remove, although extensive losses may be incurred.

*Observation and Viability of Cultures*

Cultures are observed daily with an inverted phase microscope. Healthy cultures will appear as monolayers of plump or stellate, phase-

[17] C. Nathan, *J. Immunol.* **118,** 1612 (1977).

dense, well-spread, and closely approximated cells. Unhealthy cultures will be sparsely populated with rounded, unspread, and phse-lucent cells and will contain many floating cells.

More stringent criteria for determining viability include testing the ability to phagocytose particles, to secrete lysozyme, and to hydrolyze fluorescein diacetate.[18]

### *Long-Term Cultures*

Macrophages in culture are generally nonreplicating cells except under special circumstances.[19] Murine peritoneal macrophages and human monocytes under well-controlled conditions can survive for periods of up to 2 or 3 weeks.

## Preparation of Macrophages for Study

### *Morphologic Studies*

For study of macrophages in exudates or suspension cultures, an excellent method is to prepare them in a Cytocentrifuge (Shandon Southern Instrument Co., Sewickeley, Pennsylvania) and stain the resultant smears with a commercially available Wright's stain (Diffquik, Arthur Thomas Co.). This stain is also convenient for macrophages cultivated on coverslips or in Lab-Tek chambers.

Phase microscopy is extremely useful for more detailed examinations. Macrophages are particularly advantageous to study by this method, since they spread after adherence and thus reveal their cytoplasmic contents in the thinned cytoplasm. Cultures of spread macrophages on coverslips are washed in PBS and fixed 5 min in 4% osmium tetroxide in S-Collidine buffer while *in a fume hood*. The coverslips are immersed over PBS in a trough whose *shallow* walls of silicone grease are applied to a glass slide with a syringe and needle. The resultant inverted wet mount is studied under the phase microscope and can be preserved by coating it with clear nail polish.

### *Biochemical Preparation*

Studying the intracellular contents of macrophages poses problems, since macrophages are relatively resistant to lysis. For example, mac-

[18] A. C. Allison, *in* "*In Vitro* Methods in Cell-Mediated and Tumor Immunity" (B. R. Bloom and J. R. David, eds.), pp. 395–404. Academic Press, New York, 1976.

[19] M. Virolainen and V. Defendi, *Wistar Inst. Symp. Monogr.* **7,** 67 (1967).

rophages can remain intact after multiple cycles of freezing and thawing. We generally employ the detergent Triton X-100. We add ice-cold PBS containing 0.2% Triton X-100 to washed cultures of macrophages and incubate the cultures with the detergent for 30 min. The entire procedure must be conducted on ice to prevent liberated lysosomal enzymes from destroying the desired cellular constituent. After the incubation, macrophages are completely removed from the culture vessel by scraping with a rubber policeman. The preparation of organelles from macrophages has been described in detail.[20]

The study of products secreted by macrophages is conducted on medium conditioned by growing macrophages in it.[6] Since many secreted products, e.g., neutral proteases, are inhibited by serum constituents, the cultures are frequently conducted in serum-free medium. Siliconized glassware is useful because many of the secreted products are extremely labile and adhere to glass. Daily, the cultures of macrophages are aspirated and the aspirates placed in siliconized conical centrifuge tubes on ice and immediately spun 10 min at 10,000 *g* at 4°. If the desired secretory product is present in low concentration, the conditioned medium can be lyophilized or dialized against an adsorbent such as Acquacide II (Calbiochem, San Diego, California) to concentrate the constituent. In our hands, concentration in a stirred cell against an appropriately sized UM membrane (Amicon Inc., Lexington, Massachusetts) has proven the most useful and satisfactory method.

Specific activity of a given constituent within macrophages can be expressed in relation to either protein or cell number. Determination of cell number is generally advantageous because cultivated macrophages usually produce large amounts of protein in culture and thus specific activity may fall in the face of a rising total amount of a particular enzyme.[21] Determining cell number by hemocytometer count is often inaccurate because a large but variable number of macrophages is lysed while scraping them from the culture vessel. Determination of DNA content provides an accurate, reliable, and sensitive method for quantifying cell number.[22]

### Continuous Cell Lines

Several lines of continuously replicating cells having the properties of macrophages are now available (for review, see Defendi[23]). These neo-

[20] Z. Werb and Z. A. Cohn, this series, Vol. 31, P. A, pp. 339–345.

[21] Z. A. Cohn and B. Benson, *J. Exp. Med.* **121,** 153 (1968).

[22] S. Cookson and D. O. Adams, *J. Immunol. Methods* (1978) in press.

[23] V. Defendi, *in* "Immunobiology of the Macrophage" (D. S. Nelson, ed.), pp. 275–286. Academic Press, New York, 1976.

plastic murine lines have the morphology of macrophages and are phagocytic. Depending on the particular line, these cells may express various other functions generally associated with macrophages. The macrophage-like lines are generally easy to grow and are passaged without difficulty, since they detach readily from the culture vessels. They offer the obvious advantage of providing large numbers of similar cells with minimal difficulty.

## Appendix: Protocol for Culturing Stimulated Macrophages

### *Animals*

Two C57 B1/6J mice of either sex (Jackson Laboratories, Bar Harbor, Maine) weighing 25–30 g each are used.

### *Supplies*

Small, lidded plastic bucket, containing one-half pound of Dry Ice
Small dissecting board, covered with sterile barrier
70% ethanol
Two small forceps with teeth
One 3 inch dissecting scissors
Sterile, plugged, Pasteur pipettes with rubber bulbs
Two sterile 23-gauge 1 inch needles
Two sterile 10-ml syringes
One sterile 50-ml conical *polypropylene* centrifuge tube with cap
30 ml of ice-cold washout medium [Eagle's minimal essential medium with Earle's salts (Grand Island Biological Co., Grand Island, New York, cat. no. F-11); to this, add 10 units of heparin per milliliter]
Ice bucket filled with ice
3 cluster dishes with four 60-mm wells of tissue culture plastic (Falcon Plastics, Los Angeles, California)
250 ml of sterile Eagle's MEM with Earle's salts to wash plates
Hemocytometer
Gilson or similar pipettes, dispensing 20 and 200 $\mu$l
10 ml sterile plugged pipettes
Glass slides
Coverslips
2 ml of sterile, Brewer's thioglycolate broth (Difco Manufacturing Co., Detroit, Michigan, cat. no. B-236; prepare and store according to manufacturer's instructions)

150 ml of complete tissue culture medium to culture macrophages: Eagle's minimal essential medium with Earle's salts (Gibco no. F-11) plus 10% heat-inactivated fetal calf serum plus $2.5 \times 10^4$ units of penicillin per 100 ml plus 12.5 mg of streptomycin/100 ml plus 29.2 mg of fresh L-glutamine (Sigma Chemical Corp., St. Louis, Missouri)/100 ml. Basic medium can be made in advance, but the final assembly of medium plus serum, antibiotics, and supplements should be prepared on the day of the experiment.

*Stimulation of Peritoneal Exudate Cells (PEC)*

Three days before the experiment, inject each mouse sterilely with 1 ml of thoglycolate broth intraperitoneally with a 23-gauge needle. Be sure that the injection does not perforate or enter the intestines.

*Collection of PEC*

Lay out all equipment. Load two syringes with 10 ml of chilled washout medium and place them and the centrifuge tube on ice. Note that the entire collection and culture procedure is to be done under sterile conditions.

To kill the mouse, place it in the plastic bucket containing the Dry Ice. Suspend it over the ice by its tail placed under the rim of the closed lid. After 1 min, remove the dead mouse.

Wash the skin of the abdomen of the mouse vigorously with 70% ethanol. Gently pick up the abdominal skin with a pair of toothed forceps in each hand and gently manipulate the skin up and down to separate it from the peritoneal wall. Grasp the skin with both forceps and gently tear it, making sure not to tear the peritoneal wall. Pull the torn skin caudally and rostrally until the carcass of the mouse is completely exposed. Do not tear the peritoneal wall or the thorax.

Rewash the peritoneal wall with 70% alcohol. Insert the needle into the peritoneal cavity in the midline and inject the 10 ml of washout medium. Massage the flanks of the animal for several seconds. Elevate the shaft of the needle to create a small tent just below the xyphoid process and aspirate the injected fluid. At least 9 ml should be recovered.

Remove the needle from the syringe and gently introduce the collected fluid into the centrifuge tube. Cover the centrifuge tube. Wash the second mouse similarly. Add the fluid from the second mouse to that of the first and mix gently. Remove a few drops of the fluid and prepare smears on the Cytocentrifuge, keeping the centrifuge tube on ice.

*Culture of Macrophages*

Immediately centrifuge the collected peritoneal fluid for 10 min at 250 *g* at 4°. Carefully aspirate the supernant from the resultant cell pellet. Add 10 ml of complete tissue culture medium. Gently disperse the cell button with a cotton-plugged sterile Pasteur pipette and take a small sample. Return the capped centrifuge tube to the ice bucket.

Determine the cell number in a hemocytometer, using the Gilson pipettes for dilution and counting at least 400 cells. Examine the stained cytocentrifuge smear and do a differential count. Typically, each mouse should yield 20–30 $\times 10^6$ PEC, of which approximately 90% should be large macrophages. Calculate the final dilution of the PEC. To get $3 \times 10^5$ adherent macrophages/cm$^2$, a concentration of $2.5 \times 10^6$ PEC/ml in 6 ml will be added to each 30 cm$^2$ well. ($2.5 \times 10^6$ PEC/ml $\times$ 90% macrophages $\times \frac{2}{3}$ of elicited macrophages to adhere $\times$ 6 ml $\div$ 30 cm$^2$ = $3 \times 10^5$ adherent macrophages/cm$^2$).

Make this dilution and add 6 ml of the suspension to each well. Shake the dishes laterally to disperse the cells evenly and incubate the dishes for 3 hr in a humidified $CO_2$ incubator at 37°. Aspirate the medium and wash each well vigorously 3 times with tissue culture medium containing no serum or supplements. The cultures at this time should consist of monolayers of well-spread phase-dense macrophages having prominent cytoplasmic ruffles. Add 6 ml of complete culture medium to each well and return to the incubator.

Acknowledgments

This work was supported in part by U.S.P.H.S. Grants CA-14236 and CA-16784.

# [44] Mouse Erythroleukemia Cells

*By* T. V. Gopalakrishnan and W. French Anderson

Mouse erythroleukemia (MEL) or Friend cells grow in suspension culture and have been extensively used as a model system for studying erythropoiesis *in vitro*.[1,2] When MEL cells are grown for several days in the presence of dimethylsulfoxide (DMSO) or a variety of other chemical agents,[3] they undergo changes similar to the normal maturation of red

[1] C. Friend, H. D. Preisler, and W. Scher, *Top. Dev. Biol.* **8**, 81 (1974).
[2] P. R. Harrison, *Int. Rev. Biochem.* **15**, 227 (1977).
[3] R. C. Reuben, R. L. Wife, R. Breslow, R. A. Rifkind, and P. A. Marks, *Proc. Natl. Acad. Sci. U.S.A.* **73**, 862 (1976).



ISBN 0-12-181958-2

blood cells including alterations in morphology and the production of hemoglobin, heme, carbonic anhydrase, spectrin, and red blood cell surface antigens.[4-8] However, these cells are insensitive to the hormone erythropoietin, the natural mammalian hormone involved in the *in vivo* development of erythroid cells.[9] Hence, Friend cells are considered by most observers to be erythroid cells arrested at or before the proerythroblast stage. The reason for the inability of these cells to undergo the normal maturation process in the absence of any chemical inducers remains unknown.

Because of the ease of altering conditions *in vitro*, MEL cells are a potentially useful system for studying the molecular control of hemoglobin gene expressions as well as other biochemical and genetic aspects of erythroid cell maturation.[1,2,10-12] The unique advantage of the globin system is that the immediate gene product, globin messenger RNA (mRNA), can be assayed directly and quantitatively using a specific molecular probe, globin complementary DNA (cDNA). Little or no hybridizable globin mRNA is present in uninduced MEL cells, but globin mRNA can be detected 2 days after induction with DMSO and reaches a maximum concentration 4 days after incubation.[13] Considerable studies are now underway attempting to understand the molecular events that occur inside the intact MEL cell between the time of exposure to an inducing agent and the synthesis of globin mRNA.

## Growth of MEL Cells

*A. Cells*. MEL cells can be obtained from the Institute for Medical Research, Copewood and Davis Streets, Camden, New Jersey. Strains GM-86 and GM-979 are presently available. Other isolates or mutants can be obtained from individual investigators. In addition, these cells can be

[4] C. Friend, W. Scher, J. G. Holland, and T. Sato, *Proc. Natl. Acad. Sci. U.S.A.* **68,** 378 (1971).

[5] D. Kabat, C. C. Sherton, L. H. Evans, R. Bigley, and D. Koler, *Cell* **5,** 331 (1975).

[6] P. S. Ebert and Y. Ikawa, *Proc. Soc. Exp. Biol. Med.* **146,** 601 (1974).

[7] H. Eisen, R. Bach, and R. Emery, *Proc. Natl. Acad. Sci. U.S.A.* **74,** 3898 (1977).

[8] Y. Ikawa, M. Furusawa, and H. Sugano, *Bibl. Haematol. (Basel)* **39,** 955 (1973).

[9] A. W. Nienhuis, J. E. Barker, and W. F. Anderson, *in* "Kidney Hormones" (J. W. Fischer, ed.), p. 245. Academic Press, New York, 1978.

[10] A. W. Nienhuis, J. E. Barker, A. Deisseroth, and W. F. Anderson, *Ciba Found. Symp.* **37,** 329 (New Ser.), (1976).

[11] T. V. Gopalakrishnan, E. B. Thompson, and W. F. Anderson, *Proc. Natl. Acad. Sci. U.S.A.* **74,** 1642 (1977).

[12] D. E. Axelrod, T. V. Gopalakrishnan, M. Willing, and W. F. Anderson, *Somatic Cell Genetics*, **4,** 152 (1978).

[13] J. Ross, Y. Ikawa, and P. Leder, *Proc. Natl. Acad. Sci. U.S.A.* **69,** 3620 (1975).

obtained as primaries by following the procedure of Friend.[1] Briefly, susceptile mice are infected with Friend virus complex from which they develop an erythroleukemia syndrome. Cells from the enlarged spleen are then inoculated subcutaneously into another susceptible mouse and a tumor resembling a reticulum cell sarcoma develops. Cells from this tumor can be cultured indefinitely in suspension as MEL cells.

*B. Growth Conditions*. MEL cells are grown at 37° in a humidified water-jacketed incubator equilibrated with 10% $CO_2$ in air. Growth at 37° is not greatly influenced by slight variations in the $CO_2$ concentration. Since the $CO_2$ content inside the incubator determines the pH of the medium (which is buffered with $NaHCO_3$), too much or too little $CO_2$ will lower or raise the pH and be lethal for the cells. A $CO_2$ concentration that lies between 5–10% maintains the pH of any one of the media described below in a range that is conducive to good cell growth.

*C. Medium*. MEL cells are easily grown as a suspension culture in any of the established media supplemented with 10% fetal calf serum (GIBCO). Routinely these cells have been grown in BME (Basal minimal medium, GIBCO) or in slightly richer media such as Modified Improved Minimal Essential Medium (Modified IMEM) or F14 medium (available from GIBCO as F12 modified medium 72004 and then supplemented with 3.7 g/liter of $NaHCO_3$). Depending upon the medium and growth conditions, MEL cells grow with a doubling time of 10–18 hr. It is important to maintain the cells in logarithmic growth. This is achieved by diluting the cells 1:50 twice a week; slower growing cells can be passed at 1:100 dilution once a week. These dilutions are useful for the maintenance of stock cultures, whereas cells to be used regularly for experiments can be maintained at 1:10 dilution at frequent intervals.

*D. Serum*. Fetal calf serum is used with medium at a concentration of 10% (v/v). Several different lots of serum should be tested for their ability to promote optimal growth as well as for hemoglobin induction of the MEL cells. Markedly different doubling times and percent inducible cells are obtained with different lots of serum from the same (or different) companies.

### *1. Lot Testing for Growth Support*

Growth media supplemented with 10% fetal calf serum of different lot numbers are prepared. The capacity of different lots of serum to support cell growth can be tested in two different ways: (a) by cloning efficiency or (b) by doubling time.

(a) Cloning efficiency is determined by suspending in separate tubes MEL cells in medium plus each different lot of serum, then plating in microtiter dishes. The cell suspensions are diluted to give a concentration of 1.5 cells/ml; 0.2 ml of each suspension is added to each well of a 96-well microtiter dish (Linbro or Falcon Plastics); one microtiter dish is used for each serum lot being tested. Thus, roughly 30% of the wells will have one cell. The dishes are then kept in the incubator until visible colonies appear in the wells. One can see a colony with the naked eye in 5–7 days or under low power with an inverted phase microscope earlier. If cloning efficiency were 100%, 30% of the wells (29) would have colonies; the actual cloning efficiency can be calculated from the actual number of colonies observed. The serum lot that gives the best cloning efficiency is then determined.

(b) An approximate doubling time can be determined as follows: $10^4$ MEL cells are seeded in 10 ml of growth medium prepared with different lots of serum in separate 100-mm Petri dishes in duplicate. The dishes are incubated for a fixed period of time, normally 3–5 days. At the end of this period, the total number of viable cells in each dish is determined with a hemocytometer. The number of hours of growth is divided by the number of cell doublings.

### 2. *Lot Testing for Hemoglobin Inducibility*

For testing the inducibility of hemoglobin by DMSO (or another inducing agent) using the medium prepared with different lots of serum, MEL cells are centrifuged and washed once with serum-free medium. The cells are suspended in serum-free medium and their concentration determined. The cell density is then adjusted by suitable dilution with serum-free medium to $10^6$ cells/ml. Nine milliliters each of medium prepared with different lots of serum are placed in separate 100-mm Petri dishes and to each of these 0.2 ml of DMSO is added and mixed; then 1 ml of the MEL cell suspension is added and mixed. The dishes are incubated for 5 days at 37°, at which time the percentage of benzidine-positive cells in each dish is determined as described below. The serum lot that gives the maximum percentage of benzidine-positive cells is selected. It has been observed that the ability of different lots of sera to bring about hemoglobin induction by 2% DMSO varies more widely than growth support. Therefore, a lot that gives the highest induction is usually selected over one with a slightly superior growth support but less induction potential. Once a satisfactory lot is found, it is purchased in bulk and stored frozen at -20° or below until used.

### Hemoglobin Induction

For induction experiments, cells are plated in regular growth medium at a density of $1 \times 10^5$ cells/ml with 1.5–2% (v/v) DMSO.[14] After 4 or 5 days, the cell pellet appears pink to deep red depending upon the extent of hemoglobin production. For the quantitative estimation of the percentage of cells producing hemoglobin, the cells are stained with benzidine as described below.

DMSO should not be autoclaved for sterility since this inhibits induction of hemoglobin in MEL cells. Hence, a stock solution of 15–20% DMSO in growth medium is prepared and then sterilized by filtration through Nalgene filters with 0.45-$\mu$m pores. The stock solution can be stored at 4°. This solution is diluted 1:10 with medium so as to bring the final concentration of DMSO to 1.5–2.0%. In addition to DMSO, a variety of other chemical agents bring about the induction of hemoglobin in MEL cells.[3] Of these inducers, hexamethylenebisacetamide has proved to be the most potent inducer for hemoglobin.

### Benzidine Staining

*A. Stock Solution*. The stock solution is 0.2% (w/v) benzidine dihydrochloride in 0.5 *M* acetic acid. The solution is stable for many months when stored in a brown bottle and kept refrigerated. *Note: This is a potential carcinogen and, therefore, should be handled using proper precautions.*

*B. Stain Solution*. A fresh solution of 0.4% of 30% $H_2O_2$ (v/v) in the benzidine stock solution (i.e., 10 microliters 30% $H_2O_2$ in 2.5 ml stock solution) is prepared just prior to use. One-tenth volume of the stain solution is added to a cell suspension that is to be tested for hemoglobin induction. Routinely we use 4 ml of a suspension of MEL cells that have been treated with 2% DMSO for 5–6 days; 0.4 ml of the stain solution is added and mixed. The cells that contain hemoglobin turn blue in 5–10 min. The percentage of benzidine-positive cells is determined by bright-field light microscopy; 200–500 cells are counted for an accurate estimation. It is essential to examine cells in different areas of the dish to insure that the distribution of benzidine-positive cells is uniform. A potential source of error arises from the tendency of the floating population of cells to concentrate at the center of the dish. One way of overcoming this problem is to mix the cell population very well after the addition of the stain solution. We have also noted that the benzidine-positive cells have a tendency to float and the negative cells to sink. This should be taken into account during counting.

[14] Obtained from Fisher, Fisher Scientific Co., Pittsburgh, Pa. Since different lots of DMSO also differ in their ability to bring about hemoglobin induction, DMSO should be lot tested also.

### Spectrometric Determination of Hemoglobin

Hemoglobin induction in a population of cells also can be estimated by lysing the cells and measuring the hemoglobin content spectrophotometrically.[5] This is done as follows: The cell suspension is centrifuged at 100 *g* for 3 min and the cell pellet washed once in normal saline. The cells are lysed by suspending the cells in 3 volumes of lysing buffer (50 m*M* Tris·HCl, pH 7.0; 25m*M* KCl; 5 m*M* $MgCl_2$; 1 m*M* 2-mercaptoethanol; and 0.3% Triton X-100) at 4° and vortexing briefly. The lysed cell suspension is centrifuged in an Eppendorf centrifuge at maximum speed (8000 *g*) for 10–15 min using microcentrifuge tubes. The supernatant liquid is collected, and absorbance is measured in a spectrophotometer at 420 nm using the lysing buffer as blank. The absorbance at 420 nm is multiplied by 0.0945 to give hemoglobin concentration in milligrams per milliliter. The hemoglobin content in the cell lysate is normalized per $10^6$ cells or per milligram of soluble protein.

# [45] Skeletal Myoblasts in Culture

*By* IRWIN R. KONIGSBERG

Cell culture systems have been employed for the past 17 years[1,2] to study the differentiation of skeletal muscle fibers from myoblasts, the embryonic stem cells from which this tissue develops *in vivo*. When appropriately cultured, these progenitor cells reproduce, with remarkable fidelity, the sequence of events that have been observed during normal embryonic development. Cell culture techniques provide a number of advantages unattainable in the intact organism. One can obtain a uniform, highly purified population of cells of known developmental fate. These can be maintained under rigidly controlled conditions and subjected to a wide range of experimental intervention. Finally one can manipulate culture conditions to impose a greater degree of synchrony than is ever observed in the organism.

The use of skeletal muscle cell culture as an experimental system is now widely employed. Although the same overall strategy is used, the details of the specific techniques used vary widely from one laboratory to another. At times, therefore, it becomes difficult to compare the results obtained by different investigators. This article will deal largely with the techniques and procedures currently in use in our laboratory. Where

[1] L. M. Rinaldini, *Exp. Cell Res.* **16**, 477 (1959).

[2] I. R. Konigsberg, *Exp. Cell Res.* **21**, 414–420 (1960); I. R. Konigsberg *et al.*, *J. Biophys. Biochem. Cytol.* **8**, 333 (1960).





ISBN 0-12-181958-2

alternatives procedures are employed by other investigators these will be described briefly and referenced. To the extent that we know how variations in protocols and procedures might affect results and observations, these will be discussed.

Since cell culture is more art than science, much of the technology is empirical but based on extensive testing that is, unfortunately, largely unpublished. Wherever appropriate, the results of such tests will be cited and in general the rationale of our procedures will be indicated.

## Cell Source

Embryonic muscle from a number of species is commonly employed (chick,[1,2] quail,[3] rat,[4] calf,[5] mouse,[6,7] and human[8]), but muscle from such sources as *Drosophila*,[9] the butterfly,[10] cricket,[11] and lizard[12] has also been cultured. We routinely use avian embryos, both chick (*Gallus domesticus*) as well as quail (*Coturnix coturnix japonica*), finding the latter species preferable. From the standpoint of the yield of viable cells we find 11-day chick embryos (either pectoral or leg muscle) and 9-day quail embryos (pectoral muscle) to be the most suitable stages. The choice of stage is a compromise. With increased age a larger volume of muscle tissue can be obtained per embryo. However, as development increases the ratio of myoblasts to nonproliferative nascent multinuclear fibers decreases. Some difficulty is also encountered in dissociating tissue from older embryos due, presumably, to increased cellular packing and the presence of more substantial amounts of extracellular connective tissue elements.

## Dissection and Mincing the Tissue

In establishing primary cultures all operations between sacrificing the embryo to inoculating the petri plates with single cells must be performed as quickly as possible. We use no more tissue than can be processed

[3] I. R. Konigsberg, *Dev. Biol.* **26,** 133 (1971).

[4] D. Yaffe, *Curr. Top. Dev. Biol.* **4,** 37 (1969).

[5] M. E. Buckingham, D. Caput, A. Cohen, R. G. Whalen, and F. Gros, *Proc. Natl. Acad. Sci. U.S.A.* **71,** 1466 (1974).

[6] F. Bowden- Essien, *Dev. Biol.* **27,** 351 (1972).

[7] S. D. Hauschka, C. H. Clegg, T. A. Linkhart, and R. W. Lim, *Cell. Biol.* **75,** No. 2, Part 2, 383 (1977).

[8] S. D. Hauschka, *Dev. Biol.* **37,** 329 and 345 (1974).

[9] R. L. Seecoff, *Am. Zool.* **17,** 577 (1977).

[10] T. J. Kurtti, and M. A. Brooks, *Exp. Cell Res.* **61,** 407 (1970).

[11] R. Wittmann, J. G. Moser, B. Heinkin, and B. Wolf, *Cytobiologie* **8,** 468 (1974).

[12] P. G. Cox, and S. B. Simpson, Jr., *Dev. Biol.* **23,** 433 (1970).

within 1 hr. Embryos are removed sterilely, one at a time, and transferred to a 100-mm glass petri plate. Skin and other overlying tissues are removed and discarded and the muscle cleanly dissected and transferred to a 50-mm petri dish containing a drop or two of a phosphate-buffered balanced salt solution. The pooled muscle is then thoroughly minced with a pair of sharp curved scissors. During this procedure the tissue is confined to one edge of the dish by tilting the dish and using the edge of the scissors to push the mince together. This should be done quickly to avoid desiccation. One quickly learns to gauge the proper degree of mincing (by counting scissor strokes and observing the degree of homogeneity of the mince). If the tissue is not adequately minced this will become obvious at the next step, which requires diluting and pipetting the mince to a tube (or flask).

### Cell Dissociation (Mechanical)

The mince can be dissociated into single cells either enzymically or mechanically. The two enzymes commonly employed are pancreatic trypsin (Difco 1:300) or bacterial collagenase (Worthington Type I or Type II); both preparations are relatively unpure. Crude trypsin is preferred because one or more of the contaminating enzymes (presumably DNase) digests a viscous material present in the mince;[13] with more highly purified trypsin the viscous material remains, trapping cells and clogging whatever sieve is used to remove undissociated clumps of cells. Similarly, but for different reasons, crude collagenase is preferred to the crystalline enzyme. Our own attempts to use highly purified collagenase were singularly unsuccessful, leading us to assume that the efficacy of the crude enzyme preparation was due to the contaminating proteases rather than to collagenase activity *per se*. Of the two enzymes we have found collagenase to be preferable since it yields a population of more viable cells than trypsin.

A practical procedure for the mechanical dissociation of embryonic skeletal muscle was first devised by Arnold Caplan several years ago. He communicated the technique to several laboratories, including our own. This technique has since been published[14] as well as a number of variations of the original procedure.[15,16]

We adopted mechanical dissociation for the preparation of primary cultures during a period when we had difficulty obtaining standardized

[13] M. S. Steinberg, *Exp. Cell Res.* **30,** 257 (1963).
[14] A. I. Caplan, *J. Embryol. Exp. Morphol.* **36,** 175 (1976).
[15] K. Tepperman, G. Morris, F. Essien, and S. M. Heywood, *J. Cell Physiol.* **86,** 561 (1975).
[16] J. C. Bullaro, and D. H. Brookman, *In Vitro* **12,** 564 (1976).

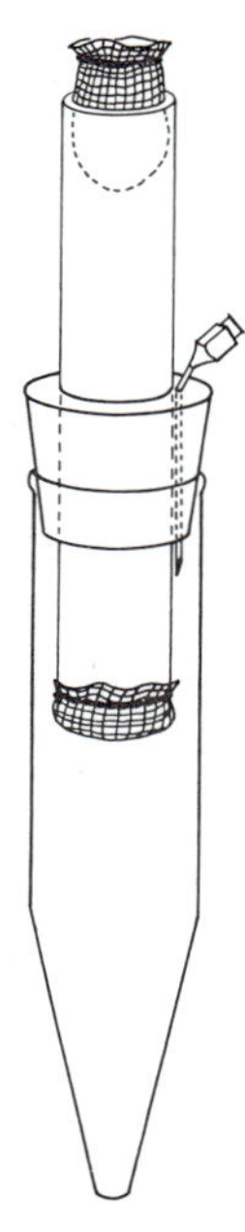

Fig. 1. Cell filter. The device centers a cotton-plugged polypropylene test tube (Nalge #3110-0180), with the bottom machined off, inverted in a 40-ml Pyrex centrifuge tube with a #5 silicone rubber stopper. The centrifuge tube is vented by the insertion of a 20-gauge hypodermic needle. Nitex or gauze filters (see text) are tightly tied to the flared lip of the inverted tube and discarded after use.

batches of collagenase. The procedure proved satisfactory and we use it routinely, as follows:

1. The mince from a pectoral muscle of four 10-day quail embryos is suspended in 2 ml of complete growth medium and transferred to a sterile 25-ml polycarbonate graduated centrifuge tube.

2. The tube and its contents are subjected to the action of a Vortex mixer (Vortex-Genie, Model K 550 G) for 30 sec at the highest speed setting.

3. Large clumps are allowed to settle by gravity, and the supernatant liquid is withdrawn with a curved-tip Pasteur pipette (Bellco Glass) with rubber bulb.

4. The cell suspension is filtered through a Nitex mesh[17] (20-$\mu$m opening) screen positioned in a 40-ml graduated conical Pyrex centrifuge tube (see Fig. 1).

[17] Tetko Inc., Elmsford, New York.

5. While the first cell suspension is filtering, an additional 2 ml of complete growth medium are added to the settled clumps that are resuspended by a 3-sec vortexing at half-maximum speed.

6. The suspension is filtered through the same Nitex filter, with continous swirling to prevent clogging the Nitex, and pooled with the first filtrate.

7. The cell number in the pooled suspension is counted in a hemocytometer, and primary cultures are established by inoculating approximately $10^7$ cells into each of several gelatin- or collagen-treated Falcon 100-mm petri plates (see Culture Surface) containing 5 ml of complete growth medium each. (Primary plates are pre-equilibrated for 1-2 hr in a humidified incubator at 36.5° in an atmosphere of 4% $CO_2$ in air before use.)

8. Unattached cells and debris are removed by replacing the medium with 10 ml of complete medium 2 hr after the primary plates are seeded.

9. After 16–18 hr of incubation these primary plates are used to prepare secondary cell suspensions (as described below), which are then used to set up experimental and control cultures.

The establishment of satisfactory primary cultures requires not only a brief (ca. 1 hr) transit time between dissection and plating but also a judicious adjustment of inoculum size. It is difficult to accurately count cells in mechanically dissociated suspensions that are not monodisperse but contain many cell clumps. With practice, however, one can obtain reproducible primaries. One should always monitor primary plates by microscopic inspection before use. Extremely sparse primaries yield poor secondary cultures. However, in overloaded primary plates, a higher proportion of multinucleated cells are found the next morning, presumably due to the early initiation of fusion. Moderate overloading thus selects against the cell type, i.e., the myoblast, of interest. Grossly overloaded primaries, even though they may not contain proportionately higher numbers of multinucleated cells to the eye, invariably yield cultures that grow poorly and contain abnormal-looking cells. Our suspicion is that in such densely seeded primaries metabolic by-products accumulate, the buffer capacity is exceeded, and the drop in pH damages the cells.

### Cell Dissociation (Enzymic)

Cell yield, by mechanical dissociation, is lower than one obtains enzymically since many of the cells are clumped and consequently lost during sieving. Although we prefer to establish experimental cultures from secondary cell suspensions, the reader's experimental design may require the use of freshly isolated primary cells. Since accurate cell counts are more readily obtained using enzymically dissociated cells, the collagenase procedure previously used by us is given here:

1. Muscle tissue is dissected and minced as described above (see "Dissection and Mincing the Tissue") using two or four embryos.

2. The mince is transferred with two 2.5-ml portions of nominally 0.1% collagenase in Puck's saline G[18] into a 25-ml Erlenmeyer flask using a serological pipette with an orifice adequate to accept the minced fragments (usually 10 ml).

3. The suspended tissue is maintained at 37° for 5 min and subjected to gentle pipetting with a 10-ml serological pipette to disperse the cells.

4. Enzymic digestion is stopped by adding 5 ml of complete growth medium at about 5°, and the suspension is filtered through three double layers of cheesecloth (see Fig. 1) into a 40-ml Pyrex conical centrifuge tube.

5. The filtered suspension is centrifuged at 800 rpm with a bench-top centrifuge (I.E.C., Model H) and the supernatant liquid removed by aspiration.

6. The cell pellet is dispersed in 2–5 ml of complete growth medium by repeated pipetting with a hypodermic syringe equipped with a 20-gauge 5-inch spinal-tap needle.

7. Following a final filtration through a Nitex (Nylon monofilament) screen with a mesh opening of 10 $\mu$m, the suspension is counted with a hemocytometer chamber.

8. At this point, primary cultures are started in the same manner as with mechanically dissociated cells and are used on the following day to prepare secondary cell suspensions that have been enriched for myoblasts.

Other investigators prefer to establish experimental and control cultures from such primary suspensions. Under these circumstances enrichment can be achieved by exposing the suspension briefly to culture surfaces to which fibroblasts attach preferentially.[19]

The same protocol can be used with trypsin as well although cell viability is somewhat poorer. During the period in which we routinely employed this enzyme we used 0.05% trypsin (N.B.C. or Difco 1:300) made up in Puck's saline G.

### Secondary Cell Suspension

We prefer to use cultures from secondary suspensions harvested from briefly cultured primary cells for the following reasons: (1) We feel that we

[18] T. T. Puck, S. J. Cieciura, and A. Robinson, *J. Exp. Med.* **108,** 949 (1958).
[19] D. Yaffe, *Proc. Natl. Acad. Sci. U.S.A.* **61,** 477 (1968).

get a more accurate count of *viable* cells (after plating such secondary suspensions we see relatively few "floaters"). (2) We can more readily enrich for myoblasts during the harvest of primary cultures.

The protocol we adopted to prepare secondary cell suspensions from mechanically dissociated primary cultures is described below. In this procedure we now use crystalline trypsin[20] since it has the singular advantage of being rapidly inactivated by crystalline soybean trypsin inhibitor[20] in stoichiometric amounts.

The minimum effective concentration of enzyme is empirically determined. Primary cultures are briefly exposed to the enzyme, and enzyme activity is stopped rapidly. We have also abandoned the practice of centrifuging down the cells, washing, and resuspending the pellet (see preceeding page, steps 5 and 6). This step, we find, lowers the cell yield (either by damaging cells or by the loss of cells in undissociable cell clumps).

1. The medium is aspirated from primary cultures incubated overnight and each petri plate rinsed with 10 ml of saline G to remove residual medium.

2. An iced solution of trypsin (Worthington, 3× crystallized), 5 ml containing 13 $\mu$g/ml in saline G, is delivered into each petri plate.

3. The petri plates are maintained at room temperature and observed periodically at a magnification of ×100 under phase-contrast microscopy.

4. Cultures are gently swirled intermittently. When approximately half of the cells have either rounded up or detached from the surface, about 5 min, the enzyme solution is gently pipetted over the culture surface.

5. The resultant cell suspension is quickly transferred to an iced 40 ml Pyrex conical centrifuge tube of complete growth medium (0.5 ml for each primary plate used) containing 0.5 mg/ml of soybean trypsin inhibitor (×3 crystalline, Worthington).

6. The suspension is mixed well, and the cells are enumerated in a hemocytometer chamber. Suspensions prepared as described above tend to be dilute and are generally counted by the procedure used for white cell counts.

Procedures for the enrichment of myoblasts in cell suspensions are based on the observation that fibroblasts stretch out and are more tenaciously bound to the culture surface [21] (see Fig. 2). In the above protocol, differential release of myoblasts is achieved by a controlled, brief trypsinization that leaves most of the fibroblasts bound to the petri plate. We

[20] Worthington Biochemical Co.

[21] I. R. Konigsberg, *Science* **140**, 1273 (1963).

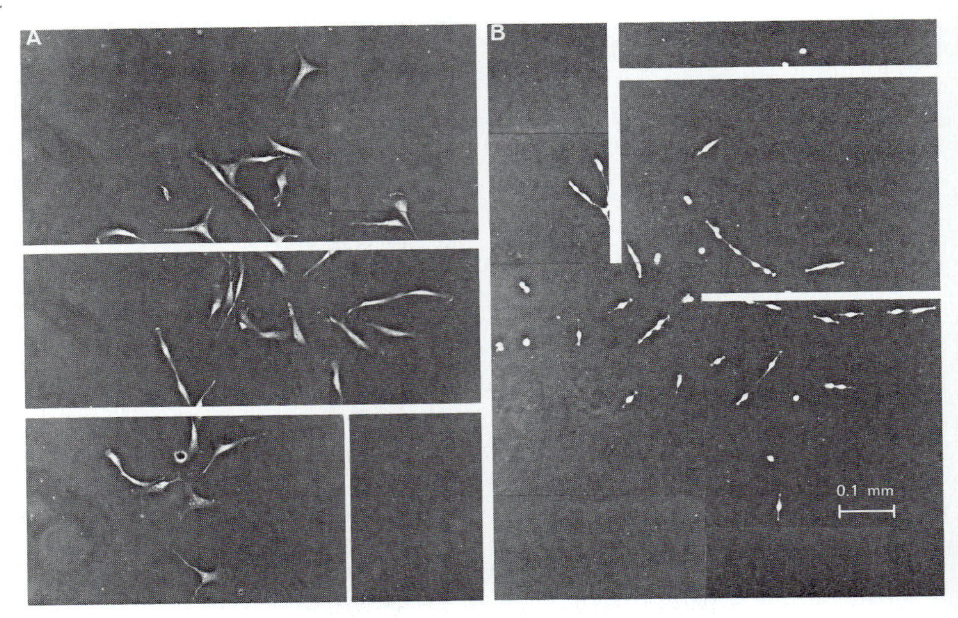


Fig. 2. Morphological differences between cells in a clone of chick muscle fibroblasts (A) and a clone of myoblasts (B) photographed on the fourth day of culture using phase-contrast optics. Since fibroblasts attach more tenaciously they appear to be larger. The remarkable similarity of cell shape predictably identifies the clonal type.[21a]

prefer to sacrifice yield for purity of cell type and can, with care, obtain cell suspensions in which a minimum of 96% of the cells are myogenic.[22]

## Inoculum Size, Medium Composition, and Synchrony

One of the chief advantages of muscle cell culture as an experimental system is the high degree of synchrony of differentiation that can be achieved. Synchrony can be imposed because cell culture is a closed system in which the proliferating cells eventually deplete the medium of heterologous, exogenous mitogens and lag in the $G_1$ phase of the cell cycle. The protraction of $G_1$ is the cue for the initiation of differentiation. By adjusting the inoculum size and the volume and composition of the growth medium, the timing of differentiation can be manipulated within limits. The limits are imposed by the establishment, in static cultures of a gradient of depletion in the immediate vicinity of the cells. This can be

[21a] I. R. Konigsberg, *Scientific American,* **211,** 61–66 (1964).

[22] P. A. Buckley, and I. R. Konigsberg, *Dev. Biol.* **37,** 186–193 (1974).

overcome to a certain extent by continuous, gentle perturbation of the cultures[3] or by scheduled refeeding. Under such conditions the period of rapid log growth can be extended, but short of a regime of continuous perfusion,[23] increasing cell density eventually exceeds the practical limits of any refeeding schedule. The protocols employed to achieve synchrony, therefore, represent compromises of one sort or another.

In order to cleanly divide the stages of culture development into a pure rapid proliferative stage and a stage in which myogenic fusion is initiated and proceeds rapidly, we employ the following protocol:

1. After determining the number of cells per milliliter in a secondary cell suspension (usually about $8 \times 10^4$; total volume per primary plate, 3 ml), an aliquot is serially diluted to provide a suspension in which 0.1 or 0.2 ml contains the inoculum per petri plate to be delivered ($2 \times 10^4$ cells).
2. Cells are inoculated into each of a series of collagen- or gelatin-treated petri plates containing 2 ml of high-growth medium. Culture dishes containing medium are equilibrated in the incubator before use.
3. Following incubation for 16–18 hr, the medium is aspirated, replaced with 3 ml of high-growth medium, and not replaced thereafter.

In cultures established on this schedule, fusion—in the form of short, stubby multinuclear cells (two to six nuclei per cell)—is first observed on day 3.[22] Following the same protocol but using a low-growth medium (10% serum, 1% embryo extract) instead of high-growth (10% embryo extract, 15% serum), fusion is initiated 24 hr earlier (Fig. 3) and still earlier by further depleting the low-growth medium (see "Conditioned Media").

Using this tactic of adjusting the rate of depletion of an initially growth-promoting medium, fusion, once initiated, occurs rapidly but certainly not at the maximum rate. However, greater synchrony can be achieved by employing a "step-down" medium change; replacing a high-growth medium with a defined, synthetic medium containing no serum or embryo extract (see "Media"), with the plate rinsed once with the defined medium before refeeding.[24]

The suggested inoculum size yields satisfactory cultures using secondary suspensions of quail myoblasts. In our experience, inocula 2–3 times larger are required to obtain comparable cultures of chick myoblasts. However, even a cursory examination of the current literature indicates that many investigators use inocula many times heavier than suggested here, ranging to $1–2 \times 10^6$ cells per 5-cm petri dish and more. This degree

[23] I. R. Konigsberg, *in* "Pathogenesis of Human Muscular Dystrophies" (L. P. Rowland, ed.), pp. 779–798. Excerpta Med. Found., Amsterdam, 1976.

[24] C. P. Emerson, Jr., *in* "Pathogenesis of Human Muscular Dystrophies" (L. P. Rowland, ed.), pp. 799–809. Excerpta Med. Found., Amsterdam, 1976.

of cell crowding in culture can hardly be considered physiological. Indeed, our experience has been that a smaller percentage of such inocula actually attach. Synchrony is completely lost in such cultures, myogenic fusion being initiated during the first 24 hr in culture, and, as one might expect, cell proliferation is diminished by cell crowding. If, as we suspect, the use of such heavy inocula reflects an inability to obtain cell survival and proliferation at lower cell density, the investigator should reexamine the cell dissociation techniques, media preparation, and incubation conditions that are being used.

The criteria for the adequacy of these parameters are the plating efficiency, colony size, and differentiation of cells plated at clonal density (200–400 cells per 5-cm dish). Plating efficiencies of about 20% (chick) and 40% (quail) are acceptable.

## Media (General Considerations)

Media composition, preparation, and sterilization are discussed elsewhere in this volume [1] [2] [5]. This section will deal with just those media requirements peculiar to freshly isolated cells and muscle cells particularly. These cells, unlike cell lines, are neither adapted to nor selected for the culture environment. They are, therefore, more sensitive to variations in quality of media components than are established cell lines.

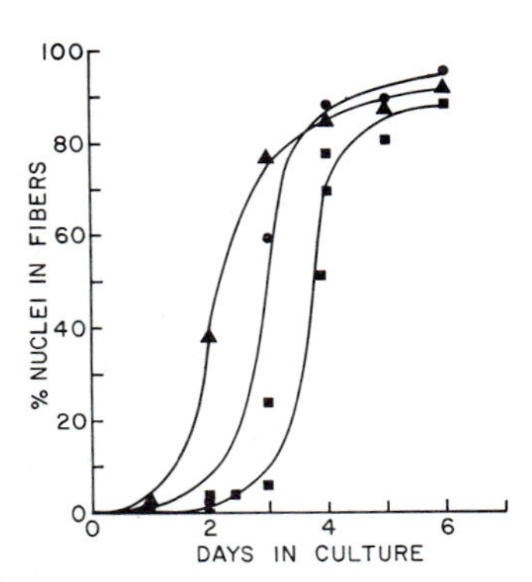


Fig. 3. The increase in percentage of nuclei in nascent multinuclear muscle fibers as a function of time in culture, using media of differing growth-supporting properties. Each point represents the average of two cultures each inoculated with $10^4$ cells of a secondary suspension of quail myoblasts. Fusion is initiated earliest in conditioned low-growth medium, CM 1-10 (▲); then in freshly prepared low-growth medium, FM 1-10 (●); and finally in high-growth medium, FM 10-15 (■). (From Wm. M. Sutherland, Doctoral Thesis, University of Virginia 1977.)

1. We have been forced for logistic considerations to use commercially prepared media components, such as MEM, F10, and F12. Although these products are satisfactory they are not quite up to the standard of these components as prepared by ourselves from reagent-grade chemicals.

2. All of the sera that we use are pretested by us, using the ability of each serum when incorporated in our "high-growth" medium to support growth and differentiation at clonal density (see above). Generally, one of two (rarely four) samples of horse serum proves satisfactory, and we order and store at −60° sufficient serum to last the year.

3. Unlike cell lines, freshly isolated myoblasts require, in addition to horse serum, the incorporation of a saline extract of chick embryos. In the absence of embryo extract, little if any proliferation occurs. The protocol (see "Appendix") for preparing this extract is the result of exhaustive empirical testing of the mincing procedures, extraction time, centrifugation speeds and times, as well as filtration procedures. We have employed this protocol, with only minor changes, over the past 10 years with good results. The extract has a useful life of only 2 weeks even when stored at -60° and therefore is prepared on a regular schedule. Alternative methods of preparing extract are employed by other investigators. These alternatives range from essentially the same procedures as ours,[25] to substituting a brief (20-min) blending step (Waring Blendor)[26] for the mincing procedure we use, to forcing the embryos through a stainless-steel screen followed by freeze–thawing and centrifugation.[27]

### Constituents of the Medium

The media used in most laboratories in which muscle cell culture is practiced consists of: (1) one of the standard synthetic mixtures of amino acids, vitamins, and cofactors (such as Eagle's MEM), (2) serum (usually equine), (3) embryo extract, and (4) a mixture of antibiotics.

Normally a $Ca^{2+}$ ion concentration of 1.4 m$M$ is employed. By reducing the calcium ion concentration to 35 $\mu M$, myoblast fusion is reversibly blocked without affecting the rate of cell proliferation.[28,29] Under these conditions the mononucleated myoblasts deplete the medium, differentiate, and withdraw from the cell cycle[29] as mononucleated myocytes.

[25] N. K. White, P. H. Bonner, D. R. Nelson, and S. D. Hauschka, *Dev. Biol.* **44**, 346 (1975).
[26] S. D. Hauschka, personal communication.
[27] M. C. O'Neill and F. E. Stockdale, *J. Cell Biol.* **52**, 52 (1972).
[28] B. Paterson and R. C. Strohman, *Dev. Biol.* **29**, 113 (1972).
[29] C. P. Emerson, Jr. and S. K. Beckner, *J. Mol. Biol.* **93**, 431 (1975).

Reduction of the calcium ion concentration is accomplished by either of two procedures:

1. Supplementing low-calcium synthetic media with dialyzed embryo extract and serum[30] (or reducing the calcium ion concentration by ion-exchange chromatography[31]).

2. Chelating calcium in the medium by the addition of EGTA. The precise concentration of EGTA (made up in the synthetic low-$Ca^{2+}$ media stock, adjusting pH to 7.2) must be determined empirically for each batch of medium since the concentration of calcium in the embryo extract varies. This is done by adding back $Ca^{2+}$ (from a 25 m*M* stock solution) to the level that minimizes cell detachment from the plate surface while still completely inhibiting cell fusion. Levels of EGTA of approximately 1.9 m*M* are generally adequate.

### The Synthetic Nutrient Component

The synthetic nutrient mixtures most frequently employed in muscle cell culture are: (1) Eagles MEM,[32] (2) Ham's F10,[33] and (3) Puck's NC (also called NCI).[34] Other synthetic formulations, as well as mixtures of two or more, have also been used. Although it is assumed that such usage is based on comparative tests, it is not obvious that any of the alternatives to the three most frequently used synthetic nutrients offer any unique advantage in cultures seeded at mass density. Indeed, one should not expect to detect differences except when comparing clonal with high-density cultures. Cells at clonal density are far more sensitive to nutritional deficiencies (as well as to inhibitory concentrations of required nutrients). Such conditions, one would expect, would be rapidly adjusted by metabolic cooperation at high cell density. On the other hand, media devised to support clonal growth may be rapidly depleted of one or more essential components present in low concentration. In fact, chick myoblasts at mass density grow well in NC-based medium but do very poorly on F10 medium, probably due to the lower $Ca^{2+}$ ion concentration in F10.[26] Conversely, growth and development at clonal density is far superior in medium containing F10 than in medium containing NC.

[30] A. Shainberg, *et al.*, *Exp. Cell Res.* **58,** 163 (1970).
[31] E. Ozawa, *Biol. Bull.* (*Woods Hole, Mass.*) **143,** 431 (1972).
[32] H. Eagle, *Science* **130,** 432 (1959).
[33] R. G. Ham, *Exp. Cell Res.* **29,** 515 (1963).
[34] P. I. Marcus, S. J. Cieciura, and T. T. Puck, *J. Exp. Med.* **104,** 615 (Sec P. 616) (1956).

Variations of Media Composition and the Control of Proliferation

The standard medium used in this laboratory was developed to support the growth and differentiation of avian myoblasts at clonal density applying the criteria of plating efficiency, colony size, and per cent differentiated clones. For that reason it is a "high-growth" medium, supporting a rapid rate of cell proliferation. In developing this medium a number of parameters were systematically varied, the most important of which, it turned out, were the proportions of serum and embryo extract in the medium. The concentration of equine serum that proved optimal was 15% (v/v). We found also that for chick myoblasts the maximum concentration of embryo extract was 5% (v/v) and that higher concentrations were actually inhibitory (see also Stockdale[35]). In adapting this medium for quail myoblast culture[3] we found that Eagle's MEM was as effective a nutrient base as F10 and that the quail cells, unlike the chick, would tolerate and do better on 10% embryo extract medium.

The high-growth ("FM 10-15") medium is currently employed for cells at both clonal and mass density (see above) and consists of:

| | |
|---|---|
| Eagle's MEM (with Earle's salts)[36] | 740 ml |
| Equine serum | 150 ml |
| Embryo extract | 100 ml |
| Penicillin-streptomycin stock[37] | 10 ml |
| Fungizone stock[38] | 2.5 ml |

By varying the ratio of serum and embryo extract, a low-growth ("FM 1-10") medium was derived[29] that limits proliferation and promotes earlier differentiation (fusion and/or the initiation of myosin synthesis). This medium, when appropriately "conditioned" to further reduce the concentration of mitogens, promotes still more limited proliferation and correspondingly earlier differentiation.

Finally, a completely defined synthetic medium (Emerson's F12 [PVP][24]) that contains neither serum not embryo extract has been devised. It is prepared as follows.

[35] F. E. Stockdale, *in* "Regulation of Cell Proliferation and Differentiation" (W. W. Nichols and D. G. Murphy, eds.), pp. 165–176. Plenum, New York, 1977.

[36] This synthetic medium is modified to contain Eagle's nonessential amino acids (see Eagle[32]) and contains 1.2 g/liter of $NaHCO_3$ rather than the concentration normally employed.

[37] P and S stock contains $10^6$ units of penicillin G (potassium) and 500 mg streptomycin sulfate per 100 ml of P-C chick (see Pannett and Compton[37a]).

[37a] C. A. Pannett and A. Compton, *Lancet* **206,** 381 (1924).

[38] Fungizone stock contains 50 mg amphotericin B (Fungizone®) in 62.5 ml of distilled $H_2O$.

1. Ham's F12 is prepared to contain, in addition to all of the normal components of F12, the following:

Eagle's essential amino acids in the same concentration as in MEM
146 mg glutamine per liter
221 mg $CaCl_2 \cdot 2\ H_2O$ per liter

2. In place of the concentration of sodium bicarbonate normally employed in F12, this solution is made to contain 1.2 g/liter. The complete medium has the following composition:

| | | |
|---|---|---|
| Ham's F12 (as modified above) | 990 | ml |
| Polyvinylpyrrolidone 360 (Sigma) | 500 | mg |
| Penicillin-streptomycin stock | 10 | ml |
| Fungizone stock | 2.5 | ml |

This defined synthetic medium does not support cell proliferation but will support rapid, synchronous fusion (or myosin synthesis in low-$Ca^{2+}$, fusion-blocking F12[PVP]) when cultures are switched from growth to defined medium.[24] This indicates that no constituent of either serum or embryo extract is required for either fusion or cell-type specific synthesis. It suggests, in fact, that these heterologous components of the medium, by reason of stimulating proliferation, prevent or delay the initiation of differentiation. This suggestion is also supported by the observation that fusion is inhibited by continuous perfusion with high-growth medium but not with F12(PVP).[23]

## Conditioned Media

Conditioning freshly prepared media by subjecting it to the metabolic activities of cultured cells could conceivably alter media composition in a number of ways, for example, by the depletion of media constituents or the accumulation of cell products. Such media were employed in developing the art of cell culture (see this volume [5]) and have also proven useful in the development of the muscle cell culture system. Specifically, conditioned media have been used: (1) to develop adequate cloning procedures,[21] (2) to study and control the time of initiation of differentiation,[3,22,23] and (3) to identify subpopulations of myoblasts in early embryonic development.[25,39] Further analysis of two of these media has led to some understanding of the nature of their alteration and eventually to better-defined substitutes for the conditioning procedures. To a large extent it is no longer necessary to employ conditioned media to clone myoblasts (from embryos of advanced stages) since a gelatin- or

[39] N. K. White and S. D. Hauschka, *Exp. Cell Res.* **67,** 479 (1971).

collagen-coated culture surface adequately substitutes for the altered medium.[40,41] In addition synchronous differentiation now can be achieved by a step-down to defined medium (see above) rather than by employing a nitrogen-depleted medium.

However, should the use of any of these conditioned media seem advantageous to the investigator's aims he would be advised to follow the published protocols in minute detail, at least initially. All of the protocols cited above were derived empirically. Media conditioned by each of the three protocols elicit quite a different response from the cultured myoblast. These protocols differ principally with respect to cell type, population size, and duration of conditioning. The different results achieved with each most probably reflect a different balance of the large number of metabolically generated changes that are occurring. Very brief exposure to moderate numbers of cells might simply promote the accumulation in the medium of low-molecular-weight biosynthetic products that leak into the medium.[42] At the other extreme of population size and duration the over riding effect may reflect the depletion of mitogenic medium components.

### Collagen Pretreatment of the Culture Dish

Myoblasts adhere and stretch out more satisfactorily to a collagen or gelatin substratum than to the bare polystyrene TC dish, as supplied by the manufacturer. Furthermore, unless provided with such an attachment surface, the percentage of clones that differentiates is low. The few clones that do differentiate are small and contain relatively few muscle fibers.[40,41,43] In mass cultures established from cell suspensions that are not enriched for myoblasts, differentiation is equally normal in collagen-treated and untreated petri plates. Although not strictly required, the use of pretreated plates is advantageous since it delays the detachment of the cell sheet that eventually occurs in long-term mass culture.

Collagen can be applied by raising the ionic strength of a concentrated solution of the protein in 0.15 $M$ acetic acid. A small measure volume is then spread on the petri plate surface, which is subsequently rinsed.[41] Solutions of gelatin in distilled water also can be spread and simply allowed to dry.[44]

[40] I. R. Konigsberg and S. D. Hauschka, *in* "Reproduction: Molecular, Subcellular and Cellular" (M. Locke, ed.), pp. 243–289. Academic Press, New York, 1965.

[41] S. D. Hauschka and I. R. Konigsberg, *Proc. Natl. Acad. Sci. U.S.A.* **55,** 119 (1966).

[42] H. Eagle and K. Piez, *J. Exp. Med.* **116,** 29 (1962).

[43] I. R. Konigsberg, *Proc. Natl. Acad. Sci. U.S.A.* **47,** 1868 (1961).

[44] S. D. Hauschka, *in* "Growth, Nutrition and Metabolism of Cells in Culture" (G. H. Rothblat and V. J. Cristofalo, eds.), Vol. II, pp. 67–130. Academic Press, New York, 1972.

To insure more uniform distribution and to avoid the tedium of spreading we now flood TC petri plates with a larger measured volume, 1.4 ml, of more dilute collagen or gelatin.[3,45] The protein is allowed to adsorb to the surface overnight (about 18 hr) at room temperature in an atmosphere equilibrated with the protein solvent used. The collagen solution applied is made up by adding 40 μl of 3.76% NaCl to 1.0 ml of a purified collagen stock containing 40 μg collagen per milliliter. Gelatin solutions contain 100 μg/ml of a good bacteriological grade of gelatin (Oxoid) in distilled water. After aspirating the excess, plates are rinsed twice with distilled water, placed in a desiccator over silica gel, and stored at room temperature until used.

Our procedure for preparing and purifying collagen has been published.[3] It is more exhaustive than need be, and our earlier procedure[43] is perfectly adequate. We find it simpler to prepare and handle collagen sterilely. However, we have sterilized collagen-treated petri plate surfaces under UV light and found them perfectly satisfactory. Since gelatin solutions can be autoclaved the investigator may prefer this alternative. We have found no difference between the native and denatured molecule in routine culturing. When cultures are to be subjected to constant[23] or intermittent[3] perturbation, however, myoblasts seem to adhere better to the collagen substratum.

## Appendix: Procedure for Preparing Embryo Extract

1. Incubate fertile chick eggs for 12 days. Candle and discard infertile eggs and eggs containing dead embryos.
2. Cut a circular hole in the blunt end of the egg with a pair of curved scissors and remove the embryo.
3. Cut off and discard that portion of the head anterior to the posterior border of the eyes. Make a longitudinal cut through the body wall from the cloaca to the base of the neck. Collect the embryos in a 150-mm petri plate.
4. When all of the embryos have been collected, decant the blood and empty the petri plates onto paper toweling. Transfer the embryos to a large square of double-thickness cheesecloth. Suspend the cheesecloth "bag" in a beaker and wash with several changes of P-C chick[37a] (until most of the blood is removed).
5. Place the cheesecloth on several thicknesses of paper toweling (to drain) and transfer the embryos to a tared beaker and weigh.

[45] I. R. Konigsberg, *in* "Chemistry and Molecular Biology of the Intercellular Matrix" (E. A. Balazs, ed.), Vol. 3, pp. 1779–1810. Academic Press, New York, 1970.

6. Run embryos through a machined Lucite Latapie mincer with a volume of P-C chick equal in milliliter to the weight in grams of the embryos. (Hold back a small quantity of this measured amount of P-C chick. Weigh out 4 mg [or 1080 U] of hyaluronidase for every 100 g of embryos and add to this P-C chick.)

7. Transfer the minced embryos to a large beaker, add the hyaluronidase, and stir at cold room temperature (4–5°) for 1 hr.

8. Distribute the extract to 90-ml ultracentrifuge tubes and centrifuge for 3 hr in the 21 rotor in the Spinco Model L ultracentrifuge at 20,500 rpm (40,000 *g*) at 1–2°.

9. Decant and collect the supernatant after centrifugation.

10. The extract is clarified by pressure filtration through a 142-mm diam fiberglass prefilter over a 0.45-$\mu$m millipore filter of the same diameter held in a large-volume "pancake" filter holder (Millipore). The supernatant collected in step 9 is transferred to an 800-ml minireservoir (Amicon) and transferred through heavy-duty silicone rubber tubing through the filtering system.

11. The filtered extract is recentrifuged as in step 8 (above) overnight (ca. 18 hr), decanted, and stored at $-60°$. Sterilization by filtration is performed after the complete medium is made up.

# [46] Clonal Strains of Hormone-Producing Pituitary Cells

*By* ARMEN H. TASHJIAN, JR.

Clonal strains of pituitary tumor cells that synthesize and secrete prolactin, growth hormone and adrenocorticotropic hormone (ACTH) have been established and serially propagated in culture for periods of up to 15 years. The rates of biosynthesis of the specific hormonal peptides by these clonal strains are high (up to 10% of total protein synthesis in appropriately stimulated cells), and they respond in culture to many of the same regulatory factors as normal pituitary cells do *in situ*. However, because certain of these regulatory factors, such as the hypothalamic peptides, affect more than one cell type in the intact pituitary gland, it is not possible to perform unambiguous mechanistic studies with hemi-pituitary fragments or mixed primary cultures of freshly dispersed pituitary cells. Therefore, strains of homogeneous populations of functional cells serve as useful model systems for determining the mechanisms of action of factors that regulate the release and synthesis of prolactin, growth hormone, and ACTH.




ISBN 0-12-181958-2

## Establishment of Pituitary Cell Strains

This summary is not intended as a review of all functional pituitary cell systems now in use, but rather as a description of cells now actively being studied in this laboratory. These include the rat GH cells and mouse AtT20/D16 cells.

A variety of cultures from a growth hormone- and prolactin-producing rat pituitary tumor, MtT/W5,[1] were established in 1965.[2] Because the organ-specific function originally characterized was growth hormone production, these cells became known in the literature as "GH cells." It was learned subsequently that certain GH cell strains also produce prolactin.[3] Cells were adapted to growth *in vitro* by alternate culture and animal passage.[2] Dispersed tumor cells were allowed to attach and grow in plastic culture dishes for periods of several days to about 1 month. The surviving attached cells were then harvested and injected into rats of the Wistar-Furth strain. When a new tumor developed, it was removed and a second generation of culture-derived tumor cells was established *in vitro*. This process of alternate passage between culture and animal was repeated several times in order to select for tumor cells that would attach and grow readily in the culture environment.[4]

Several epithelial cell strains have been cloned,[2] and some have been maintained in continuous culture for as long as 10 years without loss of hormone production. Table I gives the characteristics of several clones of GH cells.[2,3,5–15] The $GH_1$ and $GH_3$ strains are available from the American Type Culture Collection (12301 Parklawn Drive, Rockville, Maryland); the Cell Repository Numbers for the strains are CCL 82 and CCL 82.1, respectively.

[1] H. Takemoto, K. Yokoro, J. Furth, and A. I. Cohen, *Cancer Res.* **22**, 917 (1962).
[2] A. H. Tashjian, Jr., Y. Yasumura, L. Levine, G. H. Sato, and M. L. Parker, *Endrocrinology* **82**, 342 (1968).
[3] A. H. Tashjian, Jr., F. C. Bancroft, and L. Levine, *J. Cell Biol.* **47,** 61 (1970).
[4] V. Buonassisi, G. Sato, and A. I. Cohen, *Proc. Natl. Acad. Sci. U.S.A.* **48,** 1184 (1962).
[5] A. H. Tashjian, Jr., N. J. Barowsky, and D. K. Jensen, *Biochem. Biophys. Res. Commun.* **43,** 516 (1971).
[6] T. F. J. Martin and A. H. Tashjian, Jr., *Biochem. Actions Horm.* **4,** 269 (1977).
[7] A. Schonbrunn and A. H. Tashjian, Jr., *J. Biol. Chem.* **253,** 6473 (1978).
[8] H. H. Samuels, J. S. Tsai, and R. Cintron, *Science* **181,** 1253 (1973).
[9] H. H. Samuels, J. S. Tsai, J. Casanova, and F. Stanley, *J. Clin. Invest.* **54,** 853 (1974).
[10] H. H. Samuels, Z. D. Horwitz, F. Stanley, J. Casanova, and L. E. Shapiro, *Nature (London)* **268,** 254 (1977).
[11] D. K. Biswas, J. Lyons, and A. H. Tashjian, Jr., *Cell* **11,** 431 (1977).
[12] P. S. Dannies and A. H. Tashjian, Jr., *J. Biol. Chem.* **248,** 6174 (1973).
[13] P. M. Hinkle and A. H. Tashjian, Jr., *J. Biol. Chem.* **248,** 6180 (1973).
[14] L.-Y. Yu, R. J. Tushinski, and F. C. Bancroft, *J. Biol. Chem.* **252,** 3870 (1977).
[15] U. I. Richardson, *J. Cell. Physiol.* **88,** 287 (1976).

TABLE I: FUNCTIONAL CHARACTERISTICS OF CLONES OF RAT PITUITARY GH CELLS[a]

| Clone | Hormone production | | Specific receptors for and responses to | | | Comments | References |
|---|---|---|---|---|---|---|---|
| | PRL | GH | TRH | SRIH | TH | | |
| $GH_1$ | yes | yes | yes | yes | yes | TRH increases PRL release and synthesis and decreases GH production. TH increases cell growth and GH synthesis | 2,3,5–10 |
| $GH_12C_1$ | no | yes | no | no | —[b] | Synthesis of PRL induced by BrdU | 2,11 |
| $GH_3$ | yes | yes | yes | yes | yes | Similar to $GH_1$ | 2,3,5–10,12,13 |
| $GH_4C_1$ | yes | yes | yes | yes | yes | Similar to $GH_3$ but with higher rate of PRL and lower rate of GH synthesis; SRIH inhibits PRL and GH release and synthesis | 2,3,5–10,12,13 |
| GC | no | yes | yes | yes | yes | Similar to $GH_3$ in regulatory responses but no detectable PRL produced | 14 |
| $F_4C_1$ | yes | yes | no | no | — | No detectable ACTH produced | 15 |
| $BGH_12C_1$ | yes | yes | no | no | — | Subclone of $GH_12C_1$ resistant to BrdU | 11 |
| $F_1BGH_12C_1$ | yes | yes | no | no | — | Similar to $BGH_12C_1$ | 11 |

[a] Abbreviations: PRL = prolactin; GH = growth hormone; TRH = thyrotropin-releasing hormone; SRIH = somatotropin release-inhibiting hormone (somatostatin); TH = thyroid hormone (thyroxine and triiodothyronine); BrdU = 5-Bromodeoxyuridine.

[b] Not determined.

Hormone-producing mouse pituitary tumor cells have also been established in culture. The AtT20/D16 cells are a clonal strain isolated in 1962 by Yasumura[16] from a radiation-induced pituitary tumor in the mouse.[17] These cells have continued to produce large amounts of ACTH during serial propagation in culture for 15 years.[18] Table II gives the functional characteristics of this strain.[19–23] The AtT20/D16 strain is available from the American Type Culture Collection (Cell Repository Number CCL 89).

## Methods of Culture

*Type of Culture.* GH cells can be grown as monolayers on plastic or glass surfaces,[2] in roller bottles, or in suspension culture.[24]

*Medium.* GH cells are not fastidious in their medium requirements. They are usually grown in monolayer in this laboratory in Ham's nutrient mixture F10[25] supplemented with 15% horse serum and 2.5% fetal calf serum at 37 ± 0.5° in a humidified atmosphere of 5% $CO_2$ and 95% air.[2] They also can be grown in Ham's nutrient mixture F12,[26] Eagle's minimum essential medium,[27] and Dulbecco's modified Eagle's medium[28] supplemented with horse serum (5–15%) and 2.5% fetal calf serum. In addition, for experiments of relatively brief duration (1–48 hr) requiring the absence of serum, GH cells can be maintained in a functional hormon-producing and hormone-responsive state in either Neuman and Tytell's serumless medium[29] or in Ham's nutrient mixture F10 supplemented with lactalbumin hydrolysate (5 g/liter). Hayashi and Sato have reported growth of $GH_3$ cells in chemically defined medium supplemented with certain "hormones."[30]

For spinner cultures, modified Eagle's medium for suspension[31] supplemented with 15% horse serum and 2.5% fetal calf serum is used.

[16] Y. Yasumura, *Am. Zool.* **8,** 285 (1968).

[17] J. Furth, *Recent Prog. Horm. Res.* **11,** 221 (1955).

[18] U. I. Richardson, *Endocrinology* **102,** 910 (1978).

[19] R. E. Mains and B. A. Eipper, *J. Biol. Chem.* **251,** 4115 (1976).

[20] R. E. Mains, B. A. Eipper, and N. Ling, *Proc. Natl. Acad. Sci. U.S.A.* **74,** 3014 (1977).

[21] G. Giagnoni, S. L. Sabol, and M. Nirenberg, *Proc. Natl. Acad. Sci. U.S.A.* **74,** 2259 (1977).

[22] J. L. Roberts and E. Herbert, *Proc. Natl. Acad. Sci. U.S.A.* **74,** 5300 (1977).

[23] E. Herbert, R. G. Allen, and T. L. Paquette, *Endocrinology* **102,** 218 (1978).

[24] F. C. Bancroft and A. H. Tashjian, Jr., *Exp. Cell Res.* **64,** 125 (1971).

[25] R. G. Ham, *Exp. Cell Res.* **29,** 515 (1963).

[26] R. G. Ham, *Proc. Natl. Acad. Sci. U.S.A.* **53,** 288 (1965).

[27] H. Eagle, *Proc. Soc. Exp. Biol. Med.* **89,** 362 (1955).

[28] J. D. Smith, G. Freeman, M. Vogt, and R. Dulbecco, *Virology* **12,** 185 (1960).

[29] R. E. Neuman and A. A. Tytell, *Proc. Soc. Exp. Biol. Med.* **104,** 252 (1960).

[30] I. Haysahi and G. H. Sato, *Nature (London)* **259,** 132 (1976). See also [6].

[31] H. Eagle, *Science* **130,** 432 (1959).

TABLE II
FUNCTIONAL CHARACTERISTICS OF CORTICOTROPIN-PRODUCING MOUSE PITUITARY CELLS[a]

| Strain | Specialized products synthesized | | | Specific receptors for and responses to | | | References |
|---|---|---|---|---|---|---|---|
| | ACTH | β-endorphin | Enkephalin | TRH | SRIH | HC | |
| AtT20/D16 | yes | yes | yes | no | yes | yes | 19–23 |

[a] Abbreviations: ACTH = adrenocorticotropic hormone; TRH = thyrotropin-releasing hormone; SRIH = somatotropin release-inhibiting hormone (somatostatin); HC = hydrocortisone (glucocorticoid steroid hormones).

*Subculture.* GH cells are released from substrate for subculture by brief (2–5 min) incubation at 37° with Viokase (0.1%) or trypsin (0.1%) in either phosphate-buffered saline or nutrient mediun lacking serum.[2] Split ratios of 1 : 10 to 1 : 40 are used. Plating efficiencies are 30–50%.

*Growth Rates.* In monolayer culture in Ham's F10 medium, the population doubling time determined from counts of cell number or total cell protein is 45–52 hr. GH cells do not form confluent monolayers, but after a period of logarithmic growth, they reach a plateau density at which the cells cover 60–80% of the available surface.[2] Cell division continues with viable cells being shed into medium at about their rate of division. Hormone production occurs during both exponential growth and plateau density.[2,32]

Values for alkali-soluble protein and intracellular volume for GH cells grown in monolayer were 2.1 mg cell protein per $10^7$ cells and 17 μl per $10^7$ cells, respectively.[33]

## Characteristics of GH Cells

*Stability.* The rates of growth and hormone production by GH cells have generally remained unchanged over many years of continuous propagation. However, some changes have been noted: for example, $GH_3$ cells that initially produced about equal quantities of growth hormone and prolactin[3] now synthesize 5–10 times more prolactin than growth hormone, and in the $GH_4C_1$ strain derived from $GH_3$ cells,[34] the ratio of prolactin to growth hormone is even greater. It is probable that such changes in the ratio of hormone production reflect some unknown current environmental alteration, such as medium composition, because $GH_3$ cells that are thawed from liquid nitrogen change their ratio of prolactin to growth hormone synthesis over a period of several weeks from about 1 : 1 to the same high ratio (about 10 : 1) as serially propagated $GH_3$ cells.

*Karyotype.* GH cells are aneuploid with a modal number of chromosomes ranging from 69 to 76.[35] The modal chromosome number for the AtT20/D16 cells is 76.[18]

*Hormone Production.* As described in Table I, most of the GH cell strains, except $GH_12C_1$ and GC, produce both growth hormone and prolactin although the ratios of hormone synthesis among different clones varies considerably, with $GH_3$ and GC being the highest producers of

[32] O. P. F. Clausen, K. M. Gautvik, and T. Lindmo, *Virchows Arch. B* **23,** 195 (1977).

[33] T. F. J. Martin and A. H. Tashjian, Jr., *J. Biol. Chem.* **253,** 106 (1978).

[34] A. H. Tashjian, Jr., P. M. Hinkle, and P. S. Dannies, *in* "Endocrinology" (R. O. Scow, ed.), p. 648. Excerpta Med. Found., Amsterdam, 1973.

[35] C. Sonnenschein, U. I. Richardson, and A. H. Tashjian, Jr., *Exp. Cell. Res.* **61,** 121 (1970).

TABLE III
RESPONSES OF GH CELLS TO HORMONES AND DRUGS[a]

| Signal | Effect | References |
|---|---|---|
| TRH | *Acute actions* (0–30 min) | |
| | Binds to membrane receptors | 13,40–43 |
| | Quenches membrane fluorescence | 44 |
| | Increases cyclic AMP | 45–47 |
| | Releases stored PRL | 6,42 |
| | Enhances uridine uptake | 33,48 |
| | *Chronic actions* (4–48 hr) | |
| | Increases PRL synthesis | 5,6,12,49,50 |
| | Decreases GH synthesis | 5,6,51 |
| | Decreases TRH receptor number | 6,52,53 |
| | Decreases uridine uptake | 33,48 |
| | Morphologic changes | 51,54 |
| Cortisol | Increases GH synthesis | 10,14,36,55,56 |
| | Decreases PRL synthesis | 12,36 |
| | Increases TRH receptor number | 43 |
| Estrogens | 17β-estradiol increases PRL and decreases GH production | 51,57,58 |
| | Antiestrogens increase PRL synthesis | 57 |
| Thyroid hormone | Increases cell growth | 8 |
| | Increases GH synthesis | 8–10 |
| | Decreases TRH receptor number | 59 |
| Somatostatin | Binds to specific receptors | 7 |
| | Decreases GH release | 7 |
| | Decreases PRL release | 7 |
| Bromocriptine | Decreases PRL synthesis and release | 51,60 |
| Prostaglandins | Increases PRL synthesis | 61 |
| | No change in GH production | 61 |
| Cations | Role of calcium in hormone release and synthesis | 62,63 |

[a] Abbreviations: TRH = thyrotropin-releasing hormone; PRL = prolactin; GH = growth hormone; bromocriptine = 2-bromo-α-ergokryptine (CB-154); cyclic AMP = adenosine 3′:5′-monophosphate.

growth hormone and $GH_4C_1$ producing the most prolactin. No clone yet isolated synthesizes only prolactin.

The specific rates of growth hormone and prolactin synthesis in GH cells can be modulated by a variety of steroid and peptide hormones as well as by drugs added to the culture medium (see below). At maximum stimulation with cortisol and thyroid hormone, $GH_3$ or GC cells can produce as much as 50–400 μg growth hormone/mg cell protein/24 hr.[14,36,37] In comparable experiments at maximum stimulation with thyrotropin-releasing hormone, $GH_4C_1$ can produce 40–150 μg prolactin/mg cell protein/24 hr.[3,5,6] Under these conditions, the fraction of total protein synthesis represented by either growth hormone or prolactin synthesis is 2–10%.

At least 30 serial clones of $GH_3$ cells have been isolated, each derived from a single cell of the preceding clone, and each clone produces both growth hormone and prolactin. This finding suggests strongly, but does not prove, that both hormones are produced by the same cell.

*Somatic Cell Hybrids*. The $GH_12C_1$ clone was hybridized with Clone 1D mouse fibroblasts.[38] Six hybrids clones were analyzed in detail and none produced growth hormone. These hybrid lines were subsequently found to exhibit unexpected resistance to X-irradiation.[39] Subsequently, $GH_3$ and $GH_4C_1$ cells have been hybridized with mouse fibroblasts (Thompson, Dannies, and Tashjian, unpublished data), and a variety of clones have been isolated; some produce no hormone, some produce only prolactin, and some produce both growth hormone and prolactin.

*Modulation of Hormone Biosynthesis and Release in GH Cells*. Table III summarizes the more prominently studied responses of GH cells to hormones and drugs known to affect the function of mammotrophs and somatotrophs in the normal pituitary gland *in situ*.[5–10,12–14,33,36,40–63]

Although less extensive studies have been performed with AtT20/D16 cells, they have been used to examine the negative feedback of adrenoglucocorticoid hormones on ACTH release,[23,64] the ACTH release induced by corticotropin-releasing factor,[23] the ultrashort loop self-regulation of ACTH release by ACTH itself,[18] and the mechanism by which adrenoglucocorticoids enter target cells (using the related AtT20/D-1 cell strain).[65,66]

[36] F. C. Bancroft, L. Levine, and A. H. Tashjian, Jr., *J. Cell Biol.* **43,** 432 (1969).

[37] J. A. Martial, J. D. Baxter, H. M. Goodman, and P. H. Seeberg, *Proc. Natl. Acad. Sci. U.S.A.* **74,** 1816 (1977).

[38] C. Sonnenschein, U. I. Richardson, and A. H. Tashjian, Jr., *Exp. Cell. Res.* **69,** 336 (1971).

[39] J. B. Little, U. I. Richardson, and A. H. Tashjian, Jr., *Proc. Natl. Acad. Sci. U.S.A.* **69,** 1363 (1972).

[40] P. M. Hinkle, E. L. Woroch, and A. H. Tashjian, Jr., *J. Biol. Chem.* **249,** 3085 (1974).

[41] P. M. Hinkle and A. H. Tashjian, Jr., *Endocrinology* **97,** 324 (1975).

[42] P. S. Dannies and A. H. Tashjian, Jr., *Nature* (*London*) **261,** 707 (1976).

[43] A. H. Tashjian, Jr., R. Osborne, D. Maina, and A. Knaian, *Biochem. Biophys. Res. Commun.* **79,** 333 (1977).

[44] T. Imae, G. D. Fasman, P. M. Hinkle, and A. H. Tashjian, Jr., *Biochem. Biophys. Res. Commun.* **62,** 923 (1975).

[45] P. S. Dannies, K. M. Gautvik, and A. H. Tashjian, Jr., *Endocrinology* **98,** 1147 (1976).

[46] K. M. Gautvik, E. Walaas, and O. Walaas, *Biochem. J.* **162,** 379 (1977).

## GH Cells as a Model System

As summarized above, GH cells produce large amounts of growth hormone and prolactin. The biosynthesis and release of these two protein hormones are modulated by specific factors that have been found to influence the pituitary gland in the intact animal. In fact, observations initially made with GH cells were subsequently shown to operate *in vivo.*[5] Unlike the pituitary gland, however, GH cells are clonal populations that can be studied in a highly controlled environment. They therefore represent a useful tool for studies on the molecular mechanisms underlying the actions of hypothalamic peptides on growth hormone and prolactin release and synthesis as well as how peptide hormone receptors are modulated. Because GH cells are neoplastic, their functions and responses may not be identical to comparable cells in the normal animal. It is recognized that tumor cells can lose functions or express functions inappropriately, but it is not likely that they invent new functions or novel mechanisms. Nevertheless, it is essential to perform, wherever possible, control experiments with normal pituitary cells or explants to confirm that mechanisms deduced from tumor cells correspond to normal mechanisms.

---

[47] P. M. Hinkle and A. H. Tashjian, Jr., *Endocrinology* **100,** 934 (1977).

[48] T. F. J. Martin, A. M. Cort, and A. H. Tashjian, Jr., *J. Biol. Chem.* **253,** 99 (1978).

[49] P. S. Dannies and A. H. Tashjian, Jr., *Biochem. Biophys. Res. Commun.* **70,** 1180 (1976).

[50] G. A. Evans and M. G. Rosenfeld, *J. Biol. Chem.* **251,** 2842 (1976).

[51] A. H. Tashjian, Jr. and R. F. Hoyt, Jr., *in* "Molecular Genetics and Developmental Biology" (M. Sussman, ed.), p. 353. Prentice-Hall, Englewood Cliffs, New Jersey, 1972.

[52] P. M. Hinkle and A. H. Tashjian, Jr., *Biochemistry* **14,** 3845 (1975).

[53] P. M. Hinkle and A. H. Tashjian, Jr., *in* "Hormones and Cancer" (K. W. McKerns, ed.), p. 203. Academic Press, New York, 1974.

[54] R. F. Hoyt, Jr. and A. H. Tashjian, Jr., *Anat. Rec.* **175,** 374 (abstr.) (1973).

[55] P. O. Kohler, L. A. Frohman, W. E. Bridson, T. Vanha-Perttula, and J. M. Hammond, *Science* **166,** 633 (1969).

[56] R. J. Tushinski, P. M. Sussman, L.-Y. Yu, and F. C. Bancroft, *Proc. Natl. Acad. Sci, U.S.A.* **74,** 2357 (1977).

[57] P. S. Dannies, P. M. Yen, and A. H. Tashjian, Jr., *Endocrinology* **101,** 1151 (1977).

[58] E. Haug and K. M. Gautvik, *Endocrinology* **99,** 1482 (1976).

[59] M. H. Perrone and P. M. Hinkle, *J. Biol. Chem.* **253,** 5168 (1978).

[60] K. M. Gautvik, R. F. Hoyt, Jr., and A. H. Tashjian, Jr., *J. Cell. Physiol.* **82,** 401 (1973).

[61] K. M. Gautvik and M. Kriz, *Endocrinology* **98,** 352 (1976).

[62] K. M. Gautvik and A. H. Tashjian, Jr., *Endocrinology* **92,** 573 (1973).

[63] K. M. Gautvik and A. H. Tashjian, Jr., *Endocrinology* **93,** 793 (1973).

[64] H. Wantanabe, W. E. Nicholson, and D. N. Orth, *Endocrinology* **93,** 411 (1973).

[65] R. W. Harrison, S. Fairfield, and D. N. Orth, *Biochemistry* **14,** 1304 (1975).

[66] R. W. Harrison, S. Fairfield, and D. N. Orth, *Biochim. Biophys. Acta* **466,** 357 (1977).

# [47] Liver Cells

*By* H. L. LEFFERT, K. S. KOCH, T. MORAN, and M. WILLIAMS

Methods for establishing primary monolayer fetal and adult rat hepatocyte cultures are presented. Techniques are emphasized rather than rationale or methodological validation.[1–4] Media preparations (Tables I and II), plating procedures (Tables III and IV), and buffer formulations (Table V) are outlined with explanatory details provided in the text. Table VI summarizes some differentiated properties of these culture systems.

## Animals

Fetal livers are obtained from pregnant rats 19–21 days in gestation; 14–19-day-old fetuses can be used but hepatocyte yields are reduced. Adult livers are obtained from rats, 150–300 g body weight, and fed standard Purina Chow and water *ad libitum*. Animals are housed at 21° with alternating 12-hr cycles of light and darkness.

## Biological Reagents

Fetal bovine serum is purchased from standard suppliers and is pretested to ensure its suitability. Defrost, mix well, and heat the serum at 56° for 30 min. Dialyze against 0.15 *M* NaCl (1 : 80, v/v) for 24 hr at 4°, followed by two similar changes, and store aliquots at −20° for up to 12 months. Dialysis tubing ($\frac{5}{8}$ inch; Van Waters and Rogers #25225-226) is boiled 5 min in water containing 0.1% ethylenediaminetetracetic acid (EDTA), washed 20 min with deionized water, and stored at 4°. All serum used is dialyzed unless otherwise noted.

Collagenase (Sigma Type I, #C0130) also varies in effectiveness and must be pretested for its suitability. The enzyme preparation is stored desiccated at −20° for up to 6 months.

Amino acids, hydrocortisone-succinate, vitamins, inosine, bovine serum albumin (BSA), and sodium pyruvate are from Sigma; penicillin G and streptomycin-sulfate are purchased from Squibb and Pfizer, respec-

[1] H. L Leffert and D. Paul, *J. Cell Biol.* **52,** 559 (1972).

[2] H. L. Leffert and D. Paul, *J. Cell. Physiol.* **81,** 113 (1973).

[3] H. L. Leffert, T. Moran, R. Boorstein, and K. S. Koch, *Nature (London)* **267,** 58 (1977).

[4] H. L. Leffert, K. S. Koch, B. Rubalcava, S. Sell, T. Moran, and R. Boorstein, *Natl. Cancer Inst., Monogr,* **48,** 87 (1978).




ISBN 0-12-181958-2

TABLE I

CONSTRUCTION OF 8 LITERS OF TWICE-CONCENTRATED (×2) MEDIUM

| Step | Procedure |
|---|---|
| 1. | Thaw: 320 ml amino acid stock<br>160 ml vitamin stock<br>40 ml antibiotic stock |
| 2. | Dissolve 1.152 g L-tyrosine in 500 ml boiling water. Add 3.2 g $MgSO_4 \cdot 7\ H_2O$. Cool to room temperature. |
| 3. | Dissolve 0.748 g L-cystine in 200 ml water (at 60°) containing 7 ml 1 *N* NaOH. Cool to room temperature. |
| 4. | Dissolve 4.2 g $CaCl_2 \cdot 2\ H_2O$, 6.4 g KCl, and 102.4 g NaCl in 1500 ml water. |
| 5. | Add amino acid stock to #4 using stirring bar. Add:<br>24 ml 1% (w/v) phenol red<br>72 g glucose (anhydrous)<br>2.3 g $NaH_2PO_4$<br>0.112 g Inositol<br>16 ml 0.01% (w/v) $FeCl_3$<br>40 ml antibiotic stock<br>16 ml antimycotic<br>160 ml vitamin stock<br>Let stir for 5–10 min. |
| 6. | Add 9.34 g L-glutamine and 1.76 g sodium pyruvate to #5. Stir. |
| 7. | Add cooled solutions (#2, #3) to #6. Stir. |
| 8. | Bring to 3200 ml with water; bubble $CO_2$ for about 1 hr while adding 800 ml water containing 59.7 g $NaHCO_3$. |
| 9. | Filter through 0.22 $\mu$m Millipore GS filter. |
| 10. | Dispense into sterile bottles and store in the dark at 4° for up to 6 weeks. |

TABLE II

STOCK SOLUTIONS FOR ARGININE-FREE MEDIUM

| | | | |
|---|---|---|---|
| *Amino acids* (4 liters) | | *Vitamins* (2 liters) | |
| L-Glycine | 6.00 g | Choline chloride | 0.80 g |
| L-Histidine HCl | 8.40 g | Folic acid | 0.80 g |
| L-Isoleucine | 20.96 g | Nicotinamide | 0.80 g |
| L-Leucine | 20.96 g | Pantothenic acid | 0.80 g |
| L-Lysine | 29.24 g | Pyridoxal HCl | 0.80 g |
| L-Methionine | 6.00 g | Thiamine | 0.80 g |
| L-Ornithine | 16.80 g[a] | Riboflavin | 0.08 g |
| L-Phenylalanine | 13.20 g | | |
| L-Serine | 8.40 g | *Antibiotics* (330 ml) | |
| L-Threonine | 19.04 g | Penicillin G | $30 \times 10^6$ U |
| L-Tryptophan | 3.20 g | Streptomycin sulfate | 6.0 g |
| L-Valine | 18.72 g | | |
| | | *Antimycotic* (1 liter) | |
| | | *n*-Butyl-*p*-hydroxybenzoate | 0.1 g |

[a] Can be omitted or added from a standard sterile stock solution (332 mg/100 ml water; add 2.5 ml/100 ml ×1 medium).

TABLE III
FETAL RAT LIVER CELL CULTURE PLATING PROCEDURE

| Step | Manipulation |
|---|---|
| 1. | Lightly etherize pregnant rat; immobilize the animal on its back and soak entire abdomen with 95% (v/v) ETOH. |
| 2. | Make a midline incision with a sterile scissors, cutting first skin and then muscle layers; do not perforate intestines, which lie under muscle layer. |
| 3. | Remove fetuses one at a time from embryonic sacs. Cut umbilical cord and decapitate immediately. Place about 10–15 into a sterile 9-cm dish. |
| 4. | Dissect out intact livers sterilely and put about 15 into a 5-cm dish containing 5 ml ornithine-minus medium at 37° supplemented with 10% (v/v) serum. Inspect each liver to ensure absence of nonhepatic tissue. |
| 5. | Cut each liver into its four separate "lobes"; remove medium by aspiration and wash once with 5 ml of fresh medium. |
| 6. | Pool washed, cut tissue and incubate in fresh collagenase-supplemented digestion buffer (1–2 ml/liver), for 8–10 min at 37° in a 125-ml Erlenmeyer flask containing a magnetic stirring bar. Set rheostat = 35–40. |
| 7. | Tilt flask to allow undigested tissue to settle and remove cell suspension, with a sterile Pasteur pipette, placing it into a "storage" flask containing 10 ml medium (maintain on ice). |
| 8. | Repeat tissue digestion steps 6 and 7, a total of 4–5 times, pooling each subsequent harvest. Add an equal volume of fresh ice-cold medium to the final cell suspension and swirl gently. Dispense 20–40 ml (by pouring) into sterile 50-ml plastic tubes. |
| 9. | Centrifuge 10 sec at full speed and apply brake immediately thereafter. |
| 10. | Remove supernatant liquid by suction and discard; wash pellet once with 10 ml of serum-free medium (blow fluid rapidly against sidewall to disrupt pellet and then take up the cell suspension; blow it out gently against sidewall). Repeat step 9. |
| 11. | Discard second supernatant liquid. Resuspend pellets as in step 10, pool (store on ice), and titer a 0.5-ml aliquot. |
| 12. | Pipette cell suspensions into appropriate media and plate using a flow syringe. |

tively. Eastman Kodak produces *n*-butyl-*p*-hydroxybenzoate. Porcine insulin is obtained from the Eli Lilly Research Laboratory.

### Inorganic Reagents

Triple-distilled water is used unless otherwise noted. Tris base is from Sigma.

### Tissue Culture Supplies

*Equipment.* Tissue culture incubators supplied with $CO_2$ and air line filters, regulatory valves (settings are 0.5 and 4.0 for $CO_2$ and air, respec-

TABLE IV
ADULT RAT LIVER CELL CULTURE PLATING PROCEDURE

| Step | Manipulation |
|---|---|
| 1. | Lightly etherize rat. Immobilize the animal on its back and soak entire abdomen with 95% ethanol. |
| 2. | Make a midline incision extending into the thoracic cavity. Exsanguinate the animal by cardiac puncture, removing 5–8 ml blood, and push the intestinal contents aside everting them to expose portal vessels. |
| 3. | Clamp portal vein (to the left of splanchnic and gastric venous inputs) together with the celiac trunk (small pulsating arterial system dorsal to everted portal system). Clamp inferior vena cava above the renal venous inputs but below the hepatic vein. |
| 4. | Insert a 20-ml plastic syringe equipped with a 25-gauge needle (bevel up) and perfuse 20 ml hormone, purine, and collagenase-supplemented "digestion buffer" for 2–3 min through the liver. Complete tissue blanching; a light tan color indicates adequate perfusion. |
| 5. | Excise perfused liver, placing it in a 9-cm plastic petri dish and removing extraneous tissue if present. Wash liver twice with 10 ml of "complete medium." Add 10 ml of fresh medium and dissect tissue (two livers) into 5 × 8 mm pieces and wash twice. |
| 6. | Incubate cut tissue at 37° with 40 ml of hormone, purine, and collagenase-supplemented digestion buffer for 8–10 min in a 125-ml Erlenmeyer flask containing a magnetic stirring bar. Set rheostat = 35–40. |
| 7. | Tilt flask to allow undigested tissue to settle, and remove cell suspension with a sterile Pasteur pipette, placing it into a flask containing 10 ml of medium (maintain on ice). |
| 8. | Repeat tissue digestion (steps 6, 7) four times, pooling each subsequent harvest. Add 160 ml ice-cold "complete" medium (a total volume of about 320 ml) and dispense 40 ml into eight 50-ml sterile plastic centrifuge tubes. |
| 9. | Centrifuge 13 sec (full speed) and apply brake. |
| 10. | Remove supernatant liquid by suction and discard. Wash pellet once with 10 ml of "basal" medium; blow fluid rapidly against the sidewall to disrupt pellet and then take up the cell suspension and blow it out gently against sidewall. Repeat step 9. |
| 11. | Remove by suction and discard clear supernatant fluid; pellets should be "white." Resuspend pellets as in step 10 twice but blow out hard. Transfer to ice-cold sterile bottle, swirl pooled cell suspension (about 80 ml), and resuspend seven times. Titrate a 0.5-ml aliquot. |
| 12. | Redistribute in portions of 20 ml or less to centrifuge tubes and repeat step 9. |
| 13. | Resuspend pellets in 10 ml (twice, hard) of appropriate plating media. Bring up to needed volume and plate using a 10-ml plastic pipette. |

tively), and external thermometers are from Napco (#3321) or Shel-Lab (Portland, Oregon). Water trays are used (gas is bubbled through water) to maintain humidity. Calibrate temperature (use internal thermometers at different tray levels) and gas tension (use test media and check pH). Napco also supplies a small air incubator (#310) used to perform tissue digestion and trypsinizations at 37°. Cell numbers are determined electronically with a Coulter Particle Counter (Model Fn, Hialeah, Florida).

TABLE V
BUFFER SOLUTIONS

| | | |
|---|---|---|
| *$Ca^{2+}Mg^{2+}$-free concentrate* (2 liters) | | |
| NaCl | 64.0 | g |
| KCl | 3.04 | g |
| $Na_2HPO_4$ (anhyd.) | 0.08 | g |
| Tris base | 24.0 | g |
| 1 *N* HCl | 160 | ml |
| (Adjust pH to 7.4) | | |
| *$Ca^{2+}Mg^{2+}$-free Tris buffer* (2 liters) | | |
| $Ca^{2+}Mg^{2+}$-free concentrate | 500 | ml |
| Water | 1500 | ml |
| (Adjust pH to 7.4) | | |
| *Tris buffer* (3 liters) | | |
| $Ca^{2+}Mg^{2+}$-free concentrate | 750 | ml |
| Water | 2250 | ml |
| 15% (w/v) $CaCl_2 \cdot 2\ H_2O$ | 2 | ml |
| 15% (w/v) $MgCl_2$ | 2 | ml |
| (Adjust pH to 7.4) | | |
| *Digestion buffer* (1 liter) | | |
| KCl | 0.224 | g |
| NaCl | 7.587 | g |
| $NaH_2PO_4 \cdot H_2O$ | 0.138 | g |
| D-glucose | 1.802 | g |
| Phenol red | 0.0012 | g |
| Insulin[a] | 0.010 | g |
| Hydrocortisone-succinate[a] | 0.010 | g |
| Inosine[a] | 0.010 | g |
| (Adjust pH to 7.4) | | |
| *Trypsin concentrate* (1 liter) | | |
| Difco 1:250 trypsin | 25 | g |
| Water | 1.0 | liter |

*Steps:*
1. Dissolve 8 hr at 4° with stirring; let settle.
2. Filter supernatant at 21° using Whatman No. 1 paper.
3. Sterilize with 0.22 $\mu$m GS Millipore and AP2512450 prefilter.
4. Store 60 ml portions at −20°.

| | | |
|---|---|---|
| *Trypsin stock solution for cell counts* (0.6 liter) | | |
| Trypsin concentrate | 60 | ml |
| 1% phenol red | 2 | ml |
| (Filter through 0.45 $\mu$m Nalgene filter) | | |
| Sterile $Ca^{2+}Mg^{2+}$-free Tris buffer | 540 | ml |
| Store 10 ml portions at −20°. | | |

[a] Freshly added for adult liver cell isolation only.

A desk-top centrifuge containing a timer, brake, and rpm indicator is obtained from International Equipment Co. (#HN-S, Needham Heights, Massachusetts). A #958 head, adapted for IEC #325 rings and #305 buckets with rubber stoppers, is used.

The electric rheostat-controlled stirrer, from A. H. Thomas (Magne-Stir, #10), is placed into the Napco air incubator.

Aluminum tissue culture trays (12 × 30 cm) with grated open sides (1-cm high) and bottoms were constructed by this Institute's engineering shop.

Magnetic stirring bars from Van Waters and Rogers (#58948-171; 2 × $\frac{5}{16}$ inch) are replaced monthly.

TABLE VI
DIFFERENTIATED PROPERTIES OF PRIMARY MONOLAYER RAT LIVER CELL CULTURES

| | Fetal | Adult |
|---|---|---|
| *Enzymes* | | |
| Gluconeogenic | + | + |
| Glutathione S-transferase B[EC 2.5.1.18] | U[a] | +[b] |
| Glycogenolytic | + | + |
| Microsomal $P_{448,450}$ | U | + |
| Ornithine transcarbamylase [EC 2.1.3.3] | + | + |
| Pyruvate kinase [EC 2.7.1.40] | | |
| Fetal (type III) | + | +[b] |
| Adult (type I) | U | +[b] |
| Tyrosine aminotransferase [EC 2.6.1.5] | U | + |
| Urea "cycle" | + | + |
| *Secretory proteins* | | |
| Albumin | + | + |
| $Alpha_1$-fetoprotein | + | +[b] |
| Haptoglobin | + | U |
| Hemopexin | + | U |
| Very low density lipoprotein | −[c] | +[b,c] |
| *Ultrastructure*[c] | | |
| Bile "canaliculi" | + | + |
| Desmosomes | + | + |
| Glycogen granules | + | + |
| Peroxisomes | + | + |
| Tight junctions | + | + |

[a] Unknown.

[b] Growth state dependent (H. L. Leffert *et al., Proc. Natl. Acad. Sci. U.S.A.* **75,** 1834 (1978).

[c] Unpublished observations (D. Weinstein, K. Dempo, S. Sell, and H. L. Leffert).

*Plasticware and Glassware.* Falcon Plastics supplies sterile 50-ml (#2070) and polypropylene test tubes (#2059), petri dishes, 35-mm diameter tissue culture dishes (#3001), and 1-ml (#7506) and 10-ml (#7530) pipettes. NUNC (Roskilde, Denmark) supplies 5-cm (#1400) and 9-cm (#N-1415) diameter tissue culture dishes. Dish lot numbers should be noted in case of toxicity problems.

Nalgene 0.20 and 0.45-$\mu$m filters (#120-0020 and #245-0045) are purchased from Sybron (Rochester, New York) and should be checked for tightness of seal and detergent contamination (removable with warm-water washes).

Glass Erlenmeyer flasks (125 ml; #26500 or #26650) and other routine items are purchased from Kimax.

Cornwall continuous pipetting outfits (5-cc "flow syringes") are from

Becton Dickinson (#3054) and are fitted with autoclavable $\frac{1}{8}$-inch ID × $\frac{1}{4}$-inch OD silastic tubing from Dow Corning (#601-365).

### Surgery

Sterile technique is preferred but not necessary; alcohol-washed areas and instruments suffice. Animals are placed onto clean absorbent cotton pads layered over a simple flat cork board and operated on between 0900–1200 hours.

### Media and Digestion Buffer Construction

Twice-concentrated medium is prepared as described in Table I. Amino acid, vitamin, and antibiotic stock solutions (Table II) are prepared in the appropriate portions and stored in darkness at −20° until use. The antimycotic stock solution is stored in darkness at room temperature. L-Ornithine is usually present but is not necessary for fetal liver cultures[1–4]. Arginine is added after plating only under special conditions.[5–9]

One hundred milliliters of ×1 medium at pH 7.4 is prepared fresh as follows: 45 ml sterile ×2 medium; L-ornithine, if desired (see Table II); serum (0–30 ml); and sterile water to make a final volume of 100 ml. If contamination occurs, add 0.25 ml of antibiotic stock.

For the adult system (Table IV), "complete medium" contains L-ornithine, 15% (v/v) serum, and 10 μg/ml each of insulin, hydrocortisone, and inosine. "Basal medium" is complete medium minus hormone and purine supplements. Stock solutions of hormone or purine, 5 ml, are prepared fresh as 1 mg/ml solutions in polypropylene test tubes in 0.15 *M* NaCl containing 0.3% (w/v) BSA. One drop of 6 *N* HCl will dissolve insulin in the 5-ml solution. Supplements, 1 ml/100 ml medium, are added to the medium followed by sterilization through a 0.20-μm Nalgene filter.

Digestion buffer (Table V) is stored at −20° until use. Fresh collagenase, 3 mg/ml, is added to buffer at 37°; after dissolving, the solution is sterilized (as above) and preincubated at 37° for 1–4 hr prior to use. For adult liver tissue digestion only, prior to sterilization, 100 ml of buffer

[5] H. L. Leffert and S. Sell, *J. Cell Biol.* **61,** 823 (1974).
[6] H. L. Leffert, *J. Cell Biol.* **62,** 767 (1974).
[7] K. Koch and H. L. Leffert, *J. Cell Biol.* **62,** 780 (1974).
[8] H. L. Leffert, *J. Cell Biol.* **62,** 792 (1974).
[9] H. L. Leffert and D. B. Weinstein, *J. Cell Biol.* **70,** 20 (1976).

containing enzyme is supplemented with 1 ml each of the insulin, hydrocortisone, and inosine stock solutions.

## Cell Isolation and Plating

Detailed procedures appear in Table III (fetal) and Table IV (adult). All steps are performed with sterile technique. The stirring speed, i.e., rheostat knob setting, for all tissue digestion is 35–40. For fetal cultures, four animals are used and usually yield 20–40 fetuses giving rise to about 1–3 $\times$ $10^8$ cells. For adult cultures, two livers are used giving rise to about 1–2 $\times$ $10^8$ cells. For both systems, the time from surgery to plating should be about 3 hr or less. At plating, greater than 95% of isolated cells are hepatocytes[3] with 70–95% viability as judged by exclusion of 0.05% (w/v) trypan blue at 1 min.

Seeding densities, i.e., numbers of cells per milliliter medium, and volume of media per dish vary with experimental design.[1–3] For the fetal system, usual parameters for 35-, 50- and 90-mm diameter dishes are 1.25 $\times$ $10^5$ cells/ml, 2 ml; 1.5 $\times$ $10^5$ cells/ml, 5 ml; and 2 $\times$ $10^5$ cells/ml, 10 ml, respectively. For the adult system, 5 $\times$ $10^5$ cells/ml, 2 ml, are plated per 35-mm dish. "Healthy" cultures show cellular aggregates upon the surface of the dish within 24 hr and flattening (nonrefractile nuclear and cytoplasmic morphology under phase optics) within 48 hr. Medium changes are performed as needed but are not necessary. In this laboratory, functional fetal cultures plated without L-ornithine have survived for 6–8 weeks without a medium change; functional adult cultures plated with L-ornithine have survived for 2 weeks without a medium change.

## Cell Counts

At plating, titrations are performed with 0.5 ml of cell suspension added to 9.5 ml of Isoton (Scientific Products, #B3157-11) and counted electronically (threshold at 10 $\mu$m; attenuation and aperture settings of 1 and 256, respectively). Hemocytometer counts should be within $\pm$10% of corrected electronic counts using these settings.

Table V lists buffer solutions used to measure attached cell numbers. Dishes are washed twice with Tris buffer, pH 7.4 (a volume equal to initial medium volume). Then an equal volume of $Ca^{2+}Mg^{2+}$-free Tris buffer with trypsin (10 ml trypsin stock solution per 180 ml $Ca^{2+}Mg^{2+}$-free Tris chloride at pH 7.4) is added and the dishes incubated for 30–40 min at 37°. A 2 ml (#122402010) or 5 ml (#122405010) glass handpipette (Bellco) equipped with a rubber bulb (Scientific Products, #R5000-1) is useful for detaching the cells; vigorous pipetting, about 20 times, over the dish surface usually reduces cell aggregation. The cell suspension or an aliquot of it is brought up to 10 ml with Isoton and counted.

### Functional Properties

Typical differentiated properties of primary monolayer rat liver cell cultures are summarized in Table VI. Additional sources, including those cited above, should be consulted.[10-19]

[10] S. Sell, H. L. Leffert, U. Mueller-Eberhard, S. Kida, and H. Skelly, *Ann. N. Y. Acad. Sci.* **259,** 45 (1975).
[11] H. Watabe, H. L. Leffert, and S. Sell, *in* "Onco-Developmental Gene Expression" (W. Fishman and S. Sell, eds.), p. 123. Academic Press, New York, 1976.
[12] D. M. Bissell, L. E. Hammaker, and U. A. Meyer, *J. Cell Biol.* **59,** 722 (1973).
[13] R. J. Bonney, *In Vitro* **10,** 130 (1974).
[14] C. Guguen, C. Gregori, and F. Schapira, *Biochimie* **57,** 1065 (1975).
[15] G. Michalopolous and H. C. Pitot, *Exp. Cell Res.* **94,** 70 (1975).
[16] T. H. Claus, S. J. Pilkis, and C. R. Park, *Biochim. Biophys. Acta* **404,** 110 (1975).
[17] G. M. Williams, *Am. J. Pathol.* **85,** 739 (1976).
[18] D. Bernaert, J. C. Wanson, P. Drochmans, and A. Popowski, *J. Cell Biol.* **74,** 878 (1977).
[19] H. L. Leffert and K. S. Koch, *Ciba Found. Symp.* **55** (New Ser.), 61 (1978).

## [48] Liver Cells (HTC)

*By* E. BRAD THOMPSON

HTC (hepatoma, tissue culture) cells are a line of cells established from the ascites form of the rat hepatoma, Morris 7288C.[1] The original hepatoma was dissected from the liver of a male rat of the inbred Buffalo strain that had been fed the carcinogen, *N,N'*-2,7-fluorenylenebis-2,2,2-trifluoroacetamide. It subsequently was converted to the ascites form and carried serially as such. The HTC line was cultured directly from the ascitic fluid of tumor-bearing rats.

HTC cells have proved useful in studying a number of both liver-specific and general cellular processes at the cellular level. The cells have been used especially for studies of the effect of steroid hormones on enzyme induction. Glucocorticoids induce in these cells the hepatic marker enzyme tyrosine aminotransferase (L-tyrosine: 2-oxoglutarate aminotranserase, EC 2.6.1.5), as well as ornithine decarboxylase and glutamine synthetase. In HTC cells, such steroids also increase cell adhe-

[1] E. B. Thompson, G. M. Tomkins, and J. F. Curran, *Proc. Natl. Acad. Sci. U.S.A.* **56,** 296 (1966).



ISBN 0-12-181958-2

siveness; reduce the activity of cyclic 3′,5′-AMP(cAMP) phosphodiesterase and of plasminogen activator; induce an increase in one of the isoaccepting forms of phenylalanyl tRNA; induce mouse mammary tumor virus in infected cells; render cells more sensitive to cAMP analogs; and inhibit the transport of certain amino acids. Besides studies on the action of steroid hormones, HTC cells have been utilized in studying the control of fatty acid and cholesterol synthesis, the viral-like A particles, RNA viruses (C particles), growth control, phenylalanyl hydroxylase, and an extensive list of other enzymes. The cells also have been used in somatic cell genetics, through formation of somatic cell hybrids with various other cell types and through the isolation of stable cell variants cloned from the original cell line.

## Propagation of HTC Cells in Monolayer Culture

HTC cells are a hardy line that grow readily as monolayer cultures on either glass or plastic substrata. They have been successfully carried in both commercially made special glass tissue culture flasks and in simple medicine bottles laid on side. Plastic tissue culture flasks and dishes from virtually all major U.S. suppliers also have been employed to grow these cells. HTC cells in monolayer culture, being transformed cells, grow in an irregular, trabecular pattern, display both round and polygonal cell morphology, and do not exhibit density-dependent inhibition of growth. After being seeded so as to form a lawn of individual cells, they grow atop one another as well as side to side, and frequently reach stationary phase with areas of the substratum still unoccupied by cells.

### *Media*

HTC cells are not particularly fastidious with regard to medium. They were originally established and maintained for years in Swim's medium #77, (modified from 5103, with hydroxyproline omitted[2]) supplemented with 20% bovine plus 5% fetal bovine serum. The accompanying table lists the ingredients of this medium. Later it was found that they grow in a number of standard media, including Eagle's minimal essential medium,[3] Eagle's basal medium,[4] Ham's F12,[5] and improved minimal essential medium, zinc option (IMEM-ZO).[6] For routine culture, these media usually

[2] H. E. Swim and R. F. Parker, *J. Lab. Clin. Med.* **52,** 309 (1958).

[3] H. Eagle, *Science* **130,** 432 (1959).

[4] H. Eagle, *Science* **122,** 501 (1955).

[5] R. G. Ham, *Proc. Natl. Acad. Sci. U.S.A.* **53,** 288 (1965).

[6] A. Richter, K. K. Sanford, and V. J. Evans, *J. Natl. Cancer Inst.* **49,** 1705 (1972).

INGREDIENTS OF MEDIUM S77 AS USED FOR THE CULTURE OF HTC CELLS

| | Concentration (m$M$) | | Concentration (m$M$) |
|---|---|---|---|
| *Salts* | | *Amino Acids* | |
| Calcium chloride (2 $H_2O$) | 1.8 | L-Alanine | 0.3 |
| Magnesium sulfate (7 $H_2O$) | 0.8 | L-Aspartic acid | 0.15 |
| Potassium chloride | 5.5 | L-Arginine (HCl) | 0.8 |
| Sodium chloride | 116.2 | L-Cystine (HCl) | 0.05 |
| Sodium bicarbonate | 26.2 | L-Histidine (HCL · $H_2O$) | 0.05 |
| $NaH_2PO_4 \cdot H_2O$ | 1.2 | Glycine | 0.15 |
| | | L-Glutamine | 2.0 |
| *Vitamins, miscellaneous* | | L-Isoleucine | 0.2 |
| Glucose | 5.6 | L-Leucine | 0.2 |
| Choline bitartrate | 0.01 | L-Lysine (HCl) | 0.2 |
| Folic acid | 0.005 | L-Methionine | 0.1 |
| Calcium pantothenate | 0.005 | L-Phenylalanine | 0.1 |
| Mesoinositol | 0.01 | L-Proline | 0.15 |
| Nicotinamide | 0.005 | L-Serine | 0.2 |
| Pyridoxal · HCl | 0.005 | L-Threonine | 0.4 |
| Riboflavin | 0.001 | L-Tryptophan | 0.05 |
| Thiamin HCl | 0.005 | L-Tyrosine | 0.1 |
| Phenol red (sodium salt) | 0.02 | L-Valine | 0.4 |

need supplementation with only 5% fetal calf serum. Although it is not necessary, heat inactivation of the serum (56° for 60 min) to destroy complement and to reduce the chance of contamination with mycoplasma is recommended. In IMEM-ZO, the serum requirement can be reduced sharply, and the cells grow readily in as little as 0.25% fetal calf serum.

HTC cells also have been adapted to growth in totally defined, serum-free medium without macromolecular supplements. For this purpose, the medium of choice is IMEM-ZO. All these media are bicarbonate-buffered and therefore should be used in an incubator kept at 37° with a humidified atmosphere of 5% $CO_2$ in air.

### *Routine Culture*

Routine cultures are fed 3 times a week by removing the medium from the monolayer and replacing it with fresh medium at 37°. The cells double about every 24 hr under these conditions, and for best viability they should be kept in logarithmic growth phase by subculturing appropriately into fresh culture flasks or dishes. They can be subcultured from near-confluent flasks at a ratio of 1 : 10 or 1 : 20 with virtually 100% viability. After such subculturing, they should show a lag phase of no more than 24 hr, and less if handled expeditiously. For subculturing, cells can be removed from their substratum in any of several ways. The preferred methods is to use dilute trypsin or trypsin-EDTA as follows:

Remove spent growth medium by suction. Gently introduce 5–10 ml of phosphate-buffered saline (PBS) (138 m*M* NaCl, 2.7 m*M* KCl, 8.1 m*M* $Na_2HPO_4$, 1.47 m*M* $KH_2PO_4$, pH 7.4) at 25°, and tip the dish to wash over cells. Remove by suction. Introduce 1–2 ml of Puck's Saline A (containing, in grams per liter: 20 mg phenol red, 0.35 g $NaHCO_3$, 8.0 g NaCl, 0.4 g KCl, and 1.0 g dextrose), which contains 0.05% trypsin and 0.02% ethylenediaminetetracetic acid (EDTA), prewarmed to 37°. Tip to wash over cells. Remove by suction all but approximately 0.5 ml. Place in a 37° incubator and examine after 5 min. Tap the dish against the heel of the hand to dislodge cells, which should come free in 5–15 min. As soon as cells float free, add serum-containing medium or soybean trypsin inhibitor at 37° and reflux-pipette cells gently about 3 times to give a suspension of single cells. Plate appropriate portions of cell suspension into fresh dishes.

Cautions: If trypsin-EDTA is interrupted too soon or allowed to continue overlong, cell clumping will result. In addition, with overlong trypsinization, there may be loss of viability. The pH during these procedures should not be allowed to rise above 8.0; the inclusion of phenol red in the trypsin-EDTA solution allows visual monitoring of the approximate pH. If a large number of cultures are to be subcultured, it is recommended that they be carried through the procedure a few at a time, and that those waiting to be handled be kept in the air–$CO_2$ atmosphere of the incubator.

Alternate methods of cell removal for subculturing that can be used successfully are: scraping with an L-shaped glass rod, either naked or with its operative end covered by a silicone rubber tip; trypsin alone, up to 0.1 g/100 ml in saline A or PBS; scraping with a Teflon-coated magnetic bar inside the tissue culture flask and a stronger magnet outside moved to manipulate the inner magnet. EDTA alone is not as satisfactory as the other methods, although it will remove some cells.

For cells grown in serum-free medium, best results are obtained with the trypsin-EDTA technique, using soybean trypsin inhibitor to stop proteolysis. If the introduction of even this small amount of protein is to be avoided, scraping is the second-best choice.

*Cloning*

HTC cells can be cloned with high efficiency by any of several standard techniques in standard medium. If good tissue culture technique is used, cloning efficiency should be 50–100%. Methods that have produced good results include cloning in plastic tissue culture dishes or flasks, in multiwell dishes, and in soft agar. To clone in a tissue culture dish or flask, cells are trypsin-EDTA treated as described above (if grown monolayer-style) or simply taken as grown if in suspension culture (see below). In any case, for best results logarithmic-phase cells should be cloned; they are easier to obtain in monocellular suspension and give higher cloning efficiency. The original cell suspension should be counted carefully by hemocytometer or electronic counter and in either case examined by phase-contrast microscopy to be sure the closest possible approximation to a monocellular suspension is achieved. Proper attention to procedure should result in 90–95% of all cells existing independently, with the remainder appearing as doubles and only rare groups of three or more. It is worth performing preliminary experiments to be sure this degree of dispersion is achieved, since most cloning experiments are aimed at obtaining truly clonal populations, and contamination with greater than 5% with pseudoclones, derived from more than a single cell, may complicate subsequent results unnecessarily.

The dispersed cells are serially diluted with growth medium at 37°, taking care not to pipette too vigorously, until a convenient dilution is reached. If the cells are to be cloned *en masse* in large plates or flasks, a seeding concentration of 1–2 cells $cm^2$ results in a yield of colonies convenient for counting or picking for subculture. Larger numbers may be plated, of course, if selective conditions resulting in survival of only a fraction of the seeded cells are to be used. To cleanly pick colonies, each colony is first isolated in a stainless-steel ring and then trypsinized, as described by Ham and Puck.[7] Because of the possibility that multiple colonies obtained from a single dish may include either seed colonies from an initial site or cross-contaminated colonies from cells detached from other, large clones, it is recommended that attempts to obtain truly clonal populations can be carried out either in multiwell dishes or in soft agar. For the former, good results can be obtained with the standard 96-well dishes at a volume of about 0.1 ml per well. If cells are diluted to average one cell per five wells at the time of seeding, then about 95% or more of the subsequent colonies will have risen from a single cell. Seeding at higher numbers of cells per well results in significantly greater numbers of nonclonal colonies. Care must be exercised to keep the parent cell suspension well mixed during the seeding. This can be done by frequent manual mixing or by use of a sterile Teflon-coated magnet in the suspended "mother" culture, turning slowly under the control of an external magnetic stirrer. In addition, using no more than a 2-ml pipette to seed the 0.1-ml aliquots will help avoid nonrandom cell distribution. In larger pipettes or with very slow pipetting, cells may accumulate along the

[7] R. G. Ham and T. T. Puck, This series, Vol. 5, [9] p. 107.

pipette walls, resulting in irregularities in the number of cells delivered per unit volume. When sufficient cells have grown in the wells, they can be replica-plated if desired. Commercial devices for replica-plating from such dishes are available.

To clone in agar, either of two methods may be used. High cloning efficiencies result with either in a final agar concentration of 0.3–0.5%. Agarose, at 0.35%, works equally well. In the first procedure, a stock agar solution is made by autoclaving 2 g agar (Difco) in 100 ml of PBS. This is kept in solution in a 56° water bath. Cells to be cloned are diluted to 1.25 times the final desired concentration in normal growth medium at 37°. The agar is allowed to cool to near gelling temperature; then it is quickly mixed with the cell suspension, 1 part agar to 4 parts cells, and plated in appropriate dishes. The freshly plated dishes are transferred to a refrigerator for a few minutes to assist uniform gelling and then moved to the 37° incubator. The 20% dilution of medium nutrients incurred in this procedure does not seem to affect the cells adversely. An alternate procedure, however, avoids this dilution. The agar or agarose is made as an aqueous solution and autoclaved as above. A special supply of growth medium is prepared, containing twice the normal concentration of salts, nutrients, and serum. Cells grown normally are diluted down to 10–100 times the final desired concentration. These are then rapidly diluted to the final concentration in the concentrated medium, which is at once mixed 1 : 1 with the agar (or agarose) and plated as described before. When clones appear, they may be removed from the agar bed with a sterile, cotton-plugged Pasteur pipette and transferred to other vessels for propagation as monolayer cultures.

### *Serum-Free Growth of HTC Cells*

For some purposes, such as defining growth factors and amino acid and cofactor requirements, developing auxotrophs, and studying sterol synthesis and certain hormone actions, cells grown in defined medium offer many advantages. HTC cells can be adapted to growth in totally defined medium lacking macromolecular supplements.[8] For this, the medium of choice is IMEM-ZO. The cells can be adapted directly to the serum-free medium, or carried for a few weeks in intermediate levels of fetal calf serum, for example 1% followed by 0.25%. In the defined medium, the cells undergo a brief adaptation period and then grow with a doubling time only a few hours longer than that in 5% fetal calf serum. Cell transfer of the cells grown without serum can be accomplished by scraping the monolayer with an L-shaped glass rod with or without a

[8] E. B. Thompson, C. U. Anderson, and M. E. Lippman, *J. Cell. Physiol.* **86**, 403 (1975).

silicone rubber sheath, or by trypsinization. If trypsinization is used it should be noted that the serum-free cells are exquisitely sensitive to the proteolytic enzyme, and should be monitored constantly during the process. Overlong treatment results in rapid loss of cell viability; consequently the standard procedure for serum-grown cells is modified as follows. Growth medium is removed and a few milliliters of trypsin-EDTA solution introduced, washed over the cells, and all but 0.1 ml immediately removed. The cell layer is left with only this slight film of liquid, and as soon as the cells loosen from the substratum a slight excess of soybean trypsin inhibitor in PBS is added. The cells can be centrifuged for 3 min at 800 g and resuspended in growth medium or simply diluted into growth medium for transfer. They should not be subdivided at a ratio below 1 : 10, and 1 : 4 is recommended for best viability. The medium containing the trypsin and inhibitor is removed as soon as the transferred cells have adhered to the plastic substratum. The advantage of trypsin for cell transfer is that it results in a well-dispersed cell monolayer; the obvious disadvantage is that the process briefly introduces exogenous proteins into the culture system.

As with many cell lines, HTC cells grown in defined medium are more sensitive to mishandling than are serum-grown cells. Care to avoid rough pipetting, pH extremes, and toxic contaminants in medium, glassware, and plastics must be taken. The cells clone poorly in serum-free medium, with an efficiency not exceeding 1%.

## Propagation of HTC Cells in Suspension Culture

HTC cells have been adapted to grow in suspension (spinner) culture in a modified form of Swim's S77 medium, in which $CaCl_2$ is omitted, $NaHCO_3$ is reduced to 0.5 g/liter, glucose is increased to 3 g/liter, and 0.05 *M* Tricene at pH 7.6 is added.[9] (This medium works equally well with only 1 g/liter glucose.) The medium is supplemented with 10% serum, at least 5% of which should be fetal calf serum. In this medium the cells can be grown in tightly capped bottles without added $CO_2$. Suspension culture also can be carried out in the usual bicarbonate-buffered S77 monolayer medium from which $CaCl_2$ is omitted. In this case, to maintain proper pH, the cultures must be grown in 5% $CO_2$ in air. Any of several vessels have been used for this form of culture: commercial suspension culture flasks, with Teflon-coated magnetic stir bars on Teflon pivots (Belco); Wheaton bottles or screw-capped Erlenmeyer flasks in which a Teflon-coated magnetic bar simply rests on the bottom; or bottles with a stir bar suspended from a rubber stopper by a stainless-steel "bathtub" chain. All

[9] R. S. Gardner, *J. Cell Biol.* **42,** 320 (1969).

work well, although the cheapest and simplest by far is the Wheaton bottle–stir bar combination.

Initiating a suspension culture from HTC cells previously grown exclusively as monolayer cultures requires adaptation by the cells. Considerable cell death is often seen during the adaptation period, and once the number of viable cells falls below a critical level, the entire culture dies. Consequently, several cell concentrations should be set out initially. Cells in monolayer culture should be adapted to the spinner medium for several days while still grown as monolayers. The cells that come free from the substratum should be collected by centrifugation and returned to the culture, and not discarded during medium changes. After this initial adaptation, the monolayer-grown cells are collected, with those still adhering to the substratum removed by trypsin-EDTA. Suspend 1–5 $\times$ $10^5$ cells/ml in fresh, suspension-culture growth medium at 37° to a total volume of at least 100 ml. Add the cell suspension to a vessel chosen so that it will be $\frac{1}{2}$–$\frac{2}{3}$ full, containing a stir bar whose length equals 1–1.5 times the radius of the vessel. Place the vessel on a magnetic stirrer set at a speed to produce gentle mixing and no turbulence. There should be a slight vortex formed at the surface of the liquid. During the adaptation period the cells may multiply quite slowly. Medium changes, without diluting the cells, can be carried out by separating the cells by centrifugation (5 min at 800 g) or by removing the flask from the magnetic stirrer and allowing the cells to settle for 30–60 min, and then removing most of the overlying medium by suction.

Once the cells adapt to suspension culture conditions, they double approximately every 24 hr. If fed 3 times per week, and kept between $10^5$ and $10^6$ cells/ml, they will remain in logarithmic growth phase indefinitely. They have been grown successfully in volumes as low as 50 ml and as high as 50 liters. The suspension-adapted cells can be again grown as monolayers, although at first they attach less firmly and have a more spherical morphology.

## Cryopreservation of HTC Cells

HTC cells can be stored by freezing in a viable state, using standard methods. Medium containing at least 10% serum is supplemented with 10% dimethylsulfoxide (DMSO) or 10–15% glycerol. Cells are harvested and suspended in this medium at about $10^6$ cells/ml. Aliquots of 1–1.5 ml are placed in small screw-capped ampules and immediately frozen, lowering the temperature at 1–3°/min. This can be achieved several ways: by use of special instruments, by putting the vials in an alcohol–ice bath to which frozen $CO_2$ chips are added, or by simply standing the vials on a bed of frozen $CO_2$ or, wrapped in paper towels, in a −75° freezer. For greater security against contamination, sealed glass ampules may be substituted for the screw-capped type. After the cells are frozen, they are transferred to a liquid $N_2$ freezer; the lower temperature it provides is essential for viability after long storage. HTC cells have been regrown after more than 5 years of such preservation.

When frozen cells are to be grown, the vial is promptly brought to ambient temperature or to 37° in a water bath and cultured without delay. Glycerol-frozen cells can simply be diluted with fresh medium and grown. DMSO-frozen cells should be removed from the toxic freezing medium either by centrifugation as soon as the cells are thawed or by plating the thawed cells with fresh medium and then replacing the medium again after a few hours, when the cells have adhered to the substratum. Initial culture in any case should be as monolayers, not suspension cultures. Cells grown serum free can be frozen in serum-supplemented medium as above, or 15–20% polyvinylpyrrolidone can be substituted for serum.

## [49] An Established but Differentiated Kidney Epithelial Cell Line (MDCK)

*By* MARY TAUB and MILTON H. SAIER, JR.

The Maden Darby canine kidney cell line (MDCK) is one of the best characterized epithelial cell lines available for study. MDCK was derived from the kidney of a normal male cocker spaniel in 1958. Initially, the cells were used primarily for viral production.[1] However, subsequently, the differentiated properties of MDCK were examined. Although maintained in culture for almost 20 years, the cells still bear close resemblance to transporting epithelia present in the distal tubule of the kidney.[2] Thus, the cells may be used to answer the following questions about kidney epithelia: (1) What are the mechanisms involved in the regulation of epithelial cell growth? Previous studies concerned with growth regulation have been concerned primarily with fibroblast cultures,[3] and the observations made with fibroblasts do not necessarily apply to epithelial cells. (2) What are the mechanisms responsible for the biogenesis of the polarity in these cells? Epithelial cells in general have two distinct surfaces, serosal

[1] C. R. Gaush, W. L. Hard, and T. F. Smith, *Proc. Soc. Exp. Biol. Med.* **122,** 931 (1966).

[2] M. H. Saier, Jr., M. J. Rindler, M. Taub, L. Shaeffer, and L. M. Chuman, *Proc. Natl. Acad. Sci. U.S.A.* (submitted for publication).

[3] B. Clarkson and R. Baserga, eds., "Control of Proliferation in Animal Cells." Cold Spring Harbor Lab., Cold Spring Harbor, New York, 1976.



ISBN 0-12-181958-2

and mucosal, which have distinct structures and enzyme markers. The mechanisms by which these enzyme markers segregate to the two surfaces are not well understood. (3) What are the mechanisms by which solutes are transported across kidney epithelia? Studies of kidney function using MDCK are not only feasible, but have the advantage over *in vivo* studies that the cells in culture form a homogenous population that can be cultured indefinitely; they can be grown readily to sizable populations for biochemical studies and are amenable to genetic analysis using the techniques available to somatic cell geneticists.[4]

Several studies have been conducted concerning the differentiated processes in MDCK cells related to salt and water transport. That MDCK cells transport solutes in a vectorial manner, from the mucosal to the serosal surface, was first suggested by Leighton[5] who observed the formation of multicellular blisters or domes in confluent MDCK cultures. The blisters are groups of cells in the monolayer slightly raised from the tissue culture dish surface. Microscopically, the cells in the blisters are out of the focal plane of the majority of the cells in the monolayer. Blister formation can be accounted for by the transport of salt and water from the mucosal surface of the cells (facing the culture medium), to the serosal surface (facing the dish surface). Presumably, the solutes that have accumulated between the cell layer and the plastic surface give rise to hydrostatic pressure, which causes portions of the monolayer to be elevated from the dish surface, i.e., blister formation. This explanation is supported by both morphological and electrophysiological studies.

A prerequisite for vectorial transport is that the cells have the structural polarity characteristic of transporting epithelia of the kidney; indeed, electron microscopic studies indicate that microvilli are localized exclusively to the mucosal (upper) surface of the MDCK cells.[6] Adjacent cells are interconnected by tight junctions. The vectorial transport of salt and water by MDCK cells was first demonstrated by Misfeldt *et al.* employing a Ussing chamber.[6] They detected a spontaneous potential difference across the MDCK cell monolayer, as well as net water flux from the mucosal to the serosal surface of the monolayer.[6] More recent studies also indicate that the cells vectorially transport certain amino acids.[7]

In addition to studies of differentiated functions, both the *in vitro* and the *in vivo* growth characteristics of MDCK have been examined. *In vitro,* MDCK cells have a saturation density typical of transformed cells, have a

[4] L. H. Thompson and R. M. Baker, *Methods Cell Biol.* **6,** 209 (1973).

[5] J. L. Leighton, W. Estes, S. Mansukhani, and Z. Brada, *Cancer* **26,** 1022 (1970).

[6] D. S. Misfeldt, S. T. Hamamoto, and D. R. Pitelka, *Proc. Natl. Acad. Sci. U.S.A.* **73,** 1212 (1976).

[7] M. White and D. S. Misfeldt, personal communication (1977).

low serum requirement, and can grow in methocel suspensions.[8] *In vivo,* when injected into baby nude mice MDCK cells form nodules structurally very similar to the intact kidney.[2] Cells injected into adult nude mice remain viable over time, but nodule growth is not observed.[9] These observations suggest that MDCK cells respond to growth regulatory signals present in baby mice, but that the signals are deficient in adult animals.

## MDCK Cell Growth and Maintenance

MDCK cells are maintained as monolayers at 37° in a 5% $CO_2$ in air atmosphere, in Dulbecco's modified Eagle's medium with 10% fetal calf serum, penicillin ($1.9 \times 10^2$ IU/ml), streptomycin (0.2 mg/ml), and ampicillin (25 $\mu$g/ml) (DME). Under these culture conditions MDCK cells have a growth rate of one doubling per 22 hr and reach a final saturation density of $4 \times 10^5$ cells/cm$^2$. The cells are cuboidal in shape and form continuous tight junctions between adjacent cells, similar to other epithelial cells. At confluency the groups of cells have joined to form a continuous epithelial sheet on the dish surface. In such confluent cultures the multicellular blisters or domes first become apparent. The regulation of blister formation may be studied under these conditions by simply adding reagents of interest to the culture medium and microscopically determining their effects on the number and size of such blisters as a function of time.[10]

As an alternative to monolayer cultures, MDCK cells can be grown as papillary cysts in suspension, a method developed by John Valentich.[11] To initiate cyst culture development, cells are grown to late confluency (1.2–$1.4 \times 10^7$ per 32-ounce bottle) in DME. The medium is changed 24 hr prior to harvesting the cells for cyst formation. The next day fragments of the cell sheet are detached by scraping with a Pasteur pipette, and aliquots of the resulting cell suspension are placed in 25-ml Erlenmeyer flasks (aliquots are 4 ml per 50 ml DME, when using a 32-ounce prescription bottle). The flasks are shaken at 70 rpm at 37°. Twenty-four hours later cysts appear and increase in size and number over a 3–4 day period.

## Solute Transport

The uptake of labeled solutes is determined using MDCK cell monolayers in 35-mm diameter plastic dishes containing DME. For up-

[8] C. D. Stiles, W. Desmond, L. M. Chuman, G. Sato, and M. H. Saier, Jr., *Cancer Res.* **36,** 3300 (1976).

[9] C. D. Stiles, W. Desmond, L. M. Chuman, G. Sato, and M. H. Saier, Jr., *Cancer Res.* **36,** 1353 (1976).

[10] J. D. Valentich, R. Tchao, and S. Leighton, *J. Cell Biol.* **70,** 330a (1976).

[11] R. Tchao, personal communication (1976).

take studies, confluent cultures are washed 3 times with the uptake buffer [phosphate-buffered saline (PBS) or 1 m*M* Tris made isotonic to PBS using sucrose], and the cells are incubated with labeled solute at room temperature. After the uptake period, the solution is removed by aspiration, and the monolayers are washed 3 times with 3 ml of incubation buffer. The labeled material is solubilized by incubation for 1 hr in deionized water (for cation transport) or in 5% trichloroacetic acid (for sugar or amino acid transport). The solubilized material is added to scintillation vials containing a Triton–toluene based scintillation fluid, and the radioactivity is determined. Total label accumulated intracellularly is measured in duplicate and is corrected for zero time uptake, i.e., for unincorporated material not removed by the washing procedure.

The label present per cell population is standardized by either the protein content per dish, determined using the Lowry protein assay as modified for tissue culture by Oyama and Eagle,[12] or by the total cell number per dish, determined using a Coulter counter. In order to determine the concentration of solute present intracellularly, the volume is estimated by measuring the average MDCK cell size with a Coulter counter. Pollen particles are used as standards, and a 75% water content is assumed.[13] The protein content is used to determine the intracellular fluid volume per dish since a direct correlation has been made between these two factors as a function of cell density.[13]

## Growth for Studies in Ussing Chamber

The transepithelial transport of salt and water by MDCK cells has been studied with a Lucite Ussing chamber (Fig. 1) that was originally developed for the measurement of the transepithelial electric potentials generated by frog skin. For the Ussing chamber studies MDCK monolayers are grown on Millipore filters, so that the monolayers resemble a continuous epithelium. Membrane filters (Millipore, HAMK 02512) are boiled 5–10 min to remove the wetting agent and are attached to plastic tissue culture dishes while still wet using droplets of Millipore cement (Formulation No. 1). MDCK cells are distributed into the dishes containing the filters and DME. When the cells on the filters have grown to form a continuous epithelial sheet, a filter may be gently detached from the dish and placed between Lucite Ussing chambers containing either culture medium without serum or a balanced salt solution (for example, Hanks's balanced salt solution).[6]

The Ussing chamber (designed originally by Ussing and Zerahn[14]) is

[12] V. I. Oyama and H. Eagle, *Proc. Soc. Exp. Biol. Med.* **9,** 305 (1956).

[13] D. O. Foster and A. Pardee, *J. Biol. Chem.* **244,** 2675 (1969).

[14] H. H. Ussing and K. Zerahn, *Acta Physiol. Scand.* **23,** 110 (1951).

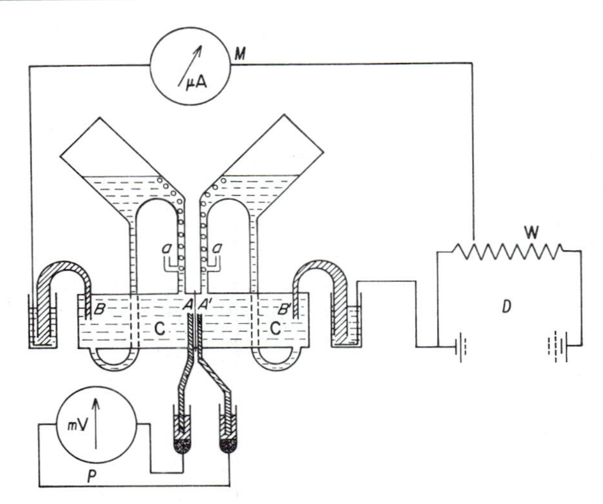


FIG. 1. Diagrammatic sketch of Ussing chamber. Abbreviations used in the figure are as follows: *C*, Celluloid or Lucite chamber containing salt solution; *S*, skin or membrane filter with MDCK cells; *a*, air inlets; *A* and *A'*, Agar-Ringer bridges, connecting outside and inside solutions with calomel electrodes and potentiometer; *B* and *B'*, Agar-Ringer bridges used for applying outside emf; *D*, battery; *W*, potential divider; *P*, tube potentiometer.

illustrated in Fig. 1. The epithelium is placed in a space between two adjacent chambers, thereby separating the solutions in these compartments. Saturated KCl–calomel electrodes are connected to the solutions in the two chambers by means of 1 *M* NaCl–4% agar bridges. A potentiometer is connected to these half cells to determine the potential difference between the solutions facing either the mucosal or serosal sides of the cell layer.

In applying this apparatus to the study of MDCK cells, a Beckman model 76 pH meter is used as a potentiometer and the potential is recorded with a Beckman model 1005 recorder.[6] The system has to be carefully balanced, i.e., the components present on either side of the bridges have to be symmetrical, in order to measure spontaneous potentials. Relatively high potentials induced by external currents may be measured if a Simpson multimeter is placed in series with the potentiometer.

The spontaneous electrical potential of the MDCK cells has been determined when identical solutions were present on either side of the Ussing chamber.[6] Immediately after initiating the readings, relatively high potentials are observed; a decline follows after 3–10 min. At this lower steady-state level a potential of $1.42 \pm 0.26$ mV with a negative serosal surface was determined. Due to the wide range of readings multiple determinations are necessary to obtain meaningful estimations of the spontaneous transepithelial electrical potential. MDCK epithelial layers are leaky and have a much lower spontaneous current than frog skin or toad bladder.

The electrical potential across the cell layer also may be determined when an external current is applied. As illustrated in Fig. 1 an external power supply is connected in series with the Ussing chamber, and the voltage is adjusted to the desired amplitude using a potential divider (*W*). When applying such an external current, Ussing and Zerahn demonstrated a linear relationship between the $Na^+$ flux and the measured potential difference across frog skin.[14] In frog skin the measured current is equivalent to the $Na^+$ influx rate when the external voltage is adjusted so that the potential is zero. However, similar relationships have not been established with the MDCK cells; a determination of the $Na^+$ influx rate from the spontaneous potential difference may actually underestimate the total rate of $Na^+$ influx, due to the simultaneous flow of cations and anions across the monolayer.

Misfeldt *et al.* have demonstrated a constant linear relationship between applied current and transepithelial potential when the serosal side is negative;[6] a similar relationship holds when the mucosal side is negative, provided that current densities are below 800 $\mu$A. Under these conditions the transepithelial electrical resistance may be determined.

The chamber apparatus also has been used to determine the rate of water flux.[6] Capillary tubes are positioned horizontally from the center to the outside of each chamber. The chambers are filled with appropriate solutions, evacuated of air, and the vectorial water flux is measured from the position of the meniscus of water present in the capillary tubes.

Finally, the Ussing chamber can be used to measure the vectorial flux of labeled solutes. Labeled material is placed in the buffer adjacent to one side of the monolayer, and the label present in the buffer in the other chamber may be measured as a function of time. To establish vectorial flux one must demonstrate a difference in the rate of movement of labeled solute from the serosal to mucosal side as compared to the mucosal to the serosal side of the Ussing Chamber.

### cAMP Measurements

cAMP production by MDCK cells is determined by measuring intracellular cAMP or the cAMP excreted into the culture medium using the Gilman receptor protein binding displacement assay.[15] To assay cAMP production confluent monolayers of MDCK (in 30-mm diameter dishes) are washed 2 times with PBS and are incubated with the compounds to be

[15] A. G. Gilman and F. Murad, this series, Vol. 38, p. 49.

tested in serum-free DME at 37° in a 5% $CO_2$ in air incubator. To facilitate studies of the effects of hormones on MDCK, the cells are incubated concomitantly with the hormone of interest and isobutyl methylxanthine. The latter agent inhibits cAMP phosphodiesterase action. At the end of the incubation period the culture medium is removed, and intracellular cAMP is extracted from the monolayers with boiling water. Extraction with boiling water is a simpler and more accurate procedure than extraction procedures utilizing trichloroacetic or perchloric acid which require subsequent extractions to remove the denaturing agent from the cell extracts.[15] To standardize cAMP measurements protein is determined by a modified Lowry procedure.[12]

The Gilman procedure assays unlabeled cAMP by its ability to compete with labeled cAMP for binding to the regulatory subunit of a cAMP protein kinase partially purified from bovine skeletal muscle. The binding is enhanced in this assay by adding a heat-stable protein inhibitor of the kinase. After the binding has reached equilibrium, free and bound nucleotides are separated by passing the solution through cellulose ester (Millipore) filters. The concentration of labeled nucleotide that the sample displaces from the binding protein is determined, which permits the concentration of unlabeled cAMP in the sample to be estimated from standard curves.

Standard curves are constructed in which log (total [$^3$H]cAMP bound) is plotted versus log (total cAMP present, labeled and unlabeled), for each concentration of [$^3$H]cAMP used. Such plots are linear at saturating cAMP concentrations. cAMP is assayed at saturating concentrations of the nucleotide to maximize reproducibility of the assays and to facilitate estimation of cAMP content. The sensitivity of the assay is enhanced when lower [$^3$H]cAMP concentrations are employed, but accuracy suffers.

Routinely, the standard, or unknown, compounds are added to the tubes (10 × 75 mm) containing 5 μl of 0.1 *M* sodium acetate at pH 4.0, 5 μl [H$^3$]cAMP (usually 1 pmol, 10,000–20,000 cpm), and 5 μl of inhibitor protein (9–15 μg). Prior to adding binding protein the solution containing these components has a total volume of 20 μl. While the tubes are on ice, 30 μl of a solution containing binding protein and bovine serum albumin (10–15 μg) are added to each tube, raising the fluid volume to 50 μl. The quantity of binding protein used is kept minimal to ensure saturation of the protein; saturation may be defined as conditions under which less than 30% of the total [$^3$H]cAMP is bound after a 1-hr incubation at 0°. The reaction mixtures are diluted with 1 ml of 20 m*M* potassium phosphate, pH 6, at 0–3° and are passed through 25-mm cellulose ester Millipore filters (previously rinsed with buffer); the filters are washed with buffer to remove unbound cAMP. To determine [$^3$H]cAMP bound, filters placed in scintillation vials are dissolved with 1 ml of methyl cellusolve, scintillation fluid is added (the ratio PPO/POPOP : methyl cellusolve is 3 : 1), and the vials are counted in a scintillation counter.

### Comments

MDCK has both the morphological and enzymic properties of epithelial cells from the distal tubule of the kidney.[1] In addition to having high $Na^+/K^+$ ATPase activity, MDCK cells possess the kidney-specific enzymes alkaline phosphatase and leucine aminopeptidase. The peptidase has been localized primarily to the mucosal membrane by immunofluorescence. While the cells lack proximal markers ($Na^+$-dependent sugar transport, *p*-aminohippurate transport; disaccharidase), they do have a distinctive distal marker, responsiveness to the antidiuretic hormones. When present at physiological concentrations, lysine and arginine vasopressin as well as oxytocin stimulate cAMP production by adenylate cyclase.[16] Similarly, in the presence of $PGE_1$, $PGE_2$, glucagon, or cholera toxin cAMP production is enhanced in MDCK. The stimulation of adenylate cyclase by vasopressin is apparently subject to negative cooperativity; however, these studies indicate that the cells have antidiuretic hormone (ADH) receptors specific to distal kidney cells, and that a single kidney epithelial cell type is responsive to multiple hormonal agents.

When ADH is present in the distal portion of the kidney, increased water flux from the kidney lumen into the mucosal surface of the cells is observed.[17] If the MDCK cells provide a good model system for the study of kidney epithelial function one would expect that the cells would respond to treatment with vasopressin (or to dibutyryl cAMP since the vasopressin response is cAMP mediated) by increased rates of water transport. That cAMP levels in MDCK cause increased vectorial fluid transport is suggested by the observation that incubation of MDCK with dibutyryl–cAMP results in increased blister formation.[10] These observations indicate that the MDCK cells provide a novel model system for the study of distal tubular cell function.

The primary function of distal cells is to transport salt and water, thereby regulating body levels of these compounds. Misfeldt's experiments indicate that in MDCK cells, as in the kidney, salt and water

[16] M. J. Rindler, L. M. Chuman, and M. H. Saier, Jr., in preparation.

[17] E. Koushanpour, "Renal Physiology, Principles and Functions," p. 336. Saunders, Philadelphia, Pennsylvania, 1976.

transport occur transepithelially, from the mucosal to the serosal surface.[5] The transport of $Na^+$ by the $Na^+$ channel (presumably located on the mucosal surface) has been examined, both kinetically and genetically.[18] $^{22}Na^+$ uptake studies indicate that a saturable $Na^+$ transport system is present in MDCK cells that is energy independent and insensitive to ouabain inhibition.[18] $Na^+$ uptake by the channel is inhibited by a number of monovalent cations including $Li^+$, $K^+$, $Rb^+$, guanidine, and amiloride. $^{22}Na^+$ uptake studies indicated that the system is subject to two exchange processes: (1) exchange transport with intracellular $Na^+$, and (2) $Na^+$/proton antiport. Calcium is implicated as an important regulatory molecule for the channel, inhibiting $Na^+$ uptake when present extracellularly, and stimulating $Na^+$ uptake when present intracellularly.

Variant clones of the MDCK line have been isolated by their resistance to killing by $5 \times 10^{-4}$ *M* amiloride.[18] Since the clones are stable over a 6-month growth period under nonselective conditions, and since their frequency is increased by mutagens, they appear to have arisen by genetic mutations. The clones exhibit reduced uptake of $Na^+$, amiloride, and $Rb^+$ (ouabain-insensitive uptake). Since both $Rb^+$ uptake by the $Na^+/K^+$ ATPase and $\alpha$-amino-isobutyrate uptakes are not affected by the mutation, the decreased ionic uptake rates may be tentatively explained as resulting from a mutation affecting the $Na^+$ channel. The reduced uptake of amiloride by the $Na^+$ channel in the resistant cells may confer resistance to killing by amiloride by decreasing the intracellular concentration of the drug below the cytotoxic level. Reduced salt transport in these cells apparently results in a decreased propensity for vectorial fluid transport, as indicated by the lack of blister formation in the variant cells.

[18] M. Taub, M. S. Rindler, and M. H. Saier, Jr., *Cell Biol.* **75**, 367a (1977).

# [50] Large-Scale Preparation of Chondrocytes

*By* MICHAEL KLAGSBRUN

Cartilage is an unusual tissue in several respects. It has neither blood vessels nor nerves and contains only one type of cell, the chondrocyte. Thus, it is relatively easy to obtain a homogeneous population of cells from cartilage without having to resort to cell fractionation techniques. Chondrocytes are useful for the analysis of proteoglycan synthesis[1] and

[1] M. Muto, M. Yoshimura, M. Okayama, and A. Kaji, *Proc. Natl. Acad. Sci. U.S.A.* **74**, 4173–4177 (1977).




ISBN 0-12-181958-2

collagen synthesis.[2] In addition, the metabolic effects of polypeptide growth factors,[3,4] hormones,[5] and vitamins[6] on chondrocytes can be studied. Chondrocytes can be readily grown in cell culture in plastic dishes and, unlike other normal diploid cells, can be maintained and grown in suspension.[7,8] Furthermore, it has been demonstrated that chick embryo chondrocytes will grow in soft agar.[9] Growth in suspension and growth in agar are properties usually ascribed to transformed cells.

Chondrocytes are obtainable in large numbers from bovine and human sources. The shoulder of a newborn calf (1–14 days old) is an excellent source of large amounts of cartilage and is readily available from slaughterhouses that prepare veal. Since sterile conditions cannot be maintained in the slaughterhouse, care must be taken to avoid further contamination of the shoulder before its arrival in the laboratory. The shoulder is placed in a sterile plastic bag containing 0.15 *M* NaCl, penicillin (200 U/ml), and streptomycin (200 $\mu$g/ml) immediately after slaughter and packed in ice. The shoulder is then transferred to a laminar flow hood and placed on a sterile cloth. Within the calf shoulder are two major sources of cartilage. One is the scapula, which has a piece of cartilage measuring about 3 cm $\times$ 8 cm. The other source is the cartilage on the articular surfaces of the bones that make up the shoulder joint.

The surrounding muscle and connective tissue must be removed in order to obtain clean cartilage. This is done by dissection with scalpels holding sterile stainless-steel #10 and #21 surgical blades (Bard-Parker). The scapular cartilage is enclosed in a fibrous connective tissue, the perichrondrium, which is removed by scraping the scapular cartilage with a #21 blade. Once the scapular cartilage is clean, it is cut away from the scapular bone. However, the region of about 0.5 cm wide that is proximal to the bone is discarded, because this section of cartilage is slightly vascular.

The articular cartilage, which does not have a perichondrium, is easily removed from the articular surfaces of the long bones by cutting off slivers using the #21 blade.

[2] K. von der Mark, V. Gauss, H. von der Mark, and P. Muller, *Nature (London)* **267,** 531–532 (1977).

[3] M. Klagsbrun, R. Langer, R. Levenson, S. Smith, and C. Lillehei, *Exp. Cell Res.* **105,** 99–108 (1977).

[4] M. T. Corvol, C. J. Malemud, and L. Sokoloff, *Endocrinology* **90,** 262 (1972).

[5] S. Meier and M. Solursh, *Gen. Comp. Endocrinol.* **18,** 89–97 (1972).

[6] M. Solursh and S. Meier, *Calcif. Tissue Res.* **13,** 131–142 (1973).

[7] K. Deshmukh and B. D. Sawyer, *Proc. Natl. Acad. Sci. U.S.A.* **74,** 3864–3868 (1977).

[8] Z. Nevo, A. Horwitz, and A. Dorfmann, *Dev. Biol.* **28,** 219–228 (1972).

[9] A. L. Horwitz and A. Dorfmann, *J. Cell Biol.* **45,** 434–438 (1970).

Both scapular and articular cartilage are diced into pieces approximately 0.5 cm × 0.5 cm and washed twice with phosphate-buffered saline (Gibco) containing penicillin (200 U/ml) and streptomycin (200 $\mu$g/ml). The diced cartilage is incubated at 37° in 25 ml of 2 mg/ml clostridial collagenase (Worthington CLS II, 140 U/mg) made up in the phosphate-buffered saline–penicillin–streptomycin mixture. The incubation is carried out in a 60-mm petri dish that is gently rotated on a rotating apparatus (Arthur Thomas Co.) for 12–24 hr or until the pieces of cartilage are completely dissociated and the chondrocytes are liberated from the matrix. The suspension of chondrocytes is passed through a 153-$\mu$m nylon sieve (Tetko) to remove debris and undissolved cartilage fragments. The cells are subsequently washed twice with the phosphate-buffered saline–penicillin–streptomycin mixture by means of centrifugation. After the second wash, cells are counted in a Coulter counter (Coulter Electronics). A typical scapular or articular chondrocyte preparation contains 1–4 × $10^8$ cells per shoulder.

A source of large numbers of human chondrocytes is the costal cartilage obtained from children undergoing surgical correction of pectus excavatum, a disorder in which rib cartilage is deformed. These chondrocytes are particularly easy to isolate because the perichondrium, muscle, and connective tissue are removed from the cartilage during surgery. Otherwise, the isolation procedure is the same as for the bovine chondrocytes.

For cell culture, cells (3 × $10^6$ per 75 $cm^2$ culture dish) are plated sparsely in the presence of 10 ml of Dulbecco's modified Eagle's medium containing 4.5 g/liter glucose (Gibco), 10% calf serum (Colorado Serum Co.), penicillin (50 U/ml), and streptomycin (50 $\mu$g/ml) and transferred to a 37° water-jacketed $CO_2$ incubator (National Appliance Co.). Chondrocytes are round cells *in situ*. They gradually flatten in the culture dish over a period of 2–4 days; growth begins after this initial lag period. In confluent cultures, two patterns of growth are observed; some cells in the dish grow in a monolayer while others pile up and form nodules. The number of nodules increases with repeated feeding of the cells. If the cells are stained with 2% aqueous toluidine blue, metachromasia may be observed in the nodules, indicating that active synthesis of proteoglycans has taken place.

Chondrocytes cannot be maintained indefinitely in culture flasks once they become confluent because the synthesis of the sulfated proteoglycans leads to a sharp decrease in the pH of the media. The cells are harvested and either subcultured in new flasks by plating sparsely or are frozen. Harvesting of chondrocytes is accomplished by detaching the cells with 0.1% trypsin (Gibco) made up in phosphate-buffered saline lacking calcium and magnesium (Gibco). The medium is removed from a

T-75 culture flask, and 10 ml of 0.1% trypsin solution are added. The cells are incubated with trypsin at room temperature for about 5 min or until they round up and are loosely attached to the flask. The trypsin is carefully removed, and 10 ml of fresh media are added. The flask is struck several times with the palm of the hand in order to detach the cells.

Chondrocytes are frozen in a solution of Dulbecco's modified Eagle's medium containing 10% calf serum and 7.5% dimethyl sulfoxide (Fisher Scientific Co.). A stock solution of dimethyl sulfoxide (15%, v/v) in Dulbecco's modified Eagle's medium containing 10% calf serum is prepared. One milliliter of the dimethyl sulfoxide solution is mixed with 1 ml of chondrocytes ($2.5 \times 10^6$ cells/ml) resuspended in Dulbecco's modified Eagle's medium containing 10% calf serum. The cells are transferred into 2-ml ampules (Wheaton Scientific) that are subsequently sealed. The ampules are gradually frozen at $-90°$ in a Revco freezer and are stored either in the Revco freezer or in a liquid nitrogen tank (Cryogenics-East Co.).

Bovine chondrocytes will remain viable in liquid suspension culture for at least 7 days. Maintenance in liquid suspension culture can be used as a technique to remove contaminating fibroblasts since these cells will not survive in suspension for more than 2 or 3 days.[8] One technique for keeping chondrocytes in liquid suspension is to coat the bottom of a petri dish with 0.5% agar. The chondrocytes are unable to plate out in the agar and remain suspended. Agar plates are prepared as follows: A 1.0% solution of agar (Difco Bacto-agar) is made up in water and sterilized by boiling for 30 min. The 1% agar is cooled to 45° and mixed with an equal volume of warm ×2 basal Eagle's medium with Earle's balanced salts (Microbiological Associates) containing 20% calf serum, penicillin (100 U/ml), and streptomycin (100 μg/ml). Two milliliters of the 0.5% agar solution are poured into a 35-mm petri dish. After the agar hardens, 3 ml of chondrocytes ($2 \times 10^5$ cells/ml) are resuspended in Dulbecco's modified Eagle's medium containing 10% calf serum and plated onto the agar. The agar plates are maintained at 37° in a water-jacketed $CO_2$ incubator.

A major problem in chondrocyte culture is the possibility of contamination by fibroblasts arising from the perichondrium. Nevo *et al.*[8] have suggested that the fibroblasts can be eliminated by maintaining the chondrocyte preparation in suspension culture for 2 or 3 days. Fibroblasts do not survive in suspension culture.

In general, chondrocyte cultures can be characterized by several criteria. These include a high level of $^{35}SO_4$ uptake into chondroitin sulfate,[8] metachromatic staining with toluidine blue,[9] and the ability of chondrocytes to survive in suspension culture.[7-9]

Cartilage *in vivo* contains Type II collagen that is genetically distinct from the Type I collagen found in skin, bone, and tendon. Thus, chondrocytes should be the only cell capable of producing Type II collagen. However, the type of collagen produced by chondrocytes in culture has been shown to be dependent on extracellular conditions.[2,7,10] In general, monolayer cultures produce Type I and suspension cultures produce Type II.[7] Therefore, care must be taken in using collagen typing as a criterion for the presence of chondrocytes.

[10] P. D. Benya, S. R. Padilla, and M. E. Nimni, *Biochemistry* **16,** 865–872 (1977).

## [51] Melanoma Cells

*By* John Pawelek

Melanocytes synthesize melanin in a reaction controlled by the enzyme monophenyl, dihydroxyphenylalanine : oxygen oxidoreductase (EC 1.14.18.8), commonly referred to as either tyrosinase or dopa-oxidase. Melanocytes in the skin of adult mammals divide infrequently, perhaps once a year, and attempts at adapting these cells to culture have met with only limited success. On the other hand, malignant melanoma cells divide as often as once a day, and several lines have been established in culture. Commonly used lines are the B16,[1] Harding-Passey,[2] and Cloudman[3] strains (mouse); the Fortner (RPM1) strain[4] (hamster); and many human strains.[5,6] Some of these lines are available commercially.[7,8]

The Cloudman S91 melanoma was established as a transplantable tumor in DBA 2/J mice in the 1930s[9] and was adapted to culture in the 1960s.[3] In 1973 it was reported that the Cloudman cells responded in culture to melanotropin (MSH) with increases in tyrosinase activity, melanin formation, and generation time, as well as changes in morphology.[10] Since the original findings, many studies have been carried out on factors that regulate pigmentation and proliferation of these and other melanoma cells in culture (for reviews, see Pawelek[11,12]).

[1] F. Hu and R. R. Cardell, *J. Invest. Dermatol.* **42,** 67 (1964).
[2] H. E. Harding and R. D. Passey, *J. Pathol.* **33,** 417 (1930).
[3] Y. Yasumura, A. H. Tashjian, and G. H. Sato, *Science* **154,** 1186 (1966).
[4] G. E. Moore, D. F. Lehner, Y. Kikuchi, and L. A. Less, *Science* **137,** 986 (1962).
[5] R. E. Gerner, H. Kitamura, and G. E. Moore, *Oncology* **31,** 31 (1975).
[6] L. A. Quinn, L. K. Woods, S. B. Merrick, M. Arabasz, and G. E. Moore, *J. Natl. Cancer Inst.* **59,** 301 (1977).
[7] The American Type Culture Collection, 12301 Parklawn Drive, Rockville, Maryland.
[8] The Jackson Laboratory, Bar Harbor, Maine.
[9] A. M. Cloudman, *Science* **93,** 380 (1941).
[10] J. Pawelek, G. Wong, M. Sansone, and J. Morowitz, *Yale J. Biol. Med.* **46,** 430 (1973).
[11] J. Pawelek, *J. Invest Dermatol.* **66,** 201 (1976).
[12] J. Pawelek, *Pigm. Cell* **4** (in press).



ISBN 0-12-181958-2

The purpose of this article is to describe techniques for culturing Cloudman S91 in monolayers and clones, and to outline some of the problems unique to melanin-producing cells that are encountered in normal culturing procedures and in some biochemical analyses.

## Cell Culture

Culture procedures are straightforward. For monolayers, cells are placed into plastic flasks or petri dishes that have been coated for cell culture (available commercially). The culture medium consists of Ham's F10 medium[13] supplemented with 2% fetal calf serum and 10% horse serum. Antibiotics can be added if desired, but it is important to first determine whether a specific antibiotic has an effect on pigmentation. For example, tetracycline and bacitracin both increase pigment formation whereas streptomycin and penicillin do not. Fresh culture medium is added 3 times per week. Subculturing of cells involves removal of the medium and addition of ethylenediaminetetracetic acid (EDTA) (1 m*M* or less) in a Ca- and Mg-free balanced salt solution. After about 10 min at room temperature, cells detach from the culture vessel and are collected by centrifugation. The use of trypsin or other proteases should be avoided in the harvest procedure, particularly if regulatory processes involving the outer cell surface are being studied.

For culturing cells in clones, we use the "dilute agar colony method" of Chu and Fischer.[14] This involves culturing cells in medium supplemented with 0.12% agar; the concentration of horse serum is increased to 15%. Culture vessels can be in a variety of sizes and shapes, but for convenience we use plastic screw-cap test tubes. The clones grow in suspension, and it is therefore unnecessary to use vessels that have been coated for cell culture. Clones are generally visible to the naked eye after 1 week and can be picked with a Pasteur pipette after 2–4 weeks.

## Isolation of Amelanotic Mutants

Pigmentation has long been used as a genetic marker as evidenced by the fact that more than 70 genes for pigmentation have been identified in the mouse.[15] Simple procedures have been developed for isolating lines of

[13] R. G. Ham, *Exp. Cell Res.* **29,** 515 (1963).
[14] M. Y. Chu and G. A. Fischer, *Biochem. Pharmacol.* **17,** 753 (1968).
[15] W. C. Quevedo, *in* "Biology of Normal and Abnormal Melanocytes" (T. Kawamura, T. B. Fitzpatrick, and M. Seiji, eds.), p. 99. Univ. of Tokyo Press, Tokyo, 1971.

EFFECTS OF ETHYLMETHANESULFONATE ON THE APPEARANCE OF AMELANOTIC CLONES[a]

| EMS concentration (mg/ml culture medium) | Total clones | Amelanotic variants | Relative survival (%) |
|---|---|---|---|
| 0 | 6110 | 0 | 100 |
| 0.02 | 6400 | 0 | 105 |
| 0.05 | 2292 | 11 (0.5%) | 38 |
| 0.10 | 1846 | 51 (2.7%) | 31 |
| 0.20 | 1438 | 27 (1.9%) | 24 |
| 0.50 | 100 | 5 (5 %) | 1.7 |
| 1.0 | 0 | 0 | 0 |

[a] Cells in monolayer were incubated with ethylmethanesulfonate (EMS) at the concentrations indicated. After 2 hr the EMS was removed, the cells were rinsed several times with fresh culture medium, and allowed to pass through three generations (6 days). Cells ($3 \times 10^4$) from each category were then cultured in petri dishes (100-mm diameter) by the method of Chu and Fischer.[13] After 4 weeks the total number of clones and the number of amelanotic variants were determined by counting with the aid of a dissecting microscope.

amelanotic melanoma cells from large populations of isogenic melanotic cells.[16] In essence, cells are treated with an agent known to induce mutations, cultured for two to three generations so that mutations can be expressed, and cloned by the "dilute agar colony method" mentioned above.[14] After 3–4 weeks, amelanotic clones are isolated with a Pasteur pipette and recloned in dilute agar. The effects of ethylmethanesulfonate, a potent mutagen, on the appearance of amelanotic clones is shown in the accompanying table. One problem with this technique is that many of the amelanotic clones prove to be unstable in that they eventually "revert" to a melanotic phenotype. Although the instability itself may prove to be an interesting problem, we have generally discarded the unstable clones and studied lines that remain faithfully amelanotic through many cell generations. It should be noted that, in addition to mutations affecting pigmentation, a number of other mutant phenotypes can be isolated by standard techniques.[17,18]

[16] J. Pawelek, J. Sansone, J. Morowitz, G. Moellmann, and E. Godawska, *Proc. Natl. Acad. Sci. U.S.A.* **71,** 1073 (1974).

[17] J. Pawelek, M. Sansone, N. Koch, G. Christie, R. Halaban, J. Hendee, A. B. Lerner, and J. M. Varga, *Proc. Natl. Acad. Sci. U.S.A.* **72,** 951 (1975).

[18] J. Pawelek, R. Halaban, and G. Christie, *Nature (London)* **258,** 539 (1975).

### Cytotoxicity of Melanin Precursors

Several years ago, Lerner suggested that in the process of melanization pigment cells produce substances that are potentially autotoxic.[19] Later it was demonstrated that pigmented cells are killed when exposed to excess tyrosine or dopa in the culture medium.[10,11] It has been shown recently that among all the naturally occurring amino acids, only tyrosine, dopa, and trytophan are toxic to pigmented cells. The toxicity of these agents is proportional to the amount of tyrosinase activity within the cells. 5,6-Dihydroxyindole is far more toxic than tyrosine, dopa, or tryptophan, and it is likely that this melanin precursor is responsible for cell death.[20] Although these findings may be potentially useful for chemotherapy of melanoma, they present a problem when culturing melanoma cells. It is important to choose a culture medium that is low in tyrosine, and it is for this reason that Ham's F10 medium (tyrosine concentration is 2 mg/1) was used for the Cloudman cells. Most other media have tyrosine concentrations that are orders of magnitude higher than that in Ham's F10, and cells with high tyrosinase activity usually do poorly in these media.

### Cyclic AMP and the Cell Cycle

Low amounts of cyclic AMP markedly stimulate the proliferative rate of Cloudman S91 cells, while higher amounts of cyclic AMP inhibit proliferation.[18] Thus melanotropin inhibits the growth of cells by raising intracellular levels of cyclic AMP, but insulin inhibits growth of cells by lowering cyclic AMP levels. The results indicate that the Cloudman S91 cells require an optimal concentration of cyclic AMP in order to pass through the cell cycle. Whereas cells exposed either to insulin or to melanotropin cease cycling, those exposed simultaneously to both hormones continue to proliferate as if they were in ordinary culture medium.[12] The basis for this phenomenon is not understood, nor is it known whether other cell types have such cyclic AMP requirements. Serum factors are known to alter intracellular levels of cyclic AMP. This might explain why we and many other have encountered variations in the cycling time of isogenic cell lines when different lots of serum are used in the culture medium.

[19] A. B. Lerner, *Am. J. Med.* **52,** 141 (1971).
[20] J. Pawelek and A. B. Lerner, *Nature,* in press (1979).

## Tyrosinase Activity

A simple and convenient assay for tyrosinase activity was developed by Pomerantz.[21] Tritiated 3,5-tyrosine is available commercially. When this molecule is oxidized to dopa by tyrosinase, [$^3$H]$H_2O$ is formed and is separated from unreacted [$^3$H]tyrosine by the addition of activated charcoal. A second purification of [$^3$H]$H_2O$ consists of passing the material that does not bind to charcoal through a Dowex-50 column.[22] Tyrosinase can be assayed in living cells by adding [$^3$H]tyrosine to the culture medium. The cells actively take up the tyrosine and release [$^3$H]$H_2O$ to the culture medium. Thus by removing small amounts of medium and measuring [$^3$H]$H_2O$, one can measure tyrosinase activity in the cells. Experiments of this sort must be interpreted with caution, however, because [$^3$H]$H_2O$ released to the culture medium is not only a function of tyrosinase activity but also of the rate of transport of [$^3$H]tyrosine into the cells, the preexisting pool of unlabeled tyrosine within the cells, and the amount of [$^3$H]tyrosine being diverted from melanin synthesis to protein synthesis. Nonetheless, it is frequently useful to measure [$^3$H]$H_2O$ produced by living cells as an assay of tyrosinase activity. To avoid the problems mentioned above, one should then compare the results in living cells with direct measurements of tyrosinase activity in cell homogenates. A typical assay in cell homogenates has been described in detail.[22]

## Normalization of Data

The colorimetric method of Lowry *et al.*[23] for measuring protein is used widely for the normalization of data. Unfortunately this technique should not be employed with crude extracts of melanin-producing cells because melanin reacts 2–3 times more strongly than bovine serum albumin (fraction V) with the Folin reagent that is used. It is difficult and time-consuming to quantitatively separate melanin from protein, and for purposes of normalizing data we have found it more convenient to estimate DNA content by using Burton's modification of the diphenylamine reaction,[24] or to measure cell number in a Coulter counter.

## Isolation of RNA

Melanin interferes with the isolation of RNA because it binds tightly to RNA molecules. We found this to be a particularly vexing problem

[21] S. Pomerantz, *Science* **164,** 838 (1969).
[22] A. Korner and J. Pawelek, *Nature* (*London*) **267,** 444 (1977).
[23] O. H. Lowry, A. L. Rosenbrough, and R. J. Randall, *J. Biol. Chem.* **248,** 190 (1951).
[24] K. Burton, *Biochem. J.* **62,** 315 (1956).

in the isolation of polyadenylic acid-containing RNA by means of oligodeoxythymidine-cellulose affinity chromatography. By combining extraction procedures from several laboratories we were able to separate melanin from RNA and isolate poly (A)-containing RNA that translated efficiently in both wheat germ and mouse ascites cell-free protein synthesizing systems.

All procedures are carried out at 4°. Cells are suspended in 4 volumes of buffer containing 0.1 *M* Tris · HCl at pH 7.5, 0.15 *M* KCl, 15 m*M* magnesium acetate, 6 m*M* 2-mercaptoethanol, 0.5% Triton X-100, and 0.1% bentonite. The mixture is homogenized with a loose-fitting Teflon homogenizer and centrifuged at 30,000 g for 15 min. The supernatant fluid is layered onto 30% sucrose dissolved in the above buffer (without Triton X-100 or bentonite) and centrifuged (100,000 g for 5 hr). The pellet that contains polysomes is dissolved in 100 volumes of a buffer containing 0.01 *M* Tris · HCl at pH 7.5, 0.15 *M*KCl, 1.5% sodium lauryl sulfate, and 0.02% pronase K. This mixture is incubated at 20° and triturated until the solution becomes clear (20–40 min). The mixture is extracted 3 times with 3 volumes each of ether. The aqueous phase, which still contains melanin, nucleic acid, and some protein, is made 70% with ethanol and stored in the cold until complete precipitation of the macromolecules (about 2 hr). The precipitate is dissolved in 5 volumes of 0.4 *M* NaCl containing 1 m*M* ethylenediaminetetraacetic acid (EDTA) at pH 7.4. An equal volume of a mixture of phenol, chloroform, and isoamyl alcohol (50 : 50 : 2) is added. The mixture is heated to 70°, agitated on a Vortex Genie mixer (5 min at top speed), and centrifuged (20,000 g for 15 min); melanin is located as a tight band at the phenol–aqueous interface. The aqueous phase containing RNA is removed and saved. The phenol phase and interface are reextracted with an equal volume of 0.4 *M* NaCl, centrifuged, and the aqueous phases pooled, extracted 3 times with ether, and precipitated at a final concentration of 70% ethanol. The precipitate, which is white and free of melanin, is dissolved with 0.4 *M* NaCl containing 1 m*M* EDTA at pH 7.5. Poly (A)-containing RNA is isolated by chromatography on oligo (dT)-cellulose. About 3% of the RNA is retained by the oligo (dT) and represents authentic mRNA since it translates efficiently in cell-free systems.

## Comments

Cultured mammalian melanoma cells provide excellent material for studying the regulation of gene expression in eukaryotic cells. The oxidation of tyrosine to melanin is well understood. A single enzyme, tyrosinase, is thought to control this pathway. A simple and rapid assay for tyrosinase activity in living cells and in homogenates is available.

Melanin in the cells can be detected visually and thus provides an easy assay for the expression of a differentiated function. Isogenic pigmented cell lines can be established by clonal techniques, and from these lines amelanotic variants can be isolated. Since at least some of these variants should have lesions in the genes controlling melanin biosynthesis, a genetic analysis of this differentiated function should be possible. The Cloudman melanoma line has been found to respond dramatically to MSH, and studies with these cells promise to provide much information concerning the mechanism of action of this hormone. It is obvious that the factors regulating pigmentation and proliferation of melanoma cells are complex. The molecules regulating these functions include hormones affecting cyclic AMP levels, the corresponding hormone receptors, adenylate cyclases, phosphodiesterases, cyclic AMP-dependent protein kinases and their phosphoprotein substrates, and possibly phosphoprotein phosphatases. Also to be considered are structural and regulatory genes that control the synthesis of these molecules. Certainly this is an abbreviated list that should expand rapidly as research continues.

# [52] Adrenocortical Y1 Cells

*By* BERNARD P. SCHIMMER

The Y1 adrenal cell line was developed by Sato and his colleagues[1] as one of a series of endocrine cell cultures that retain many of their tissue-specific characteristics.[2] This cell line originated from a transplantable, adrenocortical tumor of an $LAF_1$ mouse;[3] it was cloned in 1964 after adaptation of the tumor to culture conditions by the technique of alternate passage between culture and animal.[4] The cell line currently is available from the American Type Culture Collection (No. CCL 79).

## The Phenotype and Its Stability

The Y1 cell line has an average doubling time of 30–40 hr, has a plating efficiency[5] of 4–10%, and reaches saturation density at approximately $2.7 \times 10^5$ cells/cm$^2$. Its morphology is epithelial-like in monolayer culture, and its karyotype is nearly diploid with a modal number of 39 acrocentric

[1] Y. Yasumura, V. Buonassisi, and G. Sato, *Cancer Res.* **26,** 529 (1966).

[2] Y. Yasumura, *Am. Zool.* **8,** 285 (1968).

[3] A. I. Cohen, E. Bloch, and E. Cellozi, *Proc. Soc. Exp. Biol. Med.* **95,** 304 (1957).

[4] V. Buonassisi, G. Sato, and A. I. Cohen, *Proc. Natl. Acad. Sci. U.S.A.* **48,** 1184 (1962).

[5] Plating efficiency is defined as the percentage of single cells plated which grow into colonies.




ISBN 0-12-181958-2

or telecentric chromosomes. Y1 cells behave like normal mouse adrenal cells in several respects: (1) they synthesize $\Delta^4$-3-keto-$C_{21}$-steroids, principally 20$\alpha$-hydroxyprogesterone and 11$\beta$-20$\alpha$-dihydroxyprogesterone,[6] from cholesterol and secrete the products; (2) they increase the rate of steroidogenesis 4- to 10-fold in response to adrenocorticotropic hormone (ACTH);[1] and (3) they concentrate cholesterol and ascorbic acid from their growth medium. The cell line, in the presence of maximally effective concentrations of ACTH, produces steroids at the rate of $0.72 \pm 0.13$ $\mu$g/mg protein per hour ($N = 7$).

In our hands, the Y1 line has maintained its karyotype, growth rate, and rate of steroidogenesis for more than 3 years in continuous culture. However, the cell line shows considerable clonal variation in maximum steroidogenic capacity, with six different clonal isolates producing steroids at an average rate of $0.38 \pm 0.23$ $\mu$g/mg protein per hour. We have isolated a subclone of the Y1 line and have maintained it in continuous culture for approximately 3 years. Thirteen independent clonal isolates of this subclone produce steroids maximally at an average rate of $0.36 \pm 0.02$ $\mu$g/mg protein per hour, indicating that clonal variation in rates of steroidogenesis can be reduced by recloning the Y1 line.

A number of stable variants of the Y1 cell line have been isolated in our laboratory including clones with altered ACTH-sensitive adenylate cyclase activity,[7] altered hypoxanthine phosphoribosyl transferase activity,[8] and altered adenosine 3′,5′-monophosphate-dependent protein kinase activity.[9]

## Growth Requirements

Usually, Y1 cells are grown in 82.5% nutrient mixture F10,[10] 15% horse serum, and 2.5% fetal calf serum. The sera used are heat inactivated at 56° for 30 min. Penicillin (200 U/ml) and streptomycin (200 $\mu$g/ml) may be included as antibacterial agents. Not all of the components in this growth medium, however, are required for growth.[11] A simpler nutrient mixture, such as Eagle's minimal essential medium,[12] may be substituted readily for the F10; and, in gradually adapted cells, 5% dialyzed horse serum can replace the usual serum supplement.[11] Growth of Y1 cells is

[6] R. W. Pierson, Jr., *Endocrinology* **81,** 693 (1967).
[7] B. P. Schimmer, *J. Cell. Physiol.* **74,** 115 (1969).
[8] B. P. Schimmer, J. Tsao, and N. H. Cheung, *Nature (London)* **269,** 162 (1977).
[9] B. P. Schimmer, J. Tsao, and M. Knapp, *Mol. Cell. Endocrinol.* **8,** 135 (1977).
[10] R. G. Ham, *Exp. Cell Res.* **29,** 515 (1963).
[11] L. J. Cuprak and G. H. Sato, *Exp. Cell Res.* **52,** 632 (1968).
[12] H. Eagle, *Science* **130,** 432 (1959).

arrested in medium containing 0.2% serum and can be reinitiated upon addition of a growth factor isolated from bovine pituitary glands.[13]

Y1 cells, when grown as monolayers, require the charged surfaces of plastic tissue culture dishes; they will not attach as well to glass. The cell line is anchorage independent and can be cloned in soft agar without feeder layers or can be grown in suspension culture. In suspension, Y1 cells grow into large aggregates that can be dispersed only after treatment with a proteolytic enzyme solution.

## Methods

All operations are carried out using strict aseptic bacteriological procedures. Growth medium and other reagents are sterilized before use by filtration through sterile membrane filters (0.2 $\mu$m pore size). Cultures should be screened regularly for aerobic and anaerobic contaminants including pleuropneumonia-like organisms (PPLO).

*Initiation of Stock Cultures.* Stock cultures of Y1 cells are grown as monolayers in plastic tissue culture flasks (75 $cm^2$ area). Dispersed cells (approximately $2 \times 10^6$) are added to 15 ml of growth medium, and the suspension is gently rocked over the growing surface of the culture flask to effect a uniform distribution of cells. The inoculated flasks are capped loosely and maintained at 36.5° in a humidified atmosphere of 5% $CO_2$ in air. The cells attach and begin to spread out on the plastic surface within 2–3 hr. Culture medium is changed every third or fourth day, and the progress of the culture is monitored visually at these intervals with an inverted, phase-contrast microscope. Spent growth medium is removed by aspiration, and fresh medium is added immediately, with care taken to ensure that the cell monolayer is not disturbed. At the second medium change, the volume of growth medium is increased to 30 ml per flask. Cells reach saturation densities of $2 \times 10^7$ cells per bottle after 10–12 days. Cells can be maintained at saturation density for at least 1 week if the growth medium is replenished regularly.

*Subculture.* When the stock culture reaches saturation density, the cells are subcultured by treatment with a solution of Viokase (Viobin Corp., Monticello, Illinois). Growth medium is removed, and the monolayer is rinsed[14] twice with 2-ml portions of the Viokase solution (0.1% in phosphate-buffered saline[15] without $Ca^{2+}$ or $Mg^{2+}$ salts) and incubated at

[13] D. Gospodarowicz and H. H. Handley, *Endocrinology* **97,** 102 (1975).

[14] Rinsing is effected by gently rocking the solution over the monolayer surface and then aspirating the liquid.

[15] R. Dulbecco and M. Vogt, *J. Exp. Med.* **99,** 167 (1954).

36.5° for 10 min. The residual Viokase solution coating the cells is sufficient to detach the cells from the monolayer. Treatment with Viokase is complete when gentle tapping of the flask dislodges the cells. The cells are washed from the flask wall with 10 ml of growth medium and are dispersed by pipetting repeatedly. One milliliter of fully dispersed cells, i.e., $2 \times 10^6$ cells, is used to inoculate each new stock flask. Since the cells settle rapidly, they must be resuspended immediately before each transfer. An Erlenmeyer flask and magnetic spin bar can be used to suspend cells where a large number of uniform transfers are required. The cells are stirred at a moderate speed over a nonheating stirrer.

The size of the culture vessel and the size of the inoculum can be altered over a wide range as required by the investigator. From the data given for plating efficiency, growth rate, and saturation density of the cell line, the state of the culture can be approximated at any time after plating.

*Storage.* Y1 cells can be stored frozen in liquid $N_2$ for at least 8 years in growth medium supplemented with 10% glycerol. Cells, $2 \times 10^6$ in 1 ml of growth medium containing glycerol, are placed in sealed ampules, packed in a cardboard container, kept at −70° for 3.5 hr, and transferred quickly to liquid $N_2$. To recover cells from the frozen state, one vial is thawed rapidly in a 37° water bath, suspended by pipetting, and divided between two tissue culture flasks (75 $cm^2$) containing 15 ml of growth medium. Cells are ready for subculture after approximately 3 weeks of growth.

*Plating of Adrenal Tumors.* Y1 cells also can be grown as tumors and studied in primary cell culture. Cells, $10^5$–$10^6$ in 0.1 ml saline, injected into the hind leg muscle of an $LAF_1$ mouse will give rise to a tumor in approximately 6–8 weeks. To harvest the tumor, the mouse is killed by cervical dislocation and cleaned with 70% ethanol. The tumor and associated leg tissues are dissected free, transferred to a Petri dish, and washed several times with saline solution. The tumor capsule is cut, and the tumor tissue is removed and minced with scissors. The fine mince is triturated by pipetting with 25 ml of Viokase (0.2% solution in nutrient mixture F10) in 5-ml portions, transferred to an Erlenmeyer flask with spin bar, stirred gently for 20 min at room temperature, and centrifuged in conical centrifuge tubes at about 200 *g* for 10 min. The pellet is washed once and resuspended in complete growth medium. Residual large tissue fragments are allowed to settle, and the dispersed cells are collected for plating. Cells from one tumor distributed among 40–60 tissue culture plates (60-mm diameter) will approach confluence after 2 weeks in culture. Propagation of Y1 cells as tumors minimizes the requirements for maintaining stocks in cell culture and also provides a convenient means of generating cells in large yields. These advantages, however, are offset to some degree by the heterogeneity of the tumor material.

Caution

Although experience to date suggests that there is no biohazard associated with the culture of these mouse adrenal tumor cells, it seems prudent to adopt containment procedures similar to those of the Medical Research Council of Canada[16] when handling them.

[16] "Guidelines for the Handling of Recombinant DNA Molecules and Animal Viruses and Cells," Cat. No. MR21-1/1977. Minister of Supply Services, Ottawa, Canada, 1977.

## [53] Long-Term Culture of Dissociated Sympathetic Neurons

*By* EDWARD HAWROT and PAUL H. PATTERSON

Long-term primary culture of dissociated neurons, although a relatively new technique, has already proved to be a valuable tool in studies of neuronal development and function. Under proper conditions neurons can be grown in culture in virtual isolation, free from the influence of other cell types. Such cultures offer the potential for systematic manipulation of the fluid medium surrounding the cells. These neuron-alone cultures are particularly well suited to biochemical analysis since the properties of any function under study are directly attributable to the neurons being cultured, without complicating contamination from other cell types. Furthermore, such cultures also permit the examination of the specific effects on neurons of co-culture with a wide variety of other cell types. In this way, for example, neuronal interaction in culture with physiologically important target cells can be directly compared to what is seen with the neuron-alone cultures.

On the other hand, the major problems with primary neuronal culture are the expense in animals and supplies, the time-consuming dissection, and the relatively small amounts of material obtained for biochemical analysis. Neurons do not appear to divide in culture and thus growth consists of cellular maturation and outgrowth of neuronal processes. Furthermore, since neuron-alone cultures are considerably more fastidious than explant cultures or cell lines, they can be difficult to reproducibly maintain over long periods of time. Although these difficulties have been largely overcome with some types of neurons, extreme care and some experience with cell culture are helpful.

In fact, the particular growth conditions employed can affect not only neuronal survival and growth, but may also influence the type of synapses formed in culture as well as the neurotransmitters produced by cultured neurons. In dissociated cell culture under appropriate conditions, sympathetic neurons exhibit a normal developmental differentiation and mat-



ISBN 0-12-181958-2

uration into adrenergic neurons.[1] Differentiation occurs in the nearly complete absence of nonneuronal cells. Nevertheless, the sympathetic neurons at the time of explantation do possess a certain restricted level of plasticity consistent with their embryologic derivation from the pluripotent neural crest.[1] This flexibility permits the experimenter to obtain, from the same population of sympathetic neurons, cultures that display strikingly cholinergic functions as opposed to the normal adrenergic differentiation. Such a phenomenon was first observed upon co-culture of sympathetically derived neurons with certain nonneuronal cells.[2,3] Neurons co-cultured with these nonneuronal cells produce 1000-fold more acetylcholine than control neuron-alone cultures. However, cell-to-cell contact between the neurons and nonneuronal cells is not necessary for this effect. Medium conditioned by incubation on cultures of appropriate nonneuronal cells, when added to neuron-alone cultures produces an increase in neuronal cholinergic characteristics with a concomitant decrease in adrenergic characteristics.[4] Similarly, the frequency of nicotinic cholinergic synaptic interactions between the neurons is increased by growth in conditioned medium. Since differential survival and selection of subclasses of the neuronal population have been ruled out, an inductive phenomenon is strongly suggested.[1] It is also important to note that an *in vivo* counterpart to the alterations in differentiated fate of cultured sympathetic neurons has been observed in embryonic transplantation studies of neural crest.[1] Clearly the extracellular environment (and choice of culture conditions) is critically important in determining the differentiated fate of sympathetically derived neurons.

## Sympathetic Neuron Cultures: Standard Conditions

### *Cell Preparation*

Newborn rats (CD colony), born and shipped the same day, are obtained from Charles River Breeding Laboratories, Inc., Wilmington, Massachusetts. In general, rat pups up to 4 days old can be used for dissociated cultures. The animals are killed by a blow to the head, and the superior cervical ganglia are removed with dissecting forceps and scissors under a stream of sterile saline. Decapitation readily exposes the ganglia for this manipulation. The ganglia are placed in plating medium (see be-

[1] P. H. Patterson, *Annu. Rev. Neurosci.* **1,** 1–17 (1978).

[2] P. H. Patterson and L. L. Y. Chun, *Proc. Natl. Acad. Sci. U.S.A.* **71,** 3607 (1974).

[3] P. H. O'Lague, K. Obata, P. Claude, E. J. Furshpan, and D. D. Potter, *Proc. Natl. Acad. Sci. U.S.A.* **71,** 3602 (1974).

[4] P. H. Patterson and L. L. Y. Chun, *Dev. Biol.* **56,** 263 (1977).

low) and cleaned free of all surrounding tissue. The ganglion cells are then dissociated mechanically using very fine watchmaker's forceps (#5). Usually the ganglion sheath is first peeled away with forceps, and the ganglion cells are then gently teased apart into smaller groups of cells. The chunks and cells are triturated by several passages through a stainless-steel hypodermic needle (22 or 23G, 1.5 inch) using a 5-ml disposable plastic syringe. The resulting suspension is further agitated with a Vortex mixer for 3–5 min (avoiding any swirling action) and undissociated pieces removed by sedimentation or filtration. The suspension is either allowed to settle undisturbed for 5–10 min or filtered, e.g., with a 10 $\mu$m Nuclepore filter (Nuclepore Corp., Pleasonton, California), to remove undissociated fragments. The dissociated cells are centrifuged (3–5 min at 800 *g*), resuspended, and plated in modified culture dishes (see below) by adding 1 or 2 drops per dish of cells, suspended in plating medium (see below). In general, the overall yield of neurons in culture ranges from 1–10% based on the number of neurons found in adult rat superior cervical ganglion (approximately 25,000). Normally, 40 to 50 dishes of about 3000 to 5000 viable neurons each are prepared from 40 pups. In addition to the mechanical procedures described here, sympathetic neurons have also been dissociated with trypsin treatment.[5,6]

*Culture Dishes*

Falcon petri dishes (35 mm, #1008) are modified by boring a 1-cm hole in the bottom of the dish and affixing a polystyrene (Lux Scientific Corp., Newbury Park, California; #5407, 25 mm) tissue culture cover slip with paraffin to form a shallow well. This well protects the cells from movement of the medium, is convenient for isotopic incubations, provides better optics for inverted phase microscopy, and facilitates preparation for electron microscopy. Prior to plating, a glass ring (1.5-cm diameter microslide ring, #6705-R12, A. H. Thomas, Philadelphia, Pennsylvania) is placed in each dish surrounding the center well. The ring is added after the growth medium (see below) and serves to restrict the area to which the cells can settle to that of the coverslip. After 2 days of culture these rings are removed.

*Culture Surfaces*

Several substrata are suitable for attachment and extension of neuronal processes. Most commonly, a thin, clear film of collagen is used.

[5] S. Varon and C. Raiborn, *J. Neurocytol.* **1,** 211 (1972).

[6] K. D. McCarthy and L. M. Partlow, *Brain Res.* **114,** 391 (1976).

Fresh collagen is prepared monthly as an acid extract of rat tail tendons.[7] Good adhesion and culture longevity are dependent on using a fresh collagen extract that is stored at 0–4°. The clear collagen extract (one adult rat tail per 250 ml of 0.1% v/v acetic acid) is diluted 1 : 4 with distilled water and 3 drops placed onto each cover slip. The dishes are dried in a desiccator for 24 hr at room temperature and sterilized by ultraviolet irradiation.

Sympathetic neurons also can be grown on three-dimensional hydrated collagen gels. These can be made either by photoreconstitution with riboflavin,[8] by exposure to ammonia vapors,[9] or by salt reconstitution.[10] Neuronal processes but not somas extend into the gel matrix, firmly anchoring the neurons in place. Gels offer the advantage of obviating the need for the viscosity-increasing agent, Methocel, which is normally added to the standard growth medium (see below).

Culture surfaces treated with polylysine or other positively charged polymers also have been successfully used with nonneuronal as well as neuronal cell types.[11–13] Sympathetic neurons attach and grow well on polyornithine-treated coverslips over the short term (<14 days) but do not survive well in long-term culture (unpublished observations). Better results are obtained if the polymer is covalently linked to an underlying layer of dried gelatin which is activated by incubation with glutaraldehyde. A solution of swine skin gelatin (7.5 mg/ml) is prepared by heating in a boiling water bath and 2–3 drops are applied to the center well of the culture dish. The gelatin is then allowed to dry down and form a clear thin film. Enough glutaraldehyde (2.5% in 0.12 *M* sodium phosphate buffer, pH 7.4) is then added to the dish to completely cover the gelatin layer. After about 1 hr at room temperature, the dishes are rinsed once with distilled $H_2O$ and the polymer (5 mg/ml) is added to the gelatin layer in a solution of 0.15 *M* sodium borate buffer, pH 8.4. Good results have been obtained using poly-D-lysine (approximate molecular weight 150,000) which should be less susceptible to proteolytic digestion. After overnight incubation, the dishes are rinsed several times with distilled $H_2O$, sterilized, and rinsed several times with medium prior to use.

Sympathetic neurons can also be grown on pre-existing monolayers of

[7] M. B. Bornstein, *Lab. Invest.* **7,** 134 (1958).
[8] E. B. Masurovsky and E. R. Peterson, *Exp. Cell Res.* **76,** 447 (1973).
[9] R. L. Ehrmann and G. O. Gey, *J. Natl. Cancer Inst.* **16,** 1375 (1956).
[10] S. D. Hauschka and I. R. Konigsberg, *Proc. Natl. Acad. Sci. U.S.A.* **55,** 119 (1966).
[11] E. Yavin and Z. Yavin, *J. Cell Biol.* **62,** 540 (1974).
[12] P. C. Letourneau, *Dev. Biol.* **44,** 77 (1975).
[13] W. L. McKeehan and R. G. Ham, *J. Cell Biol.* **71,** 727 (1976).

nonneuronal cells as substrata: both primary cardiac[14,15] and skeletal muscle cells[16] have been used. In both cases nonneuronal cell proliferation is suppressed prior to neuronal plating by means of $\gamma$-irradiation ($^{60}$Co; 5000 rad in 25–30 sec). In general, the growth medium is changed immediately after exposure to irradiation. Radiation treatment effectively blocks cell division while leaving intact such functions as the spontaneous electrical activity and beating of cardiac myocytes. In some cases the cultures were irradiated again several days after the neuronal plating to block overgrowth by the ganglionic nonneuronal cells that are plated with the neurons.

### *Culture Media*

*1. L-15-Air.* Of several commercial media tested, L-15 with modified osmolarity and pH appears to be optimal for growth of sympathetic neurons.[17] One package (14.9 g) of L-15 powder (North American Biologicals, Inc., Miami, Florida) is dissolved in 1080 ml freshly distilled water. (It is very important that the water be of the highest purity for all media additions; it can be distilled from permanganate or glass distilled after purification by charcoal and ion-exchange resins.) The following ingredients are then added: 60 mg imidazole (recrystallized from acetone), 15 mg aspartic acid, 15 mg glutamic acid, 15 mg cystine, 5 mg $\beta$-alanine, 2 mg vitamin $B_{12}$, 10 mg inositol, 10 mg choline chloride, 0.5 mg lipoic acid, 0.02 mg biotin, 5 mg *p*-aminobenzoic acid, 25 mg fumaric acid, and 0.4 mg coenzyme A. The above additives, excluding imidazole, can be prepared in advance as a combined 200-fold concentrate and stored at $-20°$. Phenol red is added at a concentration of 10 mg/liter. The pH is adjusted to 7.35 with 1 *N* HCl and the medium sterilized by filtration and stored at 4°. Filtration is performed with a Millipore stainless-steel pressure filtration apparatus (142-mm filter disc holder; Millipore Corp., Bedford, Massachusetts). Nuclepore filters (0.2 $\mu$m, standard membranes) are washed extensively prior to use, first with boiling distilled water and then with ambient-temperature distilled water. Thereafter the filter holder and filter are wrapped in aluminum foil and sterilized by autoclaving. After the apparatus has cooled to room temperature, the medium is pressure filtered with compressed nitrogen under sterile conditions into clean sterile

[14] E. J. Furshpan, P. R. MacLeish, P. H. O'Lague, and D. D. Potter, *Proc. Natl. Acad. Sci. U.S.A.* **73,** 4225 (1976).

[15] P. H. O'Lague, P. R. MacLeish, C. A. Nurse, P. Claude, E. J. Furshpan, and D. D. Potter, *Cold Spring Harbor Symp. Quant. Biol.* **40,** 399 (1975).

[16] C. A. Nurse and P. H. O'Lague, *Proc. Natl. Acad. Sci. U.S.A.* **72,** 1955 (1975).

[17] R. E. Mains and P. H. Patterson, *J. Cell Biol.* **59,** 329 (1973).

glass bottles. (Glassware is never cleaned with detergent; if used for proteinaceous solutions, it is discarded after use. Bottles are cleaned by autoclaving them full of water followed by repeated rinsing while still hot.)

2. *L-15-$CO_2$*. The medium used both for sympathetic neurons and nonneuronal cells is prepared by adding 170 ml of 150 m*M* $NaHCO_3$ for every 850 ml of basal L-15-Air (see above). (The sodium bicarbonate is sterilized by filtration either before addition to sterile L-15-Air or in combination with the L-15 medium). The pH of this medium is adjusted by blowing $CO_2$ over the solution. In general, most solutions are sterilized by passage through autoclaved Nuclepore filters (0.2 $\mu$m) washed as described above.

3. *Plating Medium*. The medium used for ganglion dissection and cell dissociation consists of 100 ml of basal L-15-Air to which is added glucose (4 ml of a 30% w/v solution), glutamine (1 ml of a 200 m*M* solution; Microbiological Associates, Inc., Walkersville, Maryland), penicillin, and streptomycin (1 ml of a solution containing 10,000 U of penicillin and 10 mg of streptomycin per milliliter; obtained as lyophilized powder from Grand Island Biological Co., Grand Island, New York). Plating medium contains twice the normal glucose concentration so as to increase the density of the suspending medium and facilitate cell settling during plating.

4. *Complete Growth Media*. For routine growth of cells in either L-15-Air or L-15-$CO_2$, the following additives are combined with 100 ml of the basic formulation: glutamine, penicillin, and streptomycin (same as for "Plating Medium"), glucose (2 ml of a 30% w/v solution), fresh vitamin mix (1 ml, see below), adult rat serum (5 ml, see below), Methocel (0.6 g; see below), and nerve growth factor (NGF; final concentration of 1 $\mu$g/ml 7S, see below). Complete media are stored for no longer than 10 days at 4°. L-15-Air cultures are incubated in a humidified air atmosphere at 36° whereas L-15-$CO_2$ cultures are incubated in a 5% $CO_2$ in air humidified atmosphere, also at 36°. The growth medium (2 ml) is changed every 4 days except for cultures receiving conditioned medium (see below) which receive fresh medium every 2 days. Rat sympathetic neurons also have been grown in media based on commercial preparations other than L-15 and with additives different from those described here.[18-20]

5. *Growth Medium Additives*. The fresh vitamin mix is prepared as a 100-fold concentrate and contains 1 mg of 6,7-dimethyl-5,6,7,8-tetrahydropterine (Calbiochem, La Jolla, California), 100 mg ascorbic acid

[18] R. P. Bunge, R. Rees, P. Wood, H. Burton, and C.-P. Ko, *Brain Res.* **66,** 401 (1974).

[19] C.-P. Ko, H. Burton, and R. P. Bunge, *Brain Res.* **117,** 437 (1976).

[20] K. J. Lazarus, R. A. Bradshaw, N. R. West, and R. P. Bunge, *Brain Res.* **113,** 159 (1976).

(Sigma Chemical Co., St. Louis, Missouri), and 10 mg of glutathione (Sigma) in 20 ml of distilled water. The pH is adjusted to 6.0 with 1 *N* NaOH and the solution stored at −20° after sterilization by filtration.

Methocel (hydroxymethyl cellulose, standard grade; Dow Chemical Co., Midland, Michigan) increases the viscosity of the medium thereby protecting the cells from mechanical stresses and fluid movements in the culture dish. It does not appear to serve a nutritive function since neurons can be grown on collagen gels in medium lacking Methocel. After sterilization in the autoclave, the dry Methocel is combined with the culture medium and the mixture stirred vigorously for 10–24 hr at 4°.

Rat serum is necessary for optimal long-term growth of dissociated rat sympathetic neurons; sera of fetal calf, calf, horse, rabbit, or swine are not as effective.[17] Commercially available rat serum is also less effective than serum prepared in the laboratory using animals obtained from a commercially maintained rat colony. Adult rats (CD colony, see above) of either sex are killed by asphyxiation in a $CO_2$ chamber and exsanguinated. The blood is collected on ice, allowed to clot, and centrifuged for 30 min at 35,000 g and 4°. The serum is stored at 0° for 12–16 hr and centrifuged to remove remaining cells. Finally, the clot-free serum is filtered by pressure filtration in the same manner as the culture media. Portions of 5 ml each are stored at −20°.

Nerve growth factor (NGF) is essential for survival, growth, and maturation of sympathetic neurons from newborn rats. In the absence of this protein, only the ganglionic nonneuronal cells will survive in L-15-$CO_2$ cultures of dissociated sympathetic ganglia. NGF is prepared as the 7S form by gel filtration through Sephadex G-100[21] followed by DEAE-cellulose fractionation.[22] The pooled DEAE-cellulose fractions are concentrated by pressure filtration (Diaflo, UM-10, Amicon Corp., Lexington, Massachusetts) to 1 mg/ml and, after sterilization by filtration, individual portions are stored at −20°. NGF is stable for 6–12 months under these conditions. A new, rapid procedure for producing the low-molecular-weight form of NGF is also available.[23]

### *Removal of Nonneuronal Cells from Dissociated Sympathetic Ganglionic Cultures*

By means of a simple manipulation in the composition of the medium, the growth of ganglionic nonneuronal cells may be effectively controlled. When dissociated sympathetic ganglia are plated in L-15-Air (see above),

[21] V. Bocchini and P. U. Angeletti, *Proc. Natl. Acad. Sci. U.S.A.* **64,** 787 (1969).
[22] S. Varon, J. Nomura, and E. M. Shooter, *Biochemistry* **6,** 2202 (1967).
[23] W. C. Mobley, A. Schenker, and E. M. Shooter, *Biochemistry* **15,** 5543 (1976).

only the neurons survive; less than 10% of the cells seen by phase microscopy are nonneuronal even after several weeks in culture.[4]

When dissociated ganglion cells are plated into L-15-$CO_2$, both neurons and nonneuronal cells become established in culture. The nonneuronal cells proliferate and form a monolayer within 2–3 weeks. Growth of nonneuronal cells in L-15-$CO_2$ cultures can be prevented by the use of antimitotic poisons such as cytosine arabinoside or fluorodeoxyuridine. Both of these inhibitors reduce the proliferation of ganglionic nonneuronal cells without producing any reduction in neuronal catecholamine synthesis. Sympathetic cultures are incubated in growth medium containing 10 $\mu M$ cytosine arabinoside (Sigma) for 2 periods of 48 hr each, beginning on days 2 and 6 in culture. Fluorodeoxyuridine and uridine (Sigma) are also antimitotic when added at 10 $\mu M$ on the same schedule.

In addition to these methods, nonneuronal cells may be removed at a stage prior to plating by cell fractionation procedures. One method successfully employed for this propose in cultures of chick sympathetic neurons involves differential cell adhesiveness and takes advantage of the fact that, under appropriate conditions, chick ganglionic nonneuronal cells attach to the substratum sooner and adhere tighter than the neuronal cell population.[5,6]

## Co-Culture of Neurons with Nonneuronal Cells

### *Ganglionic Nonneuronal Cells*

As previously mentioned, use of L-15-$CO_2$ growth medium without antimitotic treatment permits proliferation of the indigenous ganglionic nonneuronal cell types from dissociated sympathetic ganglia. Among the nonneuronal elements are glial-like (satellite) cells as well as fibroblast-like cells.

### *Other Nonneuronal Cells in Mixed Culture*

As with ganglionic nonneuronal cells, L-15-$CO_2$ rather than L-15-Air growth medium is better for growth of many other types of nonneuronal cells. Sympathetic neurons can be plated either onto pre-plated irradiated nonneuronal cells as previously discussed (see "Culture Surfaces") or together with nonneuronal cells.

## Conditioned Medium

For the study of the control of the differentiated fate of sympathetic neurons by nonneuronal cells, conditioned medium (CM)[4] can be prepared from cultures of nonneuronal cells in the following manner. Dissociated primary cells are obtained from minced tissue (e.g., heart, brain, liver, skeletal muscle) that has been incubated for 20 min with continuous agitation in phosphate-buffered saline containing 1 mg/ml collagenase (Worthington Biochemical Corp., Freehold, New Jersey). Undissociated pieces are allowed to settle and are treated with additional collagenase in a second incubation. The combined dissociated cells are collected by low-speed centrifugation, washed several times, and a sample counted in a hemocytometer. In general, 5–10 million cells are plated per 75-$cm^2$ flask in L-15-$CO_2$ growth medium containing 10% fetal calf serum (Microbiological Associates) instead of rat serum. Cell lines are similarly maintained with the same medium.

Prior to preparing conditioned medium, flasks with a dense layer of nonneuronal cells are incubated for 2–12 hr with L-15-$CO_2$ growth medium containing rat serum. After removal of this wash medium, the medium to be conditioned is added to the flasks. Usually, 20–30 ml of complete L-15-$CO_2$ growth medium containing rat serum, Methocel, etc. (see above) are added to each flask and incubated for 2 days. The CM is then harvested and stored at $-20°$. Before use, the CM is thawed and mixed with fresh medium to yield the desired final concentration. Neurons receive such mixed media every 2 days.

The acetylcholine inductive effect of CM is species specific with regard to the nonneuronal cells used for conditioning.[4] Mouse and hamster cell lines as well as mouse and chick primary heart cultures are ineffective in producing CM capable of acetylcholine induction in rat sympathetic cultures grown in the presence of rat serum. Thus this additional feature of co-culture and CM must be kept in mind in planning induction experiments.

### Isolated Sympathetic Neurons Grown in Microculture

For studies requiring the growth of solitary isolated sympathetic neurons, methods are now available for maintaining and analyzing single cells in microculture. For most biochemical purposes neurons can be grown in the segregated wells of Falcon 3034 Terasaki tissue culture plates.[24] Each plate contains 60 wells with a capacity of 10 $\mu$l each. As with the standard cultures, growth in microculture wells requires a culture substratum of dried collagen. Even under standard culture conditions single neurons appear to be more fastidious than mass cultures, and the microcultures are not always successful. Although adhesion of single cells

[24] L. F. Reichardt and P. H. Patterson, *Nature* (*London*) **270,** 147 (1977).

to the culture surface is especially problematic, the use of collagen gels instead of dried collagen films appears to improve neuronal survival. Furthermore, single-cell microcultures appear to be more easily maintained in CM-containing L-15-$CO_2$ growth medium. Growth in CM from chick heart cells produces predominantly adrenergic neurons while growth in CM from rat heart cells produces predominantly cholinergic neurons.[24] Single cells in microculture wells also can be grown on pre-plated layers of nonneuronal cells where proliferation of the nonneuronal elements is inhibited by $\gamma$-irradiation.

Alternately, for purposes of electrophysiological, pharmacological, and morphological examination, single neurons can be grown on separated collagen islands formed on the standard culture coverslips.[14,25] In general, small droplets of collagen solution are allowed to dry onto nonwetting polystyrene coverslips (Lux, untreated for tissue culture) forming islands of collagen 300–500 $\mu$m in diameter. Dissociated cardiac cells (myocytes and fibroblasts) from newborn rats are plated onto such islands by incubation for about 2 hr. Cells not adhering to the collagen islands are then washed away with medium changes. After another 1–2 days, proliferation of the heart cells is suppressed by the standard $\gamma$-irradiation procedure. Dissociated sympathetic neurons suspended at a low density are plated onto these cardiac cell islands 1–5 days after irradiation. Under proper conditions, such islands receive only one or a few neurons. These microcultures are subsequently grown in L-15-$CO_2$ growth medium lacking Methocel and containing either the standard 5% adult rat serum or 10% fetal calf serum.

## Phenotypic Characteristics of Sympathetic Cultures

Several parameters can be used to monitor the development of dissociated sympathetic neurons in culture. Overall neuronal growth is measured by the increase in rates of synthesis of protein, lipid, and RNA,[26] or by the increase in total neuronal protein and lipid.[4] Process outgrowth and elongation as well as the general culture morphology can be readily followed by inverted phase microscopy. Spontaneous and evoked electrical activity as well as synaptic interaction can be examined by standard electrophysiological techniques.

Of particular relevance to studies of neuronal differentiation, the sympathetic cultures are able to synthesize and accumulate neurotransmitters from labeled precursors. These labeled neurotransmitters can be isolated

[25] S. C. Landis, *Proc. Natl. Acad. Sci. U.S.A.* **73,** 4220 (1976).
[26] R. E. Mains and P. H. Patterson, *J. Cell Biol.* **59,** 361 (1973).

and identified by high-voltage paper electrophoresis.[17] Furthermore, uptake, storage, and evoked release of neurotransmitters can be assayed.[27] The biosynthetic enzyme, choline acetyltransferase, also can be measured in extracts of sympathetic cultures, and levels of this enzyme are elevated in cultures induced for cholinergic function.[4]

### Acknowledgments

Part of the work described here was supported by the American and Massachusetts Heart Associations, the Helen Hay Whitney Foundation, and the National Institute of Neurological and Communicable Diseases and Stroke.

[27] P. H. Patterson, L. F. Reichardt, and L. L. Y. Chun, *Cold Spring Harbor Symp. Quant. Biol.* **40,** 389 (1975).

## [54] Neuronal Cells from Rodent Neoplasms

*By* DAVID SCHUBERT and WILLIAM CARLISLE

Over 70 years ago Ross Harrison used cultures of neural epithelium from amphibian embryos to examine the mechanism of neurite growth.[1] Since these initial experiments, advances in tissue culture technology have permitted the study of many cell types in defined environments that minimize the multiple interactive events found *in vivo*. Because of the extreme complexity of the nervous system there is a clear requirement for isolating individual component cells for biochemical and electrophysiological studies. Fruitful exploitation of cell culture for the study of the nervous system requires a collection of cells representative of those found *in vivo*. In addition, the biochemical and electrophysiological behavior of these cells should mimic that found in the animal. An assumption underlying all work in cell culture is that if a function of the *in vivo* cell is found in culture, then the underlying biochemical mechanisms are the same as those employed *in vivo*. The cells can then be used to study the molecular basis of these phenomena.

Clonal cell lines that possess functional characteristics of their *in vivo* counterparts have been isolated from the nervous system. However, the variety of cell types from the nervous system available for study as homogeneous (clonal) populations is very limited, and there is a continuing need for the establishment and characterization of additional cell lines. It is to this end that we describe our experience in developing cell lines from the rodent nervous system.

[1] R. Harrison, *Anat. Rec.* **1,** 116 (1907).



ISBN 0-12-181958-2

### Methods for Obtaining Neuronal Cell Lines

Since the normal differentiated nerve cell may not divide, and a dividing cell is needed to obtain cell lines, it follows that procedures are required to select for or induce cell division. Four procedures have been used successfully to obtain clonal cell lines from the nervous system: (1) spontaneous tumors, (2) normal embryonic tissue, (3) viral transformation, and (4) chemically induced tumors.

To date, only two spontaneous rodent tumors of neuronal origin have been adapted to clonal cell culture. These are the C1300 neuroblastoma[2,3] and the PC12 sympathetic ganglion-like cell.[4] Both of these tumors occurred peripherally to the central nervous system and were identified on the basis of tumor pathology as possible neuroblastoma and pheochromocytoma, respectively. Positive identification of the cell type required the establishment of clonal cell lines from the original tumor.

The successful establishment of clonal electrically excitable cells directly from embryonic mammalian central nervous system (CNS) tissue has recently been described.[5] Embryonic brain cells from specific areas were dissociated with trypsin and exposed to antiglia antiserum in the presence of complement to lyse all cells expressing glial antigens. This selection technique was repeated several times with the intent of eliminating the more rapidly dividing glial cell population. In addition, loosely adherent cells with nerve-like morphologies were selectively passaged away from the more rapidly dividing anchorage-dependent cells by ringing techniques (described below). Several of the resultant clones expressed nerve-specific surface antigens and were electrically excitable as defined by a veratridine stimulated sodium flux assay.

The third alternative successfully used to obtain dividing nerve cells is by *in vitro* viral transformation. De Vitry *et al.*[6] dissociated single cells from a 14-day embryonic mouse hypothalamus and grew the resultant cells in culture for 6 days. At this time the cultures were washed and covered with serum-free medium containing $7 \times 10^7$ plaque-forming units of SV40. After adsorption for 30 min, the cells were covered with growth medium and the transformed cells isolated from foci of growing cells by methods similar to those described below. By selecting cells primarily on the basis of morphological complexity, e.g., the most "nerve-like" in

[2] G. Augusti-Tocco and G. Sato, *Proc. Natl. Acad. Sci. U.S.A.* **64,** 311 (1969).

[3] D. Schubert, S. Humphreys, C. Baroni, and M. Cohn, *Proc. Natl. Acad. Sci. U.S.A.* **64,** 316 (1969).

[4] L. Greene and A. S. Tischler, *Proc. Natl. Acad. Sic. U.S.A.* **73,** 2424 (1976).

[5] K. Bulloch, W. Stallcup, and M. Cohn, *Brain Res.* **135,** 25 (1977).

[6] F. De Vitry, M. Camier, P. Czernichow, P. Benda, P. Cohen, and A. Tixier-Vidal, *Proc. Natl. Acad. Sci. U.S.A.* **71,** 3375 (1974).

appearance, a clone of mouse hypothalamic neurosecretory cells was obtained which made both neurophysin and vasopressin.

The final technique for obtaining dividing cell populations of nerve and glia is by the induction of tumors of these cells. The most efficient way of obtaining brain tumors is by the transplacental induction of tumors with nitrosoethyl urea (NEU).[7] Since all rat strains are not equally susceptible to this carcinogen, BD-IX rats have been used most frequently.[7] Tumors are induced by injecting NEU, freshly dissolved in saline at a does of 40 mg/kg body weight, intravenously into the femoral vein of female rats at 15 day's gestation. This dose reduces the litter size by approximately 50%. Tumors of the trigeminal nerve and spinal roots can be induced with NEU using a different strain of rat, Sprague-Dawley, and injecting 18-day-pregnant animals intravenously with 20 mg/kg of NEU.[8] After birth the pups are weaned and placed in separate cages. Between 150–250 days after birth the majority of the offspring develop symptoms of neurological disease, including loss of motor control, eyelid closure, and weight. These rats are then watched carefully and, when they appear to be near death, are anesthesized with ether, washed with 70% ethanol, exsanguinated, and carefully autopsied. Most tumor masses are found in the cranial cavity and brain stem. Using the NEU induction scheme described above, visible tumors are found in about 93% of the BD-IX offspring.

## Reagents

### *Tissue Culture Medium*

For both historical and practical reasons all of the culture work on cell lines derived from NEU-induced CNS brain tumors has employed modified Eagle's medium[9] containing 10% fetal calf serum. With one exception, the PC12 sympathetic neuron-like cell line,[4] all cell lines that we have tried will grow in this medium, even if they have been originally adapted to other types of culture media and serum. This standardization of culture conditions simplifies medium preparation and serum testing and, in addition, provides a common culture regime by which the various cell lines can be compared. Modified Eagle's medium can be purchased in powdered or liquid form, but we have found that preparing it ourselves is necessary both to maintain high quality and to provide the flexibility needed in altering its composition for experimental purposes. (The prep-

[7] H. Druckrey, R. Preussmann, S. Ivanković, and D. Schmahl, *Z. Krebsforsch.* **69,** 103 (1967).

[8] A. Hirano, J. Hasson, and H. M. Zimmerman, *Lab. Invest.* **27,** 555 (1972).

[9] M. Vogt and R. Dulbecco, *Proc. Natl. Acad. Sci. U.S.A.* **49,** 171 (1963).

aration of media is described in this volume [5].) Before each batch of medium is used it is tested for sterility, by dilution into a nutrient broth solution that is incubated at 37°, and for its ability to support cell growth and phenotypic expression. We routinely use a skeletal muscle myoblast cell line to test the medium since its ability to fuse is very sensitive to culture conditions. Freshly prepared medium is stored in 90-ml aliquots at 4° in the dark for periods of up to 1 month. The cells are grown in a humidified atmosphere at 36° in 12% $CO_2$ in air.

### *Serum*

There is a tremendous variation in the fetal calf serum obtained from different commercial sources and between lots from the same dealer. Therefore, samples from several lots are tested for their ability to support the growth and differentiation of different cell types. Once a lot is decided upon, sufficient serum from that lot is purchased to last from 6–12 months. It is stored frozen (−20°) and dispensed in 10-ml aliquots for use over a 2-week period. The dispensed serum is stored at −20° and only thawed once immediately before use.

### *Culture Dishes*

Many cell types are anchorage dependent for growth and require a tissue-culture dish surface. Although most of the nerve cell lines will grow in plastic petri dishes, where cells attach very poorly, all of the nerve cell lines are routinely passaged on tissue culture dishes; 60-mm plastic dishes are used for most purposes.

### *Viokase (Pancreatin 4XN.F., GIBCO)*

Viokase, a mixture of pancreatic enzymes, is sterilized by passage through 0.22-$\mu$m filters and stored at −20° as 1-ml aliquots.

## Establishment of Cell Lines from Neoplastic Tissue

The tumor is removed aseptically from the animal and rinsed sequentially in three tissue culture dishes containing culture medium. After the final rinse, a portion of the tumor is removed, minced with scissors, and injected subcutaneously into syngenic animals in case the initial cells do not adapt to continuous cell culture; of the approximately 150 neoplasms we have attempted to adapt to culture, only one did not grow from the initial explant. A piece smaller than 5 mm in diameter is removed from a nonnercotic area of the tumor, placed in 5 ml of complete medium, and finely minced with scissors. This procedure liberates many viable free cells as well as producing small pieces of tissue. The mixture of pieces and cells is dispensed at different concentrations into at least 24 60-mm tissue culture dishes. For example, the first four dishes receive 1 drop, the second four receive 2 drops, and so on. A few dishes receive cells in suspension only and no pieces of tissue. After about a week, the plates are carefully examined for areas of growth. If the medium in the dishes containing higher cell densities becomes acid, the cells should be removed by scraping with a rubber policeman and the cells, tissue pieces, and medium diluted by a factor of 10 into fresh medium. Some conditioned medium should be carried along with the cells until they are cloned.

## Selection and Cloning of Cells from Primary Cultures

### *Initial Selection of Cell Type*

In our experience, cells grown from all of the primary tumor explants are morphologically heterogeneous, even if histologically the neoplasm from which they were derived consists primarily of one cell type. Thus a choice must be made as to which morphologies are to be cloned. It is desirable to initiate the selection of the desired cell type as early as possible since, in the case of neuronal tumor explants, flat fibroblast-like cells tend to quickly dominate the culture. We try to select cells with the most complex morphology, e.g., with long processes, and also cells that grow loosely attached to the substratum with round, phase-bright bodies. The latter growth characteristic is shared by most clonal nerve cell lines. Such clusters of cells are usually observed in the early explant cultures associated with a small piece of tissue. To isolate a desired cell cluster, it is located with an inverted microscope and its location marked. The culture medium is removed (some should be saved), the dish carefully washed once with serum-free medium, and a sterile cloning ring (see this volume [12]) is placed over the area. A solution of 3 parts serum-free medium and 1 part Viokase (Gibco), warmed at 37°, is placed in the ring. After 5 min at 37°, the cells are removed from the dish with a sterile Pasteur pipette and placed into a 35-mm culture dish containing approximately 80% fresh medium containing 10% fetal calf serum and 20% conditioned medium from the original dish. These cells are examined regularly for growth; if cell heterogeneity persists, the above procedure should be repeated. It should be emphasized that this procedure does not yield clones, but only enriches for a particular cell type and removes it from the less desirable cells that will eventually overgrow the primary explant cultures. Many "ringed" cultures should be made, for all do not grow.

Once these "ringed" cultures begin to divide and appear somewhat homogeneous, they should be transferred at high cell densities (1 : 5 split of a confluent dish) along with conditioned medium for several transfers. Assuming the cells are growing, they are ready for cloning.

*Cloning*

There are several methods for cloning cells (see this volume [12]). Since the CNS-derived nerve and glial cells tend to grow attached to the tissue culture dish, the method most frequently employed is that in which cloning rings are used. Exponentially dividing cultures are dissociated to single cells with a 1 : 4 dilution of Viokase for 1 hr at 37°, and plated between $10^2$ and $10^4$ cells per 100-mm tissue culture dish. If the plating efficiency is low, the use of 50% conditioned medium may be helpful. After 24 hr, well-isolated single cells are located on an inverted microscope and marked from below with a felt-point pen. Between 1 and 3 weeks later, the plates are examined again for cell growth in the marked areas. If a small colony has formed and remains isolated from other cells, then it should be removed with a cloning ring and placed in 35-mm tissue culture dishes. To be sure of the clonal origin of a cell line this procedure should be repeated once. At this stage, the clones are grown in a dozen 60-mm dishes, and the cells are stored in liquid nitrogen to preserve the lines.

## Handling of Clonal Cell Lines

*Storage and Recovery of Cells*

Cells from the late experimental stage of growth are scraped with rubber policemen, and the cells from each dish are sedimented separately by centrifugation and taken up in 1 ml of ice-cold medium containing the following ingredients: 4 parts double-concentration modified Eagle's medium minus sodium bicarbonate; 3 parts water; 2 parts fetal calf serum; and 1 part filter-sterilized dimethylsulfoxide. The suspension is sealed in ice-cold plastic screw-cap vials, frozen at a rate of 1°/min to a final temperature of −80°, and stored indefinitely in liquid nitrogen. To recover the frozen cells, vials are removed from the liquid nitrogen and immediately thawed in a 37° water bath. As soon as the medium is liquid, the suspension is diluted at different final concentrations into six or eight 60-mm dishes containing modified Eagle's medium and 20% fetal calf serum. Evidence of cell growth should be visible within a few days.

## C

D

## E

## F



## G

H

## L

## M

## N

O



P

## Q

## R

## S

## Y



## Z

# Subject Index

## A

B

## G



## H

## O



## P

## Q



## R

S

## T

## U

## V



## W



## X



## Y

## Z